Amino Group Chemistry

Edited by
Alfredo Ricci

Related Titles

Arya, Dev P.

Aminoglycoside Antibiotics

2007
ISBN: 978-0-471-74302-6

Yudin, A. K. (ed.)

Aziridines and Epoxides in Organic Synthesis

2006
ISBN: 978-3-527-31213-9

Dyker, G. (ed.)

Handbook of C-H Transformations

Applications in Organic Synthesis

2005
ISBN: 978-3-527-31074-6

Yamamoto, H., Oshima, K. (eds.)

Main Group Metals in Organic Synthesis

2004
ISBN: 978-3-527-30508-7

Greenberg, A., Breneman, C. M., Liebman, J. F. (eds.)

The Amide Linkage

2002
ISBN: 978-0-471-42025-5

Togni, A., Grützmacher, H. (eds.)

Catalytic Heterofunctionalization

From Hydroamination to Hydrozirconation

2001
ISBN: 978-3-527-30234-5

Otera, J. (ed.)

Modern Carbonyl Chemistry

2000
ISBN: 978-3-527-29871-6

Greenberg, A., Breneman, C. M., Liebman, J. F. (eds.)

The Amide Linkage

2000
ISBN: 978-0-471-35893-0

Amino Group Chemistry

From Synthesis to the Life Sciences

Edited by
Alfredo Ricci

WILEY-VCH Verlag GmbH & Co. KGaA

The Editor

Professor Alfredo Ricci
Dipartimento di Chimica Organica
Universita di Bologna
Viale Risorgimento 4
040136 Bologna
Italia

Library of Congress Card No.:
applied for

British Library Cataloguing-in-Publication Data
A catalogue record for this book is available from the British Library.

Bibliographic information published by the Deutsche Nationalbibliothek
Die Deutsche Nationalbibliothek lists this publication in the Deutsche Nationalbibliografie; detailed bibliographic data are available on the Internet at <http://dnb.d-nb.de>.

Composition SNP Best-set Typesetter Ltd., Hong Kong

Printing Strauss GmbH, Mörlenbach

Bookbinding Litges & Dopf GmbH, Heppenheim

Cover Design Anne Christine Keßler, Karlsruhe

Printed in the Federal Republic of Germany
Printed on acid-free paper

ISBN: 978-3-527-31741-7

Contents

Amino Group Chemistry. From Synthesis to the Life Sciences. Edited by Alfredo Ricci

ISBN: 978-3-527-31741-7

Preface

The book *Modern Amination Methods*, I edited for Wiley-VCH in 2000, was intended to provide an almost exclusively methodological overview of several research areas in which amination plays a key role and to introduce the reader to new concepts that were at that time developed for generating new C—N bonds. The book was well received by the chemical community and indicated the need for keeping scientists aware of the progress in the field of amino group chemistry. The increasing importance of the amino function, in simple or complex molecular systems, is in fact fully acknowledged by chemists due to the presence of these molecules in the most important areas of basic and applied chemistry, such as pharmaceutical, medicinal, agricultural and natural product chemistry, and even more and more in biochemistry. Far from being exhausted, this topic seeks novel breakthroughs to face the novel challenges of third millennium chemistry. This prompted the writing of a new book, focusing not only on the C—N bond forming methodologies, but also on the role played by the amino function in those processes that are more closely related to the life sciences.

The contributions to this book are organized into interlinked sections and will include several important aspects related to amino group chemistry. The first part of the book deals with several more methodologically addressed chapters. Not only is the use of simple amination reagents and pivotal intermediates disclosed, making thus wider the already rich arsenal of conventional and unconventional amination methods, but also the potential of synthetic strategies like MCR (multicomponent reactions), or the most up-to-date metal- or organo-catalyzed approaches to the assembly of polyfunctional complex nitrogen-containing molecules is highlighted. Throughout each chapter, clear structures, schemes and figures accompany the text. Synthetic procedures, mechanisms, reactivity, selectivity and, especially, stereochemistry are addressed. An emphasis is placed, even at this stage, on target oriented synthesis with the insertion of the generated amino function into N-containing densely functionalized chiral molecules, or precursors therof, of interest in medicinal chemistry.

In the following chapters there is a greater focus on the life sciences. The relevant role played by core units containing a preformed amino functionality, many of them coming from the chiral pool, in the construction of important targets in medicinal chemistry, exhibiting among others anticancer, antibiotic and antiviral

Amino Group Chemistry. From Synthesis to the Life Sciences. Edited by Alfredo Ricci

ISBN: 978-3-527-31741-7

activity, is discussed with a rich series of examples. An even deeper insight into the field of clinically relevant drugs containing amino functionalities is provided by those chapters dealing with the synthesis and biological activity of aminated sugar and with the selective N-modification of aminoglycosides. The primary importance of the amino group in glycol structures, toning the physico-chemical properties, actively participating in recognition phenomena and, in the case of iminosugars, in enzymatic inhibition, and the role of the amino functions in RNA binding are treated in detail.

The last chapter is devoted to the industrial approach to amination reactions via transition metal catalyzed aryl amination. The progress in this field and the transformation of formerly extremely difficult processes into trivial tasks with lots of possibilities for fine tuning apt to the large scale production of modern synthetic targets, are disclosed.

This book is timely and the up-to-date reference sections together with several laboratory protocols would make it immediately useful also for those researchers not familiar with this field. It is aimed at a mixed audience including advanced students, young researchers and, more generally, people working in scientific institutions dealing with chemistry. Industrial chemists looking for a survey of well-tried fundamental concepts as well as for information on modern development in amino group chemistry, are also likely to be interested in this book considering the extensive number of industrially important targets treated.

As far as I know there are no books closely related to or similar to this book. The only exception could be the already mentioned *Modern Amination Methods* published by Wiley-VCH in 2000. This fact, instead of constituting a point of weakness, guarantees that the new book will not give rise to a substantial superimposition with the previous publication but on the contrary will be fully complementary to it.

I would like to thank all the distinguished scientists and their coauthors for their rewarding, timely and well-referenced contributions. Grateful acknowledgements are offered to the Wiley-VCH editorial staff, in particular to Dr. Manfred Koehl for proposing to me this new challenge and to Dr. Waltraud Wuest who was of precious help for the development of this project.

Bologna, July 2007 — Alfredo Ricci

List of Contributors

Cristina Airoldi
University of Milano – Bicocca
Department of Biotechnology and Biosciences
Piazza della Scienza 2
20126 Milano
Italy

Roberto Ballini
Università di Camerino
Dipartimento di Scienze Chimiche
Via S. Agostino n. 1
62032 Camerino
Italy

Shunsuke Chiba
Nanyang Technological University
School of Physical and Mathematical Sciences
1 Nanyang Walk, Blk 5 Level 3
Singapore 637616
Singapore

Armando Córdova
Stockholm University
Department of Organic Chemistry
Arrhenius Laboratory
10691 Stockholm
Sweden

Philippe Dauban
CNRS
Institut de Chimie des Substances Naturelles
Avenue de la Terrasse
91198 Gif-sur-Yvette
France

Alexander Doemling
University of Pittsburgh
Department of Pharmaceutical Sciences
10019 BST3
Pittsburgh, PA 15261
USA

Robert H. Dodd
CNRS
Institut de Chimie des Substances Naturelles
Avenue de la Terrasse
91198 Gif-sur-Yvette
France

Barbara La Ferla
University of Milano – Bicocca
Department of Biotechnology and Biosciences
Piazza della Scienza 2
20126 Milano
Italy

Amino Group Chemistry. From Synthesis to the Life Sciences. Edited by Alfredo Ricci

ISBN: 978-3-527-31741-7

André Mann
Laboratoire de Pharmacochimie de la Communication Cellulaire-UMR 7081
Faculté de Pharmacie
74, route du Rhin – BP600 24
67400 Ilkirch Cedex
France

Enrico Marcantoni
Università di Camerino
Dipartimento di Scienze Chimiche
Via S. Agostino n. 1
62032 Camerino
Italy

Koichi Narasaka
Nanyang Technological University
School of Physical and Mathematical Sciences
1 Nanyang Walk, Blk 5 Level 3
Singapore 637616
Singapore

Francesco Nicotra
University of Milano – Bicocca
Department of Biotechnology and Biosciences
Piazza della Scienza 2
20126 Milano
Italy

Marino Petrini
Università di Camerino
Dipartimento di Scienze Chimiche
Via S. Agostino n. 1
62032 Camerino
Italy

Ramon Rios
Stockholm University
Department of Organic Chemistry
Arrhenius Laboratory
10691 Stockholm
Sweden

Ulrich Scholz
Boehringer Ingelheim Pharma GmbH & Co. KG
Binger Strasse 173
55216 Ingelheim am Rhein
Germany

Floris Louis van Delft
Radboud University Nijmegen
Institute for Molecules and Materials
Toernooiveld 1
6525 ED Nijmegen
The Netherlands

1
Simple Molecules, Highly Efficient Amination

Shunsuke Chiba and Koichi Narasaka

1.1
Introduction

In the last two decades, explosive progress has been made in synthetic methods for production of amino compounds, due to their rapidly increasing applications in pharmaceutical and material sciences. Development of amination reagents for the construction of new carbon–nitrogen bonds is one of the most important and basic processes for the synthesis of amino molecules, and this chapter introduces simple and useful amination reagents classified by reaction type, such as electrophilic amination reagents, including transition metal–nitrene and nitrido complexes, radical-mediated amination reagents, and nucleophilic amination (Gabriel-type) reagents.

1.2
Hydroxylamine Derivatives

Hydroxylamine derivatives are one of the most versatile and simple amination reagents, leading the variety of nitrogen-containing compounds. This section mainly focuses on recent advances in electrophilic amination of carbon nucleophiles with various hydroxylamine derivatives.

1.2.1
O-Sulfonylhydroxylamine

Tamura has reported the synthesis of *O*-mesitylsulfonylhydroxylamine (MSH; **1**; Figure 1.1) and related compounds and has examined their reactions with various nucleophiles in detail [1]. With regard to the formation of C—N bonds by the use of MSH, however, the applicable carbanions were quite limited – to stabilized enolates only – and the product yields of the resulting amines were quite low.

The electrophilic amination of organolithium compounds with the methyllithium-methoxyamine system, long recognized as a potentially useful

Amino Group Chemistry. From Synthesis to the Life Sciences. Edited by Alfredo Ricci

ISBN: 978-3-527-31741-7

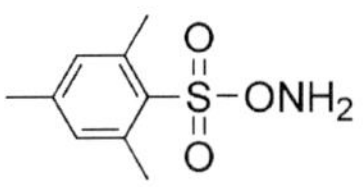

1

Figure 1.1 Tamura reagent (MSH) **1**.

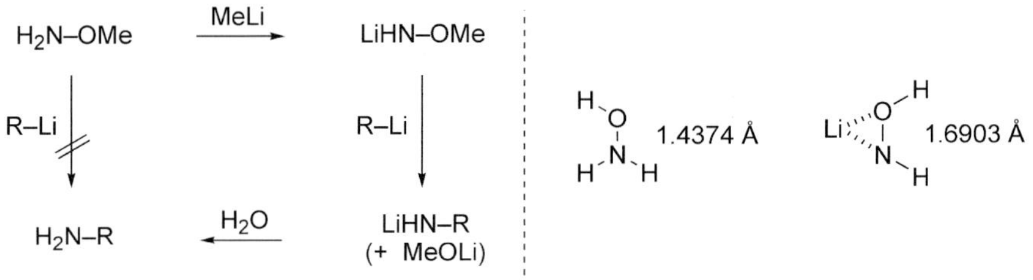

Scheme 1.1 Electrophilic amination with the methyllithium-methoxyamine system.

amination method (Scheme 1.1), was discovered by Sheverdina and Kocheskov in 1938 [2]. Overall the process is a formal displacement of the methoxy group of the lithium methoxyamide intermediate with the carbanion. Model calculations performed to provide insight into the electrophilic properties of LiHN—OMe showed that the N—O bond in LiHN—OMe is bridged by Li and is longer than the related bond in H_2N—OMe [3]. This would suggest a particular significance of the nitrenoid-like structure for the facile cleavage of the N—O bond.

These concepts have been translated into the design of some *O*-sulfonylhydroxylamines such as *tert*-butyl-*N*-tosyloxycarbamate (**2**) [4] and allyl-*N*-tosyloxycarbamate (**3**) [5], which can be easily prepared and are stable enough to handle. Actually, Boche et al. reported the crystal structure of lithium *tert*-butyl-*N*-mesityloxycarbamate and revealed that the N—O bond is longer than that in the neutral compound, *tert*-butyl-*N*-mesityloxycarbamate, which supports the calculations mentioned above [6].

Lithium *tert*-butyl-*N*-tosyloxycarbamate (**4**) and lithium allyl-*N*-tosyloxycarbamate (**5**), generated by treatment of **2** and **3** with butyllithium in THF at –78 °C, are useful for the preparation of *N*-Boc and *N*-Alloc amines. A variety of *N*-protected alkyl, aryl, and heteroaryl primary amines can be synthesized by treatment with the corresponding organolithium and -copper reagents (Schemes 1.2–1.5).

The amination of α-cuproamides and α-cuprophosphonates also proceeds effectively through the use of lithium *tert*-butyl-*N*-tosyloxycarbamate (**4**) and allyl-*N*-tosyloxycarbamate (**5**) to give α-amino acid derivatives. Asymmetric synthesis of α-amino acid derivatives is achieved by the amination of chiral amide cuprates with lithium *tert*-butyl-*N*-tosyloxycarbamate (**4**; Scheme 1.6) [7].

These methods can be applied to the amination of organoboranes [8]. Primary alkyl boranes rapidly react with lithium *tert*-butyl-*N*-tosyloxycarbamate (**4**) in a 1 : 1 molar ratio to give *N*-Boc-protected primary amines in good yield (Scheme 1.7).

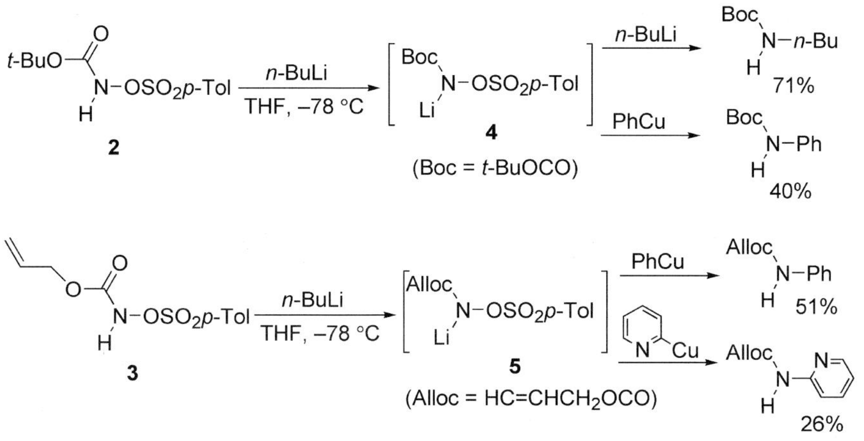

Scheme 1.2–1.5

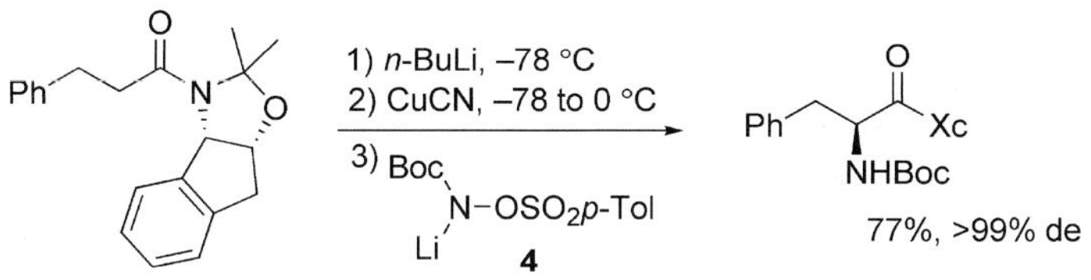

Scheme 1.6

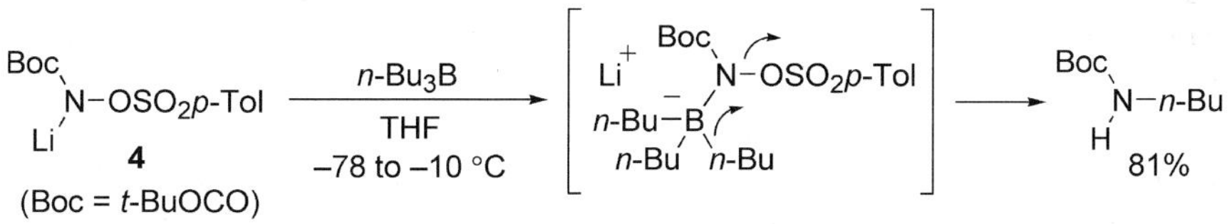

Scheme 1.7

The reaction presumably proceeds through the aniotropic rearrangement of an organoborate complex.

In addition, arylsulfonyloxycarbamates such as **6–8** can be used for aziridination of alkenes by treatment with inorganic bases such as CaO or Cs_2CO_3. The treatment of, for example, cyclohexene with ethyl *N*-arylsulfonyloxycarbamates **6** and **7** in the presence of CaO gives *N*-ethoxycarbonylaziridine along with a small amount of an allylic amination product (Scheme 1.8) [9]. As judged from the formation of an sp^3 C—H amination product, the reactive intermediate of this reaction seems to be ethoxycarbonylnitrene. Silyl enol ethers are aminated by the same procedure to afford α-amino carbonyl compounds, presumably via *N*-ethoxycarbonyl azirines (Scheme 1.9).

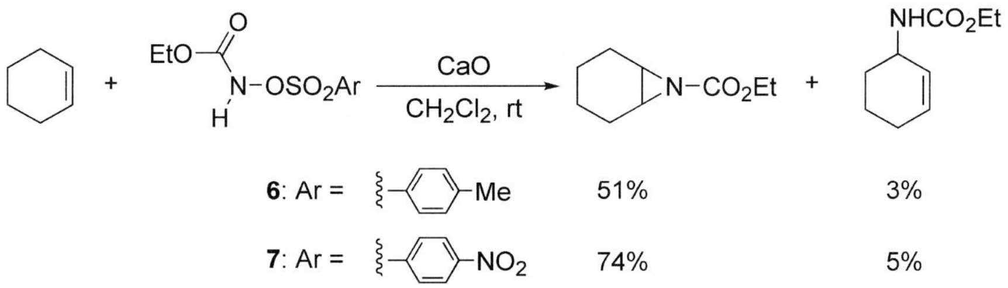

Scheme 1.8

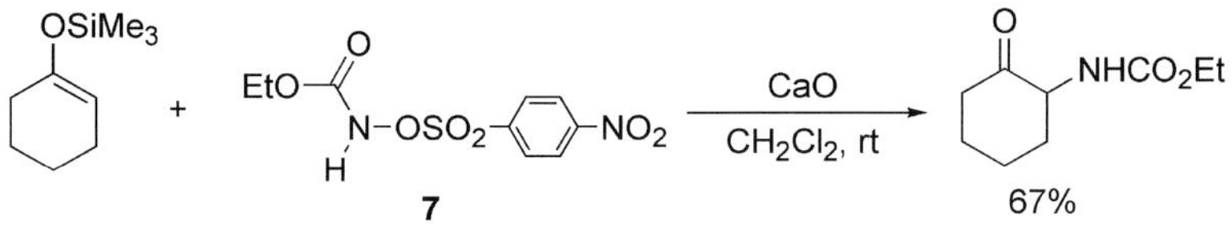

Scheme 1.9

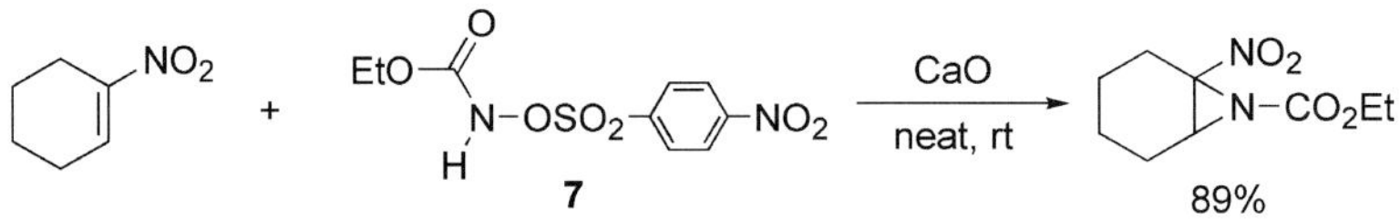

Scheme 1.10

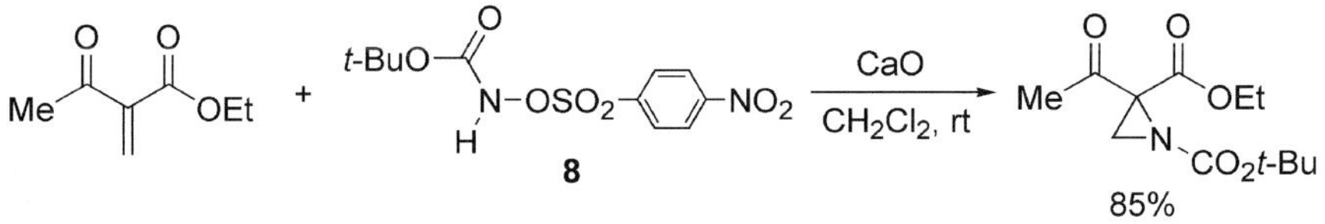

Scheme 1.11

The corresponding reactions with electron-deficient alkenes also afford aziridines, but through some other mechanism, such as an aza-Michael addition–elimination process (Schemes 1.10, 1,11) [10].

1.2.2
O-Phosphinylhydroxylamine

As well as *O*-sulfonyloximes (Section 1.2.1), *O*-(diphenylphosphinyl)hydroxylami ne (**9**) has also been utilized for the electrophilic amination of various carbanions to prepare primary amines [11]. Grignard reagents and organolithiums including enolates are aminated with **9** (Table 1.1).

The application of *O*-(diphenylphosphinyl)hydroxylamine (**9**) is limited by its low solubility in most organic solvents. Recently, Vedejs reported that *O*-di-(*p*-methoxyphenylphosphinyl)-hydroxylamine (**10**), which is soluble in THF even at

Table 1.1 Electrophilic amination with *O*-(diphenylphosphinyl)hydroxylamine (**9**).

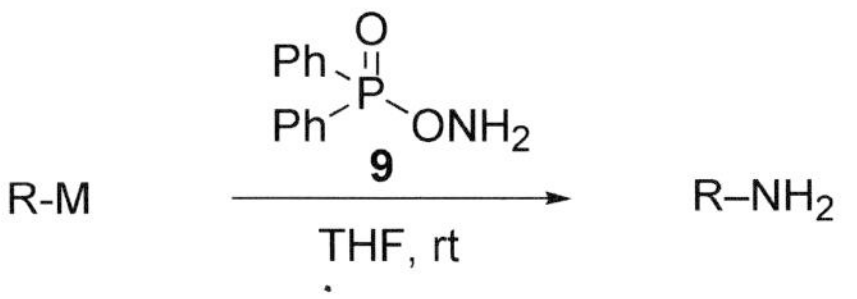

R-M	R–NH$_2$ (yield / %)	R-M	R–NH$_2$ (yield / %)
PhMgCl	Ph–NH$_2$ (35)	Ph$_3$CLi	Ph$_3$C–NH$_2$ (30)
	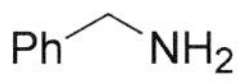 (70)	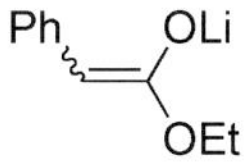	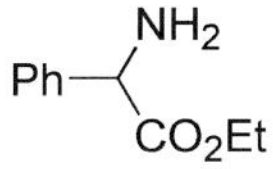 (45)

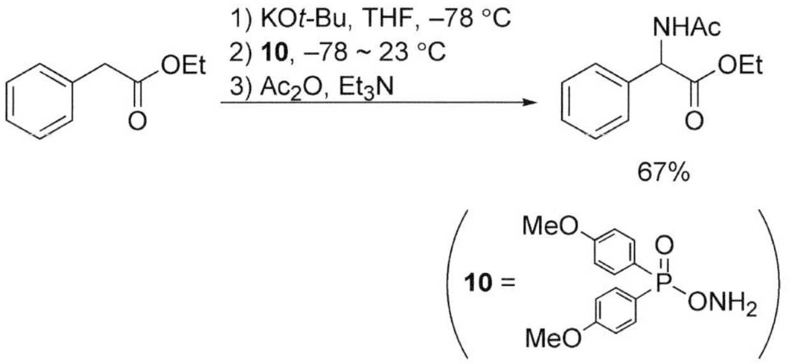

Scheme 1.12

–78 °C, reacts efficiently with stabilized sodium or potassium enolates derived from malonates, phenylacetates, and phenylacetonitrile as shown in Scheme 1.12 [12].

1.2.3
O-Acylhydroxylamine

O-Acylhydroxylamines have not been employed for electrophilic amination as extensively as *O*-sulfonyl- and *O*-phosphinylhydroxylamines [13]. Recently, though, J. S. Johnson has developed a mild and widely applicable method for the preparation of various secondary and tertiary amines through the copper-catalyzed electrophilic amination of organozinc reagents with *O*-benzoylhydroxylamines such as **11** or **12** [14]. The *O*-benzoylhydroxylamines, most of which are stable enough to be used in the subsequent amination, are prepared by the oxidation of the corresponding primary and secondary amines with benzoyl peroxide and K_2HPO_4 in DMF (Schemes 1.13 and 1.14).

Secondary and tertiary amines are synthesized by treatment of organozincs with *O*-benzoylhydroxylamines **11** and **12** in the presence of catalytic amounts of $[Cu(OTf)]\cdot C_6H_6$ (Schemes 1.15 and 1.16).

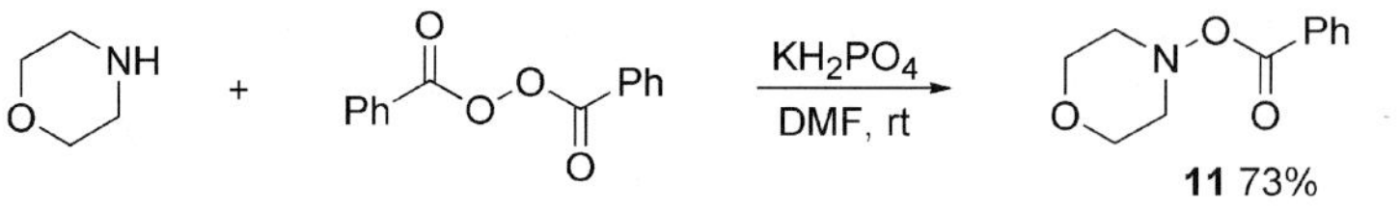

Scheme 1.13

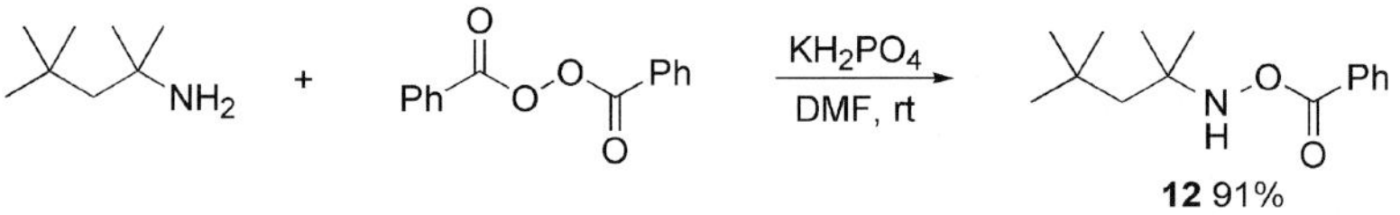

Scheme 1.14

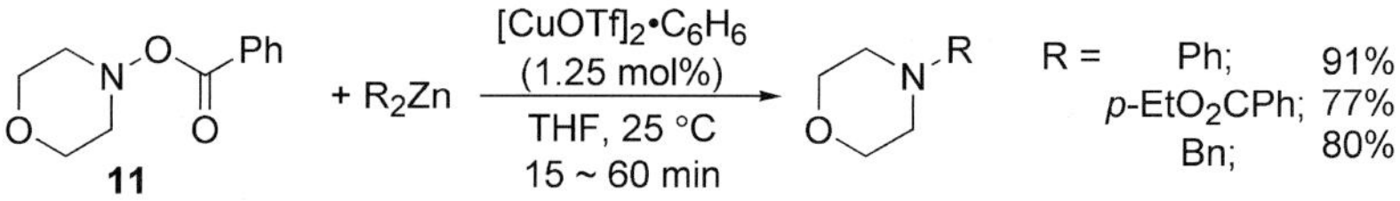

Scheme 1.15

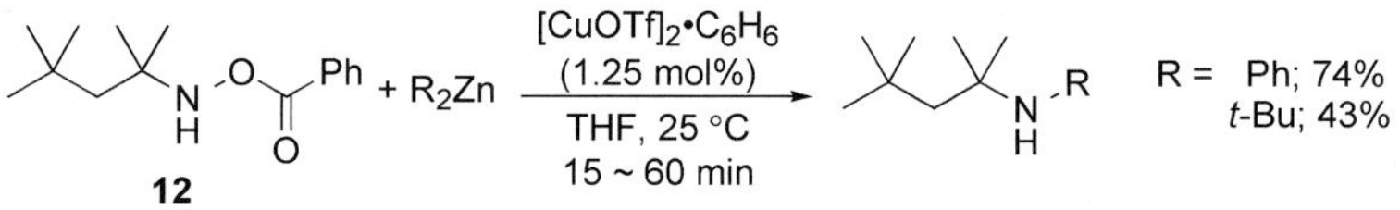

Scheme 1.16

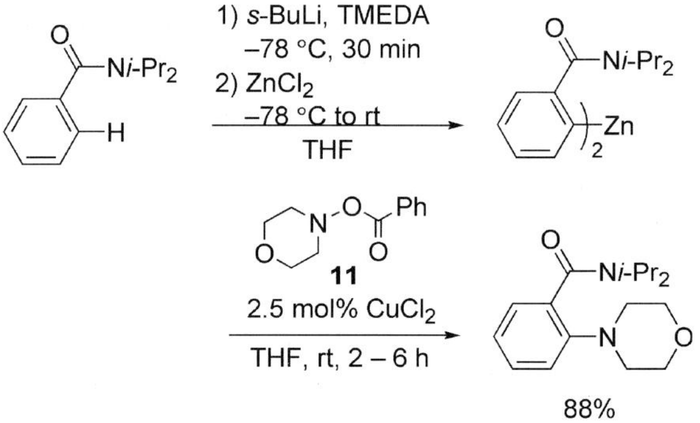

Scheme 1.17

This methodology can be used for aromatic C—H amination by combination with directed *ortho*-lithiation/transmetalation (Scheme 1.17).

1.2.4
O-Trimethylsilylhydroxylamine

Ricci developed the electrophilic amination of higher-order cyanocuprates with *N,O*-bis(trimethylsilyl)hydroxylamine (**13**), providing a suitable method for the preparation of primary amines (Scheme 1.18) [15].

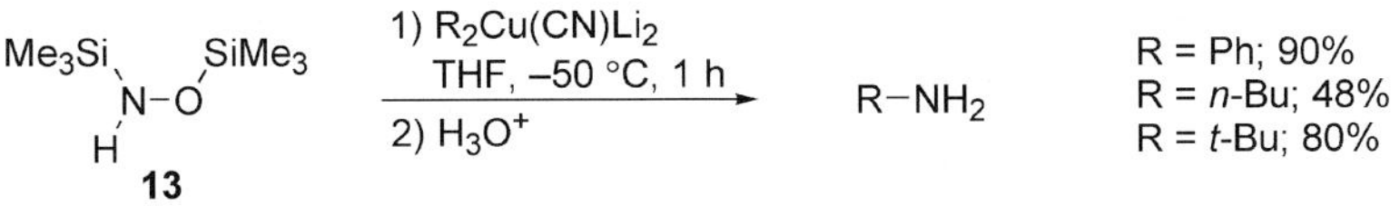

Scheme 1.18

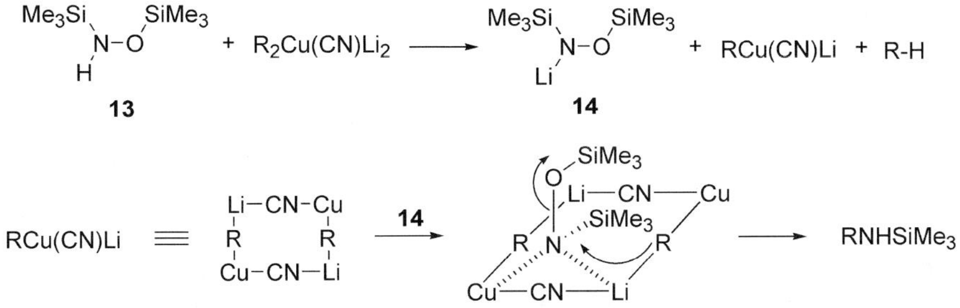

Scheme 1.19 Electrophilic amination with *N,O*-bis(trimethylsilyl)hydroxylamines (**13**).

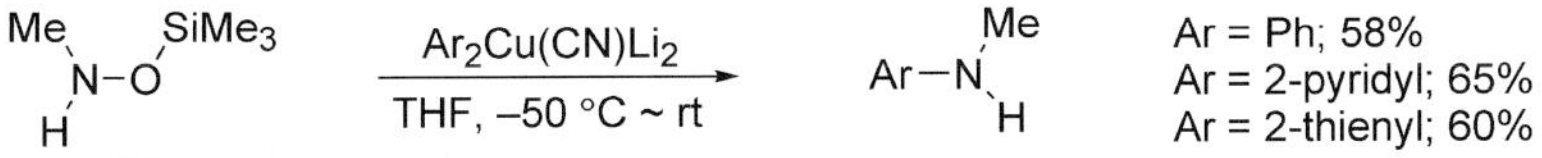

Scheme 1.20

N,O-Bis(trimethylsilyl)hydroxylamine (**13**) first reacts with the higher-order cuprate to generate lithium *N*-silyl-*N*-siloxyamide **14** and monoanionic lower-order cyanocuprate. The new C—N bond may be formed by the interception of lithium amide **14** with the thus formed cuprate via an amide–copper intermediate as shown in Scheme 1.19.

Similarly, by starting from *N*-alkyl-*O*-(trimethylsilyl)hydroxylamines such as **15**, *N*-alkyl aromatic and heteroaromatic amines are prepared by treatment with aryl- and heteroarylcyanocuprates (Scheme 1.20) [16].

1.2.5
Experimental Procedures

Representative synthesis of O-benzoylhydroxylamines Morpholine (5.2 mL. 60 mmol) was added by syringe in one portion to a mixture of benzoyl peroxide (12.1 g, 50 mmol) and dipotassium hydrogen phosphate (13.1 g, 75 mmol) in DMF (125 mL). The suspension was stirred at ambient temperature for 1 h. Deionized water (200 mL) was added, and the contents were stirred vigorously for several minutes until all solids had dissolved. The organic materials were extracted with ethyl acetate (150 mL) and the combined extracts were washed with saturated aqueous $NaHCO_3$ (100 mL × 2). The aqueous fractions were combined and extracted with ethyl acetate (100 mL × 3), and the organic fractions were combined and washed with deionized water (100 mL × 3) and brine (100 mL), and dried over $MgSO_4$. Volatile materials were removed *in vacuo* and the resulting crude mixture was purified by flash column chromatography, with elution with 50% ethyl acetate/hexane, to afford 4-benzoyloxymorpholine (**11**, 7.71 g, 37 mmol, 74%) of >95% purity by ^{1}H NMR spectroscopy.

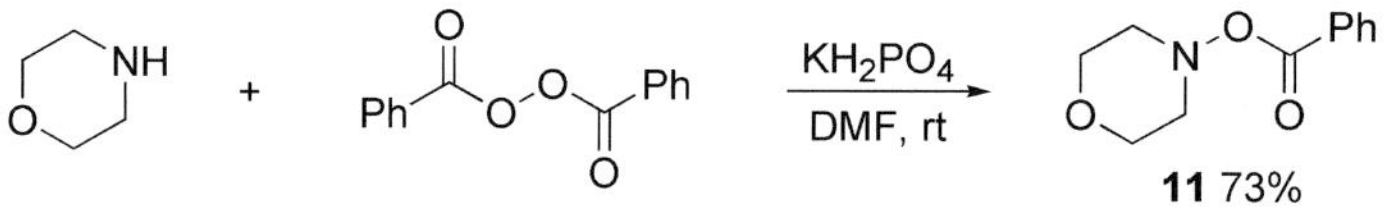

Representative electrophilic amination with O-benzoylhydroxylamines A THF solution of diphenylzinc, prepared from an ethereal solution of PhMgBr (1.0 M, 1.1 mL, 1.1 mmol) and $ZnCl_2$ (75 mg, 0.55 mmol) in THF (2.0 mL), was added by cannula in one portion to a mixture of 4-benzoyloxymorpholine (**11**, 103 mg, 0.50 mmol) and $[CuOTf]_2 \cdot C_6H_6$ (3 mg, 0.0056 mmol) in THF (5.0 mL). The resulting solution was stirred at ambient temperature for 1 h. The reaction mixture was diluted with Et_2O (10 mL) and transferred into a separating funnel. The mixture was washed with saturated aq. $NaHCO_3$ (10 mL × 3) and the amino components were extracted with 10% aqueous HCl (10 mL × 3). The aqueous extracts were basified with 10% aq. NaOH and the amino components were extracted with CH_2Cl_2 (10 mL × 3). The organic fraction was washed with brine (10 mL), dried over Na_2SO_4, and concentrated *in vacuo* to afford 4-phenylmorpholine as a white solid (80 mg, 0.49 mmol, 98%) of >95% purity by ^{1}H NMR.

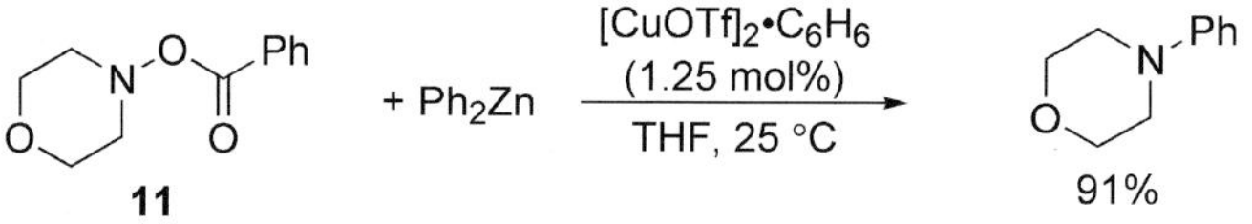

Representative electrophilic amination with N,O-bis(trimethylsilyl)hydroxylamine *N,O*-Bis(trimethylsilyl)hydroxylamine (0.426 mL, 2.0 mmol), commercially available from Aldrich Chemical Co., Inc., was added dropwise at −50 °C to a clear brown solution of $Ph_2CuCNLi_2$ (2.0 mmol) in THF After stirring for 1 h, the dark reaction mixture was hydrolyzed with 20% aq. HCl (30 mL). The aqueous layer was basified with NaOH, and aniline was extracted with Et_2O (2 × 20 mL). The organic layer was washed with brine and dried over Na_2SO_4. The solvent was removed *in vacuo* and the resulting crude materials were purified by distillation to afford aniline (167 mg, 90%) as a clear liquid.

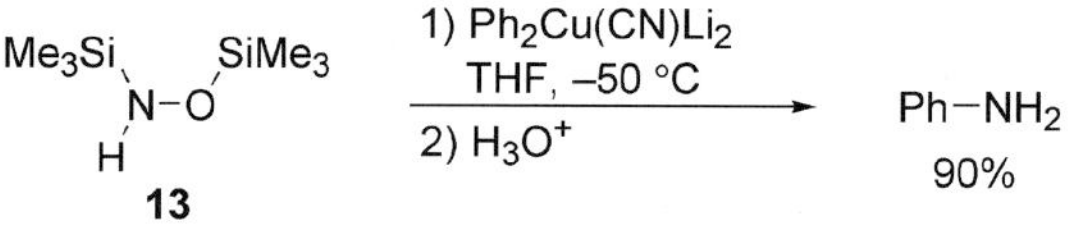

1.3
Oxime Derivatives

Oxime derivatives are readily converted into a variety of amino compounds through representative reactions such as the Beckmann rearrangement, Beckmann fragmentation, and the Neber reaction [17]. This section mainly focuses on C—N bond formation by electrophilic amination of carbanions with oxime derivatives by substitution on the sp^2 nitrogen atom.

1.3.1
Synthesis of Primary Amines by Electrophilic Amination of Carbanions

Oxime sp^2 nitrogen atoms possessing suitable leaving groups (OR^1) react with organometallic reagents (R^3–M) to afford the corresponding *N*-alkyl- or *N*-arylimines, which are readily hydrolyzed to primary amines (Scheme 1.21). To make this substitution reaction efficient, competing side reactions such as the Beckmann rearrangement and the Neber reaction have to be suppressed by suitably masking the oxime derivatives.

Murdoch reported that treatment of tetraphenylcyclopentadienone *O*-tosyloxime (**17**) with excess amounts of aryllithium and aryl Grignard reagents gives *N*-arylimines, which can be converted into primary amines and cyclopentadienone oxime by treatment with excess hydroxylamine in aqueous pyridine (Scheme 1.22) [18]. The formation of the imines probably proceeds through nucleophilic addition to the nitrogen atom of oxime **17** to generate stabilized cyclopentadienyl anions, which undergo elimination of tosylate.

Acetone *O*-(2,4,6-trimethylphenylsulfonyl)oxime (**18**) can be applied in the amination of arylmagnesium and arylzinc reagents [19]. Treatment of oxime **18** under Barbier conditions (i.e., treatment of aryl bromide with **18** and magnesium in THF at reflux temperature), followed by the hydrolysis of the resulting imines under

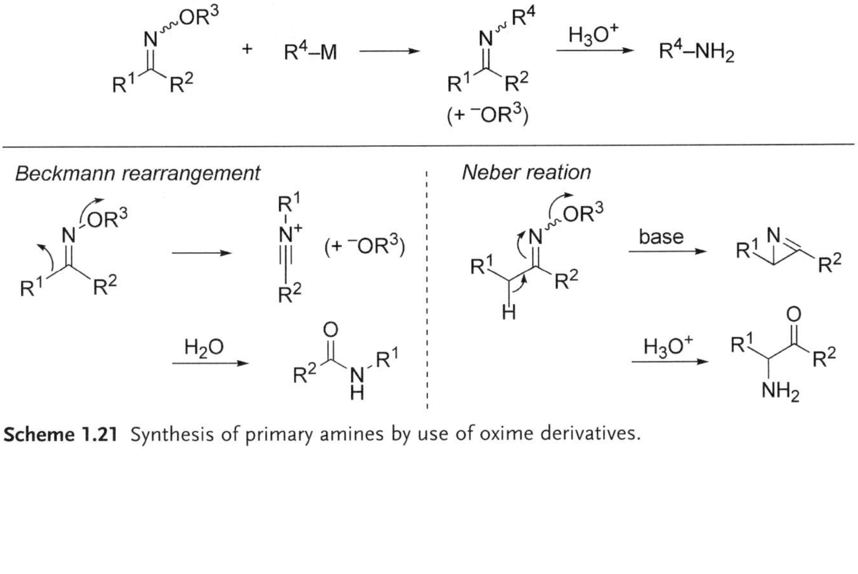

Scheme 1.21 Synthesis of primary amines by use of oxime derivatives.

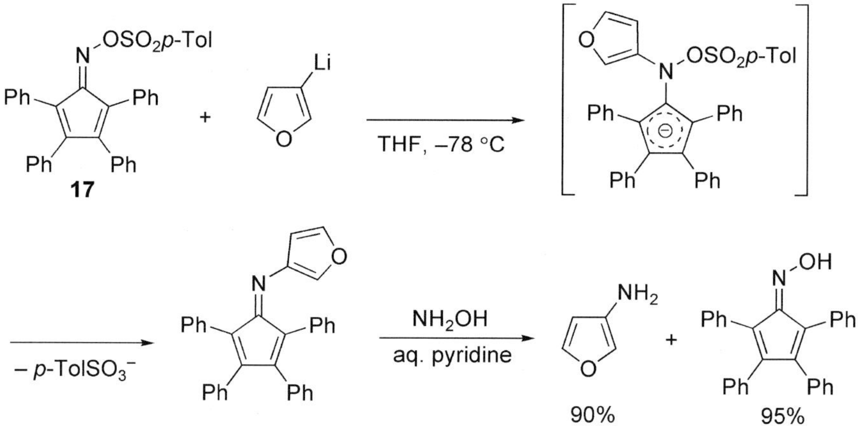

Scheme 1.22 Electrophilic amination with tetraphenylcyclopentadienone *O*-tosyloxime (**17**).

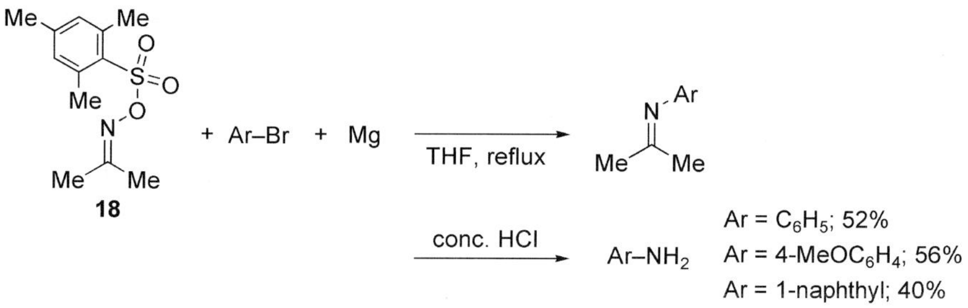

Scheme 1.23 Electrophilic amination with acetone *O*-(2,4,6-trimethylphenylsulfonyl)oxime (**18**).

acidic conditions, afforded *N*-aryl primary amines, although the yields were moderate (40–56%) (Scheme 1.23) [20].

Narasaka developed the amination of Grignard reagents with bis[3,5-bis(trifluo romethyl)phenyl] ketone *O*-tosyloxime (**19**) [21], the introduction of the electron-withdrawing trifluoromethyl groups suppressing the competing Beckmann rearrangements. Various primary amine derivatives are synthesized by the reaction with aryl and alkyl Grignard reagents, except in the case of tertiary alkyl reagents (Scheme 1.24).

A chiral secondary amine is prepared without loss of optical purity by treatment of a chiral Grignard reagent with oxime **19** (Scheme 1.25) [22]. This means that the reaction proceeds not through an electron-transfer mechanism but by nucleophilic substitution at the oxime nitrogen.

The employment of *O*-sulfonyl oximes of cyclic ureas and carbonates [23, 24] works effectively for the electrophilic amination of Grignard reagents, because they never undergo Beckmann rearrangements or Neber reactions. Among them, 4,4,5,5-tetramethyl-1,3-dioxolan-2-one *O*-(phenylsulfonyl)oxime (**20**), which can be

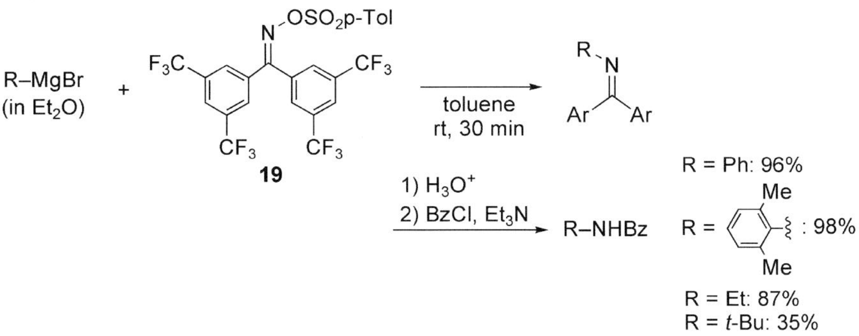

Scheme 1.24 Electrophilic amination with bis[3,5-bis(trifluoromethyl)phenyl] ketone *O*-tosyloxime (**19**).

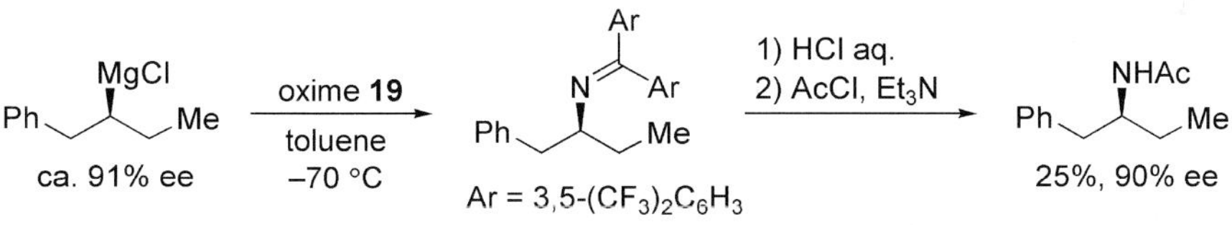

Scheme 1.25 Synthesis of a chiral secondary amine with oxime **19**.

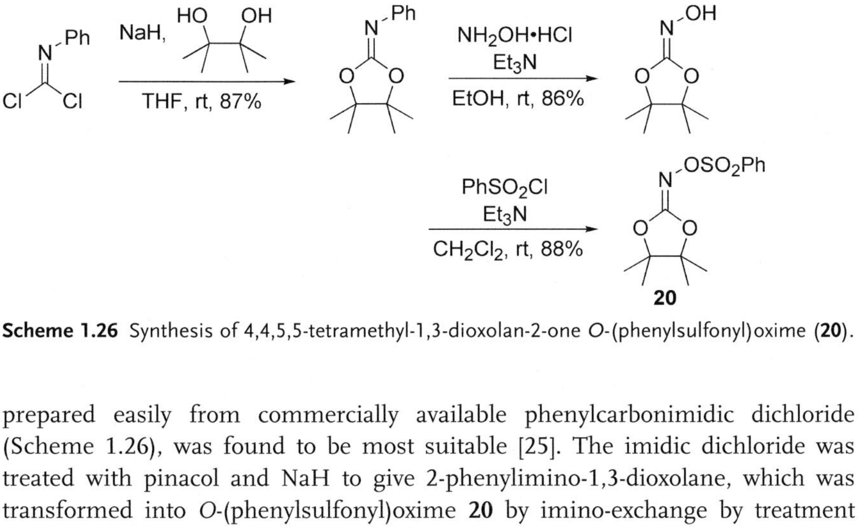

Scheme 1.26 Synthesis of 4,4,5,5-tetramethyl-1,3-dioxolan-2-one *O*-(phenylsulfonyl)oxime (**20**).

prepared easily from commercially available phenylcarbonimidic dichloride (Scheme 1.26), was found to be most suitable [25]. The imidic dichloride was treated with pinacol and NaH to give 2-phenylimino-1,3-dioxolane, which was transformed into *O*-(phenylsulfonyl)oxime **20** by imino-exchange by treatment with hydroxylamine followed by *O*-sulfonylation of the resulting oxime.

Various Grignard reagents react with oxime **20** in nonpolar solvents to afford the corresponding imines, which are easily converted into primary amines under mild acidic conditions (Table 1.2). Aryl Grignard reagents, regardless of steric congestion and the electronic effects of the substituents on the aryl group, are smoothly aminated with **20**, and anilines are obtained after hydrolysis or solvolysis of the resulting *N*-aryl imines. Primary, secondary, and tertiary alkylamines are

Table 1.2 Synthesis of primary amines by the electrophilic amination of Grignard reagents with *O*-(phenylsulfonyl)oxime (**20**).

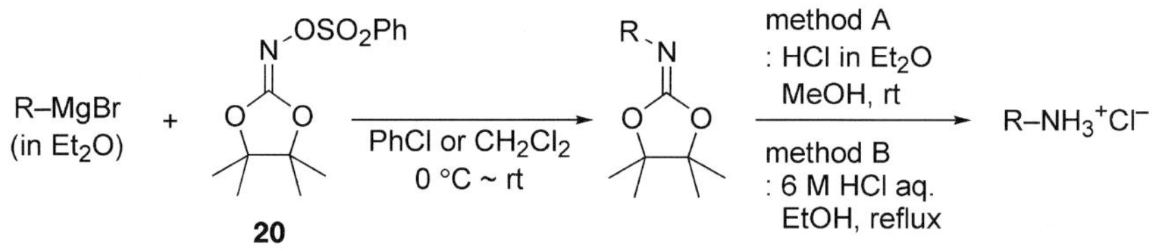

R		yield / %	R		yield / %
Ph	A	93	$PhCH_2CH_2$	B	90
p-CF_3-C_6H_4	A	94	$PhCH_2CH(CH_3)$	B	89
2,4-$(MeO)_2$-C_6H_3	A	91	1-adamantyl	B	89
2,6-Me_2-C_6H_3	B	90	CH_2=$CH(CH_3)$	—	93[a]

a) Yield of 2-aza-1,3-diene.

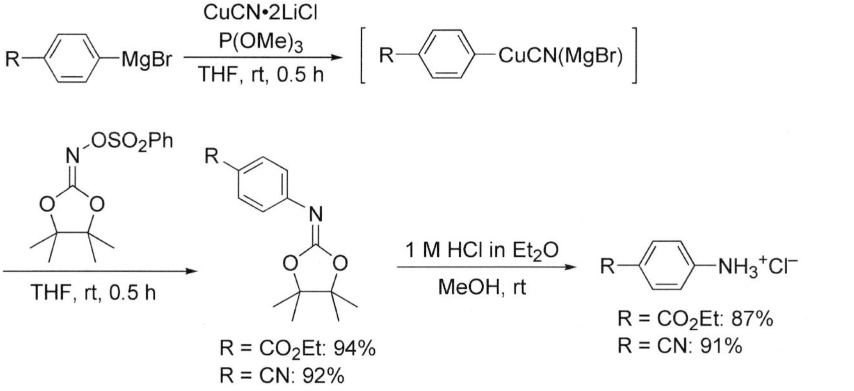

Scheme 1.27 Electrophilic amination of arylcopper reagents with *O*-(phenylsulfonyl)oxime (**20**).

prepared in high yield from the corresponding alkyl Grignard reagents, and even alkenyl Grignard reagents reacted with **20** to give 2-aza-1,3-dienes.

Aryl Grignard reagents bearing a cyano or an alkoxy carbonyl group, prepared by iodine–magnesium exchange [26], cannot be used directly for this amination procedure, because of their instability at temperatures higher than 0 °C. Arylcopper reagents generated by transmetalation of such arylmagnesium compounds with CuCN · 2LiCl in the presence of trimethyl phosphite [27] react with 4,4,5,5-tetramethyl-1,3-dioxolan-2-one *O*-(phenylsulfonyl)oxime (**20**) to afford the corresponding *N*-arylimines, which are hydrolyzed to anilines (Scheme 1.27) [28].

1.3.2
Experimental Procedures

Procedure for the preparation of 4,4,5,5-tetramethyl-1,3-dioxolan-2-one O-phenylsulfonyloxime Pinacol (15.2 g, 128 mmol) in THF (60 mL) was slowly added under argon at 0 °C to a suspension of NaH (6.36 g, 264 mmol) in THF (250 mL), after which phenylcarbonimidic dichloride (20.5 g, 128 mmol) in THF (40 mL) was added over 15 min. This mixture was stirred at room temperature for 30 min, after which the reaction was quenched with a pH 9 ammonium buffer and the mixture was extracted three times with ethyl acetate. The combined extracts were washed with brine and dried over Na_2SO_4, and the ethyl acetate was removed *in vacuo* to give an 87% yield of 4,4,5,5-tetramethyl-2-phenylimino-1,3-dioxolane (22.5 g, 111 mmol), which was used without further purification.

Triethylamine (61.7 g, 611 mmol) and $NH_2OH \cdot HCl$ (34.1 g, 491 mmol) were added to a solution of 4,4,5,5-tetramethyl-2-phenylimino-1,3-dioxolane (26.7 g, 122 mmol) in ethanol (300 mL), and this mixture was stirred at room temperature for 24 h. After the reaction had been quenched with pH 9 ammonium buffer, the mixture was extracted three times with ethyl acetate. The combined extracts were washed with water and brine and dried over Na_2SO_4, the ethyl acetate was removed *in vacuo*, and the crude materials were purified by flash column chromatography (hexane/ethyl acetate 1 : 1 to 1 : 4) to give 4,4,5,5-tetramethyl-1,3-dioxolan-2-one oxime (16.7 g, 105 mmol) in 86% yield.

Benzenesulfonyl chloride (5.59 g, 31.6 mmol) in dichloromethane (15 mL) was slowly added under argon at 0 °C to a solution of 4,4,5,5-tetramethyl-1,3-dioxolan-2-one oxime (4.49 g, 22.8 mmol) and triethylamine (5.90 mL, 42.3 mmol) in dichloromethane (100 mL), and the mixture was stirred at room temperature for 1 h. After the reaction had been quenched with ice water, the mixture was extracted three times with ethyl acetate, the combined extracts were washed with brine and dried over anhydrous sodium sulfate, the ethyl acetate was removed *in vacuo*, and the crude materials were purified by recrystallization (hexane/ethyl acetate) to give 4,4,5,5-tetramethyl-1,3-dioxolan-2-one *O*-phenylsulfonyloxime (**20**, 7.40 g, 25.9 mmol) in 88% yield.

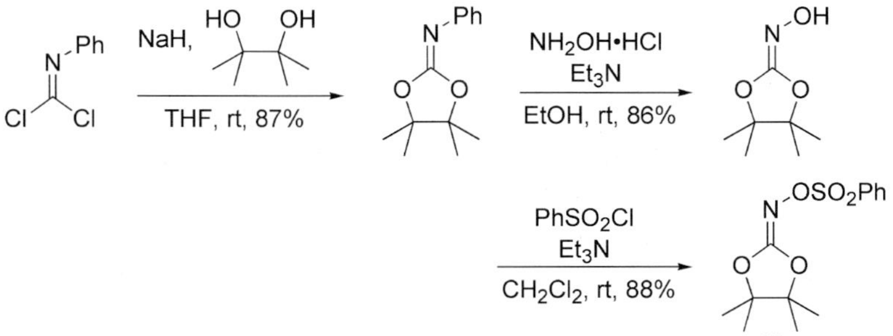

20

Representative procedure for the preparation of primary amine hydrochlorides by treatment of 4,4,5,5-tetramethyl-1,3-dioxolan-2-one O-phenylsulfonyloxime with Grignard reagents An ether solution of phenylmagnesium bromide (0.96 M, 2.3 mL, 2.2 mmol) was added dropwise under argon at 0 °C to a solution of 4,4,5,5-tetramethyl-1,3-dioxolan-2-one *O*-phenylsulfonyloxime (**20**, 593 mg, 1.98 mmol) in chlorobenzene (15 mL), and this mixture was stirred at the same temperature for 30 min. The reaction was then quenched with pH 9 ammonium buffer at 0 °C, and the mixture was extracted three times with ethyl acetate. The combined extracts were washed with brine and dried over Na_2SO_4, and the ethyl acetate was removed under vacuum. Hydrogen chloride in ether (1.0 M, 4.0 mL) was added at 0 °C to the crude imine in methanol (10 mL), and this mixture was stirred at room temperature for 1.5 h. Volatile materials were removed *in vacuo*, and anhydrous ether (40 mL) was added. The insoluble materials were collected by filtration to give aniline hydrochloride (239 mg, 1.84 mmol) in 93% yield.

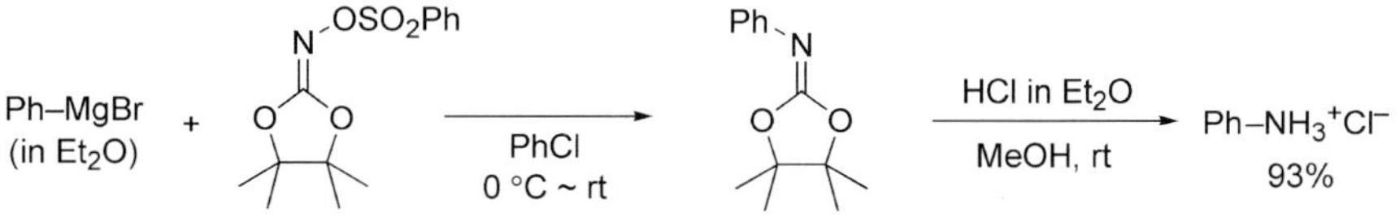

Representative procedure for the preparation of primary arylamines possessing electron-withdrawing groups by use of 4,4,5,5-tetramethyl-1,3-dioxolan-2-one O-phenylsulfonyloxime A THF solution of isopropylmagnesium bromide (1.15 M, 0.96 mL, 1.1 mmol) was slowly added under argon at −20 °C to a solution of ethyl 4-iodobenzoate (278 mg, 1.01 mmol) in THF, and this mixture was stirred at the same temperature for 30 min. A THF solution of CuCN·2LiCl (0.50 M, 2.0 mL, 1.0 mmol) was then added, the temperature again being kept below −20 °C. After completion of the addition, the reaction mixture was allowed to warm to room temperature over 30 min. Trimethyl phosphate (128 mg, 2.0 mmol) was then added and the clear solution was stirred for an additional 5 min. 4,4,5,5-Tetramethyl-1,3-dioxolan-2-one *O*-phenylsulfonyloxime (**20**, 290 mg, 0.969 mmol) in THF (3 mL)

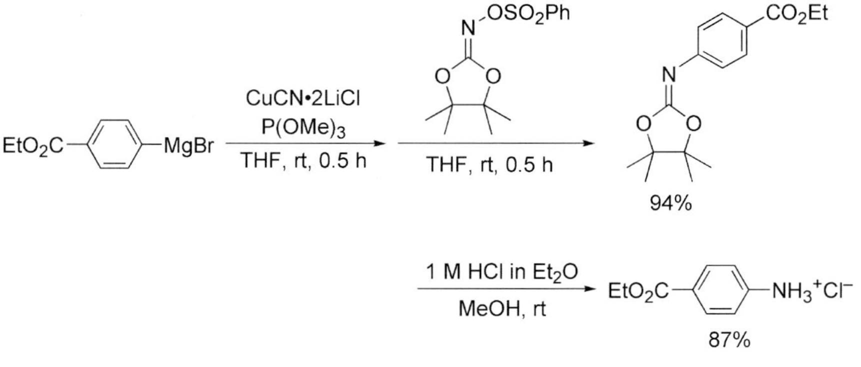

was then added dropwise and the reaction mixture was stirred at this temperature for 15 min. After the reaction had been quenched with a pH 9 buffer at 0 °C, the mixture was extracted three times with ethyl acetate, and the combined extracts were washed with brine and dried over anhydrous sodium sulfate. The ethyl acetate was removed under vacuum and the crude materials were purified by flash column chromatography (silica gel, hexane/ethyl acetate 8:2) to give 2-(4-ethoxycarbonylphenyl)imino-4,4,5,5-tetramethyl-1,3-dioxolane (266 mg, 0.913 mmol) in 94% yield. The resulting imine was converted into aniline by the same procedure as described in Section 1.3.2.

1.4 Azo Compounds

In this section some amination reactions utilizing azodicarboxylates (Section 1.4.1) and arylazo sulfones (Section 1.4.2) as nitrogen sources are illustrated. In the azodicarboxylate section, amination reactions of alkenes are discussed, because there are some reviews on the electrophilic amination of carbanions leading various hydrazine dicarboxylates [29].

1.4.1 Azodicarboxylates

1.4.1.1 Allylic Amination through Ene-Type Reactions

Ene reactions play an important role in organic transformations, and the use of azo compounds such as diethyl azodicarboxylate (DEAD; **21**) as enophiles provides a useful method for amination of alkenes (aza-ene reaction) to give allylic hydrazines (Scheme 1.28). Although thermal aza-ene reactions of various alkenes with DEAD have been reported, such reactions generally require high temperatures and are difficult to control because of the formation of bis-adducts [30].

Leblanc improved this thermal aza-ene reaction by use of bis-(2,2,2-trichloroethyl) azodicarboxylate (**22**) [31]. Reactions between alkenes and **22**

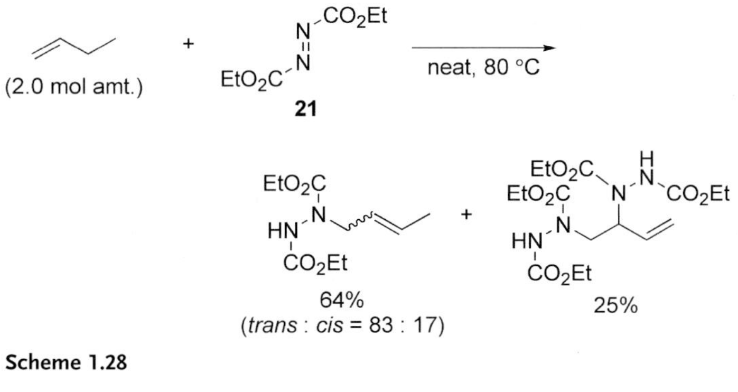

Scheme 1.28

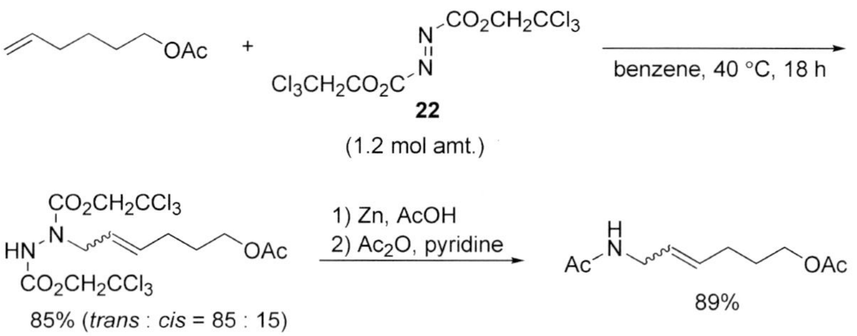

Scheme 1.29 The aza-ene reaction with bis(2,2,2-trichloroethyl) azodicarboxylate (**22**).

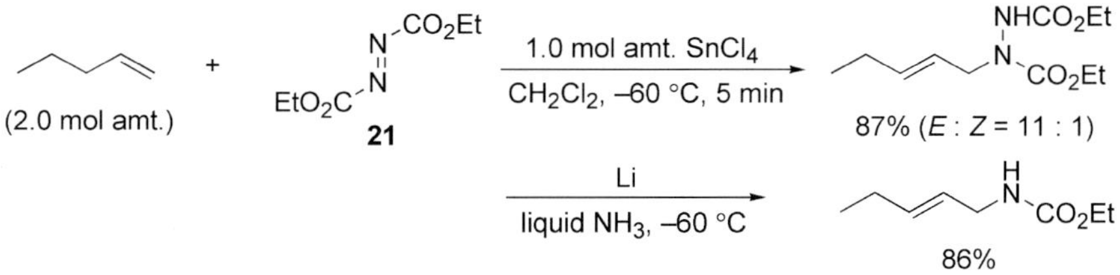

Scheme 1.30 Lewis acid-mediated allylic amination of alkenes with DEAD (**21**).

proceed under milder conditions to give the corresponding ene adducts in good yield without the formation of bis-adducts (Scheme 1.29). In addition, by treatment with Zn dust and acetic acid, the reductive cleavage of the N—N bonds in the resulting allyllic hydrazines proceeds to give allylic amine derivatives.

Heathcock developed Lewis acid-mediated allylic amination of alkenes with DEAD (Scheme 1.30) [32]. The use of $SnCl_4$ in dichloromethane promotes the reaction at –60 °C, affording the ene adducts in good yield with excellent selectivity for the formation of (*E*)-alkenes. The allylic hydrazines can be converted into carbamates by treatment with lithium in liquid ammonia. In addition, $LiClO_4$ was also able to catalyze aza-ene reactions of azodicarboxylate derivatives [33].

1.4.1.2 **Hydrohydrazination of Alkenes**

Carreira has recently developed the Co- and Mn-catalyzed hydrohydrazination of alkenes with azodicarboxylates, which enables the preparation of various alkylhydrazines from a broad range of alkenes [34].

Mono-, di-, and trisubstituted alkenes, including vinyl heterocycles, react with azodicarboxylates such as **23** in the presence of phenylsilane and the Co(III) catalyst **24**, bearing Schiff base ligands, to give alkylhydrazines in good yields (Table 1.3). Monosubstituted, 1,1-disubstituted, and trisubstituted alkenes give exclusively the Markovnikov-type hydrohydrazination products with broad functional group tolerance. In the reaction behavior of 1,2-disubstituted alkenes, the selectivity is governed by electronic effects. Phenyl substitution results in the formation of the benzylic hydrazine, while the presence of an ethoxycarbonyl group produces an α-hydrazinyl ester.

Table 1.3 Co-catalyzed hydrohydrazination of alkenes with azodicarboxylate **23**.

alkene	product (yield / %)	alkene	product (yield / %)
Ph~~~⚯	BocHN–NBoc Ph⁀⁀Me (85)	Me Ph⚯	BocHN–NBoc Ph–C(Me)(Me) (88)
Br~~~⚯	BocHN–NBoc Br⁀⁀Me (90)	Me⚯OH	BocHN–NBoc Me₂C–CH₂–OH (70)
(thiophene-vinyl)	BocHN–NBoc (thiophene)–CH(Me) (84)	Ph⚯Me	BocHN–NBoc Ph–CH(Me) (88)
(indole-vinyl)	BocHN–NBoc (indole)–CH(Me) (82)	Me⚯CO₂Et	BocHN–NBoc Me–CH–CO₂Et (66)

Reaction conditions: alkene + t-BuO$_2$C–N=N–CO$_2t$-Bu (**23**, 1.5 mol amt.), 2.5~5 mol% **24**, 1 mol amt. of PhSiH$_3$, EtOH, 23 °C, 4 h → alkylhydrazide. Co(III) cat. **24** (L = MeOH).

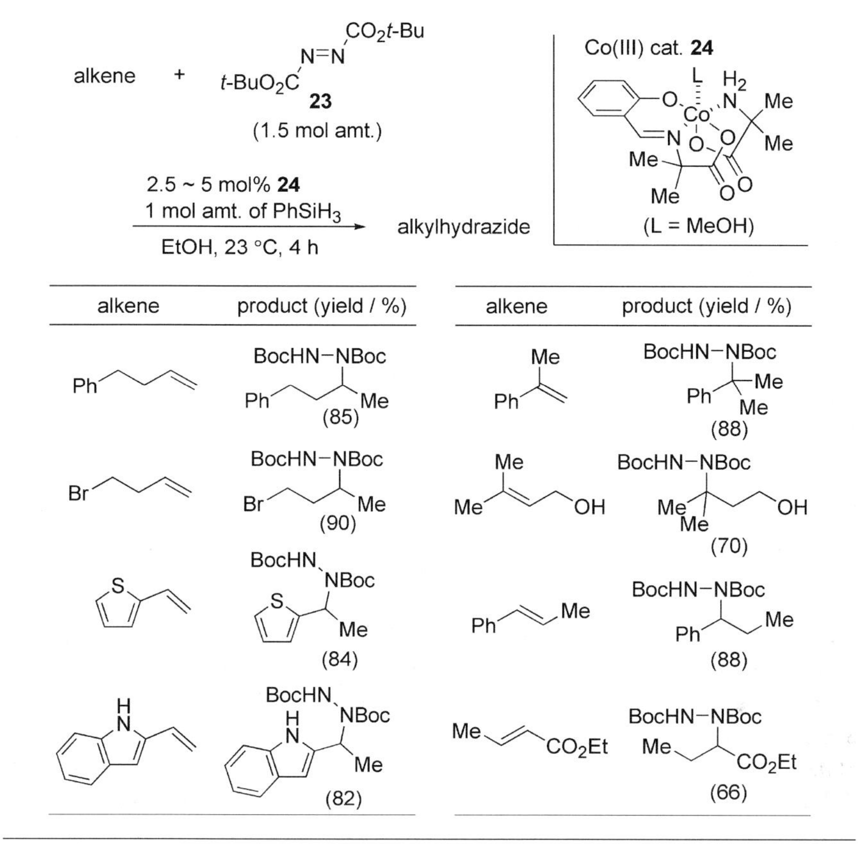

The above cobalt catalyst could not be employed for the hydrohydrazination of tetrasubstituted alkenes. As an alternative, Mn(III) complex **25** exhibits high catalytic reactivity even for the hydrohydrazination of hindered alkenes such as tetrasubstituted ones (Table 1.4).

The proposed mechanism of the Co-catalyzed reaction is shown in Scheme 1.31. The first step is the formation of the active Co(III)-hydride complex **I**. From **I**, hydrocobaltation of an alkene proceeds to form Co-alkyl complex **II**. It is believed that this step is rate-determining, whereas the following amination step is fast. The crucial amination step from **II** to **III** could proceed either by radical addition to the N=N double bond (path A) or by direct insertion of the N=N double bond into the Co-alkyl complex **II** (path B). The thus generated Co-hydrazido complex **III** reacts with a silane to regenerate Co-hydride complex **I** with the formation of silylated hydrazine derivatives, which are readily transformed into the alkylated hydrazines after ethanolysis.

This Co(III) catalyst was successfully applied to the hydrohydrazination of dienes and enynes. Although a simple reaction between di-*tert*-butyl

Table 1.4 Mn-catalyzed hydrohydrazination of tetrasubstituted alkenes with azodicarboxylate **23**.

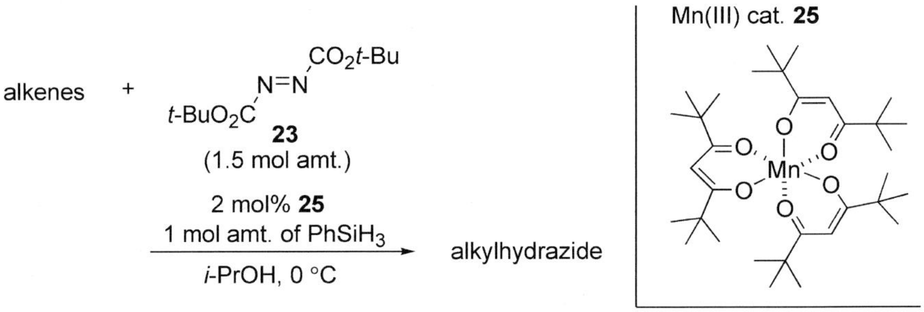

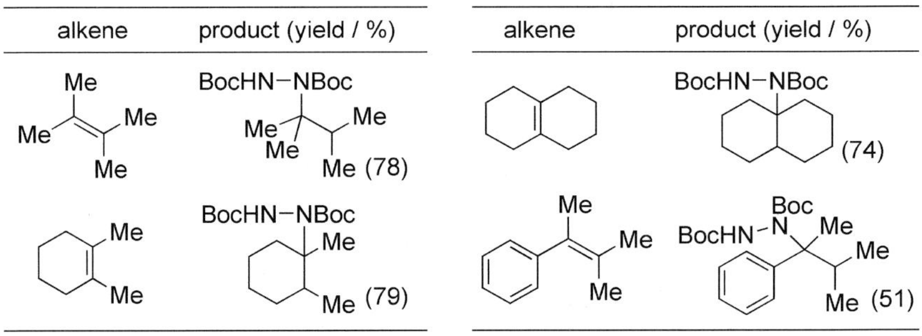

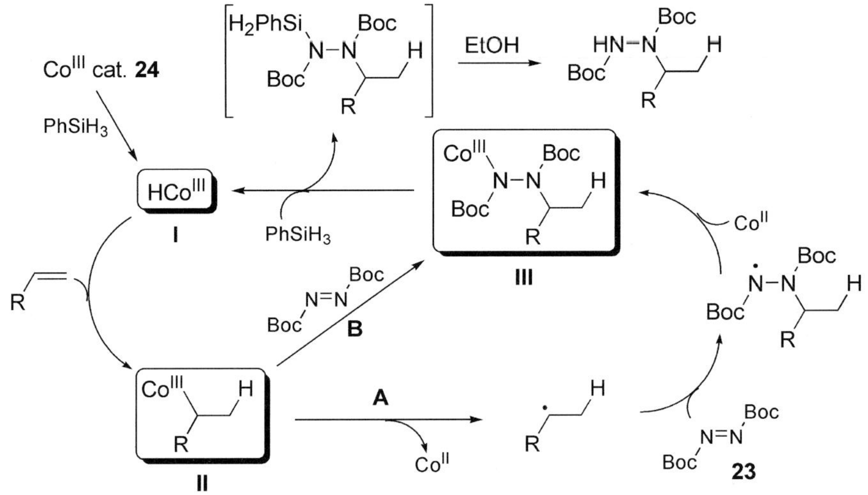

Scheme 1.31 Proposed mechanism of the hydrohydrazination of alkenes catalyzed by the Co(III) catalyst **24**.

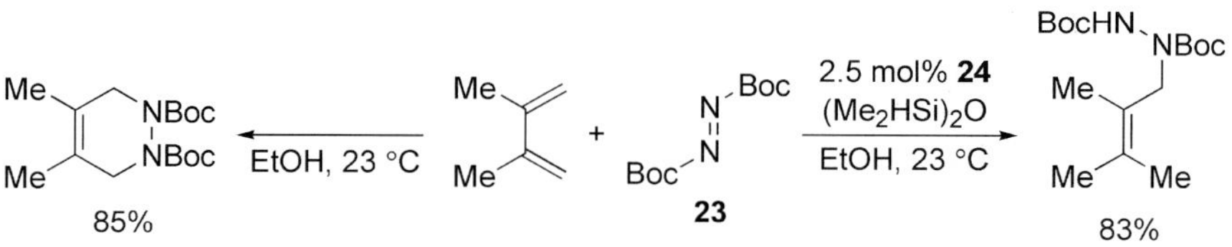

Scheme 1.32 Hydrohydrazination of dienes catalyzed by the Co(III) catalyst **24**.

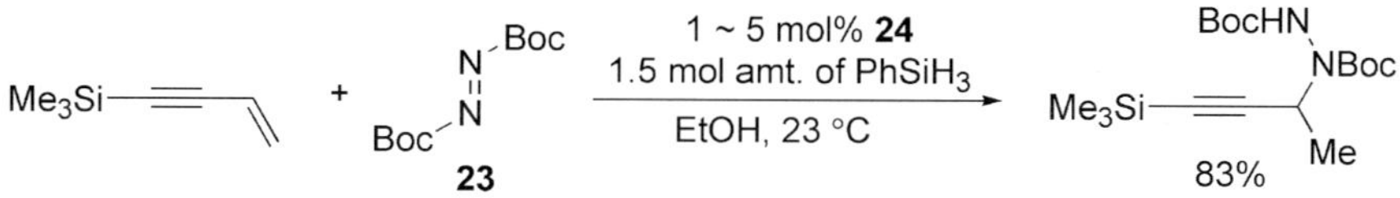

Scheme 1.33

azodicarboxylate (**23**) and 2,3-dimethyl-1,3-butadiene results in the formation of a [4+2] adduct in 85% yield at ambient temperature [35], the reaction in the presence of Co(III) catalyst **24** and tetramethyldisiloxane results in the preferential formation of allylic hydrazine in 83% yield (Scheme 1.32). The selective formation of the primary hydrazine derivative contrasts with the same reaction with alkenes, in which the formation of Markovnikov-type products was observed.

Propargylic hydrazines can be obtained from enynes through the selective amination of the double bonds (Scheme 1.33).

1.4.2
Arylazo Sulfones

Arylazo *p*-tolyl sulfones **26** (Scheme 1.34) are readily prepared from aromatic amines in a two-step sequence consisting of the formation of the corresponding

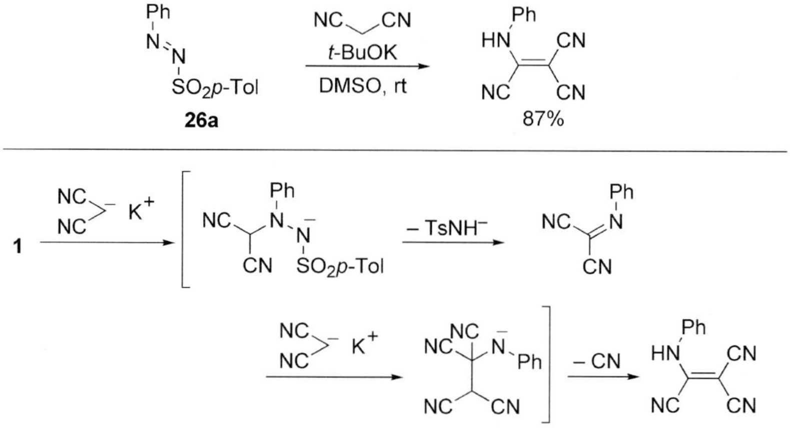

Scheme 1.34 Synthesis of tetrasubstituted ethylenes.

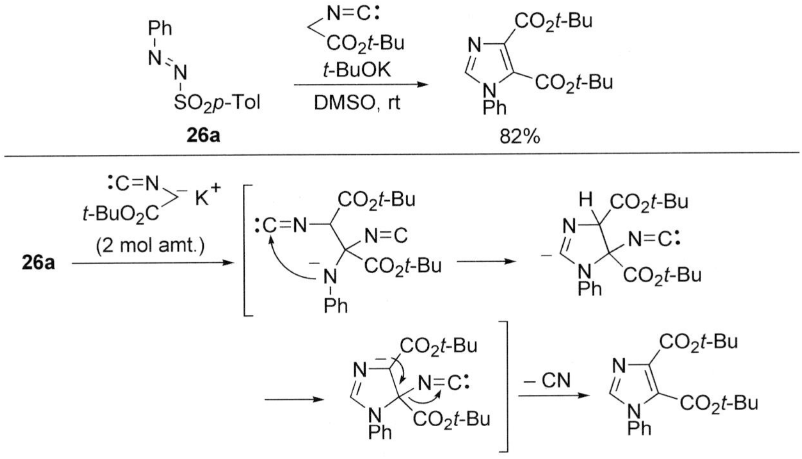

Scheme 1.35 Synthesis of tetrasubstituted 1-arylimidazoles.

arene diazonium salts and subsequent treatment with sodium *p*-tolyl sulfinate [36]. There are some reports on the synthesis of amino compounds by treatment of **26** with carbanions.

Potassium salts of active methylene compounds such as malononitrile react with phenylazo *p*-tolyl sulfones (**26a**) in DMSO to afford tetrasubstituted ethylenes bearing arylamino moieties (Scheme 1.34) [37]. Nucleophilic attack of the carbanion at the N=N double bond of **26a** and subsequent elimination of a tosylamide anion gives *N*-arylimines, on which a second nucleophilic attack by the carbanion proceeds to give tetrasubstituted ethylenes.

Substituted 1-arylimidazoles can be synthesized by treatment of phenylazo *p*-tolylsulfone (**26a**) with (*tert*-butoxycarbonyl)methyl isocyanides (Scheme 1.35) [38]. After double attack of the nucleophiles as described above (Scheme 1.34), intramolecular attack at the electrophilic isocyano group carbon, aromatization through proton transfer, and elimination of cyanide ion proceed successively to give imidazoles.

Knochel identified the utility of various arylazo *p*-tolyl sulfones **26** as synthetic equivalents of *N*-positively charged arylamine synthons. Arylazo *p*-tolyl sulfones **26** react under mild conditions with various polyfunctional arylmagnesium halides, and allylation of the resulting addition products, followed by treatment with zinc, provides polyfunctionalized diarylamines in good yield as shown in Table 1.5. Aliphatic magnesium halides are also aminated [39].

1.4.3 Experimental Procedures

Representative procedure for the Lewis acid-mediated allylic amination of alkenes with DEAD $SnCl_4$ (0.41 mL, 3.56 mmol) was added at –60 °C to a solution of DEAD (**21**, 620 mg, 3.56 mmol) and pent-1-ene (0.78 mL, 7.12 mmol) in CH_2Cl_2. After

Table 1.5 Synthesis of diarylamines by treatment of organomagnesium compounds with aryl *p*-tolyl sulfones **26**.

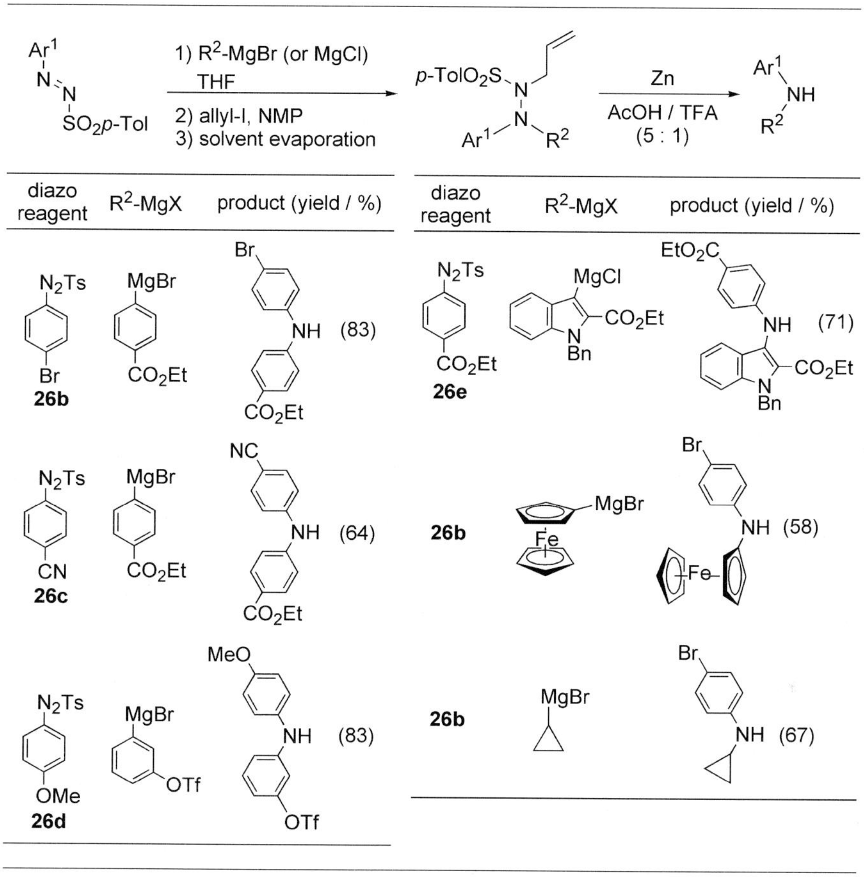

stirring for 5 min, the yellow solution had turned colorless and water (15 mL) was added. The organic materials were extracted with CH_2Cl_2 (3 × 50 mL), and the combined extracts were dried over Na_2SO_4. The solvents were concentrated under vacuum to afford crude materials, which were purified by flash column chromatography (silica gel, hexane/ethyl acetate 2 : 1) to give *N*-(pent-2-enyl)-*N*′-(ethoxycar bonyl)hydrazinecarboxylic acid ethyl ester (760 mg, 3.11 mmol) in 87% yield.

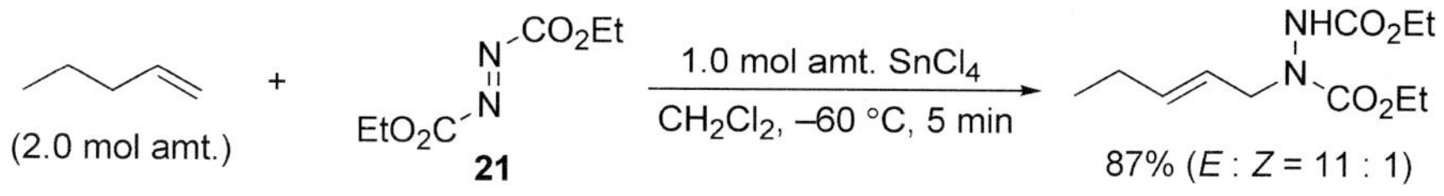

Representative procedure for the Co-catalyzed hydrohydrazination of alkenes The alkene (75 μL, 0.5 mmol) and phenylsilane (65 μL, 0.52 mmol) were added under argon at 23 °C to the Co catalyst **24** (10 mg, 0.025 mmol) in ethanol (2.5 mL). Di-*tert*-butyl azodicarboxylate (**23**, 0.17 g, 0.75 mmol) was then added in one portion, and the resulting solution was stirred at 23 °C for 4 h. The reaction mixture was quenched with water (1 mL) and brine (5 mL) and extracted with ethyl acetate (3 × 10 mL). The solvents were removed *in vacuo* to afford crude materials, which were purified by flash column chromatography (silica gel, hexane/ethyl acetate 10 : 1) to give *N*-(3-phenyl-1-methylpropyl)-*N*′-(*tert*-butoxycarbonyl)hydrazinecarboxylic acid *tert*-butyl ester (155 mg, 0.425 mmol) in 85% yield.

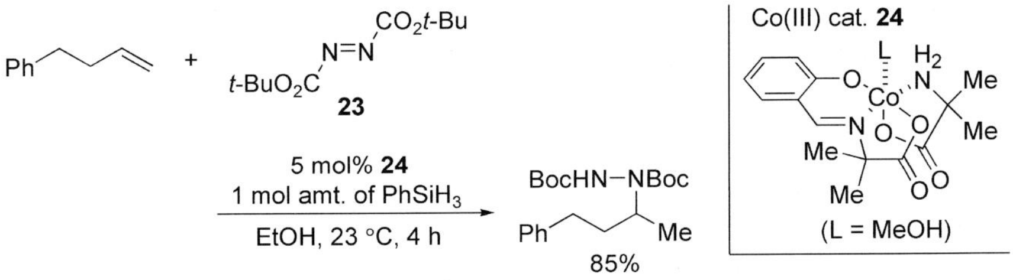

Representative preparation of arylazo tosylates: 4-bromophenylazo p-tolyl sulfone 4-Bromoaniline (1.72 g, 10 mmol) was dissolved in an aqueous HBF_4 solution (50% in water, 15 mL) and cooled to 0 °C, and then a solution of $NaNO_2$ (760 mg, 11 mmol) in water (5 mL) was added dropwise. After stirring for 30 min, the reaction mixture was allowed to warm to room temperature. The resulting white precipitate was filtered off and washed with aqueous HBF_4 solution (10 mL), ethanol (10 mL), and Et_2O (20 mL). The white crystalline powder was dissolved in CH_2Cl_2, TsNa (2.14 g, 12 mmol) was then added, the mixture was stirred overnight, and the resulting salts were removed by filtration. The solvent was removed under vacuum, and the resulting crude materials were purified by crystallization from ethanol to give 4-bromophenylazo *p*-tolyl sulfone (**26b**; 2.71 g, 8.0 mmol) in 80% yield (Scheme 1.36).

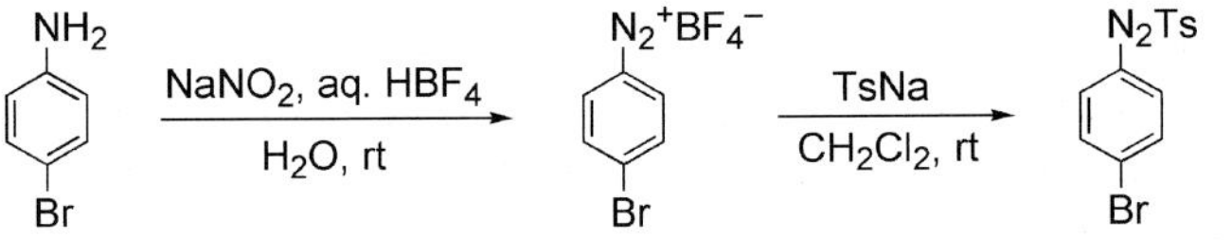

26b 80% (2 steps)

Scheme 1.36

Representative procedure for the amination of arylmagnesium reagents with arylazo tosylates A THF solution of *i*-PrMgCl (0.95 M, 1.15 mL, 1.1 mmol) was added dropwise at –20 °C to a solution of ethyl 4-iodobenzoate (306 mg, 1.1 mmol) in THF (5 mL). After the mixture had been stirred for 30 min, a solution of 4-bromophenylazo tosylate (**26b**, 339 mg, 1 mmol) in THF (3 mL) was added dropwise to the solution of the Grignard reagent, and the reaction mixture was stirred for 1 h at –20 °C. The mixture was treated with allyl iodide (510 mg, 3 mmol) and NMP

(*N*-methyl-2-pyrrolidone; 2 mL), and stirred for 2 h at room temperature. After the solvent had been removed under vacuum, the resulting residue was dissolved in glacial acetic acid (10 mL). Zn powder (10 mmol) and trifluoroacetic acid (2 mL) were added to the mixture, which was then heated at 75 °C until no starting material was evident by TLC analysis (2 h). After cooling to room temperature, the mixture was poured into crushed ice (ca. 30 g) and aqueous NaOH (2 M, 20 mL). The organic materials were extracted three times with Et_2O (30 mL) and the combined extracts were washed with saturated aqueous $NaHCO_3$ and brine. The solvent was removed *in vacuo*, and the resulting residue was purified by flash column chromatography (silica gel, pentane/Et_2O 9 : 1) to give ethyl 4-(4-bromophenylamino)benzoate (265 mg, 0.83 mmol) in 83% yield as a colorless solid.

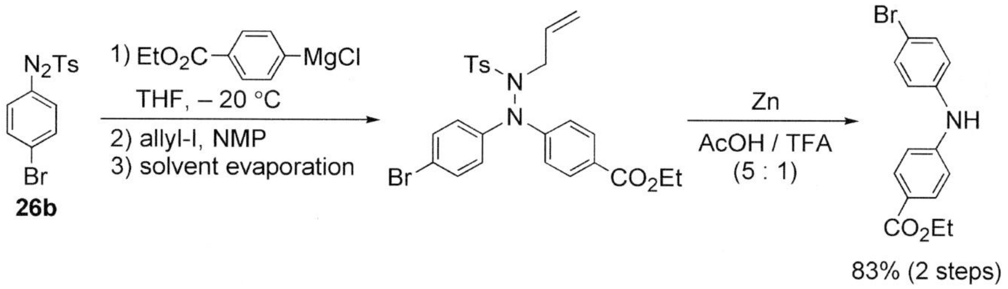

1.5 Oxaziridine Derivatives

Oxaziridines exhibit unique reactivity as a result of their ring strain and their relatively weak N—O bonds. They are utilized either as amination or as oxygenation reagents of nucleophiles. The site of nucleophilic attack (at the N or the O atom) in an oxaziridine is governed by the substituent at the nitrogen [40, 41].

1.5.1 Electrophilic Amination of Carbon Nucleophiles

N-Alkoxy- or -aminocarbonyl oxaziridines, easily prepared by treatment of the corresponding imines with *m*CPBA/*n*-BuLi, are used as aminating reagents of enolate anions [42]. *N*-Carboxamide oxaziridine **27**, for example, is used for the α-amination of various enolate anions in good to moderate yields (Table 1.6) [43]. These *N*-transfer reactions contrast sharply with those of *N*-sulfonyloxaziridines, which give α-hydroxylated product exclusively [40].

1.5.2 Amination of Allylic and Propargylic Sulfides by Use of a Ketomalonate-Derived Oxaziridine

Armstrong found that amination of sulfides proceeded with the oxaziridine **28**, derived from 2-oxomalonate, to afford a wide range of sulfimides (Scheme 1.37) [44].

Table 1.6 Electrophilic amination of lithium enolates with oxaziridine **27**.

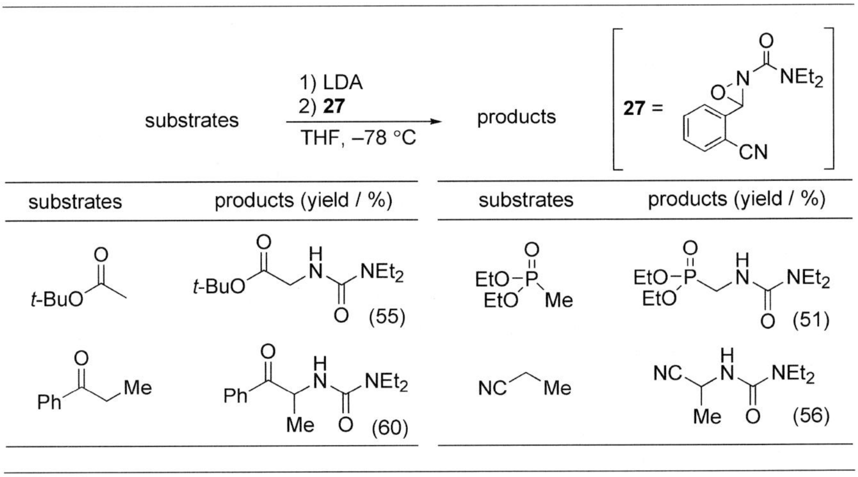

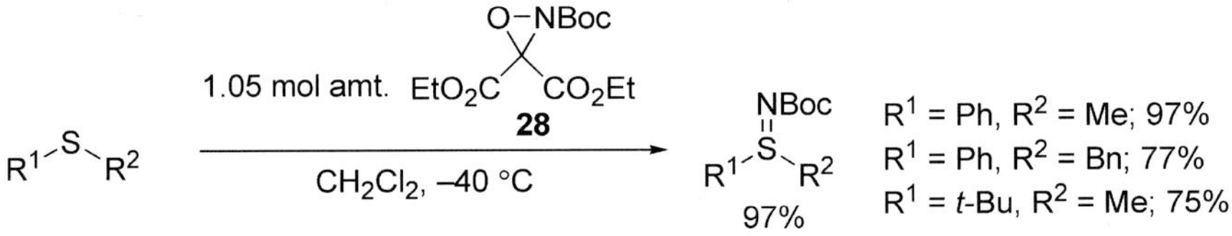

Scheme 1.37

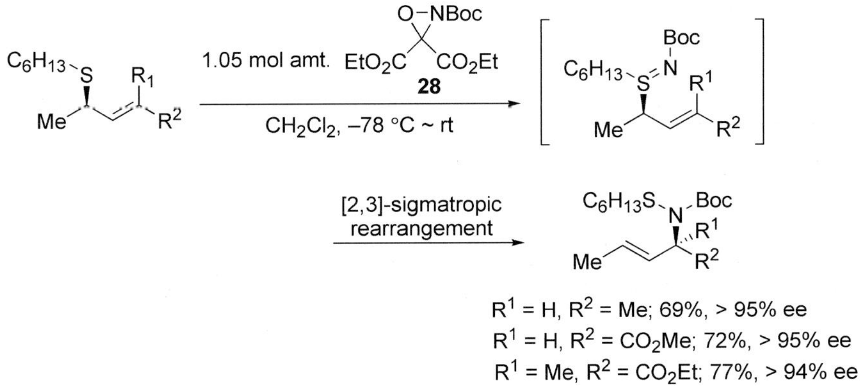

Scheme 1.38 Synthesis of allylic amine derivatives by the [2,3]-sigmatopic rearrangement of allylic sulfides.

By this method, allyl amine derivatives are prepared from allylic sulfides through the rapid [2,3]-sigmatropic rearrangement of the resulting sulfimides (Scheme 1.38). A high level of chirality transfer is observed in this rearrangement and a quaternary stereocenter is successfully constructed [45].

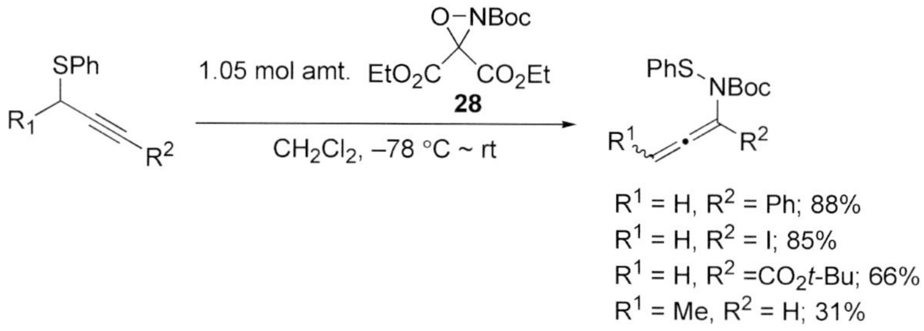

Scheme 1.39

Amination of propargylic sulfides with oxaziridine **28** gives aminoallene derivatives (Scheme 1.39) [46].

1.5.3
Experimental Procedures

Representative electrophilic amination of an enolate with *N*-carboxamido oxaziridine
n-BuLi (2.5 M, 0.21 mL, 0.53 mmol) was added at 0 °C to a solution of diisopropylamine (77 µL, 0.55 mmol) in THF (0.85 mL), and the mixture was stirred for 30 min and then cooled to –78 °C. A solution of *tert*-butyl acetate (67 µL, 0.50 mol) in THF (0.85 mL) was slowly added to the mixture, which was then stirred at –78 °C for 1 h. A solution of oxaziridine **27** (123 mg, 0.50 mmol) was added in a single portion, and the mixture was stirred at –78 °C for 3 h before being allowed to reach room temperature over 90 min. The reaction was quenched with saturated aqueous Na_2CO_3, and diluted with CH_2Cl_2. The organic materials were extracted with CH_2Cl_2, and the combined extracts were washed with saturated aqueous Na_2CO_3 and brine. The solvents were removed under vacuum, and the resulting residue was purified by flash column chromatography to give the desired product (63.5 mg, 0.28 mmol) in 55% yield.

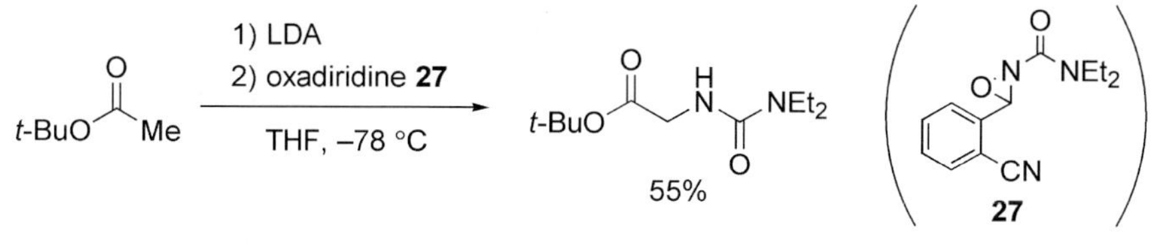

Representative procedure for the [2,3]-sigmatropic rearrangement of an allylic sulfide
The allylic sulfide (190 mg, 0.823 mmol) in CH_2Cl_2 (0.16 M) was added dropwise to a solution of oxaziridine **28** (250 mg, 0.86 mmol) in CH_2Cl_2 (0.18 M). The resulting solution was allowed to warm to room temperature over 30 min, and the solvent was removed under vacuum. The resulting residue was purified by flash column chromatography (petroleum ether/ethyl acetate 20 : 1) to give the desired product (205 mg, 0.593 mmol) in 72% yield (Scheme 1.40).

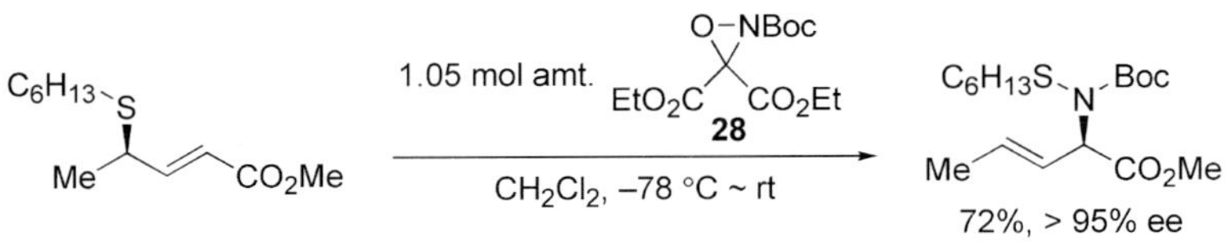

Scheme 1.40

1.6 Chloramine-T

Chloramine-T (*N*-chloro-*N*-sodio-*p*-toluenesulfonamide; **29**), a well known commercially available oxidizing reagent, serves as a source of chloronium cation and nitrogen anion [47]. Chloramine-T has been synthetically applied to a wide variety of C—N bond-forming reactions.

1.6.1 Aminochalcogenation of Alkenes

Diphenyl disulfide and diphenyl diselenide react with chloramine-T (**29**) to afford the reactive species **30**, which gives aminosulfenylated (or selenylated) intermediates (Scheme 1.41). Subsequent reduction with $NaBH_4$ gives phenylthio(or seleno)-*N*-tosylamines [48]. The insensitivity of this reaction to molecular oxygen and the exclusive formation of the *trans* products are strongly suggestive of ionic addition via sulfo(seleno)nium species.

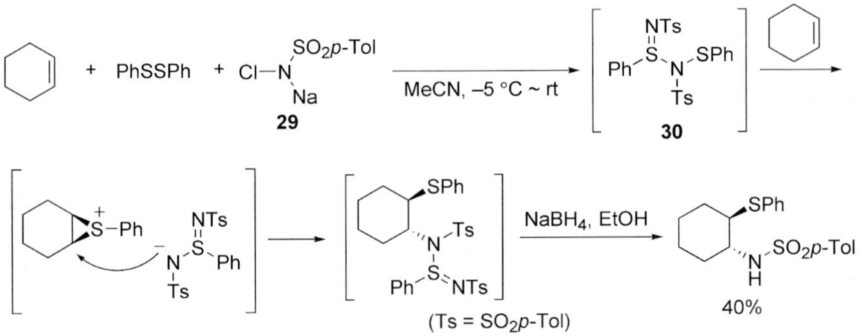

Scheme 1.41 Aminochalcogenation of alkenes by chloramine-T (**29**).

1.6.2 Aminohydroxylation of Alkenes

Treatment of alkenes with chloramine-T (**29**) in the presence of a catalytic amount of osmium tetroxide provides a convenient and general method for vicinal oxyamination of alkenes (Scheme 1.42) [49], which represents a significant improvement

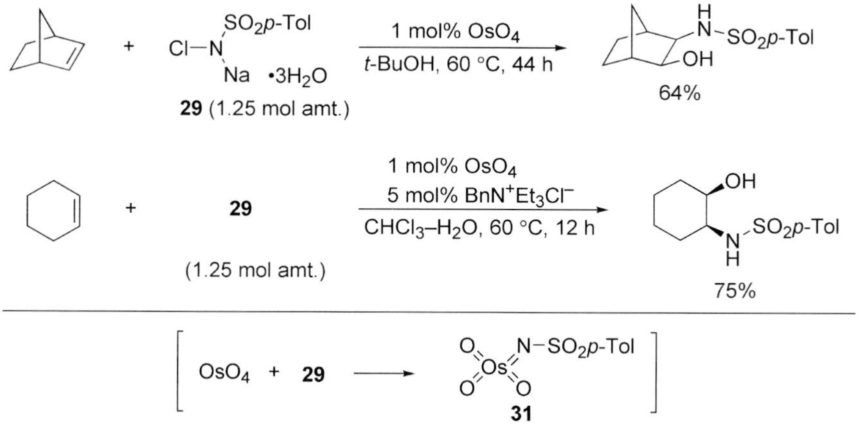

Scheme 1.42 OsO_4-catalyzed aminohydroxylation of alkenes by chloramine-T (**29**).

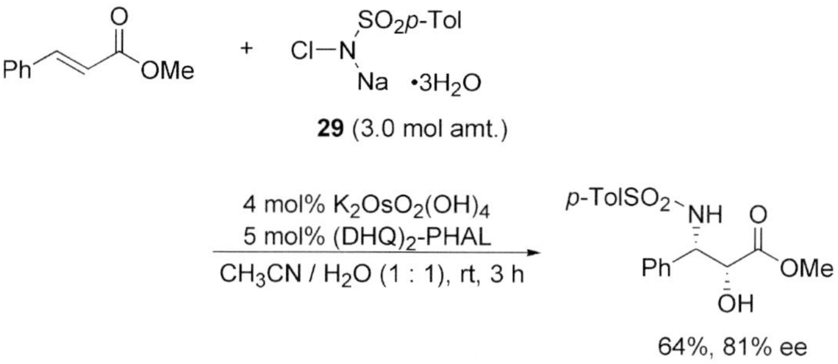

Scheme 1.43

over the stoichiometric oxyamination reactions [50]. This reaction is an aza analogue of the catalytic dihydroxylation of alkenes. The sulfonylimido osmium **31** is proposed as the key species.

This method has been extended to the practical asymmetric version of oxyamination through the use of dihydroquinine (DHQ) alkaloids and dihydroquinidine (DHQD) alkaloids as chiral ligands (Scheme 1.43) [51].

N-Halocarbamate salts such as **32** can be employed in place of chloramine-T (**29**) for oxyamination reactions of alkenes (Scheme 1.44). The *N*-chloro benzyloxycarbamate salt **32** is prepared in situ from benzylcarbamate, NaOH, and *tert*-butyl hypochlorite [52].

This method has been applied to the syntheses of nitrogen-containing natural products [53] such as ustiloxin D (Scheme 1.45) [53a].

1.6.3
Aziridination of Alkenes

Aziridines can be used as synthetic intermediates in the preparation of nitrogen-containing compounds through ring-opening and ring-expansion reactions. In

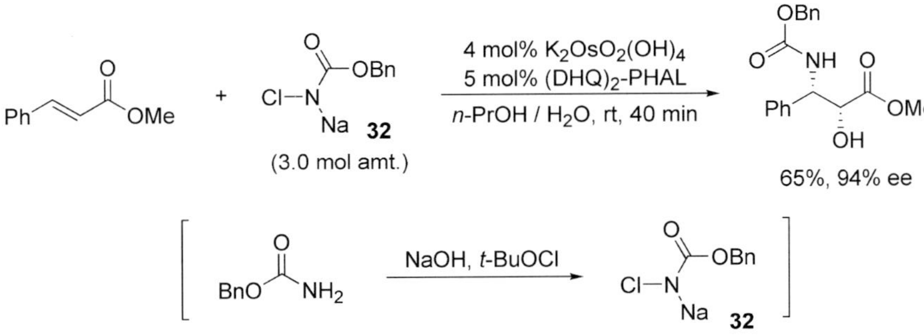

Scheme 1.44 Aminohydroxylation of alkenes with *N*-halocarbamate salt (**32**).

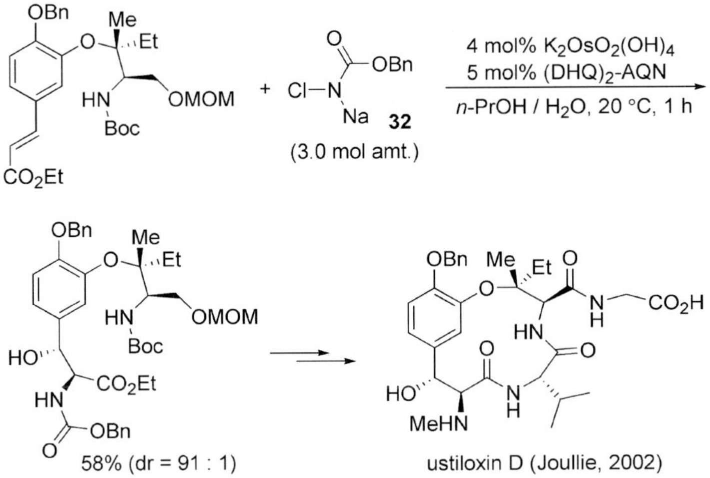

Scheme 1.45 Application of aminohydroxylation with *N*-halocarbamate salt **32** to the synthesis of ustiloxin D.

addition, aziridines are often found in natural products, most of which show potent and diverse biological activities. There have been many reports on the synthesis of aziridines from alkenes by the use of chloramine-T (**29**).

In 1998, Komatsu reported the first example of aziridination of alkenes with chloramine-T (**29**) by use of CuCl as a catalyst [54]. When anhydrous chloramine-T (**29**) was added to CH_3CN solutions of various alkenes in the presence of catalytic amount of CuCl and MS (5 Å), the corresponding aziridines were obtained in moderate to good yields (Scheme 1.46). The same types of aziridination also proceed on employment of a CuOTf–*N*-(2-pyridinylmethylene)-1-pentanamine complex (Scheme 1.47) [55] or an iron(IV) corrole catalyst [56].

The combination of chloramine-T (**29**) and $AgNO_3$ is utilized in the synthesis of aziridines from various alkenes, including electron-deficient alkenes such as α,β-unsaturated ketones and esters (Table 1.7) [57]. The following experimental evidence suggests that the reaction might involve a nitrogen radical species

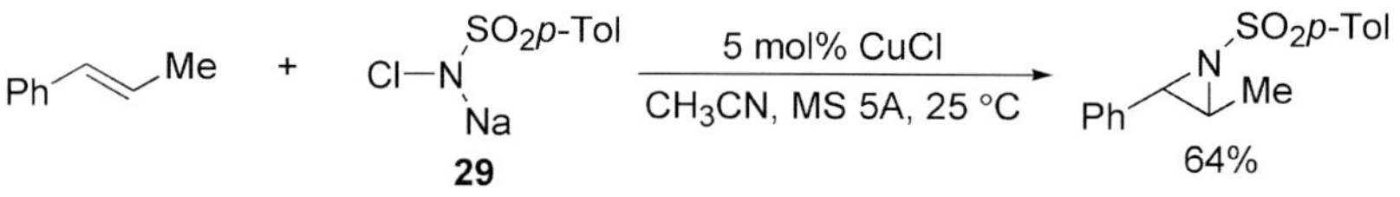

Scheme 1.46

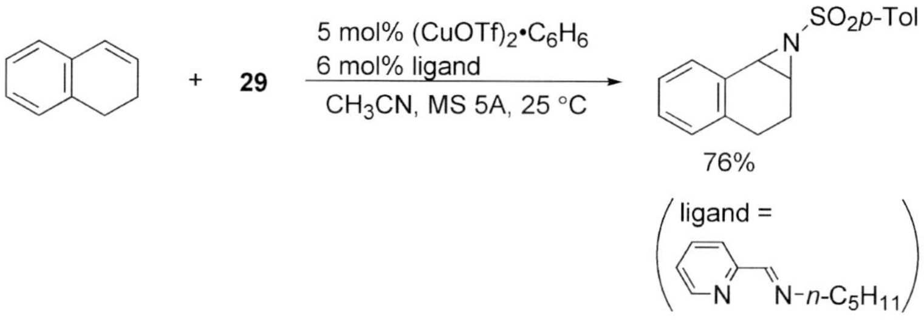

Scheme 1.47

Table 1.7 $AgNO_3$-mediated aziridination of alkenes with chloramine-T (**29**).

alkenes	aziridines (*cis* : *trans*)	alkenes	aziridines (*cis* : *trans*)
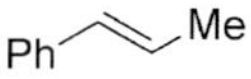	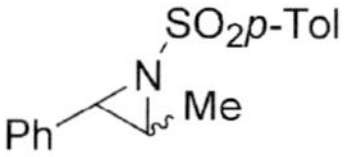 92% (6 : 94)	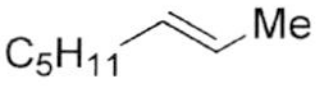	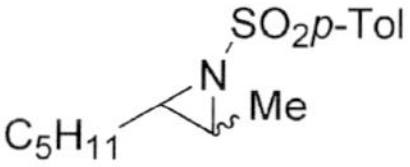 43% (38 : 62)
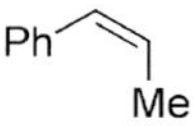	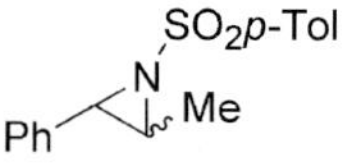 89% (7 : 93)	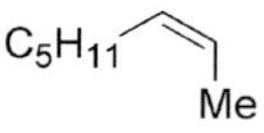	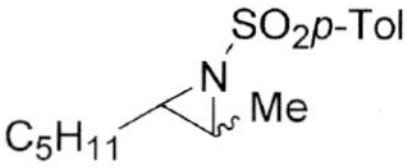 53% (43 : 57)

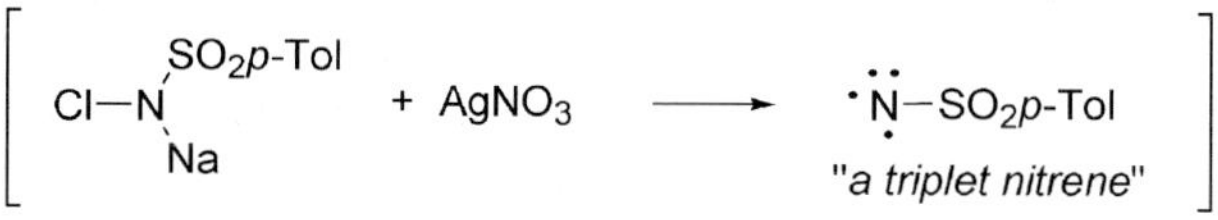

(a triplet nitrene). 1) This aziridination did not proceed at all under an oxygen atmosphere (Scheme 1.48). 2) Treatment of 1,4-dioxane with chloramine-T (**29**) and $AgNO_3$ provided the C-H amination product (Scheme 1.49). 3) The aziridination did not proceed with the retention of the stereochemistry of the starting alkenes (Table 1.7).

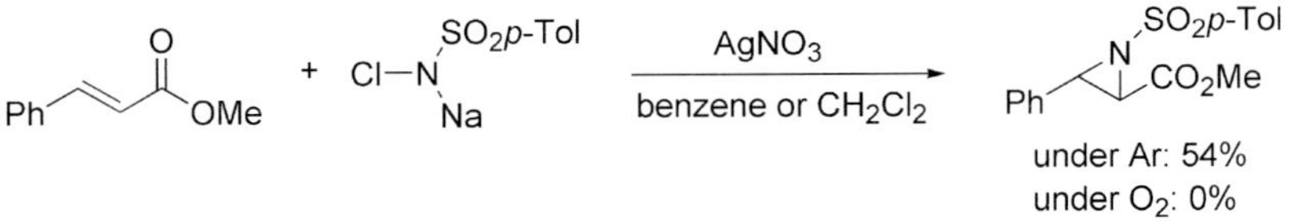

Scheme 1.48

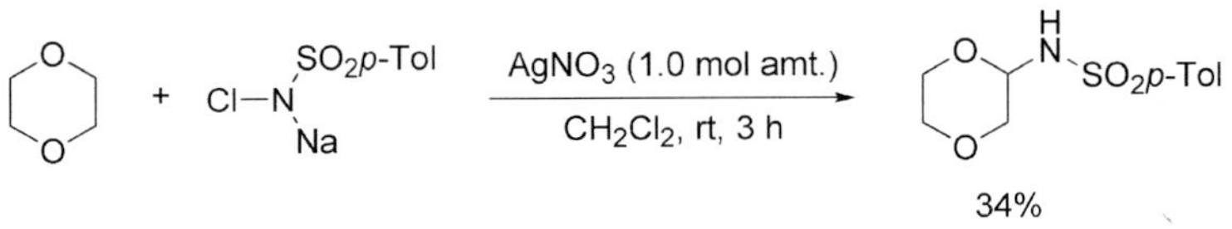

Scheme 1.49

Table 1.8 Bromine-catalyzed aziridination of alkenes with chloramine-T (**29**).

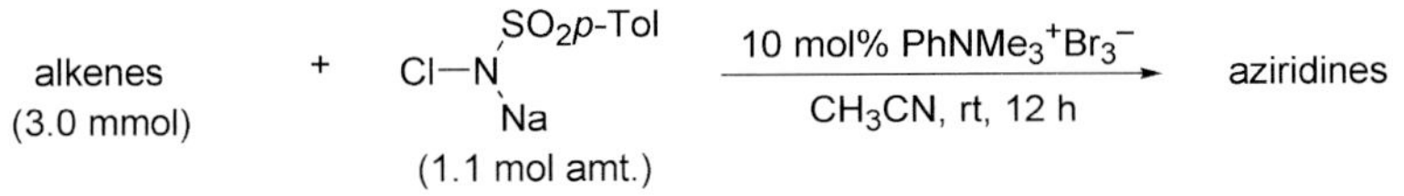

alkenes	aziridines	alkenes	aziridines
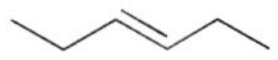	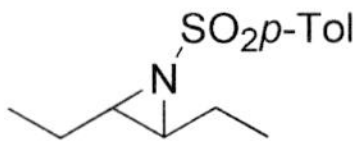 93%	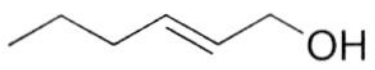	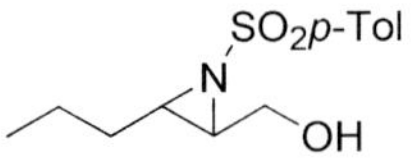 97%
	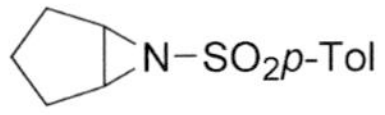 86% (80%)[a]	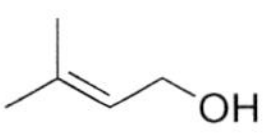	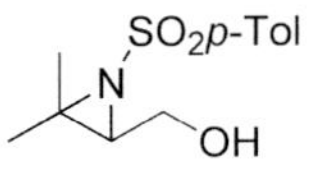 94%

a) 0.5 mol scale

Bromine (ammonium tribromide) and iodine efficiently catalyze the aziridination of alkenes with chloramine-T (**29**). Sharpless found that catalytic use of phenyltrimethylammonium tribromide ($PhNMe_3^+Br_3^-$, PTAB) with chloramine-T (**29**) provides good to excellent yields of aziridines from various alkenes (Table 1.8) [58]. This simple operation can be applied in large-scale syntheses (up to 0.5 mol scales). Pyridinium hydrobromide perbromide ($Py \cdot HBr_3$) has also been employed instead of PTAB for the chloramine-T-mediated aziridination [59].

The mechanism proposed for this bromine-catalyzed aziridination is shown in Scheme 1.50, with *cis*-β-methylstyrene **I** as a specific alkene. Initially the styrene reacts with a Br^+ source to give the bromonium cation **II**, and nucleophilic attack by TsN^-Cl then affords a bromoaminated intermediate **III**. Attack either of Br^- or of TsN^-Cl on the N–Cl group in **III** gives sulfonamide anion **IV**, which cyclizes intramolecularly to afford aziridine **V**. The regenerated Br–X initiates another catalytic cycle.

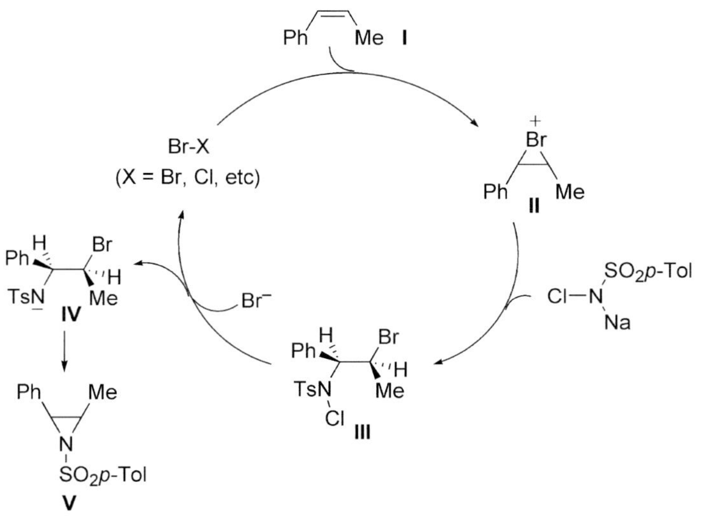

Scheme 1.50 Catalytic cycle of bromine-catalyzed aziridination.

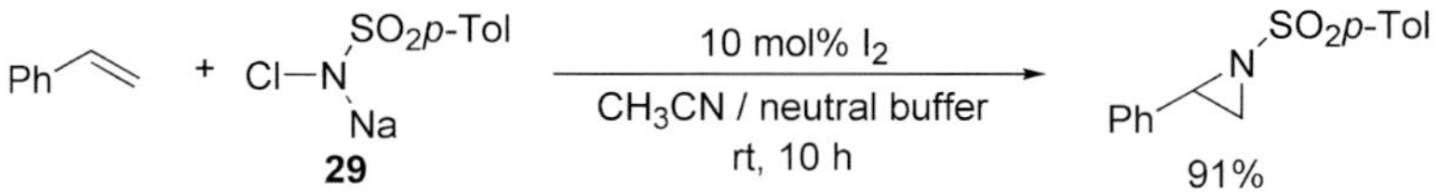

Scheme 1.51

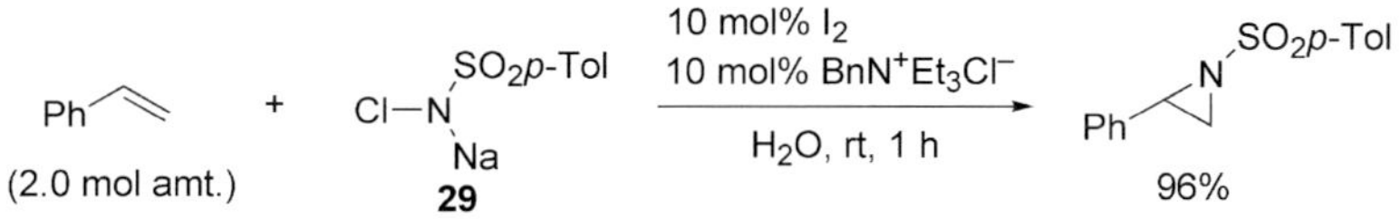

Scheme 1.52

Komatsu reported an iodine-catalyzed aziridination of alkenes (Scheme 1.51) [60]. When 2 molar amounts of styrene are added to chloramine-T (**29**) in the presence of a catalytic amount of iodine in a 1 : 1 mixture of CH_3CN and a neutral buffer, for example, the corresponding aziridine is obtained in 91% yield. A similar reaction mechanism based on the above bromine-catalyzed aziridination is proposed.

This chloramine-T/iodine system has been applied to an organic solvent-free aziridination of alkenes in combination with a phase-transfer catalyst (Scheme 1.52) [61]. No reaction is observed when alkenes are treated with chloramine-T (**29**) and a catalytic amount of iodine in water, while the addition of a catalytic amount of benzyltriethylammonium chloride (BTEAC) results in a dramatic acceleration of the reaction, to yield aziridines in good yield. The silica gel/water combination as the reaction medium was also found to be effective for the iodine-catalyzed aziridination with chloramine-T (**29**) [62].

1.6.4
Other Applications

Chloramine-T (**29**) can be used as a nitrogen source in the synthesis of azaheterocycles. A chloramine-T/$AgNO_3$ system, which is believed to generate a nitrogen radical species (a triplet nitrene), has been applied to the synthesis not only of aziridines from monoenes (Section 1.6.3), but also of bicyclic pyrrolidines from 1,6-dienes (Scheme 1.53) [57]. The bicyclic pyrrolidine derivatives were obtained by tandem radical cyclization. The formation of *trans*-substituted cyclopentane derivatives as by-products would be one piece of evidence of the formation of biradical intermediate **A**.

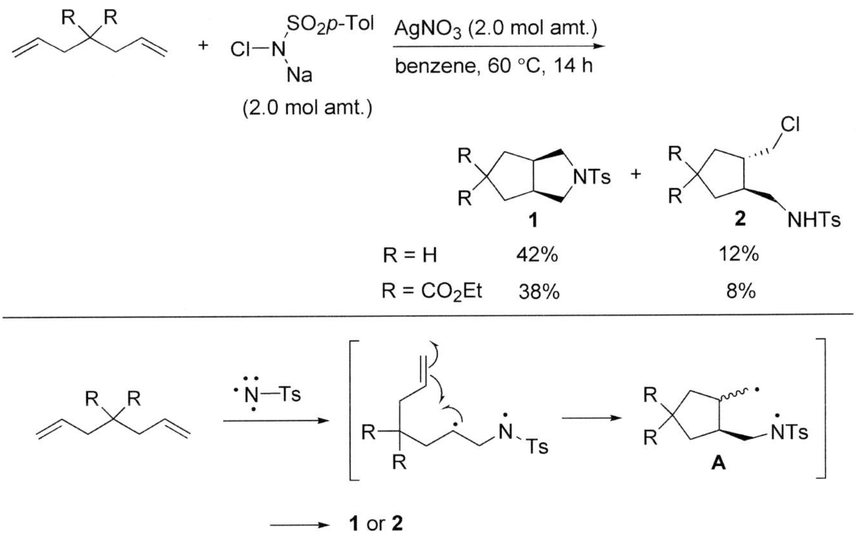

Scheme 1.53 Synthesis of bicyclic pyrrolidines from 1,6-dienes by use of the $AgNO_3$/chloramine-T system.

2-Iodomethyl pyrrolidine derivatives have been synthesized from γ-iodo alkenes with chloramine-T (**29**; Table 1.9) [63]. The cyclization proceeds with high stereoselectivity, via a cyclic iodonium intermediate. The iodo group in the substrate plays multiple roles, as: (1) a leaving group for the substitution with chloramine-T, (2) a Lewis base for the abstraction of the Cl atom, (3) an activator of the alkenyl moiety, and (4) a functional group in the product (Scheme 1.54).

Recently, Minaleata and Komatsu have reported a new type of aminochlorination of various alkenes by use of chloramine-T (**29**) and CO_2 [64]. When styrene was treated with chloramine-T under CO_2 (10 atm) at 70 °C, aminochlorination occurred regioselectively to afford the β-chloro amine in 80% yield (Scheme 1.55). In the aminochlorination of a diene such as cycloocta-1,3-diene under the same conditions, a 1,4-adduct is formed as the sole product in 70% yield without the formation of any 1,2-adduct (Scheme 1.56).

Table 1.9 Synthesis of 2-iodomethyl pyrrolidine derivatives from γ-iodo alkenes with chloramine-T (**29**).

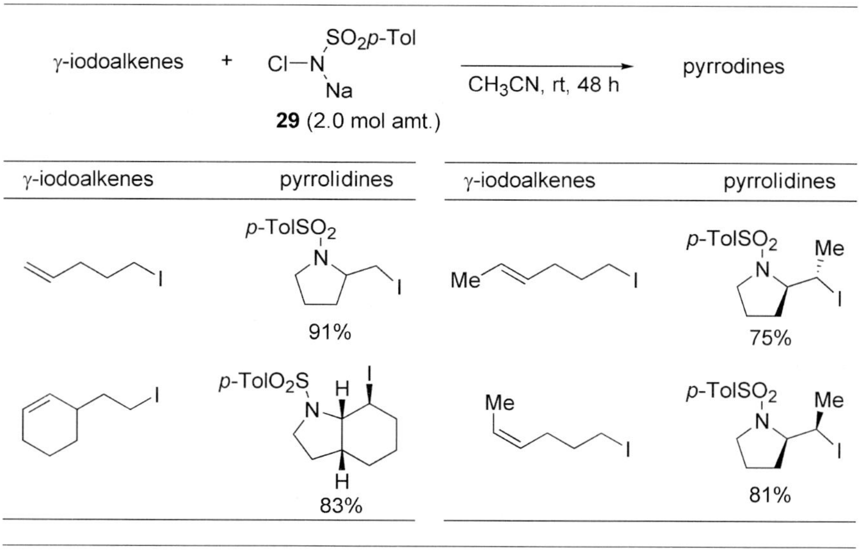

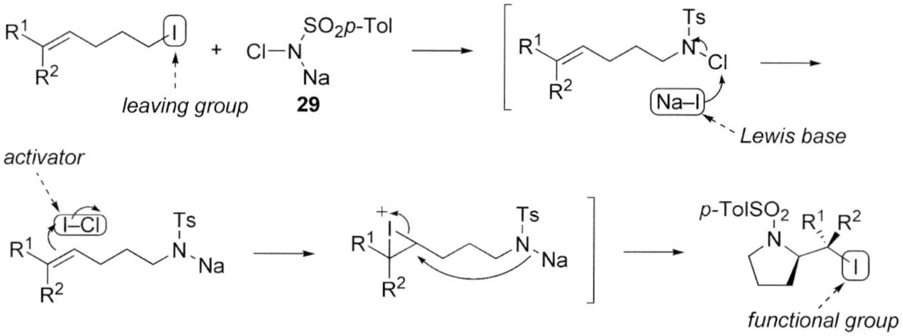

Scheme 1.54 Proposed mechanism of the stereoselective formation of 2-iodomethyl pyrrolidine derivatives.

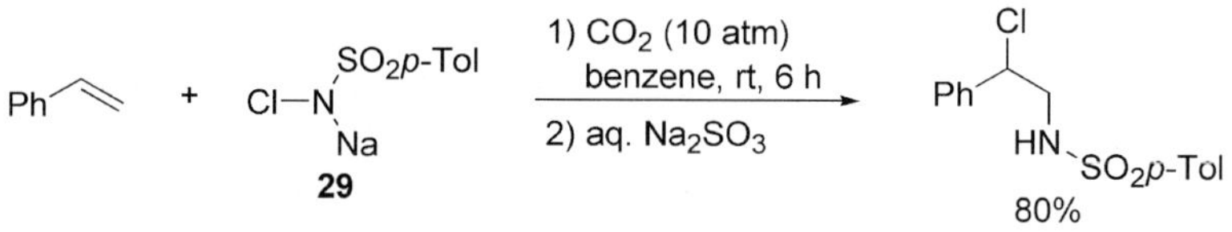

Scheme 1.55

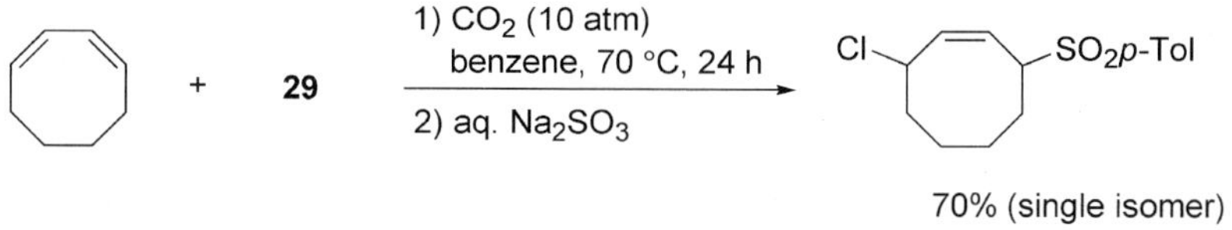

Scheme 1.56

1.6.5
Experimental Procedures

Asymmetric oxyamination of an alkene with an N-halocarbamate salt A freshly prepared water (7.5 mL) solution of NaOH (0.122 g, 3.05 mmol) was added to a solution of benzyl carbamate (0.469 g, 3.10 mmol) in *n*-PrOH (4 mL), followed by a freshly prepared solution of *t*-BuOCl (0.331 g, 3.05 mmol). Next, a solution of the ligand $(DHQ)_2$PHAL (40 mg, 0.05 mmol) in *n*-PrOH (3.5 mL) was added. After the mixture had been stirred for a few minutes, methyl *trans*-cinnamate (0.162 g, 1 mmol) and $K_2OsO_2(OH)_4$ (14.7 mg, 0.04 mmol) were added. The mixture was stirred for 40 min at room temperature, and the light green color of the solution had changed to light yellow at the end. Ethyl acetate (7 mL) was added to the mixture and the aqueous phase was extracted with ethyl acetate (5 mL × 3). The combined extracts were washed with water and brine and dried over $MgSO_4$. Volatile materials were concentrated *in vacuo* to afford the crude mixture, which was purified by flash column chromatography (hexane/chloroform/methanol 6:4:1 *v*/*v*/*v*) to provide the desired compound (0.204 g, 65%, 94% *ee*) as a colorless solid.

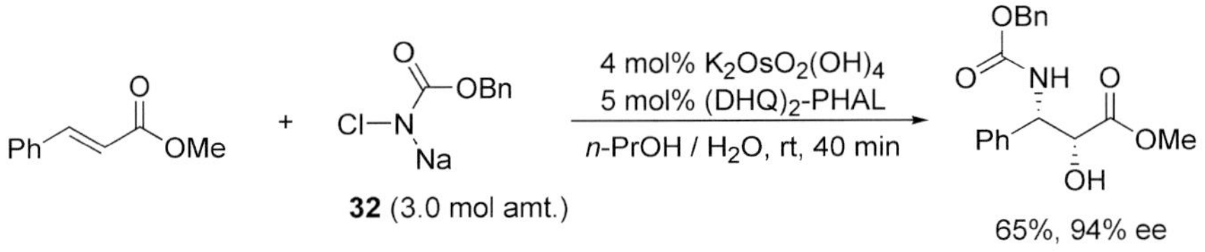

Copper(I)-catalyzed aziridination of alkenes Anhydrous chloramine-T (**29**, 114 mg, 0.50 mmol) and MS (5 Å, 50 mg) were added to a solution of CuCl (2.5 mg, 0.25 mmol) and *trans*-β-methylstyrene (295 mg, 2.5 mmol) in acetonitrile (5 mL). The reaction mixture was stirred at 25 °C for 3 h under nitrogen and the reaction mixture was then passed through a 3 cm plug of silica gel with elution with CH_2Cl_2. The solvent was removed *in vacuo* to give a crude oil, which was purified by flash column chromatography to give *trans*-2-methyl-3-phenyl-1-tosylaziridine (92 mg, 0.32 mmol) in 64% yield.

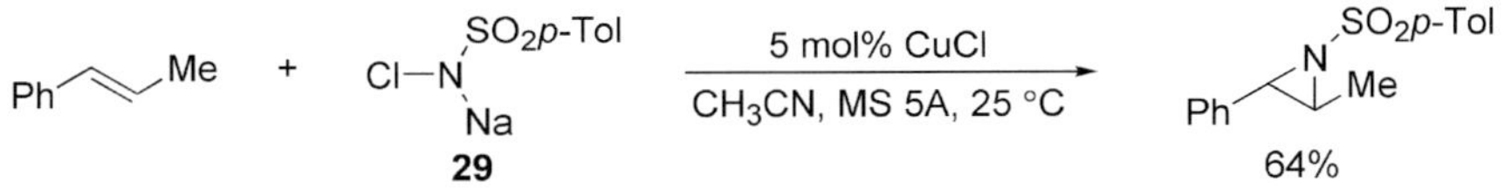

Bromine-catalyzed aziridination of alkenes Trimethylphenylammonium tribromide (113 mg, 0.3 mmol) was added at 25 °C to a mixture of *trans*-hex-3-ene (252 mg, 3 mmol) and anhydrous chloramine-T (**29**, 751 mg, 3.3 mmol) in CH_3CN (15 mL). After vigorous stirring for 12 h, the reaction mixture was concentrated (to about 1/10 volume) and filtered through a short column of silica gel (10% EtOAc in hexane). After evaporation of the solvent, the resultant solid was purified by crystallization from hexane to give *trans*-2,3-diethyl-1-[(4-methylphenyl)sulfonyl]aziridine (710 mg, 2.8 mmol) in 93% yield as colorless crystals.

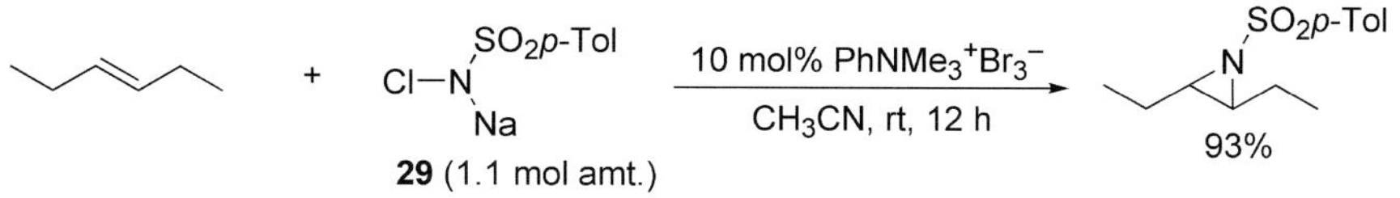

Synthesis of iodomethylated pyrrolidine derivatives 5-Iodopent-1-ene (196 mg, 1.0 mmol) was added to a suspension of chloramine-T (**29**, 455 mg, 2.0 mmol) in CH_3CN (6.0 mL). The mixture was stirred under nitrogen in the dark at room temperature for 48 h. After the addition of Et_2O (40 mL), the organic layer was washed with H_2O (60 mL), the aqueous phase was extracted with Et_2O (2 × 20 mL), and the combined extracts were washed with brine (30 mL), dried over K_2CO_3, and concentrated *in vacuo*. The resulting crude materials were purified by flash column chromatography (silica gel, 10% EtOAc in hexane) to give 2-iodomethyl-1-(*p*-tolylsulfonyl)pyrrolidine (332 mg, 0.91 mmol) in 91% yield.

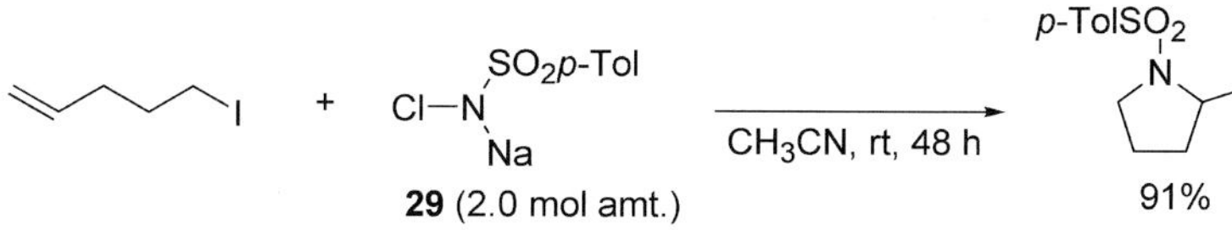

CO_2-promoted aminochlorination of alkenes Chloramine-T (281 mg, 1 mmol), styrene (104 mg, 1 mmol), and benzene (3 mL) were placed in a 50 mL stainless steel autoclave lined with a glass liner. The autoclave was closed, purged three times with carbon dioxide, pressurized with CO_2 (10 atm), and then stirred at room temperature for 6 h. After discharging of excess CO_2, aqueous $Na_2S_2O_3$ (0.5 M, 20 mL) was added to the reaction mixture, the organic materials were extracted with CH_2Cl_2 (10 mL × 3), and the combined extracts were dried over Na_2SO_4 and concentrated to give the crude products. Purification by flash column chromatography (silica gel, 30% EtOAc in hexane) gave 1-chloro-1-phenyl-2-[(4-methylphenyl)sulfonamido]ethane (247 mg, 0.797 mmol) in 80% yield.

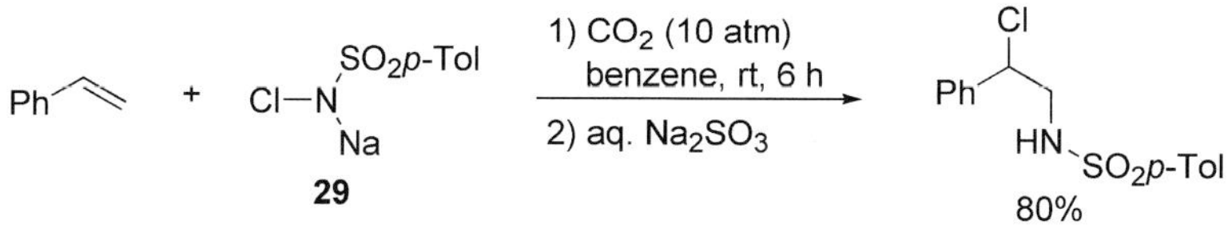

1.7
N-Sulfonyliminophenyliodinane

N-Sulfonyliminophenyliodinane [65] derivatives are employed for the generation of transition metal-nitrene complexes, which are then applied to the catalytic amination of saturated C—H bonds and alkenes. Since amination of saturated C—H bonds is summarized elsewhere in this book (Chapter 2), only the transition metal-catalyzed amination of various alkenes is introduced in this section.

1.7.1 Transition Metal-Catalyzed Amination of Alkenes

Various transition metals such as Fe [66], Mn [66], Ru [67], Rh [68], Cu [69], and Ag [70] have been used for aziridination of alkenes with *N-p*-tolylsulfonyliminophenyliodinane (**33**) (Scheme 1.57). Treatment with silyl enol ethers affords α-amino ketones through hydrolysis of the resulting 2-siloxy-1-sulfonylaziridines (Scheme 1.58).

Recently, efficient aziridination reactions of alkenes have been reported to occur on combined use of $AgNO_3$ and a tridentate ligand, 4,4′,4″-tri-*tert*-butyl-2,2′:6′,2″-terpyridine (*t*-Bu_3tpy) (Table 1.10) [70]. A unique disilver(I) complex is assumed to be the reactive intermediate; its structure has been verified by single-crystal X-ray structural analysis (Scheme 1.59).

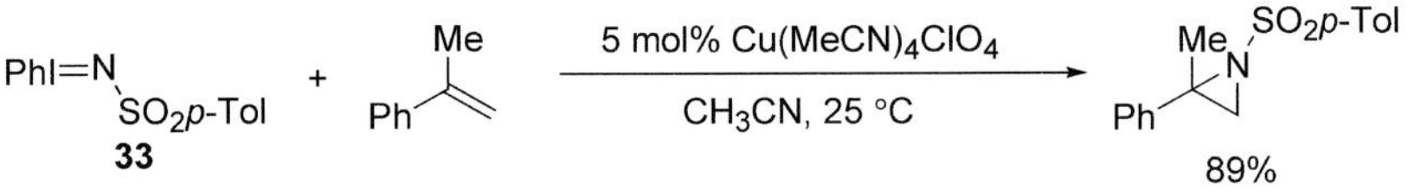

Scheme 1.57

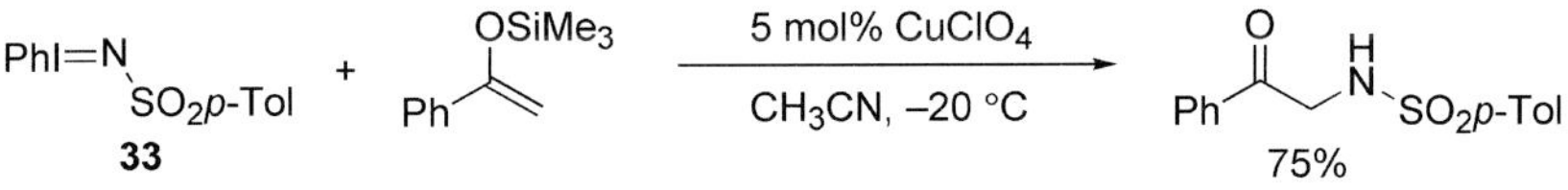

Scheme 1.58

Table 1.10 Ag(I)-catalyzed aziridination of alkenes with *N*-sulfonyliminophenyliodinane (**33**).

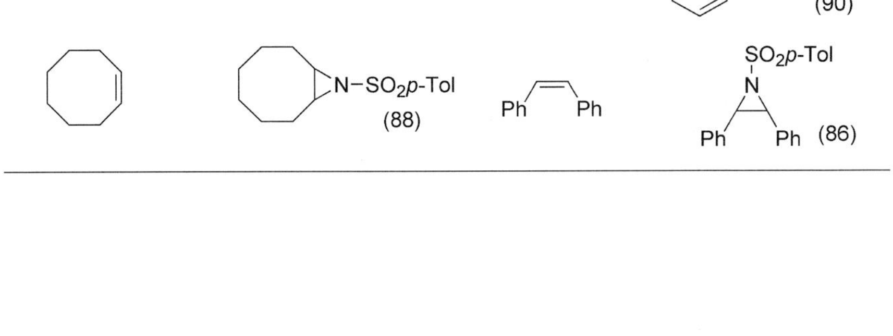

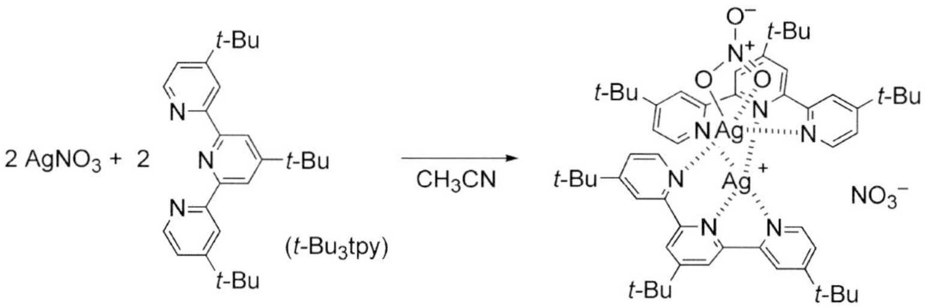

Scheme 1.59

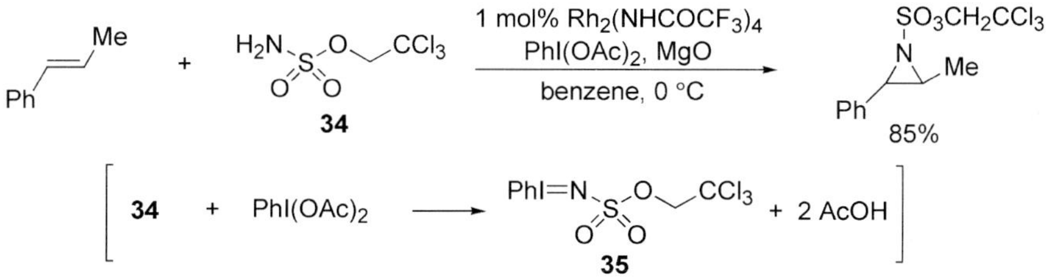

Scheme 1.60 Rh-catalyzed aziridination of alkenes by the use of sulfamate ester **34** and $PhI(OAc)_2$.

Many catalytic asymmetric aziridinations of alkenes with **33** in the presence of metal complexes with chiral bis(oxazoline), salen, and porphyrin ligands have been reported [71].

Du Bois developed Rh(II)-catalyzed aziridination of alkenes by the combined use of trichloroethylsulfamate ester **34** and $PhI(OAc)_2$ in the presence of MgO [72], in which *N*-trichloroethylsulfonyliminophenyliodinane (**35**) is generated in situ as a nitrogen source (Scheme 1.60). Au(I) complexes also catalyze the same type of aziridination of alkenes [73].

1.7.2
Experimental Procedures

Representative procedure for Ag(I)-catalyzed aziridination of alkenes A mixture of $AgNO_3$ (1.7 mg, 0.01 mmol) and 4,4′,4″-tri-*tert*-butyl-2,2′:6′,2″-terpyridine (*t*-Bu_3tpy) (4.0 mg, 0.01 mmol) was stirred in CH_3CN (2 mL) for 5–10 min. *N*-*p*-Tolylsulfonyliminophenyliodinane (**33**, 86.5 mg, 0.5 mmol), together with MS (4 Å, 0.5 g), was added to the mixture. Hex-1-ene (210 mg, 2.5 mmol) was added and the solution was stirred at 0 °C for 0.5 h, and further stirred at room temperature for 20 h. The reaction mixture was diluted with CH_2Cl_2 (10 mL) and filtered through a Celite pad. The filter cake was washed with CH_2Cl_2 (2 × 5 mL) and the combined extracts were concentrated under reduced pressure. The resulting crude mixture was purified by flash column chromatography (silica gel, hexane/EtOAc 4:1) to give the desired aziridine (89.9 mg, 0.36 mmol) in 71% yield.

Representative procedure for Rh(II)-catalyzed aziridination of alkenes by combined use of trichloroethylsulfamate ester 34 and PhI(OAc)$_2$

trans-β-Methylstyrene (59 mg, 0.50 mmol), MgO (46 mg, 1.15 mmol), and $Rh_2(NHCOCF_3)_4$ (3 mg, 5.0 mmol) were added sequentially to a solution of trichloroethylsulfamate ester **34** (126 mg, 0.55 mmol) in benzene. The resulting purple mixture was cooled to 0 °C and $PhI(OAc)_2$ (209 mg, 0.65 mmol) was added. The suspension quickly turned orange after the addition of $PhI(OAc)_2$ and was allowed to warm slowly to 25 °C over 2 h. After 6 h, the reaction mixture was diluted with CH_2Cl_2 (10 mL) and filtered through a Celite pad. The filter cake was washed with CH_2Cl_2, the combined extracts were concentrated under reduced pressure, and the resulting crude materials were purified by flash column chromatography (silica gel hexane/EtOAc 9:1) to give *trans*-2-methyl-3-phenyl-1-trichloroethylsulfonyl aziridine (140 mg, 0.43 mmol) in 85% yield.

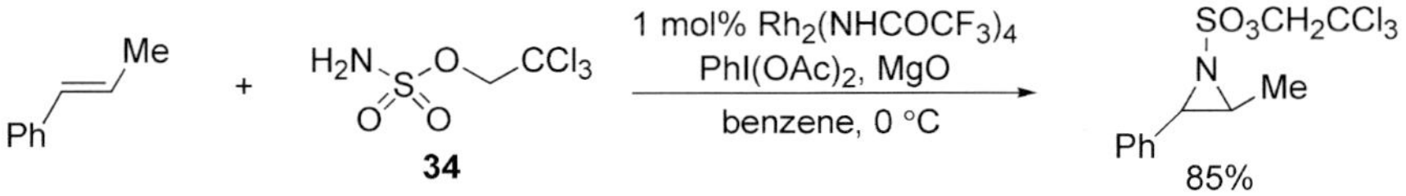

1.8
Transition Metal-Nitride Complexes

There are many reports on the generation and the characterization of transition metal/nitrido complexes [74], some of which have been applied to stoichiometric nitrogen atom transfer reactions to organic molecules.

1.8.1
Nitrogen Atom Transfer Mediated by Transition Metal-Nitride Complexes

Carreira developed new methods to prepare manganese Schiff base-nitrido complexes **35**, which were used in the amination of alkenes including silyl enol ethers and glycals [75]. The Mn(V)-nitrido complex of H_2saltmen [saltmen = (1,1,2,2-tetr amethylethylene)bis(salicylideneaminato)] **35** is prepared by the following two step-procedure at scales of up to 20 g as explained in the Experimental section (Scheme 1.61).

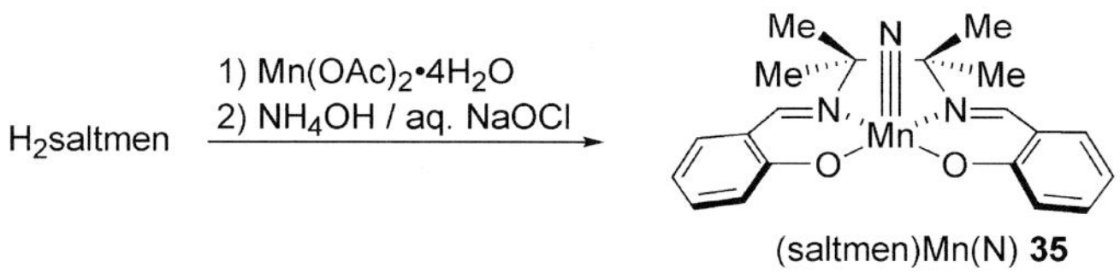

Scheme 1.61

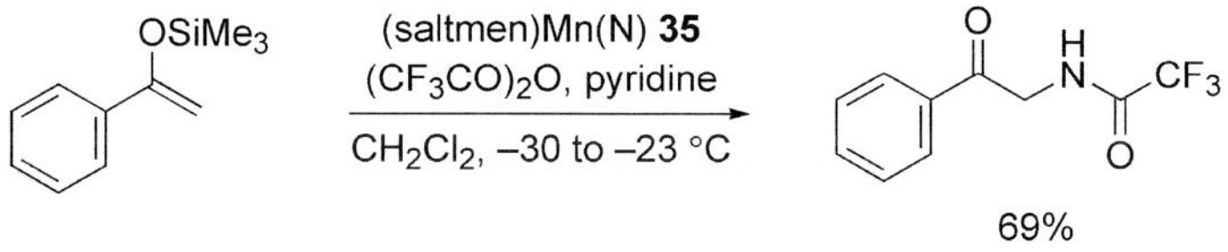

Scheme 1.62

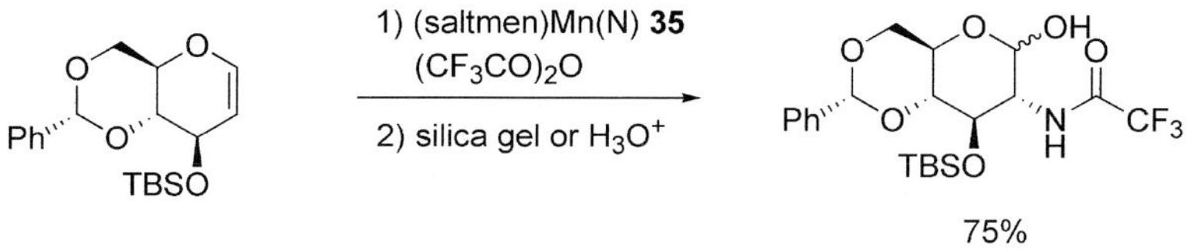

Scheme 1.63

Treatment of a solution of nitrido complex **35**, silyl enol ether, and pyridine in CH_2Cl_2 with trifluoroacetic anhydride at −30 °C provides *N*-trifluoroacetyl α-amino ketone (Scheme 1.62). This reaction may proceed through the initial formation of a reactive *N*-trifluoroacetylimidomanganese species, with the subsequent transfer of CF_3CON group to an alkene.

By this procedure the stereoselective synthesis of 2-aminosaccharides has been performed starting from glycals (Scheme 1.63).

Recently, nitrido ruthenium porphyrin complexes have been found to effect amination of silyl enol ethers or hydrocarbons [76].

Development of chiral nitridomanganese complexes such as **36** and their use for asymmetric preparation of aziridines and oxazolines have been reported by Komatsu [77]. The complex **36** reacts with alkenes in the presence of *p*-toluenesulfonic anhydride or some sulfonyl chlorides to give optically active *N*-sulfonylaziridines (Scheme 1.64). When acid chlorides are used instead of sulfonyl chlorides, oxazolines are formed, presumably via *N*-acylaziridines as intermediates (Scheme 1.65).

1.8.2 Experimental Procedures

Preparation of H_2saltmen Mn(V)/nitrido complex H_2saltmen [saltmen = (1,1,2,2-tetramethylethylene)bis(salicylideneaminato)] (10.0 g, 30.8 mmol) was suspended

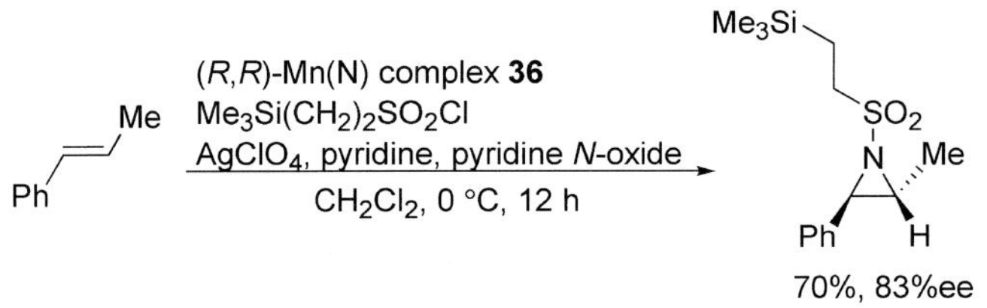

Scheme 1.64

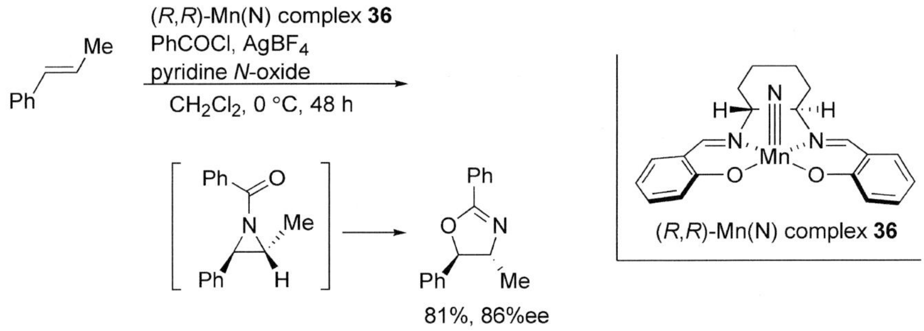

Scheme 1.65

in MeOH (400 mL) and the mixture was heated at 50–60 °C. $Mn(OAc)_2 \cdot 4H_2O$ (7.90 g, 32.4 mmol) was added portionwise to the yellow solution. The resulting dark brown solution was heated at reflux for 1 h and at 23 °C for 0.5 h. Conc. NH_4OH (15 M, 31.0 mL, 465 mmol) was then added dropwise over 5 min, after which aqueous NaOCl (0.7 M, 280 mL, 196 mmol) was added to the vigorously stirred mixture over 40 min. When the addition was complete, the mixture was cooled to 0 °C and diluted with CH_2Cl_2 (400 mL), and the resulting biphasic mixture was warmed to 23 °C and stirred for 15 min. The contents were transferred to a separating funnel with H_2O (200 mL), the organic phase was isolated, and the aqueous layer was extracted with CH_2Cl_2 (200 mL). The combined extracts were washed with H_2O (6 × 300 mL) and concentrated *in vacuo* to afford 12 g of a dark green solid. The solid materials were dissolved in CH_2Cl_2 (50–75 mL) and filtered through a 70 × 200 mm plug of Brockmann activity-IV basic Al_2O_3 with CH_2Cl_2 as the eluent. A dark green band was collected as a single fraction and the solvent was removed *in vacuo*. The resulting green solid was suspended in EtOAc at reflux (150 mL), to which hexane (300 mL) was added. The contents were cooled to 23 °C and then placed in a freezer at −20 °C for 10 h. The dark green microcrystalline precipitate was collected and rinsed with an ice cold hexane/EtOAc mixture (2 : 1, 3 × 50 mL). The product **35** (10.2 g, 85%) was dried under vacuum at 1 Torr for 5 h.

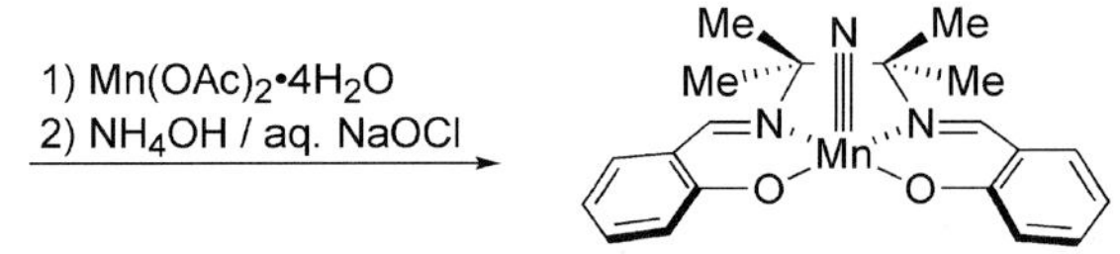

Representative procedure for the stereoselective synthesis of 2-aminosaccharides mediated by the Mn(V)/nitrido complex Trifluoroacetic anhydride (200 µL, 1.4 mmol) was added to a solution of glycal (139 mg, 0.40 mmol) in CH_2Cl_2 (0.5 mL), and a solution of (saltmen)Mn(N) **35** in CH_2Cl_2 (0.4 M, 10 mL, 0.4 mmol) was slowly added by syringe pump over 7 h. After the addition of (saltmen)Mn(N), silica gel (500 mg) and Celite (500 mg) were added to the resultant dark brown solution, together with pentane (15 mL). The dark brown slurry was stirred vigorously for 30 min before being filtered through a 20 × 50 mm plug of silica gel with Et_2O (2 × 10 mL) as an eluent. Concentration of the filtrate under vacuum afforded a pale yellow residue, which was purified by flash column chromatography (silica gel) to give the desired 2-amino sugar (143 mg, 0.30 mmol) in 75% yield.

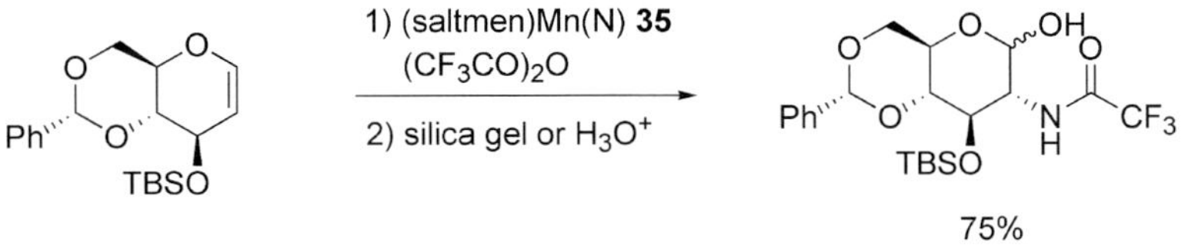

Representative procedure for the asymmetric synthesis of 2-oxazolines from alkenes with the aid of chiral nitridomanganese complexes Pyridine (12 mg, 0.15 mmol), *trans*-β-methylstyrene (354 mg, 3.0 mmol), and benzoyl chloride (51 mg, 0.36 mmol) were added at 0 °C to a mixture of an (*R,R*)-Mn(N) complex **36** (117 mg, 0.3 mmol), $AgBF_4$ (70 mg, 0.36 mmol), and pyridine *N*-oxide (34 mg, 0.36 mmol) in CH_2Cl_2 (3 mL). After the mixture had been stirred at 0 °C for 48 h, pentane (20 mL) was added. The mixture was passed through a 3 cm pad of silica gel with Et_2O (125 mL) as the eluent. The filtrate was concentrated under vacuum, and the resulting residue was purified by flash column chromatography (silica gel, hexane/EtOAc) to afford (4*R*,5*R*)-4-methyl-2,5-diphenyl-2-oxazoline (58 mg, 0.24 mmol) in 81% yield. The enantiomeric excess of the oxazoline was determined by chiral HPLC analysis (81% *ee*).

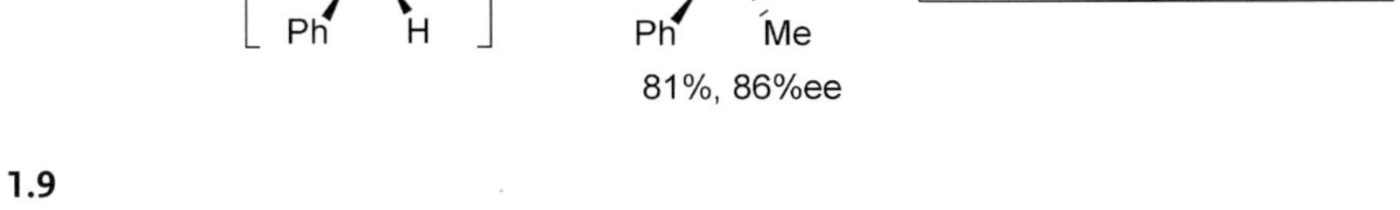

1.9 Azido Derivatives

Aliphatic azides are readily prepared by nucleophilic substitution of alkyl halides or sulfonates with sodium azide; the resulting alkyl azides are readily reduced to primary amines, while 1,3-dipolar cycloadditions of azide derivatives with dipolarophiles such as alkenes, alkynes, and nitriles [78] give various kinds of azaheterocyclic compounds. These transformations have been reported in detail in some recent reviews [79] and so are not discussed in this section.

1.9.1 Electrophilic Amination of Organometallic Reagents with Organic Azides

Some organic azido derivatives are utilized for the synthesis of primary amines through electrophilic amination of organometallic reagents.

Trost reported that phenylthiomethyl azide (**37**) can be used as a synthon for $^{+}NH_2$ by treatment with a Grignard reagent and subsequent hydrolysis of the resulting triazene (Scheme 1.66) [80]. This reagent can be applied to the preparation of a chiral secondary amine by treatment with a chiral Grignard reagent [22].

Allyl azide [81], trimethylsilylmethyl azide [82], or diphenylphosphoryl azide [83] are also used for the preparation of aromatic primary amines by treatment with aryl Grignard reagents.

For amination of enolates, Evans developed the electrophilic azidation of amide enolates with bulky 2,4,6-triisopropylbenzenesulfonyl azide **38** (Scheme 1.67) [84]. The resulting α-azido carboxiimides can be converted into α-amino acid derivatives.

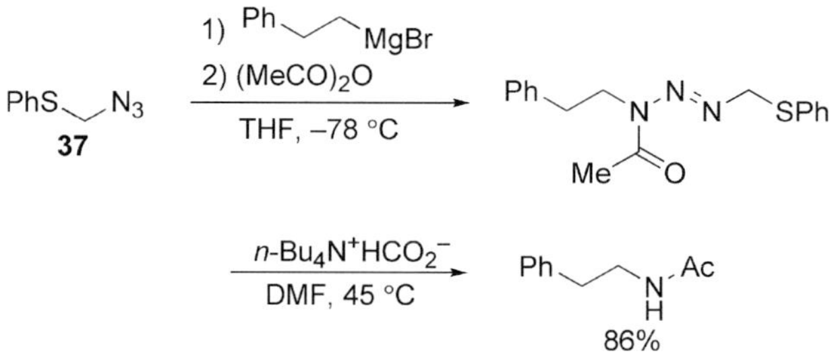

Scheme 1.66 Electrophilic amination of Grignard reagents with azide **37**.

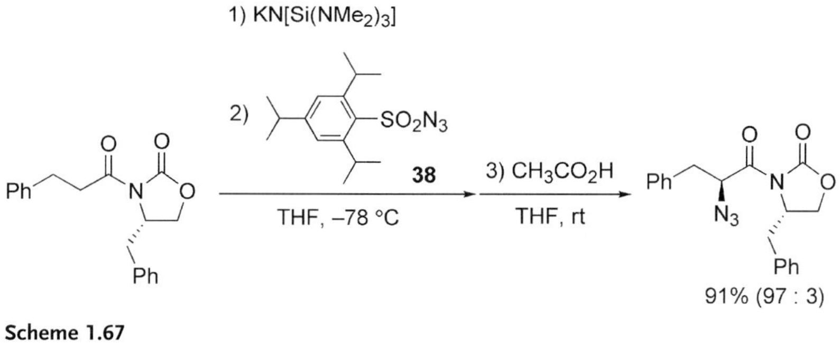

Scheme 1.67

1.9.2
Radical-Mediated Amination with Sulfonyl Azides

Organic azides have been investigated as radical traps for C—N bond formation [85]. Sulfonyl azides are suitable for azidation of nucleophilic radicals such as secondary and tertiary alkyl radicals. Renaud developed ditin-mediated radical azidation with phenylsulfonyl azide (**39**) [86]. When a mixture of alkyl iodides (Scheme 1.68) or dithiocarbonates (Scheme 1.69) and phenylsulfonyl azide (**39**) is treated with hexabutylditin as a chain transfer reagent, the radical reaction is initiated by photoirradiation or thermal decomposition of di-*tert*-butylhyponitrite to give the corresponding alkyl azides.

This radical azidation is applied to domino C—C bond formation–azidation sequences (Schemes 1.70 and 1.71) [87].

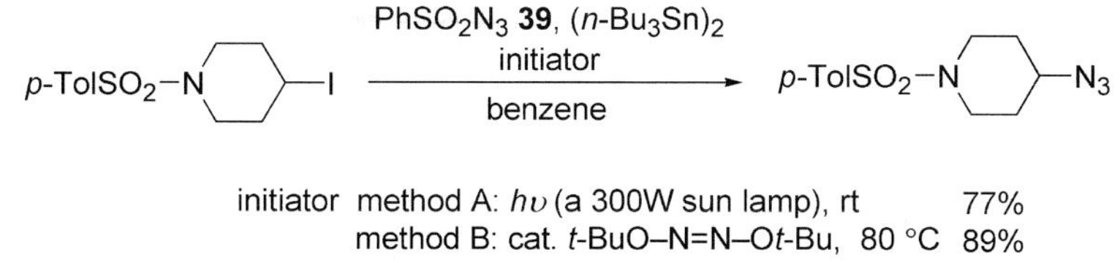

Scheme 1.68

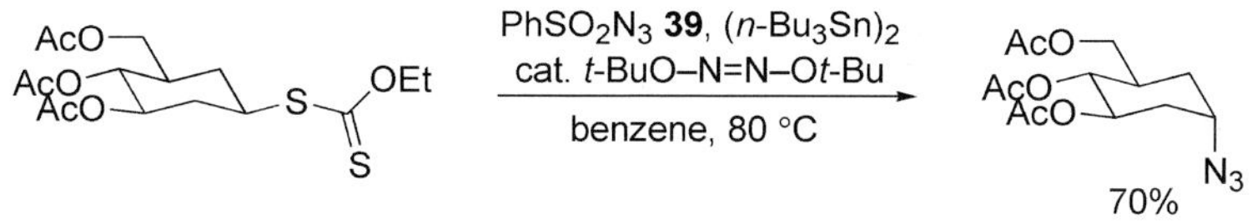

Scheme 1.69

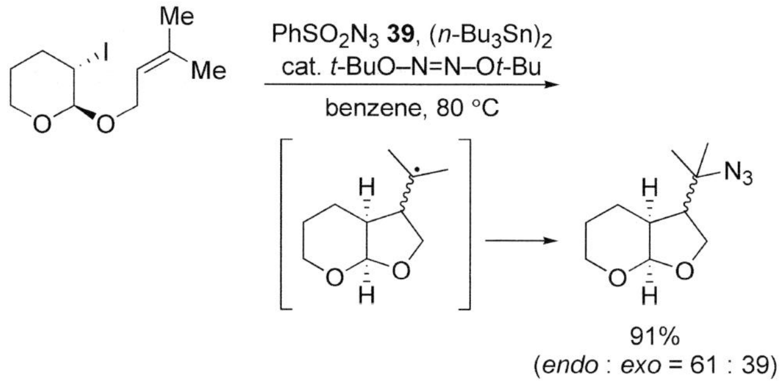

Scheme 1.70

1.9.3
Hydroazidation of Alkenes with Sulfonyl Azides

Carreira developed the cobalt-catalyzed hydroazidation of alkenes with sulfonylazides, which allows the synthesis of secondary and tertiary alkyl azides [34d].

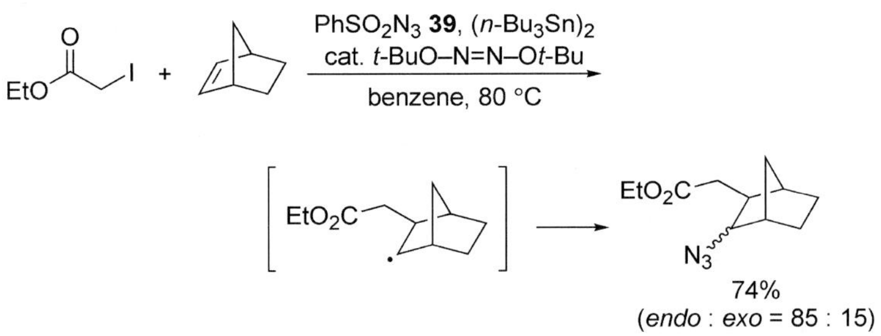

Scheme 1.71

Table 1.11 Co-catalyzed hydroazidation of alkenes with sulfonyl azide **40**.

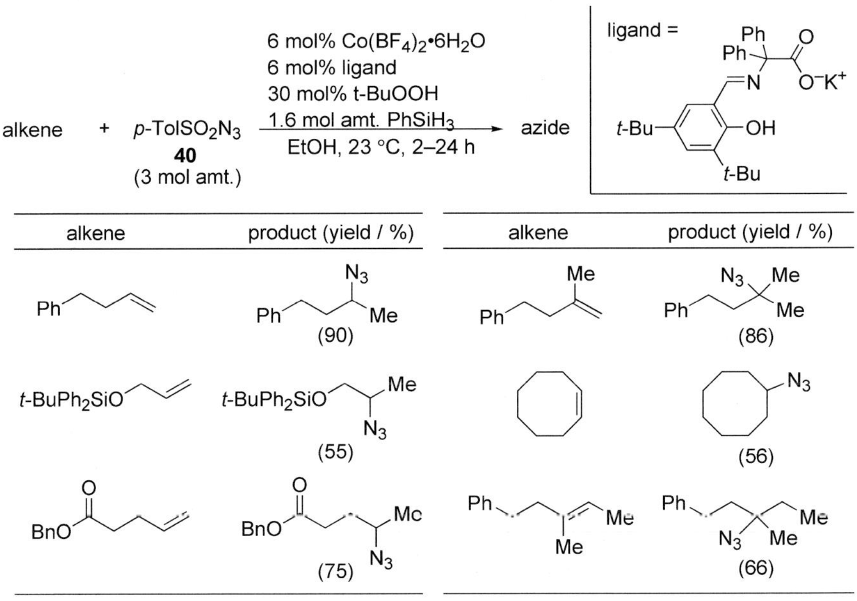

Treatment of mono-, di-, and trisubstituted alkenes with *p*-tolylsulfonyl azide (**40**) in the presence of phenylsilane and *tert*-butyl hydroperoxide (30 mol%) is catalyzed by $Co(BF_4)_2 \cdot 6H_2O$ and the Schiff base ligand to give alkyl azides in good yield (Table 1.11). The Markovnikov-type products are formed exclusively, with broad functional group tolerance.

1.9.4
Experimental Procedures

Representative procedure for electrophilic azidation of amide enolates by utilization of 2,4,6-triisopropylbenzenesulfonyl azide A toluene solution of potassium hexamethyldisilazide (KHMDS, 0.48 M, 2 mL, 0.960 mmol) was added at −78 °C to THF (3 mL). A precooled (−78 °C) solution of the acyloxazolidinone (269 mg, 0.87 mmol) in THF (3 mL) was added to the solution by cannula, and the mixture was stirred for 30 min. A precooled (−78 °C) solution of triisopropylbenzenesulfonyl azide (**38**, 330–340 mg, 1.07–1.10 mmol) in THF (3 mL) was added by cannula to the solution of the resulting potassium enolate. After 1 min, the reaction was quenched with glacial acetic acid (0.23 mL, 4.0 mmol), the cooling bath was removed, and the mixture was immediately warmed to 25–30 °C for 30 min with a water bath. The solution was partitioned between CH_2Cl_2 and brine (40 mL), the aqueous phase was washed three times with CH_2Cl_2, and the combined extracts were washed with saturated aqueous $NaHCO_3$, dried over anhydrous $MgSO_4$, and evaporated under vacuum. The resulting crude materials were purified by medium-pressure column chromatography (50 g of silica gel, 1 L linear gradient from CH_2Cl_2/hexane 6 : 4 to CH_2Cl_2) to give [3(2*S*),4*S*]-3-(2-azido-3-phenyl-1-oxopropyl)-4-(phenylmethyl)-2-oxazolidinone (278 mg, 0.79 mmol) in 91% yield.

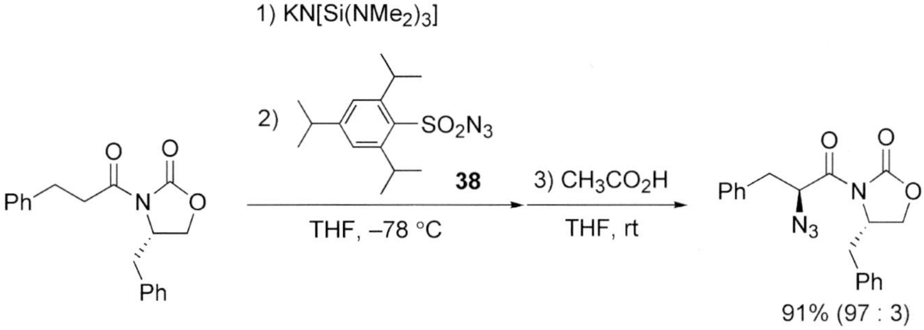

Representative procedure for azidation of alkyl radicals with phenylsulfonyl azide Di-*tert*-butylhyponitrite (10 mg, 0.06 mmol) was added to a solution of the iodide (365 mg, 1.0 mmol), phenylsulfonyl azide (**39**, 550 mg, 3.0 mmol), and $(Bu_3Sn)_2$ (0.76 mL, 1.5 mmol) in benzene (2 ml) at reflux under N_2. The reaction was monitored by TLC. Upon completion, the solvent was removed under reduced pressure, and the resulting crude materials were filtered through silica gel (hexane, then hexane/AcOEt 85 : 15). After the combined fractions had been concentrated under vacuum, the residue was purified by flash column chromatography (silica gel. hexane/AcOEt 85 : 15) to give 4-azido-1-[(4-methylphenyl)sulfonyl]piperidine (249 mg, 0.89 mmol) in 89% yield.

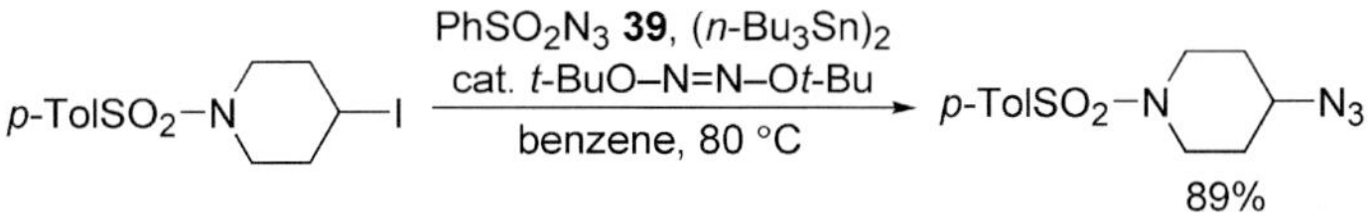

Representative procedure for hydroazidation of alkenes with sulfonyl azides $Co(BF_4)_2 \cdot 6H_2O$ (10 mg, 0.030 mmol) and the ligand (14 mg, 0.030 mmol)

were dissolved in ethanol (2.5 mL) at 23 °C under argon. After 10 min, 4-phenylbut-1-ene (75 µL, 0.50 mmol), followed by *p*-tolylsulfonyl azide (**40**, 0.23 mL, 1.5 mmol) and *tert*-butyl hydroperoxide (5.5 M in decane, 25 µL, 0.14 mmol), were added to the solution. After 5 min, phenylsilane (0.10 mL, 0.80 mmol) was added dropwise. The resulting brown solution was stirred at 23 °C. After completion, the reaction was quenched with H_2O (2 mL), and saturated aqueous $NaHCO_3$ (2 mL) and brine (5 mL) were added to the mixture. The organic materials were extracted with EtOAc (3 × 10 mL), and the combined extracts were dried over anhydrous Na_2SO_4. The solvents were removed under reduced pressure, and the resulting crude materials were purified by flash column chromatography to give (3-azidobutyl)benzene (79 mg, 0.45 mmol) in 90% yield.

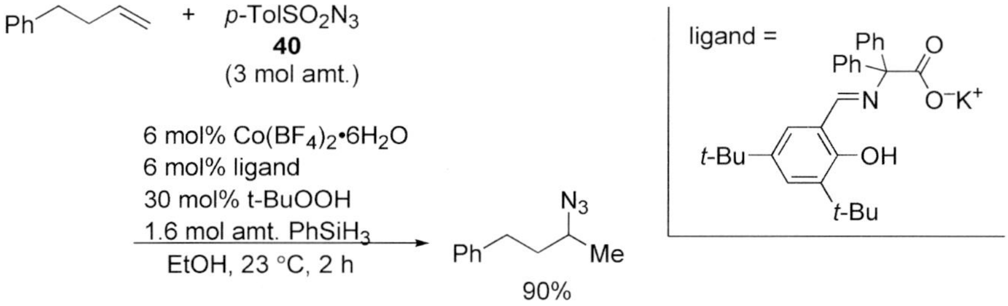

1.10
Gabriel-Type Reagents

1.10.1
Nucleophilic Amination Reactions

The Gabriel synthesis is a classical, but still very useful nucleophilic amination method for the synthesis of simple primary amines [88]. The first step of the Gabriel method, the formation of *N*-alkylphthalimide, includes substitution reactions of potassium phthalimide with alkyl halides and alkyl sulfonates. The second procedure, the conversion of the *N*-alkylphthalimides into primary amines, requires strongly acidic or basic conditions, or can alternatively be carried out by treatment with hydrazine. These severe reaction conditions prevent the application to polyfunctionalized compounds containing carbonyl groups and halogen equivalents.

To solve these problems, a number of alternatives to phthalimide have been developed in order to allow the second deprotection step to be carried out under milder conditions. Some of these reagents were reviewed by Ragnarsson in 1991 and are shown in Figure 1.2 [89], whereas other Gabriel-type reagents developed recently are covered in the following examples.

Bis(2-trimethylsilylethanesulfonyl)imide (SES_2NH, **48**), easily prepared by alkylation of bis(methanesulfonyl)imide by treatment with LHMDS and (iodomethyl)t rimethylsilane, is used in the synthesis of protected primary amines (Scheme

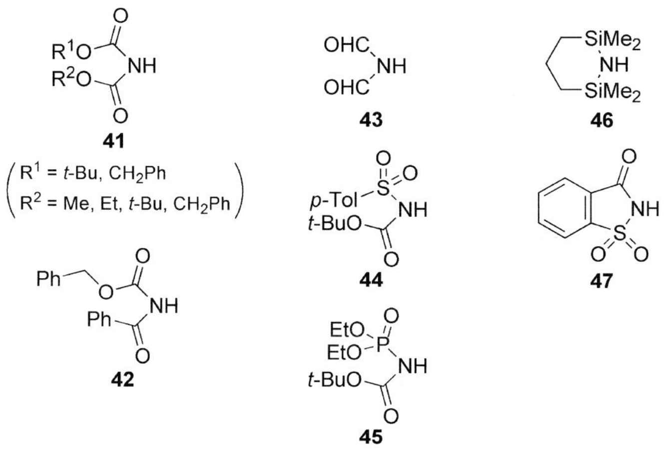

Figure 1.2 Gabriel-type reagents.

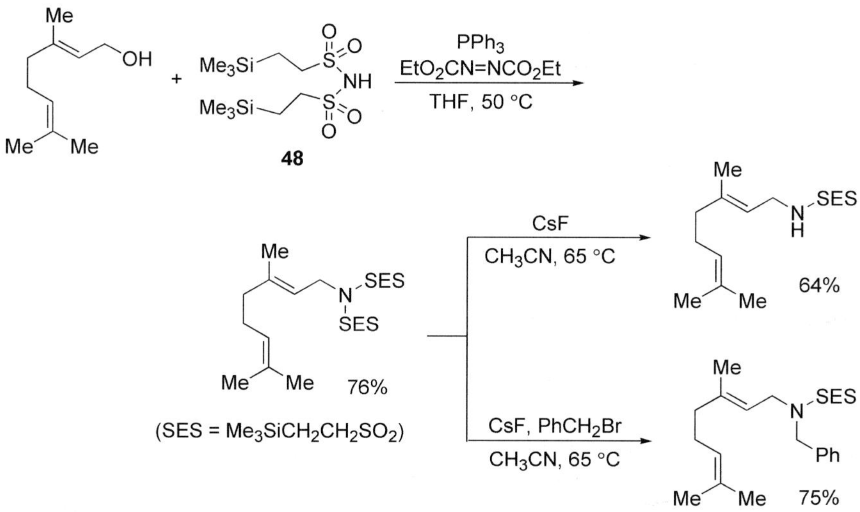

Scheme 1.72 Synthesis of protected primary amines by Mitsunobu alkylation of **48**.

1.72). This reagent undergoes Mitsunobu alkylation with both primary and secondary alcohols to afford the corresponding bis-SES imides. These imides can be selectively cleaved to the mono-SES-protected amines by treatment with CsF. In addition, one-pot monodeprotection/*N*-alkylation can be carried out by successive treatment with CsF and an alkylating agent such as benzyl bromide, affording *N*-alkyl mono-SES amino derivatives [90].

1,2,4-Dithiazolidine-3,5-dione (**49**) [91] is used for nucleophilic *N*-alkylation with alkyl halides and for Mitsunobu reactions with alcohols (Scheme 1.73) [92]. The resulting *N*-alkyl-1,2,4-dithiazolidine-3,5-diones are readily converted into urethanes via isocyanates by treatment with triphenylphosphine and alcohol.

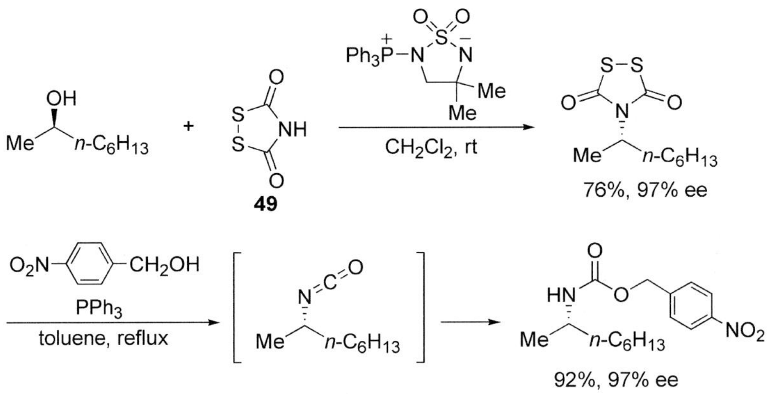

Scheme 1.73 Nucleophilic *N*-alkylation of **49** with alkyl halides.

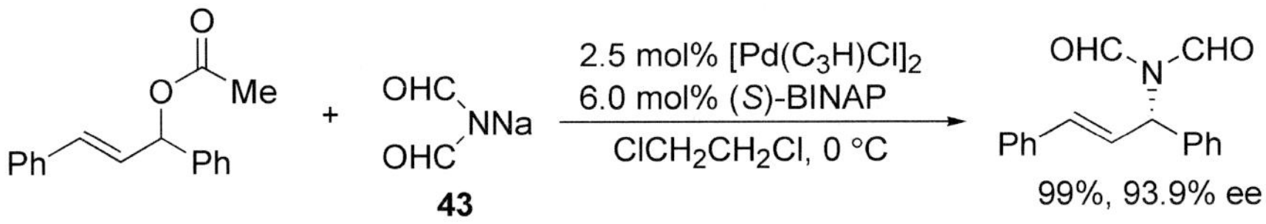

Scheme 1.74

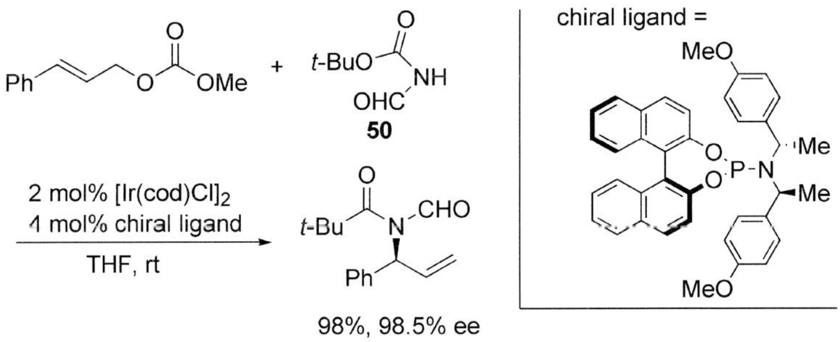

Scheme 1.75

Some Gabriel-type reagents have been applied to transition metal-catalyzed allylic amination. Ding, for example, reported the asymmetric synthesis of allylic amines from allylic acetates with sodium *N,N*-diformamide (**43**), catalyzed by allylpalladium chloride dimer and BINAP (Scheme 1.74) [93].

Iridium complexes are also suitable catalysts for the synthesis of chiral allylic amines [94]. When a mixture of allylic carbonate and *tert*-butyl formylcarbamate (**50**) is treated with a catalytic amount of $[Ir(cod)Cl]_2$ and a chiral amino phosphine ligand, the corresponding allylic amine derivative is obtained with good regioselectivity favoring the branched product, as well as with high enantioselectivity (Scheme 1.75).

1.10.2
Experimental Procedure

Representative nucleophilic amination with 1,2,4-dithiazolidine-3,5-dione (*R*)-Octan-2-ol (120 µL, 0.76 mmol) was added at room temperature to a solution of 1,2,4-dithiazolidine-3,5-dione (**49**, 110 mg, 0.82 mmol) and the betaine condensation reagent (345 mg, 0.84 mmol) in CH_2Cl_2 (2 mL). After the mixture had been stirred for 18 h, the solvent was evaporated under vacuum, and the resulting residue was purified by flash column chromatography (silica gel, petroleum ether/EtOAc 9:1) to give (*S*)-4-(1-methylheptyl)-1,2,4-dithiazolidine-3,5-dione (142 mg, 0.58 mmol) in 76% yield.

4-Nitrobenzyl alcohol (60 mg, 0.39 mml) was added to a solution of (*S*)-4-(1-methylheptyl)-1,2,4-dithiazolidine-3,5-dione (120 mg, 0.49 mmol) and triphenylphosphine (130 mg, 0.50 mmol) in toluene (2 mL), and the reaction mixture was stirred at reflux for 48 h. The solvent was removed under reduced pressure, and the resulting residue was purified by flash column chromatography (silica gel, petroleum ether/EtOAc 9:1) to give (*S*)-(4-nitrobenzyl)-1-methylheptyl carbamate (111 mg, 0.36 mmol) in 92% yield based on 4-nitrobenzyl alcohol and with 97% *ee* (determined by chiral HPLC).

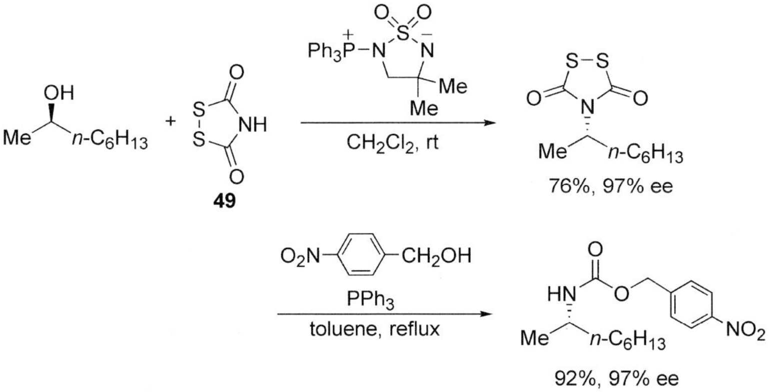

Representative palladium-catalyzed allylic amination with sodium N,N-diformylamide 1,3-Diphenylprop-2-en-1-yl acetate (46.8 mg, 0.186 mmol) and Et_3N (26 µL, 0.186 mmol) were added to a solution of allylpalladium chloride dimer (1.7 mg, 0.0047 mmol) and (*S*)-BINAP (6.7 mg, 0.0112 mmol) in 1,2-dichloroethane (3 mL), and the mixture was stirred for 10 min. After the mixture had been cooled to 0 °C, sodium *N,N*-diformamide (**43**, 106 mg, 1.116 mmol) was added. The reaction mixture was stirred at 0 °C for 6 h. After filtration, the solvent was evaporated under vacuum, and the resulting residue was purified by flash column chromatography (silica gel, hexane/EtOAc 4:1) to give (*S*)-*N,N*-diformyl-1,3-diphenyl-2-propenoylamide (49.0 mg) in 99% yield and with 99% *ee*.

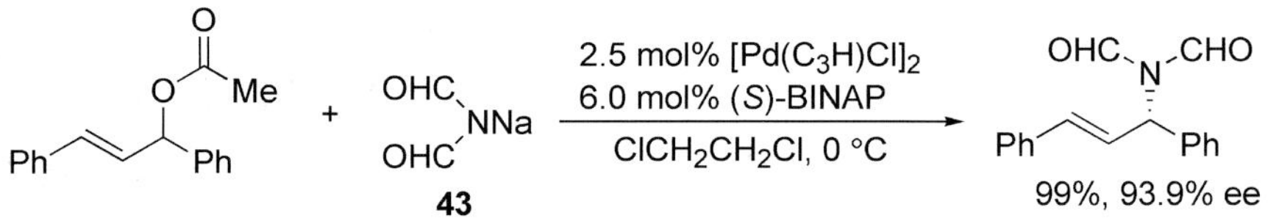

1.11 Conclusion

This chapter reviews the recent advances in the field of simple amination reagents, which have allowed synthetically efficient introduction of various nitrogen units into organic molecules, with the formation of new carbon–nitrogen bonds. Other widely used nitrogen-containing small molecules such as nitroso compounds [95], nitrites [96], or nitrogen monoxide [97] are outside the scope of this chapter and are discussed elsewhere. Nitrogen atom transfer to organic molecules through cleavage of dinitrogen by the action of transition metal complexes has also recently been reported [98] and the further development of metal-catalyzed amination reactions with dinitrogen will certainly become a challenging research area in the future.

References and Notes

1 Y. Tamura, J. Minamikawa, M. Ikeda, *Synthesis* 1977, 1–17.
2 N. L. Sheverdina, Z. Kocheskov, *J. Gen. Chem. USSR* 1938, *8*, 1825–1829.
3 G. Boche, H. H. Wagner, *J. Chem. Soc., Chem. Commun.* 1984, 1591–1592.
4 a) J. P. Genêt, S. Mallart, C. Greck, E. Piveteau, *Tetrahedron Lett.* 1991, *32*, 2359–2362. b) C. Greck, L. Bischoff, A. Girard, J. Hajicek, J. P. Genêt, *Bull. Soc. Chim. Fr.* 1994, *131*, 429–433.
5 C. Greck, L. Bischoff, F. Ferreira, J. P. Genêt, *J. Org. Chem.* 1995, *60*, 7010–7012.
6 G. Boche, C. Boie, F. Bosold, K. Harms, M. Marsch, *Angew. Chem. Int. Ed.* 1994, *33*, 115–117.
7 N. Zheng, J. D. Armstrong, III, J. C. McWilliams, R. P. Volante, *Tetrahedron Lett.* 1997, *38*, 2817–2820.
8 J. P. Genêt, J. Hajicek, L. Bischoff, C. Greck, *Tetrahedron Lett.* 1992, *33*, 2677–2680.
9 a) S. Fioravanti, M. Antonietta, L. Pellacani, P. A. Tardella, *Tetrahedron Lett.* 1993, *34*, 4353–4354. b) M. Barani, S. Fioravanti, L. Pellacani, P. A. Tardella, *Tetrahedron* 1994, *50*, 11235–11238.
10 a) S. Fioravanti, L. Pellacani, S. Stabile, P. A. Tardella, *Tetrahedron Lett.* 1997, *38*, 3309–3310. b) S. Fioravanti, L. Pellacani, S. Stabile, P. A. Tardella, *Tetrahedron* 1998, *54*, 6169–6176. c) S. Fioravanti, A. Morreale, L. Pellacani, P. A. Tardella, *Synthesis* 2001, 1975–1978. d) S. Fioravanti, A. Morreale, L. Pellacani, P. A. Tardella, *J. Org. Chem.* 2002, *67*, 4972–4974. e) S. Fioravanti, A. Morreale, L. Pellacani, P. A. Tardella, *Eur. J. Org. Chem.* 2003, 4549–4552. f) S. Fioravanti, M. Gabriella Mascia, A. Morreale, L. Pellacani, P. A. Tardella, *Eur. J. Org. Chem.* 2002, 4071–4074.
11 a) G. Boche, Encyclopedia of Reagents for Organic Synthesis; L. A. Paquette, Ed.; J. Wiley & Sons: New York, 1995; Vol. 4, p 2240–2242. b) G. Boche, M. Bernheim, W. Schrott, *Tetrahedron Lett.* 1982, *23*, 5399–5402.
12 J. A. Smulik, E. Vedejs, *Org. Lett.* 2003, *5*, 4187–4190.
13 G. Boche, *Encyclopedia of Reagents for Organic Synthesis*, L. A. Paquette, Ed.,

J. Wiley & Sons, New York, 1995, Vol. 5, pp. 3270–3271.

14 a) A. M. Berman, J. S. Johnson, *J. Am. Chem. Soc.* 2004, *126*, 5680–5681. b) A. M. Berman, J. S. Johnson, *J. Org. Chem.* 2005, *70*, 364–366. c) A. M. Berman, J. S. Johnson, *J. Org. Chem.* 2006, *71*, 219–224. d) A. M. Berman, J. S. Johnson, *Org. Syn.* 2006, *83*, 31–37.

15 A. Casarini, P. Dembech, D. Lazzari, E. Marini, G. Reginato, A. Ricci, G. Seconi, *J. Org. Chem.* 1993, *58*, 5620–5623.

16 P. Bernardi, P. Dembech, G. Fabbri, A. Ricci, G. Seconi, *J. Org. Chem.* 1999, *64*, 641–643.

17 For recent reviews on the synthetic applications of oxime derivatives, see: a) J. P. Adams, *J. Chem. Soc., Perkin Trans 1*, 2000, 125–139. b) E. Abele, E. Lukevics, *Heterocycles* 2000, *53*, 2285–2336. c) K. Narasaka, M. Kitamura, *Eur. J. Org. Chem.* 2005, 4505–4519. d) M. Yamane, K. Narasaka, *Science of Synthesis*, Vol. 27: *Carbons with Two Carbon–Heteroatom Bonds: Heteroatom Analogues of Aldehydes and Ketones*, A. Padwa, Ed., Thieme, Stuttgart, 2004, Chapter 15.

18 R. A. Hagopian, M. J. Therien, J. R. Murdoch, *J. Am. Chem. Soc.* 1984, *106*, 5753–5754.

19 a) E. Erdik, M. Ay, *Synth. React. Inorg. Met.-Org. Chem.* 1989, *19*, 663–668. b) E. Erdik, T. Daskapan, *Synth. Commun.* 1999, *29*, 3989–3997. c) E. Erdik, T. Daskapan, *J. Chem. Soc., Perkin Trans. I* 1999, 3139–3142. d) E. Erdik, *Encyclopedia of Reagents for Organic Synthesis*, L. A. Paquette, Ed., J. Wiley & Sons, New York, 1995; Vol. 1, p 41–42.

20 E. Erdik, T. Daskapan, *Tetrahedron Lett.* 2002, *43*, 6237–6239.

21 H. Tsutsui, T. Ichikawa, K. Narasaka, *Bull. Chem. Soc. Jpn.* 1999, *72*, 1869–1878.

22 R. W. Hoffmann, B. Hölzer, O. Konpff, *Org. Lett.* 2001, *3*, 1945–1948.

23 M. Kitamura, S. Chiba, K. Narasaka, *Bull. Chem. Soc. Jpn.* 2003, *76*, 1063–1070.

24 N. Baldovini, M. Kitamura, K. Narasaka, *Chem. Lett.* 2003, *32*, 548–549.

25 M. Kitamura, T. Suga, S. Chiba, K. Narasaka, *Org. Lett.* 2004, *6*, 4691–4693.

26 L. Boymond, M. Rottlander, G. Cahiez, P. Knochel, *Angew. Chem. Int. Ed.* 1998, *37*, 1701–1703.

27 W. Dohle, D. M. Lindsay, P. Knochel, *Org. Lett.* 2001, *3*, 2871–2873.

28 S. Chiba, K. Narasaka, unpublished results.

29 a) J. P. Genêt, C. Greck, D. Lavergne, *Modern Amination Methods*, A. Ricci, Ed., Wiley–VCH, Weinheim, 2000, Chapter 3. b) E. Erdik, *Tetrahedron* 2004, *60*, 8747–8782.

30 W. A. Thaler, B. Franzus, *J. Org. Chem.* 1964, *29*, 2226–2235.

31 Y. Leblanc, R. Zamboni, M. A. Bernstein, *J. Org. Chem.* 1991, *56*, 1971–1972.

32 M. A. Brimble, C. H. Heathcock, *J. Org. Chem.* 1993, *58*, 5261–5263.

33 a) A. G. Davies, W. J. Kinart, *J. Chem. Soc., Perkin Trans. II* 1993, 2281–2284. b) W. J. Kinart, *J. Chem. Res., Synop. 1994*, 486–487.

34 a) J. Waser, E. M. Carreira, *J. Am. Chem. Soc.* 2004, *126*, 5676–5677. b) J. Waser, E. M. Carreira, *Angew. Chem. Int. Ed.* 2004, *43*, 4099–4102. c) J. Waser, J. C. González-Gómez, H. Nambu, P. Huber, E. M. Carreira, *Org. Lett.* 2005, *7*, 4249–4252. d) J. Waser, B. Gaspar, H. Nambu, E. M. Carreira, *J. Am. Chem. Soc.* 2006, *128*, 11693–11712.

35 S. M. Weinreb, *Comprehensive Organic Synthesis*, B. M. Trost, I. Fleming, L. A. Paquette, Eds., Pergamon: Oxford, 1991, Vol. 5, 401–449.

36 M. F. Ahern, A. Leopold, J. R. Beadle, G. W. Gokel, *J. Am. Chem. Soc.* 1982, *104*, 548–554.

37 C. Dell'Erba, M. Novi, G. Petrillo, C. Tavani, *Tetrahedron* 1995, *51*, 3905–3914.

38 C. Dell'Erba, M. Novi, G. Petrillo, C. Tavani, *Tetrahedron* 1997, *53*, 2125–2136.

39 I. Sapountzis, P. Knochel, *Angew. Chem. Int. Ed.* 2004, *43*, 897–900.

40 For a review, see: F. A. Davis, A. C. Sheppard, *Tetrahedron* 1989, *45*, 5703–5742.

41 J. Vidal, S. Damestoy, L. Guy, J. C. Hannachi, A. Aubry, A. Collet, *Chem. Eur. J.* 1997, *3*, 1691–1709.

42 J. Vidal, L. Guy, S. Stérin, A. Collet, *J. Org. Chem.* 1993, *58*, 4791–4793.

43 a) A. Armstrong, M. A. Atkin, S. Swallow, *Tetrahedron Lett.* 2000, *41*, 2247–2251. b) A. Armstrong, I. D. Edmonds, M. E. Swarbrick, N. R. Treweeke, *Tetrahedron* 2005, *61*, 8423–8442.

44 A. Armstrong, R. S. Cooke, *Chem. Commun.* 2002, 904–905.

45 A. Armstrong, L. Challinor, R. S. Cooke, J. H. Moir, N. R. Treweeke, *J. Org. Chem.* 2006, *71*, 4028–4030.

46 A. Armstrong, R. S. Cooke, S. F. Shanahan, *Org. Biomol. Chem.* 2003, *1*, 3142–3143.

47 R. R. Goehring, *Encyclopedia of Reagents for Organic Synthesis*, L. A. Paquette, Ed., J. Wiley & Sons, New York, 1995, Vol. 2, p 1054–1056.

48 D. H. R. Barton, M. R. Britten-Kelly, D. Ferreira, *J. Chem. Soc., Perkin Trans. 1* 1978, 1090–1100.

49 a) K. B. Sharpless, A. O. Chong, K. Oshima, *J. Org. Chem.* 1976, *41*, 177–179. b) E. Herranz, K. B. Sharpless, *J. Org. Chem.* 1978, *43*, 2544–2548.

50 K. B. Sharpless, D. W. Patrick, L. K. Truesdale, S. A. Biller, *J. Am. Chem. Soc.* 1975, *97*, 2305–2307.

51 a) G. Li, H. T. Chang, K. B. Sharpless, *Angew. Chem. Int. Ed.* 1996, *35*, 451–454. b) A. E. Rubin, K. B. Sharpless, *Angew. Chem. Int. Ed.* 1997, *36*, 2637–2640.

52 G. Li, H. H. Angert, K. B. Sharpless, *Angew. Chem. Int. Ed.* 1996, *35*, 2813–2817. The original report on carbamate-based osmium-catalyzed aminohydroxylation of alkenes appeared in 1978, see: E. Herranz, S. A. Biller, K. B. Sharpless, *J. Am. Chem. Soc.* 1978, *100*, 3596–3598.

53 a) B. Cao, H. Park, M. M. Joullie, *J. Am. Chem. Soc.* 2002, *124*, 520–521. b) W. Jiang, J. Wanner, R. J. Lee, P. Y. Bounaud, D. L. Boger, *J. Am. Chem. Soc.* 2003, *125*, 1877–1887. c) B. M. Crowley, Y. Mori, C. C. McComas, D. Tang, D. L. Boger, *J. Am. Chem. Soc.* 2004, *126*, 4310–4317. d) W. Kurosawa, H. Kobayashi, T. Kan, T. Fukuyama, *Tetrahedron* 2004, *60*, 9615–9628.

54 T. Ando, S. Minakata, I. Ryu, M. Komatsu, *Tetrahedron Lett.* 1998, *39*, 309–312.

55 D. P. Albone, P. S. Aujla, P. C. Taylor, S. Challenger, A. M. Derrick, *J. Org. Chem.* 1998, *63*, 9569–9571.

56 L. Simkhovich, Z. Gross, *Tetrahedron Lett.* 2001, *42*, 8089–8092.

57 S. Minakata, D. Kano, R. Fukuoka, Y. Oderaotoshi, M. Komatsu, *Heterocycles* 2003, *60*, 289–298.

58 J. U. Jeong, B. Tao, I. Sagasser, H. Henniges, K. B. Sharpless, *J. Am. Chem. Soc.* 1998, *120*, 6844–6845.

59 S. L. Ali, M. D. Nikalje, S. Sudalai, *Org. Lett.* 1999, *1*, 705–707.

60 T. Ando, D. Kano, S. Minakata, I. Ryu, M. Komatsu, *Tetrahedron* 1998, *54*, 13485–13494.

61 D. Kano, S. Minakata, M. Komatsu, *J. Chem. Soc., Perkin Trans. 1* 2001, 3186–3188.

62 S. Minakata, D. Kano, Y. Oderaotoshi, M. Komatsu, *Angew. Chem. Int. Ed.* 2004, *43*, 79–91.

63 S. Minakata, D. Kano, Y. Oderaotoshi, M. Komatsu, *Org. Lett.* 2002, *4*, 2097–2099.

64 S. Minakata, Y. Yoneda, Y. Oderaotoshi, M. Komatsu, *Org. Lett.* 2006, *8*, 967–969.

65 Synthesis of *N*-sulfonyliminophenyliodinane, see: a) Y. Yamada, T. Yamamoto, M. Okawara, *Chem. Lett.* 1975, 361–362. b) M. J. Sodergren, D. A. Alonso, A. V. Bedekar, P. G. Andersson, *Tetrahedron Lett.* 1997, *38*, 6897–6900.

66 a) D. Mansuy, J.-P. Mahy, A. Duresult, G. Bedi, P. Battioni, *J. Chem. Soc., Chem. Commun.* 1984, 1164–1163. b) J.-P. Mahy, G. Bedi, P. Battioni, D. Mansuy, *Tetrahedron Lett.* 1988, *29*, 1927–1930. c) J.-P. Mahy, G. Bedi, P. Battioni, D. Mansuy, *J. Chem. Soc., Perkin Trans. II* 1988, 1517–1524.

67 S.-M. Au, W.-H. Fung, M.-C. Cheng, C.-M. Che, S.-M. Peng, *Chem. Commun.* 1997, 1655–1656.

68 a) P. Müller, C. Baud, Y. Jacquier, *Tetrahedron* 1996, *52*, 1543–1548. b) P. Müller, C. Baud, Y. Jacquier, *Can. J. Chem.* 1998, *76*, 738–750.

69 a) D. A. Evans, M. M. Faul, M. T. Bilodeau, *J. Org. Chem.* 1991, *56*, 6744–6746. b) D. A. Evans, M. M. Faul, M. T. Bilodeau, *J. Am. Chem. Soc.* 1994, *116*,

2472–2753. c) P. Dauban, R. H. Dodd, *J. Org. Chem.* 1999, *63*, 5304–5307.

70 Y. Cui, C. He, *J. Am. Chem. Soc.* 2003, *125*, 16202–16203.

71 For a review, see: P. Müller, C. Fruit, *Chem. Rev.* 2003, *103*, 2905–2919.

72 K. Guthikonda, J. Du Bois, *J. Am. Chem. Soc.* 2002, *124*, 13672–13673.

73 Z. Li, X. Ding, C. He, *J. Org. Chem.* 2006, *71*, 5876–5880.

74 a) W. P. Griffith, *Coord. Chem. Rev.* 1972, *8*, 369–396. b) K. Dehnicke, J. Strähle, *Angew. Chem. Int. Ed.* 1992, *31*, 955–978. c) R. A. Eikey, M. M. Abu-Omar, *Coord. Chem. Rev.* 2003, *243*, 83–124.

75 a) J. Du Bois, J. Hong, E. M. Carreira, M. W. Day, *J. Am. Chem. Soc.* 1996, *118*, 915–916. b) J. Du Bois, C. S. Tomooka, J. Hong, E. M. Carreira, *J. Am. Chem. Soc.* 1997, *119*, 3179–3180. c) J. Du Bois, C. S. Tomooka, J. Hong, E. M. Carreira, *Acc. Chem. Res.* 1997, *30*, 364–372. d) J. Du Bois, C. S. Tomooka, J. Hong, E. M. Carreira, M. W. Day, *Angew. Chem. Int. Ed.* 1997, *36*, 1645–1647. e) E. M. Carreira, J. Hong, J. Du Bois, C. S. Tomooka, *Pure & Appl. Chem.* 1998, *70*, 1097–1103.

76 S. K.-Y. Leung, J.-S. Huang, J.-L. Kiang, C.-M. Che, Z.-Y. Zhou, *Angew. Chem. Int. Ed.* 2003, *42*, 340–343.

77 a) S. Minakata, T. Ando, M. Nishimura, I. Ryu, M. Komatsu, *Angew. Chem. Int. Ed.* 1998, *37*, 3392–3394. b) M. Nishimura, S. Minakata, S. Thongchant, I. Ryu, M. Komatsu, *Tetrahedron Lett.* 2000, *41*, 7089–7092. c) M. Nishimura, S. Minakata, T. Takahashi, Y. Oderaotoshi, M. Komatsu, *J. Org. Chem.* 2002, *67*, 2101–2110.

78 a) R. Huisgen, *Angew. Chem. Int. Ed.* 1963, *2*, 565–598. b) R. Huisgen, R. Knorr, L. Möbius, G. Szeimies, *Chem. Ber.* 1965, *98*, 4014–4021.

79 a) E. F. V. Scriven, K. Turnbull, *Chem. Rev.* 1988, *88*, 297–368. b) S. Brase, C. Gil, K. Knepper, V. Zimmermann, *Angew. Chem. Int. Ed.* 2005, *44*, 5188–5240.

80 a) B. M. Trost, W. H. Pearson, *J. Am. Chem. Soc.* 1981, *103*, 2483–2485. b) B. M. Trost, W. H. Pearson, *J. Am. Chem. Soc.* 1983, *105*, 1054–1056.

81 G. W. Kabalka, G. Li, *Tetrahedron Lett.* 1997, *38*, 5777–5778.

82 K. Nishiyama, N. Tanaka, *J. Chem. Soc., Chem. Commun.* 1983, 1322–1323.

83 A. V. Thomas, *Encyclopedia of Reagents for Organic Synthesis*, L. A. Paquette, Ed., J. Wiley & Sons: New York, 1995; Vol. 4, pp. 2242–2245, and references therein.

84 D. A. Evans, T. C. Britton, J. A. Ellman, R. L. Dorow, *J. Am. Chem. Soc.* 1990, *112*, 4011–4030.

85 P. Panchaud, L. Chabaud, Y. Landais, C. Ollivier, P. Renaud, S. Zigmantas, *Chem. Eur. J.* 2004, *10*, 3606–3614.

86 C. Ollivier, P. Renaud, J. *Am. Chem. Soc.* 2001, *123*, 4717–4727.

87 a) P. Renaud, C. Ollivier, P. Panchaud, *Angew. Chem. Int. Ed.* 2002, *41*, 3460–3462. b) P. Panchaud, C. Ollivier, P. Renaud, S. Zigmantas, *J. Org. Chem.* 2004, *69*, 2755–2759.

88 a) S. Gabriel, *Ber. Dtsch. Chem. Ges.* 1887, *20*, 2224–2236. b) M. S. Gibson, R. W. Bradshaw, *Angew. Chem. Int. Ed.* 1968, *7*, 919–930. c) H. de Koning, W. Nico Speckamp, *Encyclopedia of Reagents for Organic Synthesis*, L. A. Paquette, Ed., J.Wiley & Sons, New York, 1995, Vol. 6, pp. 4141–4143.

89 For a review, see: U. Ragnarsson, L. Grehn, *Acc. Chem. Res.* 1991, *24*, 285–289.

90 D. M. Dastrup, M. P. VanBrunt, S. M. Weinreb, *J. Org. Chem.* 2003, *68*, 4112–4115.

91 G. Zumach, E. Kühle, *Angew. Chem. Int. Ed.* 1970, *9*, 54–63.

92 a) D. J. Cane-Honeysett, M. D. Dowle, M. E. Wood, *Synlett* 2000, 1622–1624. b) M. E. Wood, D. J. Cane-Honeysett, M. D. Dowle, *J. Chem. Soc., Perkin Trans. 1* 2002, 2046–2047. c) M. E. Wood, D. J. Cane-Honeysett, M. D. Dowle, S. J. Coles, M. B. Hursthouse, *Org. Biomol. Chem.* 2003, *1*, 3015–3023. d) D. J. Cane-Honeysett, M. D. Dowle, M. E. Wood, *Tetrahedron* 2005, *61*, 2141–2148.

93 Y. Wang, K. Ding, *J. Org. Chem.* 2001, *66*, 3238–3241.

94 R. Weihofen, O. Tverskoy, G. Helmchen, *Angew. Chem. Int. Ed.* 2006. *45*, 5546–5549.

95 For a recent review, see: H. Yamamoto, N. Momiyama, *Chem. Commun.* 2005, 3514–3525.

96 K. Kato, T. Mukaiyama, *Chem. Lett.* 1992, 1137–1140.

97 a) E. Hata, K. Kato, T. Yamada, T. Mukaiyama, *J. Syn. Org. Jpn.* 1996, *54*, 728–739. b) E. Hata, T. Yamada, T. Mukaiyama, *Bull. Chem. Soc. Jpn.* 1995, *68*, 3629–3636.

98 J. J. Curley, E. L. Sceats, C. C. Cummins, *J. Am. Chem. Soc.* 2006, *128*, 14036–14037, and references therein.

2
Catalytic C—H Amination with Nitrenes

Philippe Dauban and Robert H. Dodd

2.1
Introduction

Found in a plethora of natural products such as peptides, alkaloids, or nucleosides, nitrogen is also a key element in biology since, among its many properties, it can behave either as a hydrogen bond donor or as an acceptor. The paramount importance of nitrogen is illustrated by the large number of existing methodologies for the synthesis of amino compounds [1, 2] and particularly by the more recently developed enantioselective reductive amination of carbonyl derivatives [3], hydroamination [4], diamination [5] or aminohydroxylation of olefins [6], ring-opening of aziridines [7], and Buchwald–Hartwig amination [8], to name but a few. All these methods, however, share the need to incorporate some functionality (e.g., a carbonyl, a halogen, or a double bond) prior to the introduction of the nitrogen atom.

Within the context of finding ever more efficient synthetic strategies, a more straightforward process would be the design of new reactions that would allow the direct transformation of a C—H bond into a C—N bond. Such reactions would be highly attractive from a synthetic point of view, particularly if they could selectively discriminate a particular hydrogen atom among the numerous ones found in every organic compound. The last decade has witnessed the successful outcome of several investigations dedicated to transition metal-catalyzed regioselective C—H bond functionalization [9–12]. New tools have now become available to synthetic chemists for the formation of C—C, C—O, C—B or C—X (X: halogen) bonds starting from the C—H bonds previously considered to be too poorly reactive. From a mechanistic point of view, these processes may be classified into two groups depending on whether C—H bond cleavage occurs in the first step by C—H activation (inner-sphere [11] or organometallic [13] mechanisms) or in the last step through a C—H insertion (outer-sphere [11] or coordination [13] mechanisms).

In the case of C—N bond formation, the most notable achievements have arisen from recent efforts aimed at the development of efficient transition metal-catalyzed nitrene transfers [14–17]. In particular, the use of hypervalent iodine(III) reagents [18, 19] for the production of nitrenes has led to several breakthroughs

Amino Group Chemistry. From Synthesis to the Life Sciences. Edited by Alfredo Ricci

ISBN: 978-3-527-31741-7

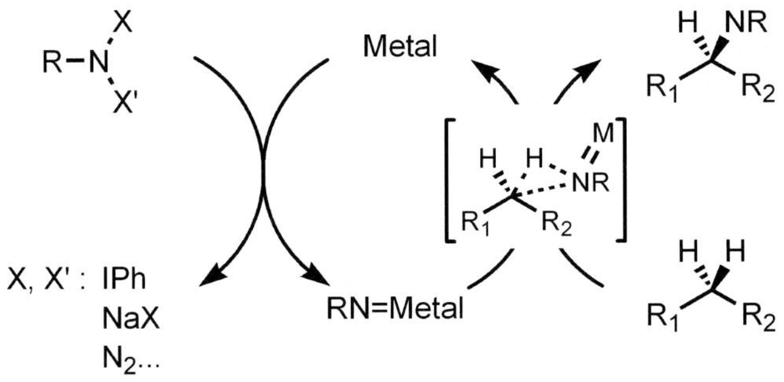

Scheme 2.1 Mechanistic principle of transition metal-catalyzed C—H amination.

in this field, although other nitrene precursors such as haloamines or azides have emerged in parallel as alternative reagents. Generally, C—N bond formation is believed to occur through the insertion into a Csp^3—H bond of a metallanitrene generated by reaction between the nitrene source and the transition metal complex (Scheme 2.1). The C—H insertion generally takes place stereospecifically through a concerted pathway, although the intermediacy of radicals has been demonstrated in some particular cases [14, 17]. The efficiency of the overall process, however, is such that it has been successfully applied to the total synthesis of various natural products.

This chapter focuses mostly on the use of nitrenes for catalytic amination of Csp^3—H bonds; other reactions such as electrophilic α-amination of carbonyl compounds [20, 21] or allylic amination based on ene reaction-like processes [22] that allow the substitution of a hydrogen atom by a nitrogen group are beyond its scope.

2.2 Historical Overview

Nitrenes, first mentioned by Tiemann in 1891 [23], were initially mostly generated from azides [24, 25], allowing the first Csp^3—H amination to be described in 1955 by Saunders [26]. Since then, seminal reports by Smolinsky [27, 28], Lwowski [29, 30], Breslow [31, 32], and Anastassiou [33, 34] demonstrated that alkane functionalization could be performed starting from azidobenzenes, azidoformates, sulfonyl azides or cyanogen azide, respectively. Even acyl azides could be used, although the efficiency of the C—H amination was rather low because of the competing Curtius rearrangement [25]. The nitrenes were generated by thermal or photochemical decomposition, while the C—H insertion generally occurred stereospecifically with retention of configuration, as elegantly demonstrated by Smolinsky in the cases of carbamate **1** and aryl azide **3** (Scheme 2.2) [35]. The synthetic interest of the intermolecular process was limited, however, due to the use of an alkane (generally cyclohexane) as the solvent, but intramolecular aminations can

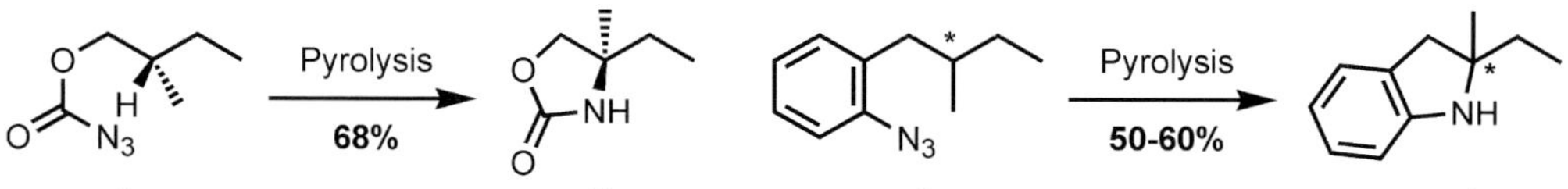

Scheme 2.2 Stereospecific C—H aminations with use of thermally generated nitrene from azide.

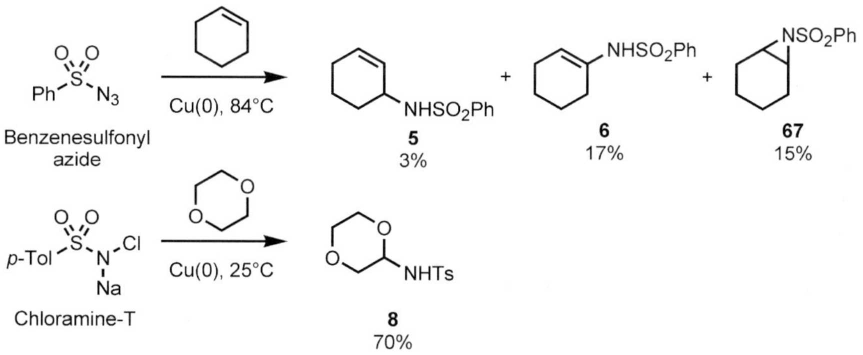

Scheme 2.3 Early copper-mediated C—H aminations of cyclohexene and 1,4-dioxane.

take place in good yields and have found some applications in total synthesis [25].

A major drawback of these free nitrenes is their high reactivity, associated with their ability to relax to their favored triplet states of radical character, resulting in numerous side reactions. Control over these reactive species was clearly needed and the use of transition metals appeared to be a possible way to achieve this goal. Consequently, metal-mediated C—H aminations were described at the end of the 1960s by three groups. The first example is generally attributed to Kwart and Kahn, who reported the decomposition of benzenesulfonyl azide in cyclohexene solution in the presence of copper, to afford allylic sulfonamide **5** (3%), enamine **6** (17%), and aziridine **7** (15%) (Scheme 2.3) [36]. A low yield of 6% was also obtained by Breslow in the amidation of cyclohexane by treatment with dichloramine-T and zinc [37]; however, the combination of copper powder and chloramine-T in dioxane at room temperature proved to give more significant results, the C—H insertion product **8** being isolated in 70% yield (Scheme 2.3) [38].

Chloramine-T was later investigated in conjunction with selenium [39] and ferrous chloride [40] for allylic amination and the functionalization of adamantane. Real advances in the field, however, occurred with the emergence of iminoiodanes, discovered independently by two groups in the mid-1970s [41, 42]. These hypervalent iodine reagents of general formula PhI=NR – nitrogen analogues of iodosylbenzene – were found to be suitable for the inter- and intramolecular amination of alkanes catalyzed by iron(III)- and manganese(III)-porphyrins [43–45]. Even some isoforms of cytochrome P-450 were able to afford such C—H amination products, although contaminated by compounds deriving from hydrolysis of the presumed iron-nitrene intermediate [46]. In the case of allylic substrates, mixtures of aziridine and allylic sulfonamides were isolated and the involvement of radicals was invoked in order to explain the observed selectivities [47]. All these studies dedicated to the modeling of cytochrome P-450 catalysis were of limited synthetic interest at the time – this was due to their low efficiency and selectivity, cyclohexane being aminated with yields of around 10% while heptane gave a mixture of four isomers in low amounts [47] – with the exception of a striking result obtained in the case of the intramolecular process. Use of an iron(III)-tetraphenylporphyrin catalyst allowed the formation of the cyclic sulfonamide **10** from iminoiodane **9** in 77% yield together with limited amounts of by-products, a result significantly improved by the use of rhodium(II) tetraacetate (Scheme 2.4) [44].

A few years later, a pioneering paper by Evans demonstrated the synthetic utility of iminoiodanes as nitrene precursors in the copper-catalyzed aziridination of olefins (i.e., nitrene insertion into a C=C bond) [48]. Various types of olefins can be transformed into the corresponding aziridines in very good yields and, in the presence of chiral ligands, with high enantioselectivities [14, 15]. Good chemoselectivity was also observed, since copper(I) and (II) complexes have a low capacity to catalyze allylic C—H insertion (Scheme 2.5) except in some rare cases [49, 50]. Rhodium(II) complexes were also found to efficiently mediate the same nitrene transfer onto several olefins, as demonstrated in a paper by Müller, but with a more pronounced tendency to induce C—H amination [51]. Thus, while the use of catalytic $Cu(OTf)_2$ in acetonitrile exclusively affords aziridine **16** from cyclohexene (**13**) and PhI=NTs (**14**) in 60% yield, the combination of $Rh_2(OAc)_4$ and PhI=NNs (**15**) in dichloromethane gives the allylic sulfonamide **18** as the major product.

In a seminal paper, Müller then studied the scope and limitations, as well as the mechanism, of this rhodium catalyzed C—H amination [52]. The reaction was shown to be highly substrate-dependent with respect to substitution and ring size, with the best yields (up to 84%) being obtained in the case of secondary benzylic and allylic positions. The overall efficiency, though, was hampered by the need for

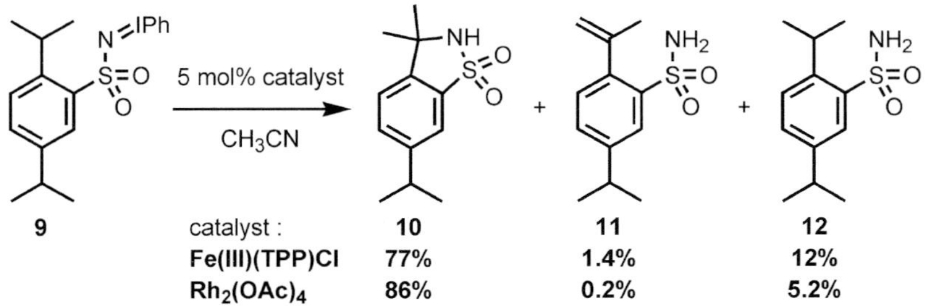

Scheme 2.4 Intramolecular metal-catalyzed C—H insertion starting from iminoiodanes.

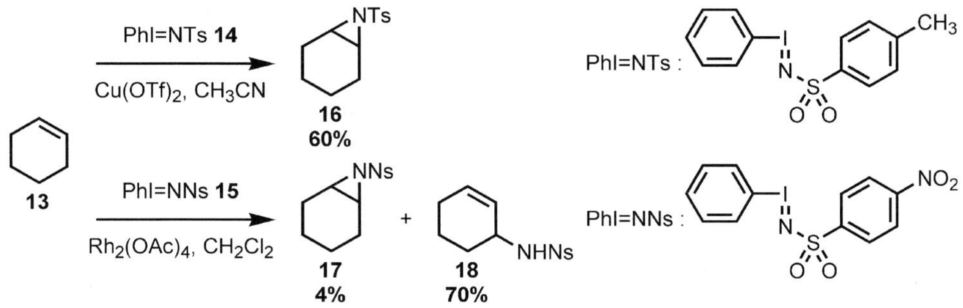

Scheme 2.5 Transition metal-catalyzed aziridination versus C—H amination.

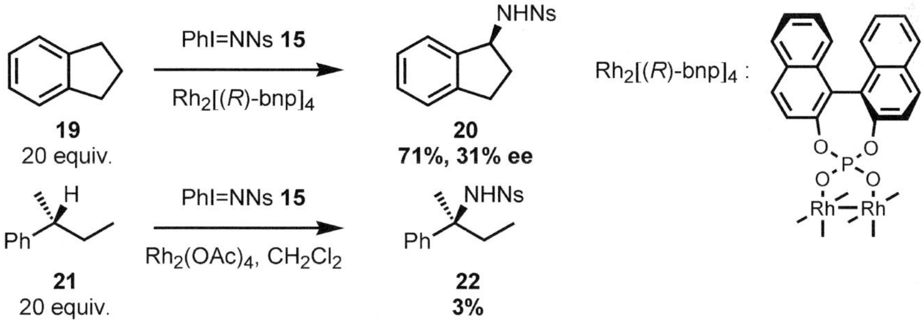

Scheme 2.6 Asymmetric and stereospecific rhodium-catalyzed C—H aminations.

a large excess (20 equivalents) of the substrate. More interestingly, it was demonstrated that the use of chiral ligands could induce asymmetric C—H amination of indan (**19**), albeit with only modest enantioselectivity of up to 31% (Scheme 2.6). Finally, numerous physical data including isotope effects, Hammett analysis, and study of a radical clock were all consistent with a mechanism involving a direct concerted C—H insertion of a nitrene in the singlet state. This was confirmed by the amidation of (*R*)-2-phenylbutane (**21**), which took place stereospecifically with retention of configuration.

This study paved the way for the development of more efficient metal-catalyzed C—H amination for synthetic applications. This goal could be achieved after the troublesome preparation and handling of iminoiodanes were resolved, thereby improving the quality of these nitrene precursors and the efficiency of the subsequent transfer. Thus, at the beginning of the 2000s, three groups independently reported key methodologies for the in situ generation of iminoiodanes [53–55] that had a major impact in the field of nitrene transfer. These improvements and their applications are presented in the following chapters.

2.3
Hypervalent Iodine-Mediated C—H Amination

2.3.1
Intramolecular C—H Amination

The sole – but striking – result obtained by Breslow in the area of intramolecular catalytic nitrene C—H insertion (see Scheme 2.4) [44] was a source of inspiration for Du Bois, who designed several elegant intramolecular C—H functionalizations using nitrenes generated from iminoiodanes [56]. A key feature of his studies was the discovery of conditions suitable for preforming these species in situ from miscellaneous nitrogenous functionalities.

2.3.1.1 From NH_2 Carbamates

In the first of his two pioneering papers, Du Bois demonstrated that a combination of iodosylbenzene diacetate [PhI(OAc)$_2$], magnesium oxide (MgO), and 5 mol% of a rhodium(II) complex in dichloromethane is optimal for the generation of a nitrene from a carbamate of general structure ROC(=O)NH_2. This undergoes intramolecular C—H insertion exclusively at the β-position of the carbon bearing the carbamate functionality, resulting in an oxazolidinone (**26–28**; Scheme 2.7) with yields of up to 84% [54]. No example of formation of the analogous six-membered ring has been reported so far. Carbamates derived from every class of alcohol are suitable substrates but those prepared from a secondary alcohol can afford a ketone resulting from a competitive α-C—H abstraction [17]. The reaction preferentially occurs at allylic and benzylic secondary C—H centers, as well as at tertiary positions, although functionalization of unactivated secondary sites is possible. In the case of cyclic substrates, *cis*-oxazolidinones are formed almost exclusively. These trends have been very helpful for the design of synthons for the application of C—H amination to total synthesis (see Section 2.6).

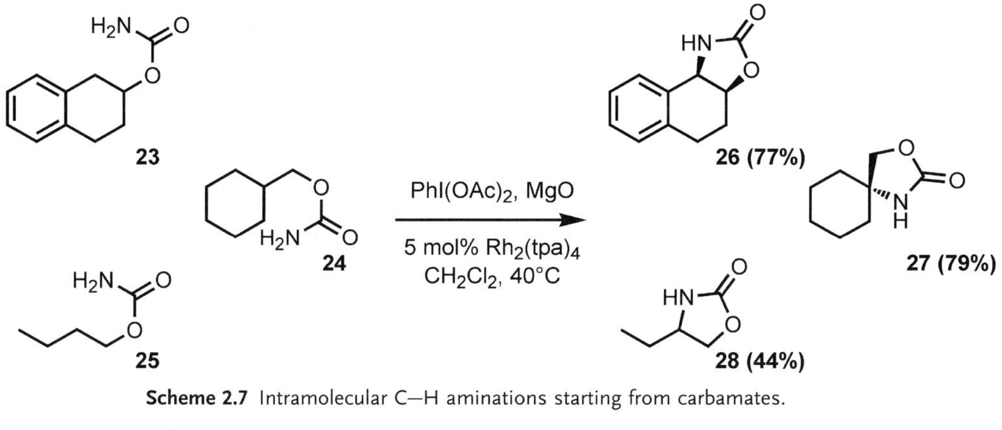

Scheme 2.7 Intramolecular C—H aminations starting from carbamates.

Typical experimental procedure MgO (2.3 equiv), $PhI(OAc)_2$ (1.4 equiv), and the rhodium (II) catalyst (0.05 equiv) are successively added to a solution of carbamate (1.0 equiv) in CH_2Cl_2 (6 mL). The reaction mixture is stirred vigorously and heated at 40 °C for 12 hours and, after cooling to 25 °C, is diluted with CH_2Cl_2 (10 mL) and filtered through a pad of Celite. The filter cake is rinsed with CH_2Cl_2 (2 × 10 mL). The combined filtrates are evaporated under reduced pressure and the isolated residue is purified by chromatography on silica gel to afford the expected oxazolidinone.

The same procedure can be applied to sulfamates with 0.02 equiv. of the rhodium (II) complex.

From a mechanistic point of view, the involvement of an iminoiodane remains hypothetical, since no such species has been isolated or spectroscopically observed. However, the stereospecific C—H amination with retention of configuration observed in the case of the carbamate derived from the (*S*)-2-methylbutan-1-ol derivative **29** (Scheme 2.8) illustrates a concerted direct C—H insertion of a singlet nitrene.

A disilver(I) complex generated in the reaction between $AgNO_3$ and 4,4′,4″-tri-*tert*-butyl-2,2′:6′,2″-terpyridine (*t*Bu_3tpy) has also been found to catalyze the intramolecular C—H amination starting from carbamates with equal efficiency and the same selectivity trends (Scheme 2.9) [57]. In some cases, reactivity was improved by addition of a catalytic quantity of 4-*tert*-butylpyridine. Involvement of a silver-nitrene intermediate has been suggested, the C—H insertion once more having been demonstrated to occur stereospecifically with retention of configuration.

Surprisingly, while the use of transition metal complexes in combination with chiral ligands offers unique opportunities to induce asymmetric reactions, no

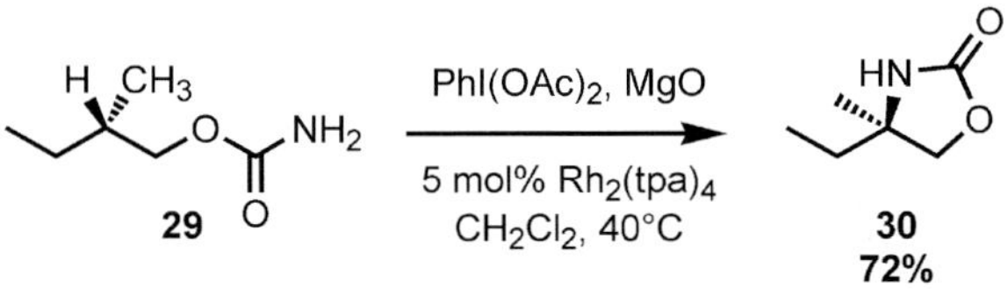

29 30 72%

Scheme 2.8 Stereospecific intramolecular C—H amination.

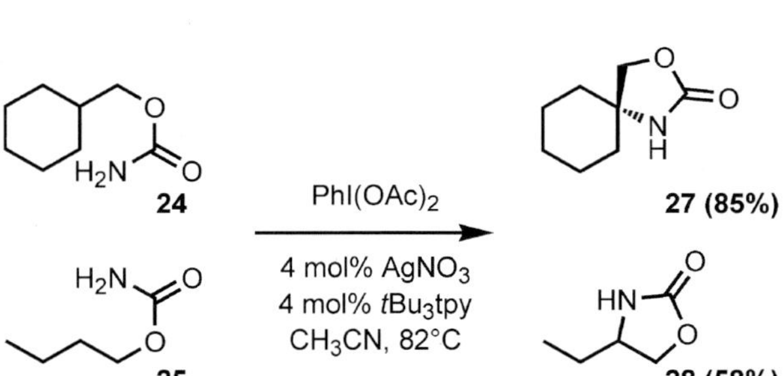

24 25 27 (85%) 28 (58%)

Scheme 2.9 Silver-catalyzed intramolecular C—H aminations.

enantioselective intramolecular C—H amination starting from carbamates has been reported so far.

2.3.1.2 From NH_2 Sulfamates

In parallel with this study, Du Bois published a second pioneering paper in which the methodology was applied to $ROSO_2NH_2$-type sulfamates [58], easily obtained from the corresponding alcohols by treatment with sulfamoyl chloride $ClSO_2NH_2$, generated in situ from chlorosulfonyl isocyanate and formic acid [59]. By application of the same conditions but with an amount of rhodium catalyst reduced to 2 mol%, reflecting the higher reactivity of sulfamates, intramolecular C—H amination occurs in very good yields of up to 91%. Unlike with carbamates, the nitrene C—H insertion occurs at the γ-position, affording six-membered cyclic sulfamates (**35–38**; Scheme 2.10), probably as the consequence of greater ring strain induced by longer S—O and S—N bonds and the obtuse N-S-O angle. Nevertheless, in the absence of a γ-C—H center, β-functionalization may take place, to afford a five-membered sulfamidate as in the case of **39**. Sulfamates derived from primary and secondary alcohols are suitable for this purpose, but benzylic and tertiary analogues have proven to be too unstable [17]. As far as the reactivity of C—H bonds is concerned, the most favored positions are, in decreasing order, tertiary, benzylic, secondary, and primary centers [56].

Recent mechanistic investigations [17] have indeed revealed the intermediacy of iminoiodanes in this process with use of sulfamates. Moreover, physical organic

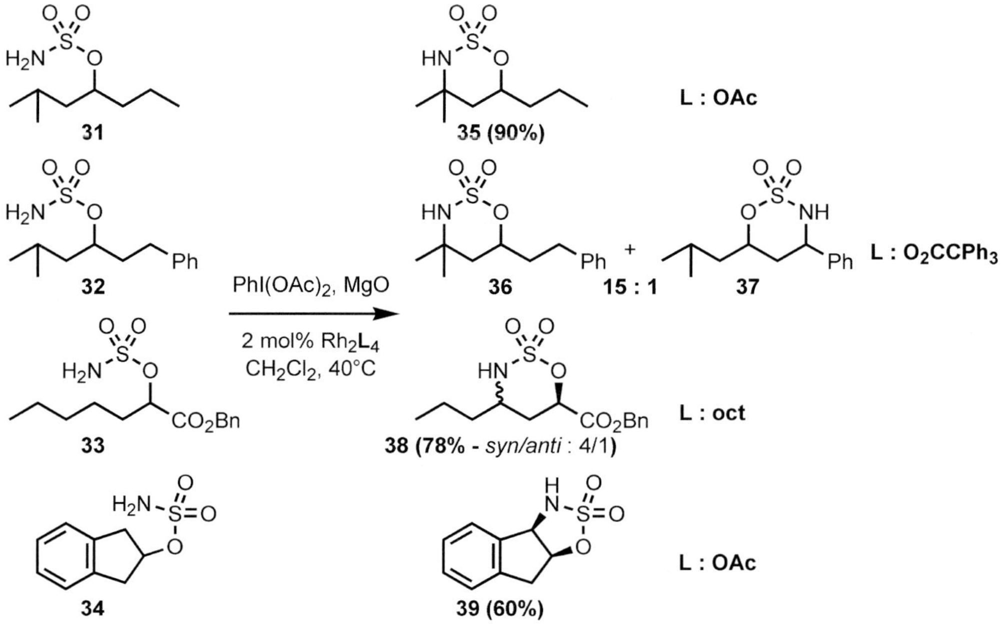

Scheme 2.10 Intramolecular C—H aminations starting from sulfamates.

experiments analogous to those performed by Müller [52] – that is, kinetic isotope effects, use of a radical clock, and Hammett analysis – have provided a wealth of evidence pointing to the concerted C—H insertion of an electrophilic rhodium-nitrene species. As in the case of carbamates, a similar chiral substrate was used as a probe for stereospecific C—H insertion that also takes place with retention of configuration.

Additionally, the other enantiomerically pure sulfamates **40** and **41** (Scheme 2.11) have been prepared with the aim of promoting diastereoselective C—H aminations [60]. A general trend has been observed, with α,γ- and β,γ-substituted oxathiazinanes **42** and **43** being preferentially obtained with *syn* and *anti* selectivity, respectively. In order to explain these results, a hypothetical model suggests a chair-like transition state with minimized *gauche* interactions in which torsional effects favor intramolecular amination of the equatorial C—H bond. From a synthetic point of view, this stereochemical preference could be very helpful for directing regioselective C—H amination from sulfamates, depending on their substitution patterns.

At this point, it can be concluded that 1,2- and 1,3-amino alcohols are both easily accessible from the same starting material by application of intramolecular C—H amination with use of carbamates and sulfamates, respectively. However, the sulfamate methodology offers wider synthetic opportunities, since the products can be seen as nitrogen analogues of cyclic sulfates and can therefore undergo nucleophilic substitution [61]. Thus, by prior carbamoylation of the free nitrogen atom, it has been demonstrated that a large variety of nucleophiles can be introduced at the carbon center bearing the oxygen functionality (see Section 2.6) [17, 58, 62]. Even more interestingly, intramolecular C—H amination from sulfamates has been shown to occur very efficiently at ethereal α-C—H positions, with yields of up to 91% [63]. This reactivity is comparable to that of a tertiary C—H center, since it has been shown with substrate **44** that a benzylic C—H bond is less reactive, a

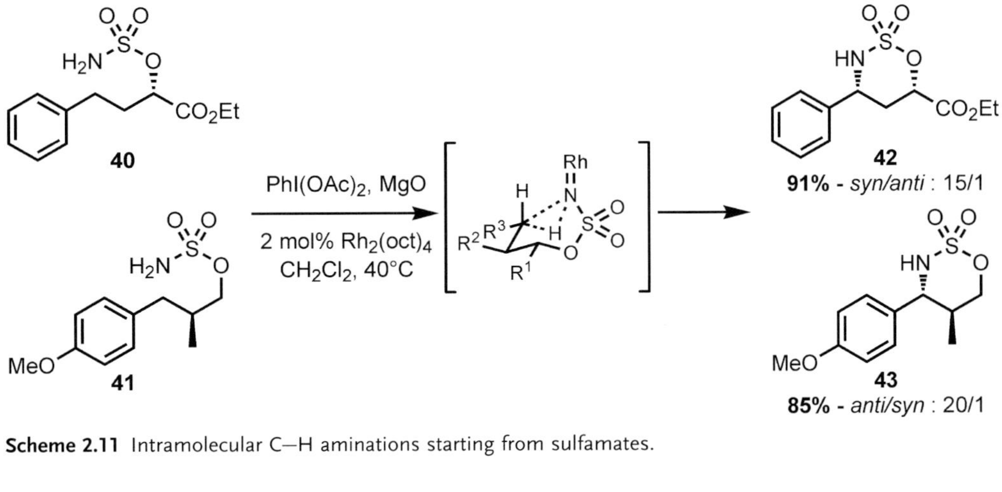

Scheme 2.11 Intramolecular C—H aminations starting from sulfamates.

result highly dependent, however, on the rhodium catalyst ligand (Scheme 2.12) [64]. The resulting crude *N,O*-acetals can then function as iminium ion equivalents, which can be trapped by various nucleophiles with good yields and stereoselectivities.

As in the case of carbamates, intramolecular C—H amination starting from sulfamates has been found to be catalyzed by disilver(I) complexes [57]. More importantly, however, Che has demonstrated that an electron-deficient ruthenium porphyrin [Ru(tpfpp)(CO)] is also an effective catalyst in this process [65, 66]. In a key paper, this group was the first to demonstrate that iminoiodanes could be generated in situ under suitable conditions and then used as hydrocarbon amidating agents (see Section 2.3.2) [53]. They then applied this methodology to the formation of six-membered ring sulfamidates; optimized yields of up to 88% were obtained when Al_2O_3 was added to the reaction mixture (**49**; Scheme 2.13) and regioselectivities comparable to those induced by rhodium complexes were observed.

The reaction was found to involve iminoiodanes that are thought to react with the porphyrin complex to afford a bis(imido)ruthenium(VI) species. Unlike in all

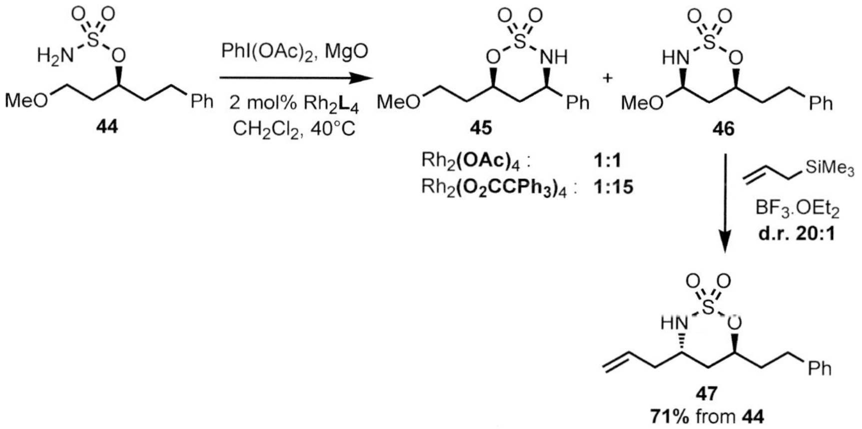

Scheme 2.12 Intramolecular C—H amination at an ethereal α-C—H bond.

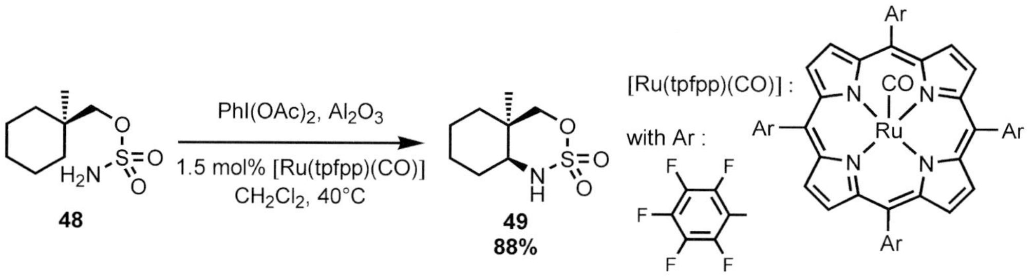

Scheme 2.13 Ruthenium porphyrin-catalyzed intramolecular C—H amination.

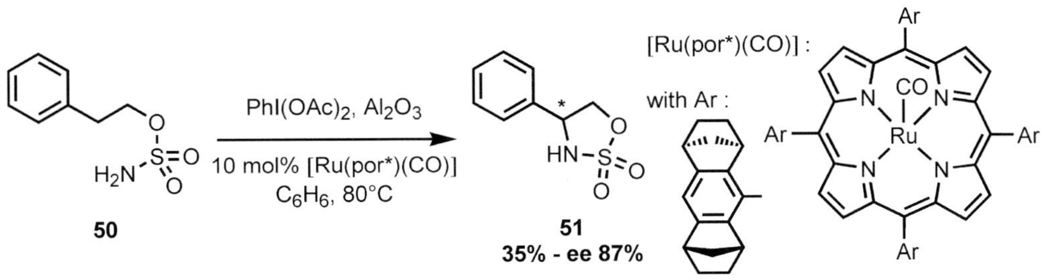

Scheme 2.14 Enantioselective porphyrin-catalyzed intramolecular C—H amination.

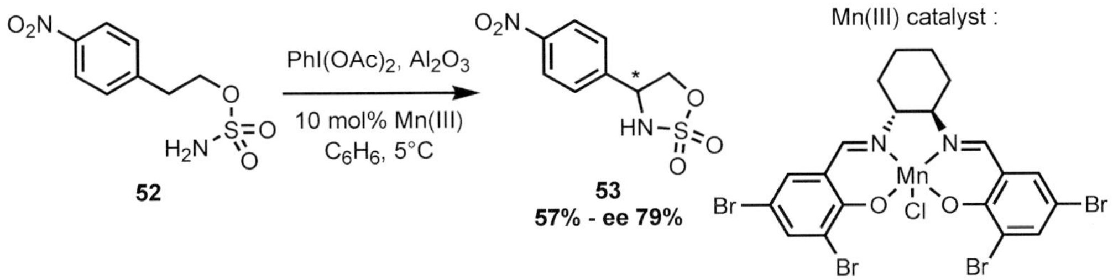

Scheme 2.15 Enantioselective manganese-catalyzed C—H amination.

the processes related so far, however, the C—H insertion may occur through an asynchronous concerted mechanism with significant hydrogen atom abstraction and involving short-lived radical species. Modification of the catalysts has also been envisaged and it has recently been shown that even a polymer-supported metalloporphyrin, prepared by covalent attachment of [Ru(tpfpp)(CO)] to a poly(ethylene)glycol chain, is effective in mediating such C—N bond formation [67]. Another modification, involving the introduction of a chiral auxiliary at each methine in the porphyrin, has resulted in the first asymmetric intramolecular C—H amination, with enantioselectivities of up to 87% (**51**; Scheme 2.14) [65].

Other enantioselective intramolecular amidations starting from sulfamates have been described. Use of chiral rhodium complexes in this context has been disappointing, with a moderate asymmetric induction of up to only 52% being obtained with $Rh_2((S)\text{-nttl})_4$ [68, 69]. Chiral manganese(III) Schiff base complexes appear to be more effective. Putative chiral manganese(V)-imido species, inducing concerted nitrene C—H insertions with *e.e.*s of up to 79%, have been suggested (**53**; Scheme 2.15) [70].

2.3.1.3 From Other Nitrogen Functionalities

As demonstrated early on by Breslow in his pioneering paper [44], intramolecular nitrene C—H insertion takes place efficiently from sulfonamides. Since then, however, these substrates have only rarely been studied for this purpose, more attention having been given to their unsaturated analogues in the context of their transformation into aziridines [50, 55, 66, 71–73]. In this reaction, products of C—H amination have often been regarded as side-products although in some cases they have been formed exclusively (**55**; Scheme 2.16). Nevertheless, in one of the few studies dedicated to intramolecular amidation from sulfonamides, use of chiral rhodium catalysts was found to induce asymmetric inductions better (up to 66%) than those obtained with sulfamates [68].

Recently, the discovery by Du Bois of a highly reactive rhodium catalyst has enhanced the variety of nitrogen substrates that can be engaged in amidation [74]. Thus, while sulfamates can be efficiently transformed into the corresponding sulfamidates with only 0.15 mol% of $Rh_2(esp)_2$, 1 mol% of this catalyst is able to induce C—H amination of sulfamide **59** and urea **60** in good yields (Scheme 2.17), with the notable difference that toluene is the solvent of choice. In the case of ureas, a follow-up paper underscored the crucial role of the secondary nitrogen protecting group – that is, the trichloroethoxysulfonyl (Tces) moiety – which alone is effective for the oxidative amination and can be removed by use of reductive conditions such as Zn in a 1 : 1 mixture of acetic acid and methanol [75]. The scope

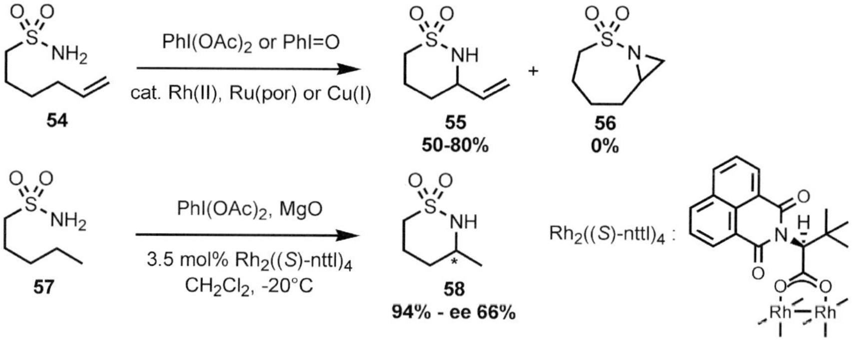

Scheme 2.16 Intramolecular C—H aminations from sulfonamides.

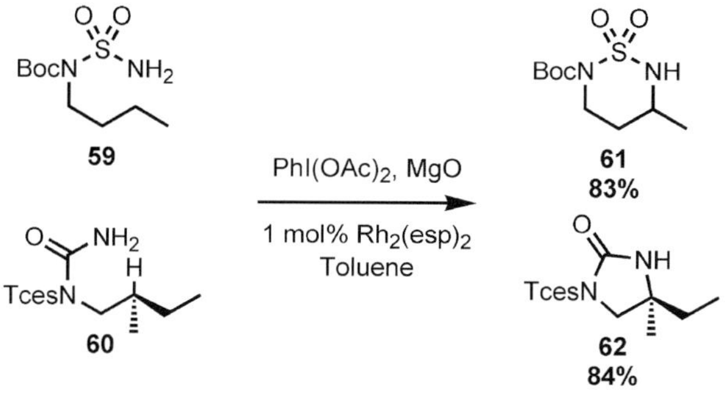

Scheme 2.17 Intramolecular C—H aminations from sulfamide and urea.

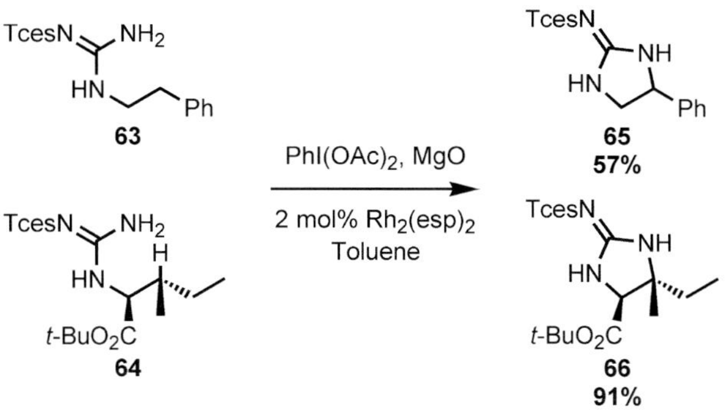

Scheme 2.18 Intramolecular C—H amination from guanidine.

and the regio- and chemoselectivity of C—H amination with ureas are comparable to those observed with carbamates and sulfamates.

The same highly effective catalyst has also been found to mediate nitrene insertion from guanidines (**63**, **64**; Scheme 2.18) [75]. Once more, the Tces protecting group appeared to be optimal. These examples clearly illustrate the great compatibility of $Rh_2(esp)_2$ with various functional groups such as basic guanidines. A general trend suggests that tertiary positions are more reactive than benzylic and secondary C—H centers. Surprisingly, secondary amine-derived guanidines do react under these conditions, while analogous carbamates were able to give rise to ketones (see Section 2.3.1.1.). Given the increasing number of natural products derived from the 2-aminoimidazole moiety [76], this methodology will surely find numerous applications in total synthesis in the very near future.

2.3.2
Intermolecular C—H Amination

2.3.2.1 General Scope and Limitations

As described in the preceding section, intramolecular C—H amination is a process mediated by several catalysts that allows efficient functionalization of various C—H bonds with good regio- and stereoselectivities. The analogous intermolecular process, however, has so far been found to be more difficult to develop. Highly reactive nitrene species can be generated for this purpose but their low stability often prevents C—H insertion from taking place in good yields, a feature less important with intramolecular reactions, due to the closer proximity of the reacting centers. To circumvent this problem, most studies have been conducted with an excess of the C—H bond-containing substrates, as described by Müller in his seminal paper [52].

Before the publication of his paper relating to the in situ generation of iminoiodanes [53], Che demonstrated that a stoichiometric amount of a previously prepared bis(amido)ruthenium(III) complex could effect the amidation of various hydrocarbons (~6 equiv) in the presence of $AgClO_4$ with yields of up to 75% [77].

Secondary allylic and benzylic positions appeared to be privileged for such functionalization. In an elegant following paper, it was then reported that bis(imido)ruthenium(VI) porphyrins are also suitable reagents for intermolecular amidation, with yields of up to 88% being obtained in the presence of 20–40 equivalents of the substrate [78]. A noteworthy feature of this work is the compilation of evidence for an H-atom abstraction mechanism involving a radical intermediate under these conditions. Che has also recently highlighted the influence of the sulfonyl substituents of the imido function on the reactivity; in particular, the thermodynamic reactivity order of tertiary versus secondary/primary positions has been found to be reversed by simply replacing a tosyl by a *p*-nitrobenzenesulfonyl group [79]. Finally, a chiral bis(imido)ruthenium(VI) porphyrin was also prepared and this reagent was shown to induce enantioselective C—H aminations with up to 58% *e.e.* on use of 10 equivalents of the substrate [80].

However, the fact that bis(imido)ruthenium intermediates could be generated with the use of a catalytic quantity of the aforementioned porphyrins Ru(tpfpp)(CO) [53] and Ru(por*)(CO) [80] has greatly improved the synthetic value of the subsequent amidation starting from a stoichiometric amount of the C—H bond-containing substrate [53]. Analogous manganese(III) porphyrins have been found to be effective for this purpose, giving even better yields of up to 86% [53], while cyclic amine or bipyridine ruthenium complexes are also suitable [81]. Once more, the reaction occurs efficiently with substrates containing allylic, benzylic, or α-ethereal CH_2 centers. Such intermolecular porphyrin-catalyzed C—H aminations were achievable either from PhI=NTs or from a combination of $PhI(OAc)_2$ and *p*-toluenesulfonamide (Scheme 2.19). The latter conditions offer several advantages, the first being the simplicity of the experimental procedure. Generally, similar or slightly better yields are obtained as in the case of ethylbenzene (**67**). Moreover, as illustrated with indan (**19**), this procedure allows in situ formation of iminoiodanes from various sulfonamides and also from carboxamides, although no analytical proof for the involvement of an iminoiodane has been provided in the latter case. It should indeed be underscored that the C—H insertion product **73** was really unexpected, since carboxamides are known to undergo Hofmann rearrangements in the presence of hypervalent iodine(III) reagents [82–84].

Modification of the porphyrin once more allowed use of polymer-supported catalysts of comparable efficiency for C—H amination [67], while incorporation of chiral auxiliaries in order to induce asymmetric amidations has been envisaged [85]. However, low conversions (up to 34%) and modest enantioselectivities (up to 56%) were observed with use of methanesulfonamide as the nitrene precursor.

In the context of enantioselective intermolecular nitrene C—H insertion, more significant results were obtained with a chiral (salen)manganese(III) complex [86] and a dirhodium(II) catalyst [87], the former inducing asymmetric C—H amination with an *e.e.* of 89% in the case of 1,1-dimethylindan (**74**), while the latter exhibited an enantioselectivity of 84% when starting from 1,1-dimethyltetralin (**75**; Scheme 2.20). In all cases, however, the efficiency of the process is hampered by the need to use an excess of the substrate (2.35 and 5.0 equivalents, respectively) in order to obtain good results. The combination of a chiral rhodium(II) catalyst and a

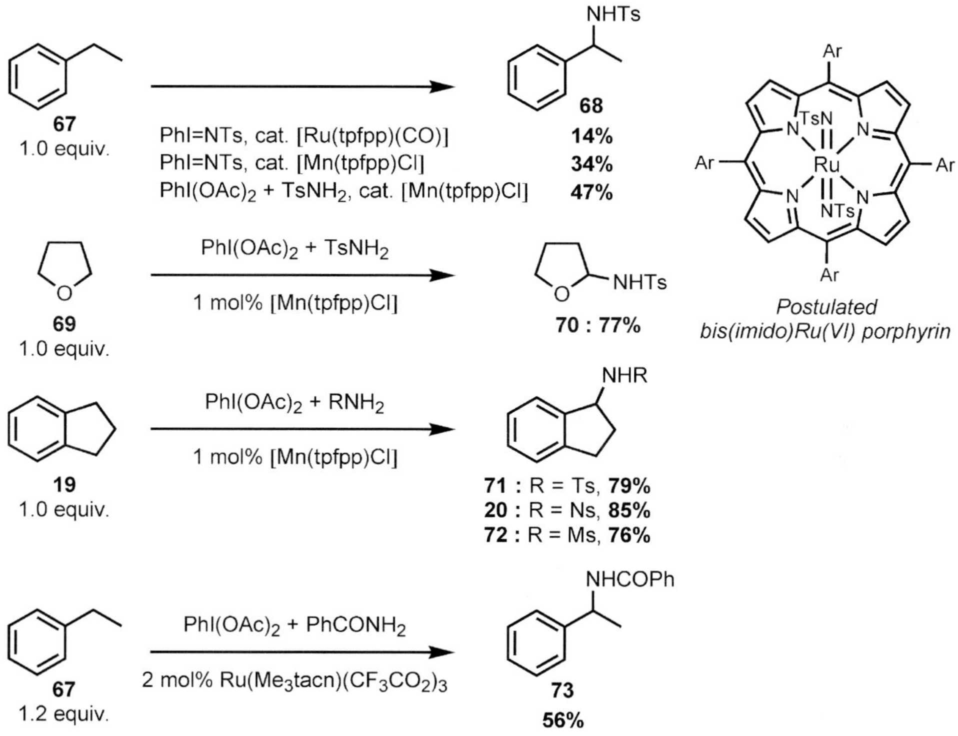

Scheme 2.19 Intermolecular porphyrin-catalyzed C—H aminations.

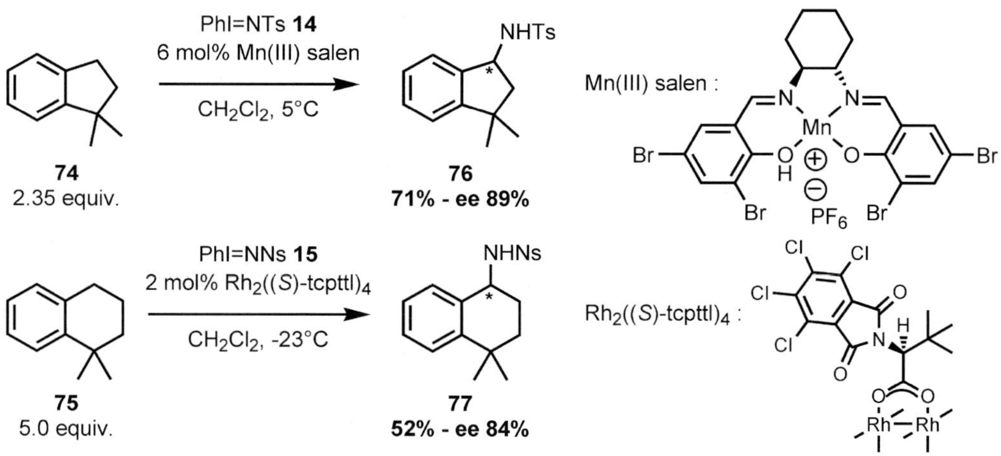

Scheme 2.20 Intermolecular enantioselective C—H aminations.

sulfamate as the nitrene precursor has also proven to be interesting, with yields of C—H amination products of up to 97% [69].

2.3.2.2 **Recent Major Improvements**

The results described above clearly indicate the limitations of intermolecular transition metal-catalyzed C—H amination: that is, low conversions, modest stereoselectivities, and functionalizations confined to activated secondary allylic and benzylic positions. Recent papers, however, have demonstrated that solutions might arise from the development of more reactive catalysts and/or nitrene precursors.

An early advance in the field was achieved with the use of copper(I) homoscorpionate complexes, which have found several applications in carbene transfer [88]. $Tp^{Br3}Cu(NCMe)$ appeared to be the catalyst of choice with which to perform efficient amination of toluene (**78**) and cyclohexane (**79**), therefore extending the scope of the reaction to primary benzylic positions and unactivated secondary sites (Scheme 2.21) [89]. More recently, the same authors have reported that all the alkyl C—H bonds of *p*-ethyltoluene (**80**) and *p*-cymene (**81**) could be functionalized under these conditions [90], but much work still remains to be done in order to induce the selective formation of *one* of the regioisomers **84–86** and **87–89**. It should also be mentioned that the C—H bond-containing substrates are used as the solvent in each reaction.

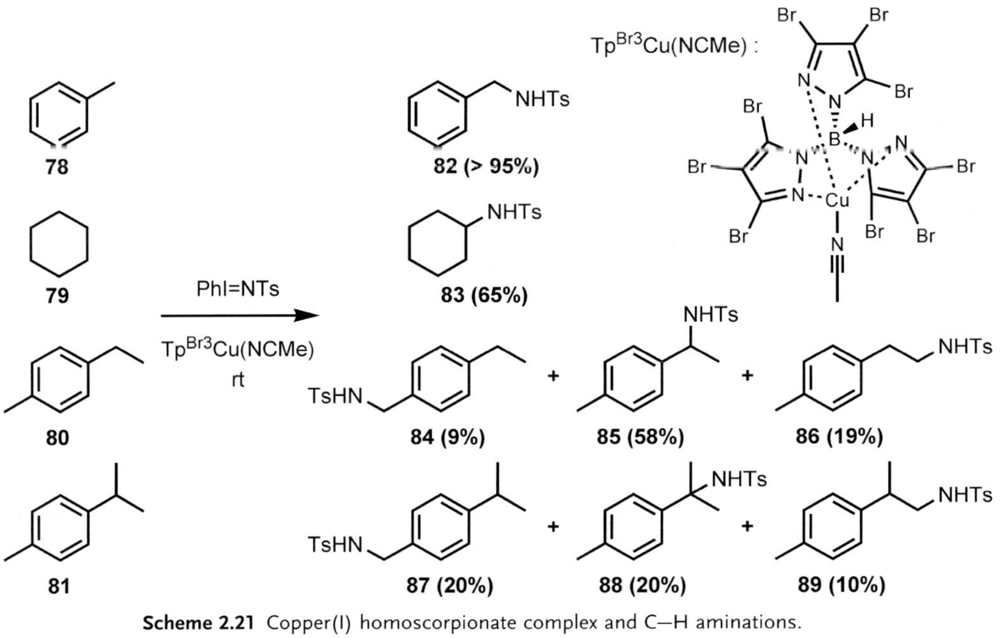

Scheme 2.21 Copper(I) homoscorpionate complex and C—H aminations.

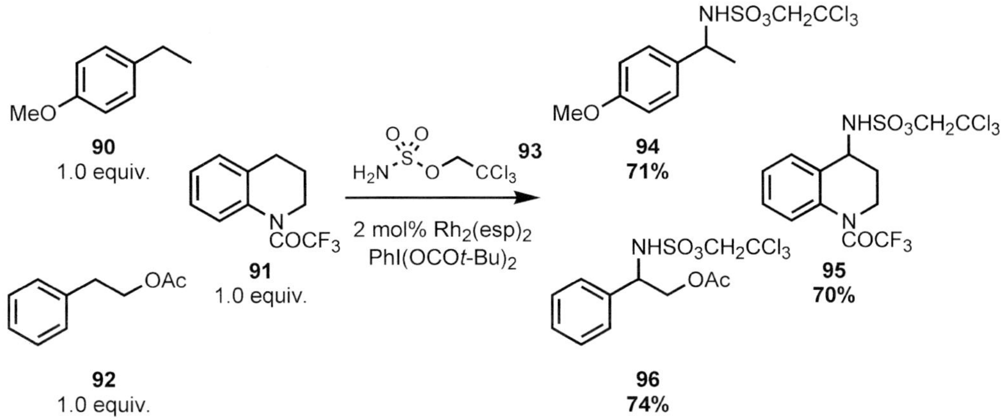

Scheme 2.22 Intermolecular C—H aminations catalyzed by $Rh_2(esp)_2$.

The high reactivity of the aforementioned rhodium complex $Rh_2(esp)_2$ has also proven to be useful in the intermolecular process. On employment of only 2 mol% of this catalyst in combination with trichloroethylsulfamate **93**, good yields of C—H insertion products **94–96** could be obtained from stoichiometric amounts of the substrates **90–92** (Scheme 2.22) [56, 74].

Finally, an intermolecular C—H amination efficient in terms both of reactivity and selectivity has recently been discovered, the process involving the generation of a chiral nitrene starting from an enantiomerically pure precursor: *N*-(*p*-toluenesulfonyl)-*p*-toluenesulfonimidamide (**99**) [91]. Such chiral sulfur(VI) reagents were first described by Levchenko over forty years ago [92], but since then have very rarely been used in organic synthesis [93]. Among these cases, though, is a publication by a Japanese group that describes diastereoselective allylic aminations mediated by a chiral selenium diimide reagent derived from **99**, albeit with a low *d.e.* of up to 42% [94]. Recently, it has been demonstrated that the reaction between a sulfonimidamide and a hypervalent iodine(III) reagent gives rise to a highly reactive nitrene species, which can react with alkenes in the presence of achiral copper or rhodium catalysts to afford the corresponding aziridines in very good yields and in some cases with good diastereoselectivities [95–97]. While use of chiral complexes with enantiomerically pure sulfonimidamides met with limited success in aziridination, such a combination has appeared to improve the efficiency of the rhodium-catalyzed intermolecular C—H amination greatly, this being correlated, moreover, with a dramatic matched effect [91]. Thus, by starting from a stoichiometric amount of tetrahydronaphthalene (**97**), use of chiral Rh_2((*S*)-nttl)$_4$ with (*S*)-**99** affords the resulting (*R*)-amino compound **98** in 80% yield and with 96% *d.e.*, whereas the reaction between (*R*)-**99** and Rh_2((*R*)-ntv)$_4$ affords (*S*)-**98** in 83% yield and with 99% *d.e.* (Scheme 2.23). The sulfonimidoyl moiety being easily removable with sodium naphthalenide at room temperature, this

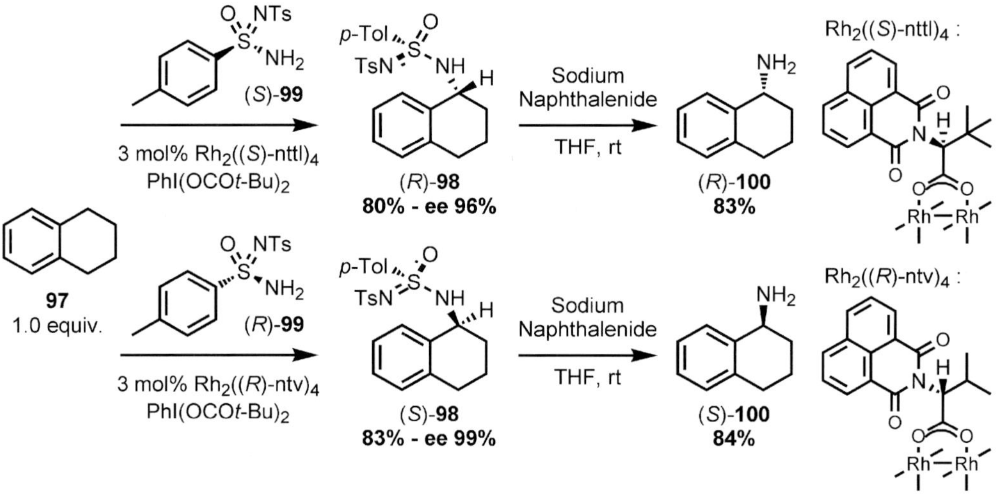

Scheme 2.23 Diastereoselective intermolecular C—H aminations using sulfonimidamide.

methodology allows isolation of both enantiomers of the benzylic amine **100** derived from **97**.

Typical experimental procedure Activated molecular sieves (4 Å, 100 mg), the chiral rhodium(II) complex (0.03 equiv), and (–)-*N*-(*p*-toluenesulfonyl)-*p*-toluenesulfonimidamide (1.2 equiv) are placed in an oven-dried tube, which is capped with a rubber septum and purged with argon. 1,1,2,2-Tetrachloroethane (3.75 mL $mmol^{-1}$) and methanol (1.25 mL $mmol^{-1}$) are added under argon and the mixture is stirred for 5 min before addition of the substrate (1.0 equiv). The tube is cooled to –35 °C, and bis(*tert*-butylcarbonyloxy)iodobenzene (1.4 equiv) is added. The mixture is stored in the freezer (–35 °C) for 3 days. After dilution with CH_2Cl_2, the molecular sieves are removed by filtration and the filtrate is evaporated to dryness under reduced pressure. The oily residue is purified by flash chromatography on silica gel to afford the C—H insertion product.

Application of these reaction conditions to several benzylic CH_2-containing substrates gave very good yields (up to 93%) and diastereoselectivities (*d.e.*s up to 99%). Secondary allylic positions were also found to be reactive, albeit to a lesser extent, whereas with a fivefold excess of cyclohexane nitrene C—H insertion occurred in 65% yield (Scheme 2.24). The key to this success is the use of methanol as a co-solvent, an unexpected observation since protic solvents have for a long time been known to induce hydrolysis of the intermediate iminoiodanes [48, 98].

It thus appears evident that these latest developments in the field of intermolecular amination should pave the way for the application of this C—H functionalization in total synthesis.

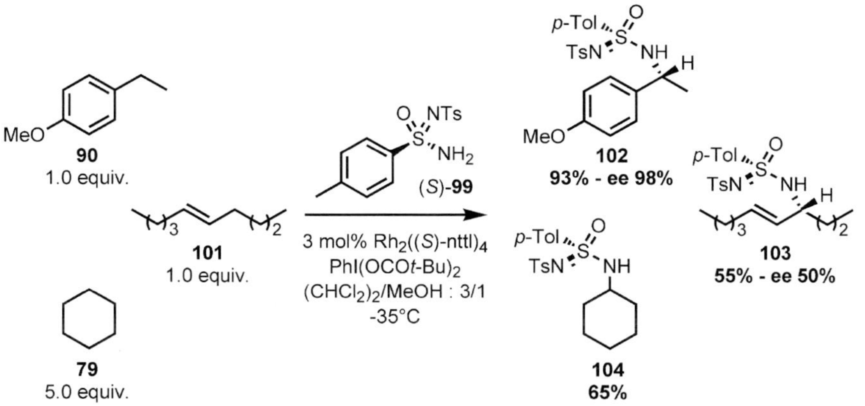

Scheme 2.24 Scope of the intermolecular C—H amination using sulfonimidamide.

2.4 Other Nitrene Precursors for C—H Amination

Noteworthy publications focusing on the use of other nitrene precursors have also appeared in parallel with these numerous studies dedicated to C—H nitrene insertion mediated by hypervalent iodine(III) reagents. In particular, these have largely been aimed at developing atom-economical methodologies, since an obvious drawback of the processes described so far is the generation of iodobenzene as an undesirable by-product.

2.4.1 Azides

Although the first reported transition metal-catalyzed generation of nitrenes was from benzenesulfonyl azide [36], few studies have since been dedicated to the use of this species in C—H amination [99]. Cenini demonstrated that ruthenium(II) and, especially, cobalt(II) porphyrins are suitable catalysts for effecting intermolecular C—H insertions from aryl azides at allylic [100] and benzylic positions [101]. However, the general interest of the process is once more diminished by the necessity to use the substrate in excess and even sometimes as the solvent. Moreover, in the former case the reaction produces either allylic amines or aziridines depending on the starting materials [100], while with primary and secondary benzylic compounds the reaction proceeds further, affording the corresponding imines [101]. As far as the mechanism is concerned, the C—H amination product could be formed from a cobalt(II)-porphyrin-azide complex; the decomposition of this, however, generates a nitrene species, which preferentially reacts with azide to give a diazene by-product.

A procedure with more synthetic potential, in which a ruthenium(salen)(CO) complex was employed, was then described [102]. Aziridination of olefins with this

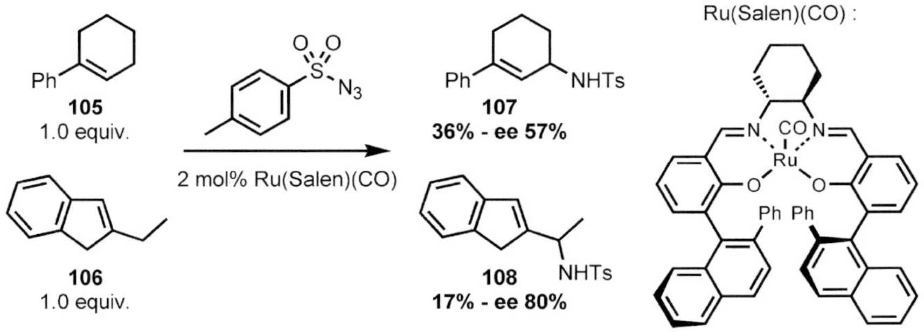

Scheme 2.25 Intermolecular C—H amination using azides.

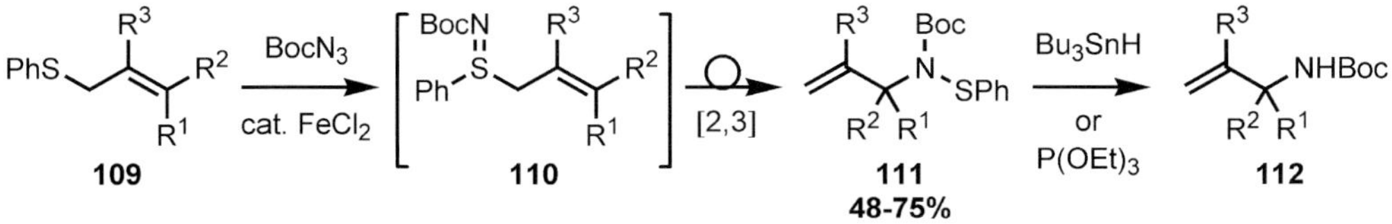

Scheme 2.26 Allylic C—H amination via imidation of sulfides.

catalyst has been reported, but alkenes **105–106**, each with a substituent *trans* to an aryl group, exclusively undergo allylic C—H amination with good enantioselectivities, albeit with low reactivity (Scheme 2.25). This low reactivity was found to be the consequence of a competitive intramolecular C—H amination that occurs on the salen ligand [103] and could be prevented with a fluorinated analogue [104].

Finally, an indirect way to mediate allylic amination is to perform the imidation of an allyl sulfide with *N-tert*-butyloxycarbonyl azide ($BocN_3$) in the presence of a catalytic quantity of $FeCl_2$ [105]. Sulfimines **110** are generated from sulfides **109** by addition of the postulated Fe(IV)-nitrene intermediate and then undergo a [2,3]-sigmatropic rearrangement to afford *N*-Boc-*N*-allylsulfenamides **111**, desulfurization of which can be achieved either with Bu_3SnH or with $P(OEt)_3$, depending on the substitution pattern of the allyl moiety (Scheme 2.26). Chirality transfer has been found possible when starting from chiral allyl sulfides but the sluggish reaction gives the sulfenamides with selectivities only in the 36–39% range. It should be pointed out that a similar indirect C—H amination had previously been devised by employing an asymmetric copper-catalyzed sulfimidation using PhI=NTs **14** that gave sulfonamides with *ees* of up to 58% [106].

2.4.2
Haloamines

While N_2 is the only by-product theoretically generated from the decomposition of an azide, innocuous and easily removable sodium halides are expected to be the sole side-products released during transformations of haloamines into nitrenes.

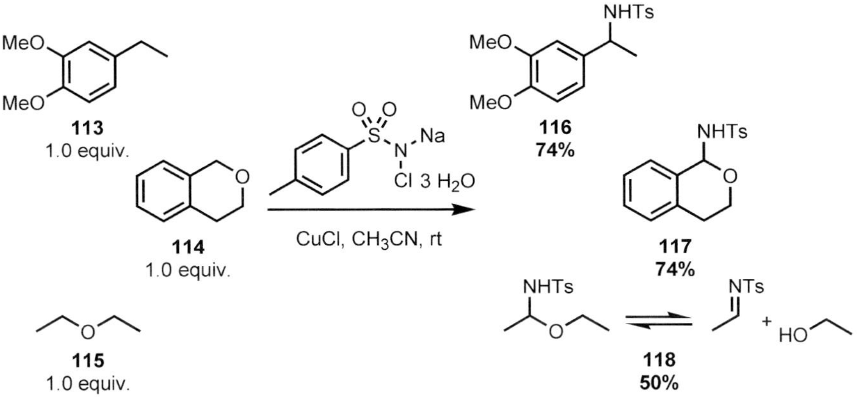

Scheme 2.27 Intermolecular C—H amination with chloramine-T.

The first study dedicated to C—H amination with chloramine-T was thus performed by Taylor [107]. Use of a copper(I) complex was found to catalyze the amination of cyclohexene and tetralin in 22 and 46% yields, respectively. A noteworthy feature of the reaction is the possibility of employing the commercially available and safer chloramine-T trihydrate. The process has recently been extended to other benzylic substrates and ethers [108]. While the former compounds give good yields of aminated products (**116**; Scheme 2.27), the latter have been found to afford hemiaminals contaminated with the imine resulting from elimination of alcohol except in some isolated cases such as with isochromane (**114**). It is also proposed that this C—H amination which can be also catalyzed by rhodium complexes, involves a concerted C—H insertion of an electrophilic metallanitrene, as often postulated with iminoiodanes [17, 52]. Very recently, a similar amination of ethers, as well as of toluene and mesitylene, with chloramine-T has been reported with a copper(I) homoscorpionate complex, but with a lower efficiency, since the substrates are employed in large excess [90]. Moreover, while bromamine-T has been found to be superior to chloramine-T in the metal-catalyzed aziridination of olefins [109, 110], this is not so clear in rhodium-catalyzed C—H amination, in which, despite the use of microwave activation, yields in the 33–70% range are obtained in the presence of 2.5 equivalents of tetralin, indan or ethylbenzene [111].

The ability of chloramine-T to induce aziridination of olefins under various conditions [112–115] has recently been applied to the α-amination of aldehydes via enolates generated in situ [116]. Such a stepwise strategy for amination of carbonyl compounds had previously been described in the context of the copper-catalyzed aziridination of olefins with iminoiodanes [48, 117, 118] and of nitrogen transfer from nitridomanganese(V) complexes [119]. In the present case, and in a one-pot process, it is postulated that use of a catalytic amount of L-proline generates an enolate **120** rather than an enamine from aldehyde **119**, since the reaction affords racemic products. Enolate **120** undergoes aziridination by reaction with chloramine-T, the resulting hemiaminal **121** then spontaneously rearranging to

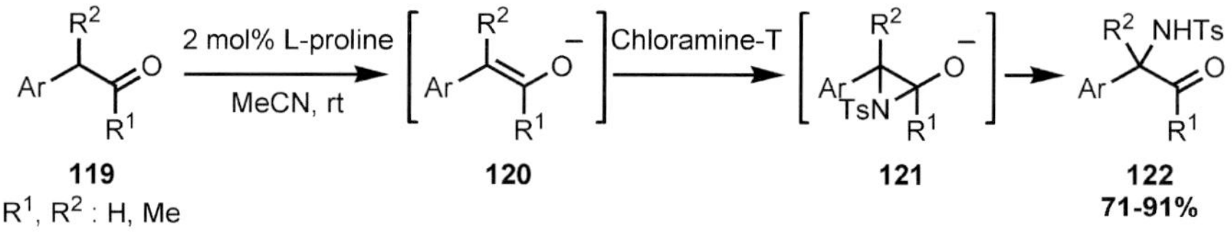

Scheme 2.28 α-Amination of aldehydes with chloramine-T.

give the expected amino carbonyl derivative **122** (Scheme 2.28). Unlike all previously described bromine-mediated aziridinations [112, 113], the reaction works even in the absence of the brominating agent while use of microwave activation allows reduction of reaction times, but its scope is limited to benzylic positions.

2.4.3 Carbamate Derivatives

Over forty years ago, Lwowski demonstrated that *N*-(sulfonyloxy)carbamates can undergo α-elimination to afford nitrene species [120]. Since then, these compounds have been particularly used by Tardella and Pellacani for the amination of either electron-rich or electron-poor olefins, although the involvement of a nitrene is unlikely in the latter case [121–124]. During the course of their studies, they designed a chiral nosyloxycarbamate **124** (derived from Helmchen's alcohol **128**), which effects allylic amination of 1-methylcyclopentene (**123**) in 45% yield and with 30% *d.e.* (Scheme 2.29) [125]. This strategy for C—H amination nevertheless turned out to be more appealing when conducted in an intramolecular manner; Lebel has recently disclosed an efficient preparation of oxazolidinones **127**, which could be isolated in very good yields from *N*-tosyloxycarbamate starting materials of type **126** in the presence of a catalytic quantity of $Rh_2(tpa)_4$ [126]. As previously observed with hypervalent iodine(III) reagents [54], the reaction occurs

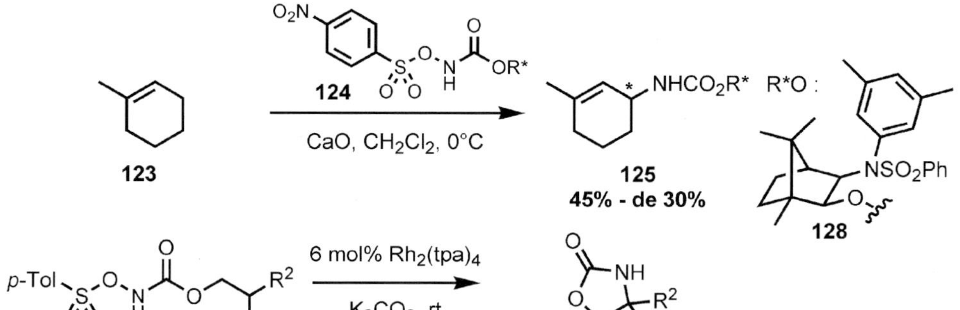

Scheme 2.29 *N*-(Sulfonyloxy)carbamates and C—H amination.

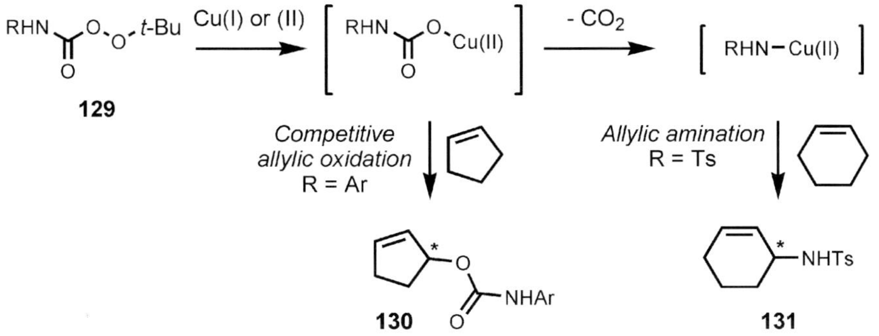

Scheme 2.30 Allylic amination with peroxycarbamates.

preferentially at tertiary and benzylic positions, through a concerted stereospecific metallanitrene C—H insertion.

"Isomeric" *N*-tosylperoxycarbamates of type **129** (R = Ts) have also proven to act as aminating agents in the intermolecular process. By analogy with the Kharasch–Sosnovsky allylic oxidation [127], based on the equilibrium between carbamic acid, amines, and CO_2, Katsuki has devised a copper(II)-catalyzed amination of benzylic and allylic derivatives that occurs in 27–55% yields [128]. Once more, use of an excess (4.0 equivalents) of the substrate is necessary while the asymmetric version does not give satisfactory results. A more recent publication has thus demonstrated that the outcome of the reaction depends strongly on the peroxycarbamate employed [129]. A competitive allylic oxidation occurs in the case of *N*-arylperoxycarbamates (**129**, R = Ar), this pathway being particularly favored in the presence of bis-oxazoline ligands. Enantioselectivities of up to 70% were then obtained when the *N*-tosylperoxycarbamate (**129**, R = Ts) was used (Scheme 2.30). It should, however, be pointed out that radical species rather than nitrenes are believed to be involved.

2.5 Amination of Aromatic C—H Bonds

While considerable attention has been devoted to the amination of sp^3 C—H bonds, some studies relating to the transformation of an aromatic C—H bond into a C—N bond have also recently emerged. The already described $Tp^{Br3}Cu(NCMe)$ [89, 90], as well as a copper(I)-phenanthroline complex [130], have been found to catalyze such functionalizations with benzene and 1,3-dimethoxybenzene, respectively, with use of iminoiodanes as nitrene precursors. In parallel, a more systematic report dedicated to the same Csp^2—H amidation of heterocycles has appeared [131]. In the presence of a catalytic quantity of a ruthenium(II) porphyrin complex and under ultrasound activation, iminoiodanes (preformed or generated in situ) effected the C—H amidation of various heteroarenes such as furan (**132**), thiophene (**133**), and the pyrroles **134** and **135** with good regioselectivities (Scheme

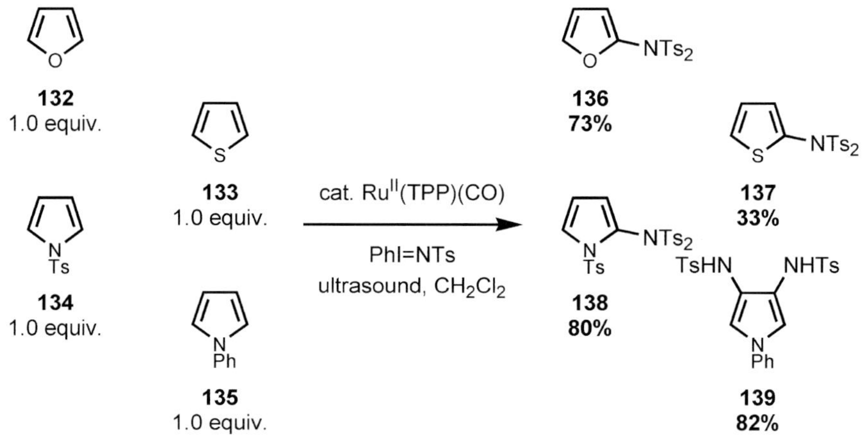

Scheme 2.31 Ru(porphyrin)-catalyzed Csp^2-H aminations of heterocycles.

2.31). Unexpectedly, the sole isolated products obtained from compounds **132–134** were the *N,N*-di(sulfonyl)amidated products **136–138**, the formation of which is poorly understood. Moreover, in the case of pyrroles the regioselectivity depends on the nitrogen protecting group, since *N*-(tosyl)pyrrole **134** afforded the C—2 substituted derivative **138**, but *N*-(phenyl)pyrrole **135** gave the 3,4-disubstituted product **139**.

More recently, the amidation of arenes through the employment of an original methodology based on a palladium-catalyzed C—H oxygenation previously described by Sanford and involving PhI$(OAc)_2$ as the oxygen atom donor has been reported [132–134]. By analogy, iminoiodanes – which can be regarded as the aza derivatives of PhI=O and PhI$(OAc)_2$ – have been tested in the closely related C—H amidation, but the combination of an amide and $K_2S_2O_8$ in the presence of 5 mol% of Pd$(OAc)_2$ has turned out to afford more effective nitrene transfers [135]. Sulfonamides, carbamates, and even acetamides can be employed for this purpose, while the amidated products are formed with complete regioselectivity. The latter result is the consequence of the pre-coordination of the catalyst to an appropriate nitrogen functionality, a pyridine ring in the case of compound **140** or an *O*-methyl oxime in that of **141**, that directs the functionalization *ortho* to the tether, a strategy often applied in C—H activation processes [11, 12, 136, 137] (Scheme 2.32). Thus, unlike in all the nitrene C—H insertions described above, the mechanism first involves a C—H activation by the palladium catalyst to afford cyclopalladated intermediates **142** and **143**. The Pd—C bond then reacts with a nitrene of still uncertain nature to afford the amidation products **144** and **145**.

The scope of this efficient C—H functionalization has been extended with equal success to Csp^3—H bonds [135]. Good to very good yields in the 58–93% range are obtained (Scheme 2.33). More interestingly, in contrast to previously described intramolecular processes (see Sections 2.3.1.1. and 2.3.1.2.), a reversed reactivity

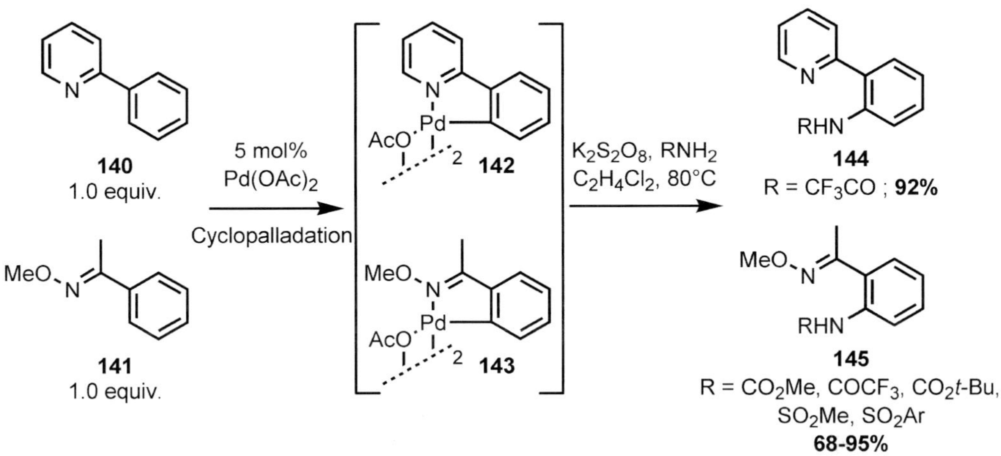

Scheme 2.32 Palladium-catalyzed Csp^2-H bond amidations by C—H activation.

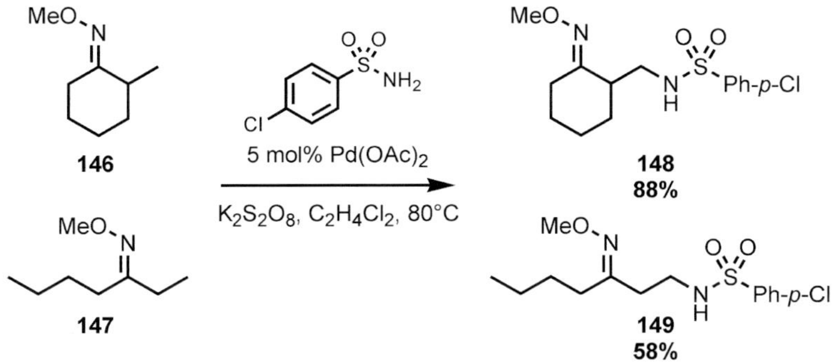

Scheme 2.33 Palladium-catalyzed Csp^3-H bond amidations by C—H activation.

is observed, amidation occurring preferentially at primary rather than secondary positions, a result correlated to steric effects.

Finally, a recent paper has reported the application of chelation-driven C—H activation to C—N bond formation not involving nitrene as the nitrogen source [138]. Buchwald has demonstrated that the transformation of 2-acetamidobiphenyl (**150**) into *N*-(acetyl)carbazole (**153**) can be catalyzed by palladium(II) complexes in the presence of stoichiometric quantities of $Cu(OAc)_2$ (Scheme 2.34). Various substitutions on both aromatic groups are tolerated well, and in particular, very good regioselectivity is observed in the case of the substrate **151**, in which the ring undergoing functionalization is already substituted. The reaction is believed to take place via a six-membered palladacycle **152**, reductive elimination of which would release the expected carbazole and a palladium(0) species that would be reoxidized to palladium(II) by the copper(II) additive.

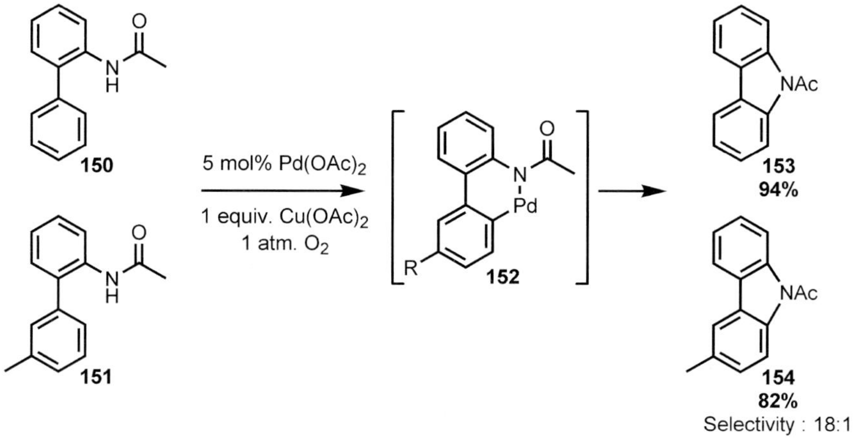

Scheme 2.34 Palladium-catalyzed C—N bond formation via C—H activation.

More recently, similar intermolecular *ortho*-C—H functionalizations of 2-phenylpyridine with anilines [139] and *p*-toluenesulfonamide [140] have been described; both reactions occur in the absence of any palladium(II) catalysts but stoichiometric amounts of copper(II) salts are needed.

2.6 Applications in Total Synthesis

Although these breakthroughs in C—H amination have been made only recently, the usefulness of C—H functionalization has already been validated by its application to the total synthesis of molecules, some of them complex. It is moreover of no surprise to note that all the examples described below almost exclusively involve intramolecular transfer of nitrene derived from iminoiodanes, either preformed or generated in situ [17, 141].

2.6.1 Application of Intramolecular C—H Amination with Carbamates

The ability of carbamates to afford exclusively five-membered ring products was first utilized for the preparation of the deoxyamino sugar methyl-L-callipeltose (**160**). The structure of this unusual carbohydrate, part of the cytotoxic glycoside macrolide callipeltoside A, makes this compound an ideal target in this context. Two groups have independently reported applications of intramolecular C—H amination to form compound **160**, starting from the same precursor **159** obtained from different starting materials (Scheme 2.35) [142, 143]. In both cases reaction conditions were adjusted to improve the yields. Trost found 2,6-di-*tert*-

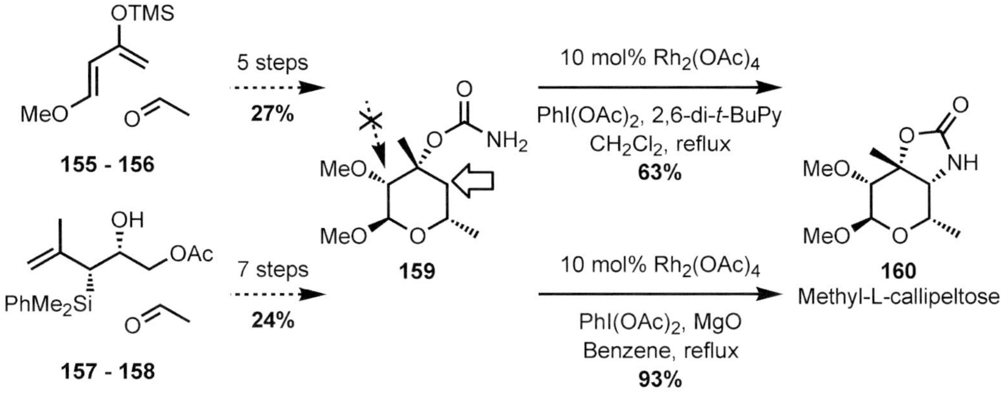

Scheme 2.35 Synthesis of methyl-L-callipeltose.

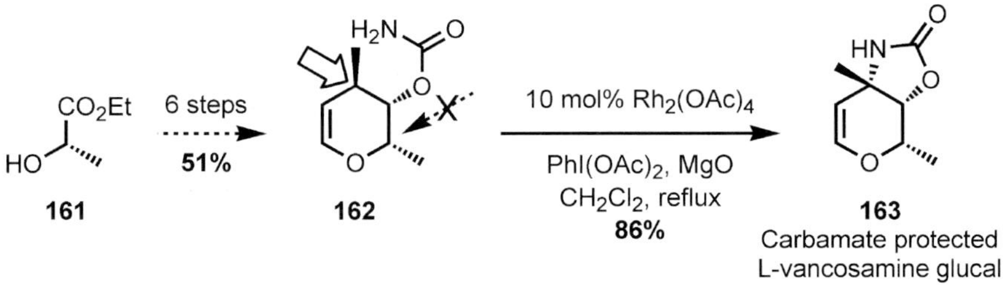

Scheme 2.36 Synthesis of a L-vancosamine derivative.

butylpyridine to be a better base than MgO, affording the expected product **160** in 63% yield with 80% conversion [142], while Panek demonstrated that complete conversion could be achieved either by using 20 mol% of $Rh_2(OAc)_4$ or, more interestingly, by carrying out the transformation in benzene at reflux [143]. Both strategies then allowed isolation of methyl-L-callipeltose in six steps with 17% overall yield and in eight steps with 23% overall yield, respectively. These routes appear to be shorter and more versatile than preceding syntheses based on the chiral pool since they can provide both enantiomers of **160**.

In parallel, a closely related synthetic target, the carbamate-protected L-vancosamine glucal **163**, was prepared by application of the same procedure to the suitably substituted precursor **162** (Scheme 2.36) [144]. C—H amination once more occurred very efficiently in the presence of 10 mol% of $Rh_2(OAc)_4$ to afford the expected oxazolidinone **163** in 86% yield (44% overall yield in seven steps from (*S*)-ethyl lactate **161**).

The same group then reported a related study of glucals in which the regio- and stereochemical outcome of the transformation could not be so easily predicted [145]. Unlike the carbamates **159** and **162**, compounds **164** and **165** both possess sterically accessible C3–H and C5–H bonds, each differently activated and situated allylic and α, respectively, to the pyranoside oxygen (Scheme 2.37). It was found

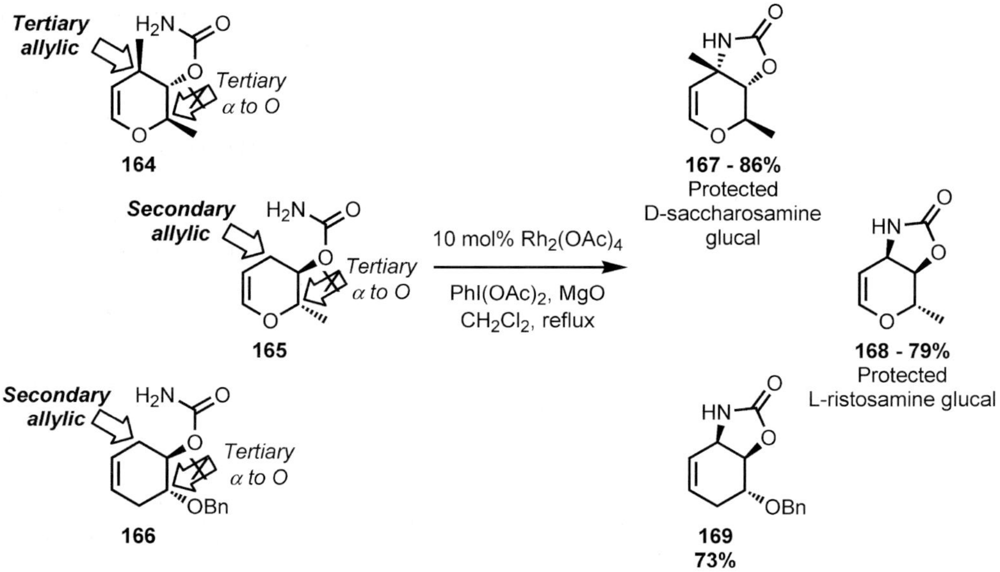

Scheme 2.37 Regioselective formation of 3,4-*cis*-3-amino glucals.

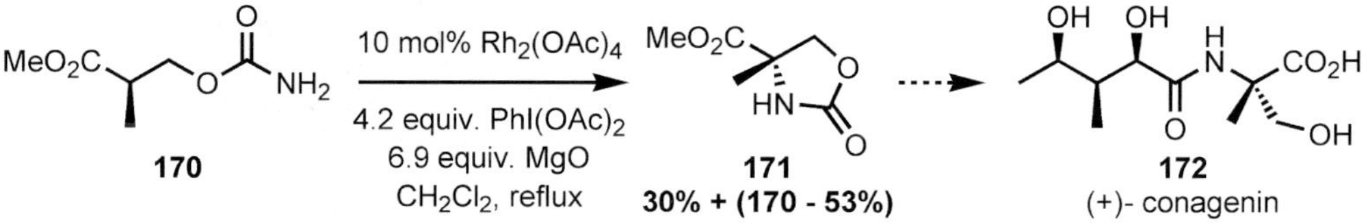

Scheme 2.38 Synthesis of (+)-conagenin.

that intramolecular rhodium-catalyzed carbamate nitrene insertion occurred selectively at the allylic C–H bonds even when these were secondary, as in the case of **165**. Since this allylic position could be regarded as a vinylogous position α to the pyranoside oxygen, a competitive experiment was devised with substrate **166**, which displays two C–H bonds activated, either by an oxygen atom or by a simple allylic system. Once more, the C–H amination was found to be chemoselective for the allylic site. It should finally be mentioned that in all these examples involving glucals, the reaction does not give the competitive α-C–H abstraction likely to afford a ketone, as is sometimes observed with carbamates derived from secondary alcohols [17] (see Section 2.3.1.1.).

Intramolecular rhodium-catalyzed nitrene insertion starting from carbamates has also been applied to the synthesis of the α-methylserine moiety **171** found in the immunomodulator (+)-conagenin **172** (Scheme 2.38) [146]. Although the substrate appears to be very simple, the reaction turned out to be rather sluggish, since 30% of product **171** together with 53% of recovered starting material were

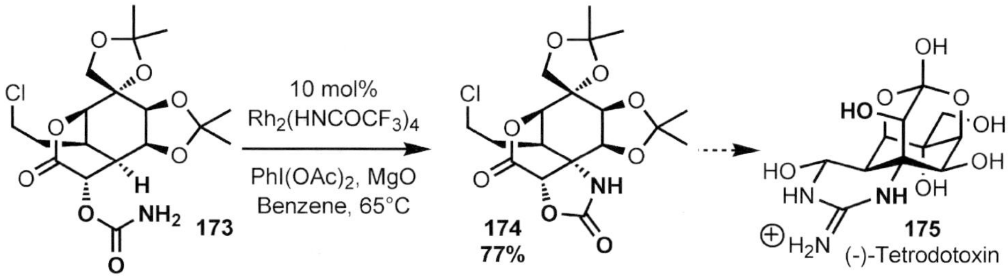

Scheme 2.39 Synthesis of (–)-tetrodotoxin.

obtained even after 40 hours at 40 °C in the presence of an excess of reagents. This result might be the consequence of the presence of the deactivating ester group, making the α C—H bond poorer in electrons.

On the other hand, the intramolecular rhodium-catalyzed C—H amination with carbamates has won acclaim with its application to the total synthesis of the much more densely functionalized (–)-tetrodotoxin (**175**) [147]. In his retrosynthetic strategy, Du Bois proposed to employ this methodology to introduce the C-8a amino function, precursor of the guanidine moiety of **175**, at a late stage of the synthesis. Thus, after the intermediate **173** had been prepared from D-isoascorbic acid in 25 steps, it was engaged in this key transformation to afford the expected oxazolidinone **174** in 77% yield in the presence of the more robust $Rh_2(HNCOCF_3)_4$ catalyst (Scheme 2.39). Given the large number of functional groups, the very good yield obtained in the intramolecular C—H amination illustrates the power of this C—H functionalization process. The clever design of compound **173** should also be underscored, since it displays only one sterically accessible tertiary position.

2.6.2
Application of Intramolecular C—H Amination with Sulfamates

While the scope of intramolecular C—H amination starting from carbamates is strictly limited to the formation of oxazolidinones likely to afford 1,2-amino alcohols, the same reaction conducted with sulfamates is more versatile, as their ability to give six-membered rings can be used to prepare 1,3-amino alcohols. As briefly mentioned in Section 2.3.1.2, however, these analogues of electrophilic cyclic sulfates [148] can smoothly undergo further functionalization by reaction with oxygen, nitrogen, sulfur, and carbon nucleophiles at the C—O bond, provided that the nitrogen atom has previously been activated by acylation [58, 61, 62]. In an early example, Du Bois described an efficient synthesis of (*R*)-*N*-Cbz-β-isoleucine (**178**), starting from (*S*)-3-methylpentan-1-ol (**176**) and involving the nucleophilic attack of water at the carbon atom bearing the oxygen of the cyclic sulfamidate **177** (Scheme 2.40) [58].

The total syntheses of manzacidins A and C (**184** and **185**; Scheme 2.41), however, are more representative examples of this synthetic opportunity [149]. In the case of **184**, application of the intramolecular C—H amination to the sulfamate

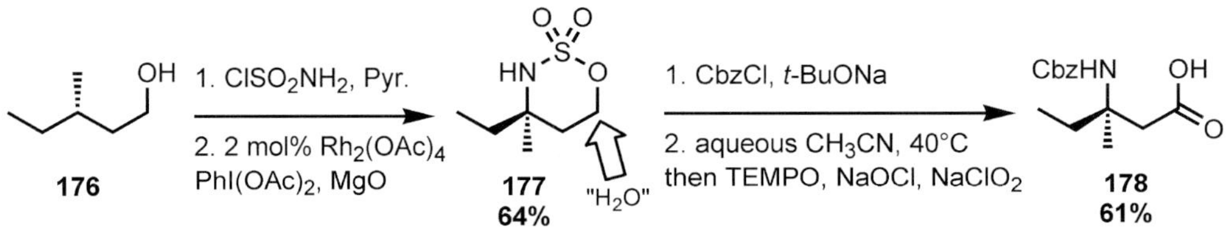

Scheme 2.40 Synthesis of (*R*)-*N*-Cbz-β-isoleucine.

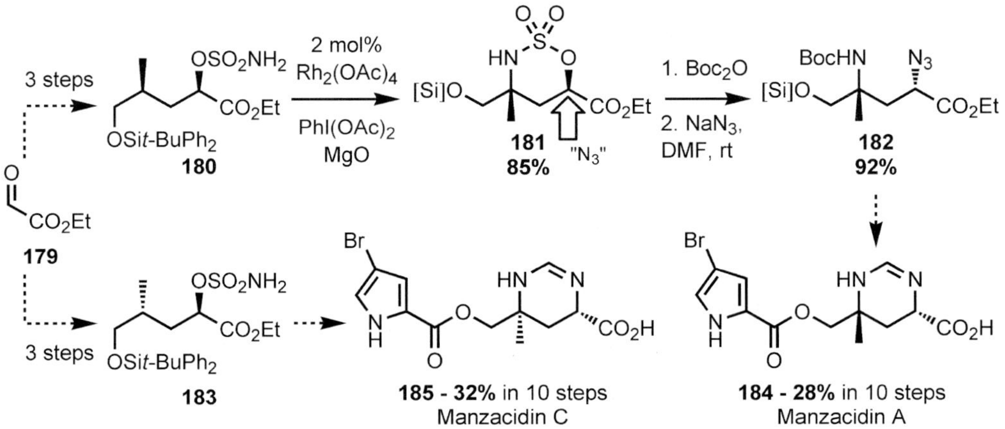

Scheme 2.41 Total syntheses of manzacidins A and C.

180 (readily obtained in three steps from ethyl glyoxylate **179**) afforded the expected oxathiazinane **181** in 85% yield. After introduction of a Boc group on the nitrogen atom, nucleophilic displacement with sodium azide was found to occur smoothly at room temperature to give product **182**, the 1,3-diamino moiety of which is the direct precursor of the tetrahydropyrimidine ring of manzacidin A. This natural bromopyrrole alkaloid could thus be obtained in 10 steps with an overall yield of 28%, while the isomeric manzacidin C was isolated through an identical pathway involving the sulfamate **183** in 32% overall yield.

A similar synthetic scheme to prepare enantiomerically pure aminodiols was also devised, these being further used for the synthesis of N^2-deoxyguanosine adducts **190** in the context of a study of their mutagenic effects. The sulfamate **186** was efficiently transformed into the corresponding oxathiazinane **187** with a good diastereoselectivity of 9 : 1 (Scheme 2.42) [150]. Compound **187** underwent, after activation, nucleophilic displacement with potassium acetate to give compound **188**, deprotection of which provided the expected 1,3,5-dihydroxyamino derivative **189**.

Recently, Du Bois has demonstrated that benzo-fused oxathiazinanes of type **192** can be synthesized from *ortho*-substituted phenolic sulfamates **191** (Scheme 2.43) [151]. It was then shown that these rarely studied heterocyclic structures

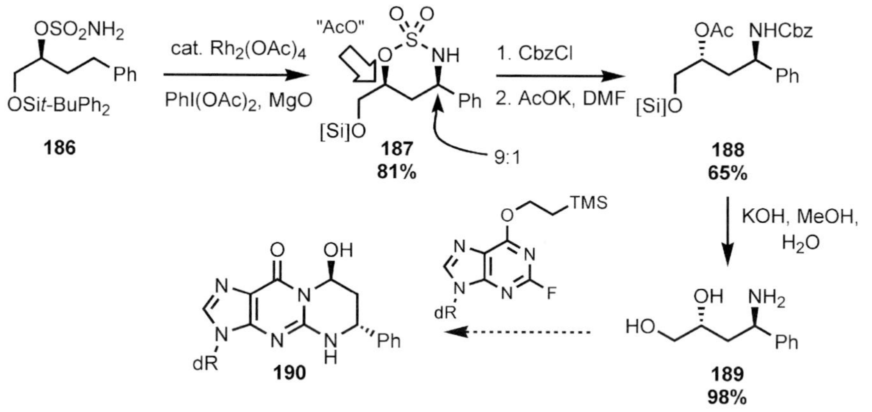

Scheme 2.42 Synthesis of enantiomerically pure aminodiols.

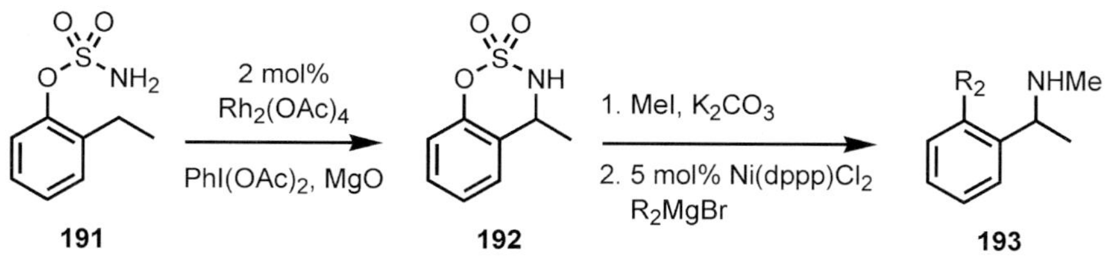

Scheme 2.43 Synthesis of arylalkylamines.

could be engaged in cross-coupling reactions analogously to triflates and tosylates. The *N*-alkyl cyclic sulfamates were thus efficient partners in nickel-catalyzed coupling reactions with aryl- and alkylmagnesium salts, affording various substituted arylalkylamines **193**.

Finally, the synthetic potential of sulfamates is further enhanced by their ability to undergo efficient intramolecular C—H aminations at ethereal α-C—H positions. Application of this strategy at the pseudo-anomeric C—H bond of the *C*-glycoside **194** afforded original spiro glycomimetics **195** (Scheme 2.44) [152]. This process occurs either with carbamates or with sulfamates, but in the latter case the reaction has been found to depend on the anomeric configuration, an observation interpreted in terms of conformational factors [153]. Interestingly, in this recent work the same authors also reported the unexpected but efficient formation of the oxathiazepane **197**, the aminal moiety of which could then be trapped by allylsilane with excellent diastereoselectivity. The regenerated sulfamate could then be used for further C—H amination.

Similarly, and as briefly mentioned in Section 2.3.1.2, intramolecular C—H aminations at ethereal α-C—H positions afford *N,O*-acetals that also function as iminium ion equivalents. These have been shown to react with several carbon nucleophiles such as alkynylzinc reagents [63], allylsilanes, silyl enol ethers, and

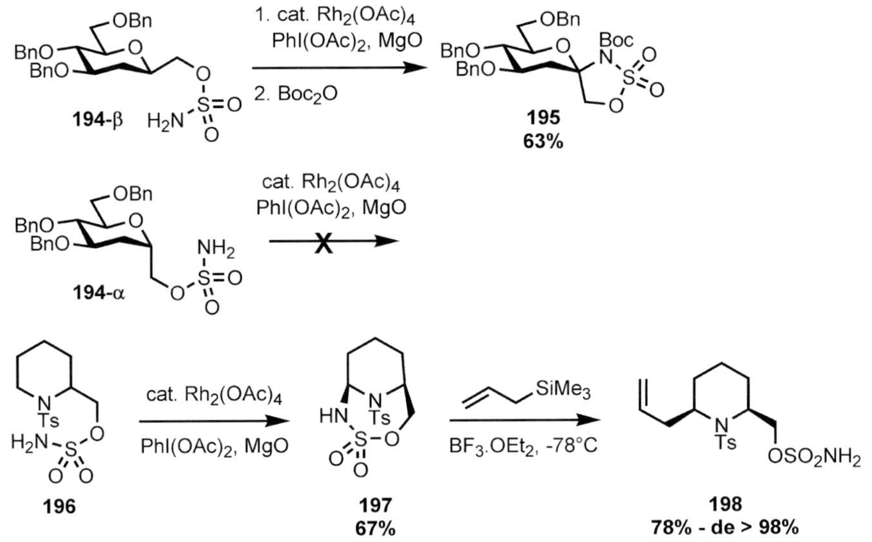

Scheme 2.44 Syntheses of glycomimetics and a substituted piperidine.

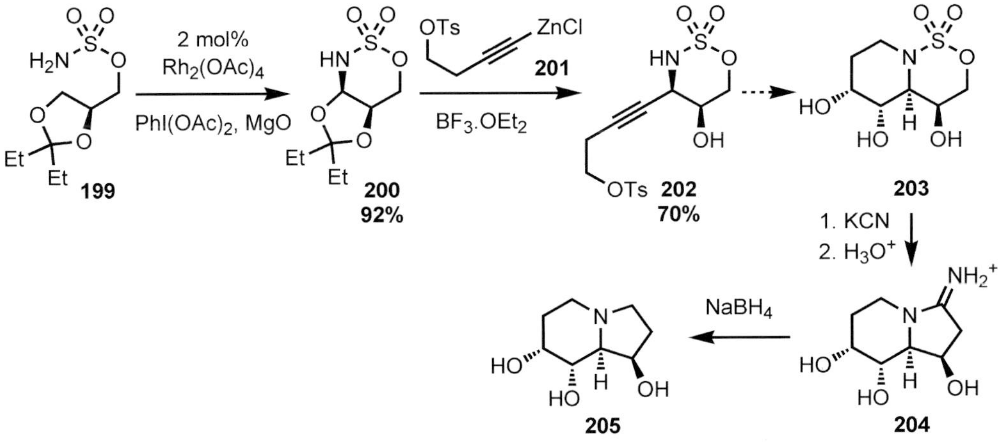

Scheme 2.45 Synthesis of an indolizidine.

ketene acetals [64]. The additions occur with very good diastereoselectivities that have been explained in terms of Stevens [154] and Felkin–Anh models. A pioneering application of this strategy was the total synthesis of the indolizidine **205** (Scheme 2.45), involving as the key steps: 1) intramolecular C–H amination, 2) treatment of the resulting *N,O*-acetals **200** with the suitably substituted alkynylzinc reagent **201**, and 3) ring-opening of the oxathiazinane **203** with potassium cyanide [63].

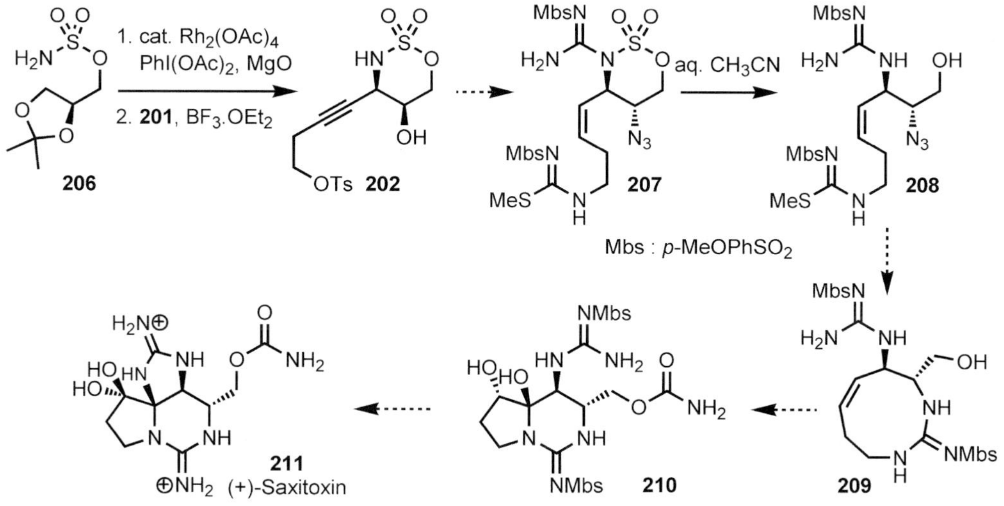

Scheme 2.46 Synthesis of (+)-saxitoxin.

A more impressive application of cyclic sulfamates, however, clearly illustrating that these are highly useful synthons for the elaboration of complex natural products, is the recently published synthesis of (+)-saxitoxin (**211**). The intermediate polyfunctionalized amine **208** could be obtained in a limited number of steps from the sulfamate **206** (Scheme 2.46) [155]. As in the preceding example, C—H amination and two nucleophilic displacements occurred efficiently under nearly identical conditions. Activation of oxathiazinane **207** toward ring-opening was achieved through the guanidinyl function present in the targeted natural product. Finally, (+)-saxitoxin (**211**) was obtained in only six steps from the key intermediate **208** by an original strategy involving the formation of the nine-membered guanidine ring **209** and transformation of this into the 5,6-bicyclic derivative **210** under finely tuned oxidative conditions.

2.6.3
Application of Intermolecular C—H Amination

The need to use an excess of substrates in order to obtain good yields of C—H amination products has hampered the use of intermolecular nitrene C—H insertion in total synthesis and such examples are therefore for the moment limited to the functionalization of steroids. While equilenin acetate (**212**) can be selectively amidated with PhI=NTs (**14**) in the presence of a manganese(III) porphyrin catalyst in 47% yield [156], it has been reported that rhodium(II) [72] and other manganese(III) porphyrin [85] complexes are able to catalyze amidation of cholesteryl acetate (**213**), albeit with modest conversions of up to 49% and moderate to good $\alpha:\beta$ selectivities (Scheme 2.47).

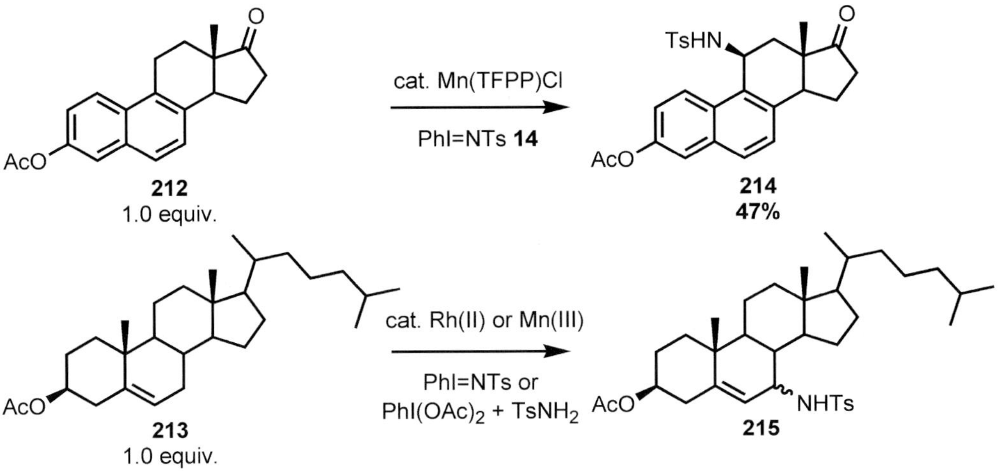

Scheme 2.47 Amidation of steroids.

2.7 Conclusions

In today's highly competitive search for efficient C—H functionalization methodologies likely to find applications in total synthesis, C—H amination is an area that has considerably expanded in the last five years, with the emergence of practical methods for the generation of iminoiodanes in situ. These hypervalent iodine(III) reagents can be simply formed from various nitrogenous functions to produce, in the presence of transition metal catalysts, nitrenes, the reactivity and the delivery of which can be particularly well controlled through intramolecular processes. Intermolecular C—H aminations have also been shown to take place with very good yields and excellent regio- and stereocontrol, although improvements are still needed in order to enhance their scope, presently limited to secondary benzylic and, to a lesser extent, allylic positions.

By comparison, it is rather surprising that fewer studies have been dedicated to azides and haloamines as nitrene precursors. Breakthroughs with regard to the use of these derivatives could offer interesting solutions to the problem of the environmentally unfriendly iodobenzene generated from iminoiodanes. The emergence of new methods involving catalytic amounts of hypervalent iodine reagents could also present interesting alternatives to this problem [157, 158]. Further studies should also be aimed at designing "smart" catalysts capable of discriminating particular C—H bonds in simple alkanes, as well as at finding chiral ligands likely to induce better enantioselectivities. To this end, in comparison with the nitrene C—H insertion, the formation of C—N bonds through C—H activation processes is a potentially valuable reaction class that has remained relatively unexplored until now.

Though C—H amination is still in its early days, the synthetic organic chemist now nonetheless has access to methodologies by which the direct transformation of a C—H moiety efficiently into a C—N bond can be efficiently achieved, as witnessed by the recent elegant total syntheses of complex natural products described above.

References

1 B. M. Trost, I. Fleming (Eds.), *Comprehensive Organic Synthesis*; Pergamon Press, Oxford, 1992.
2 A. Ricci (Ed.), *Modern Amination Methods*; Wiley-VCH, Weinheim, 2000.
3 V. I. Tararov, A. Börner, *Synlett* 2005, 203.
4 K. C. Hultzsch, *Adv. Synth. Catal.* 2005, *347*, 367.
5 J. Streuff, C. H. Hövelmann, M. Nieger, K. Muñiz, *J. Am. Chem. Soc.* 2005, *127*, 14586.
6 J. A. Bodkin, M. D. McLeod, *J. Chem. Soc., Perkin Trans. 1* 2002, 2733.
7 X. E. Hu, *Tetrahedron*, 2004, *60*, 2701.
8 J. F. Hartwig, *Synlett* 2006, 1283.
9 J. A. Labinger, J. E. Bercaw, *Nature* 2002, *417*, 507.
10 A. S. Goldman, K. I. Goldberg, in *Activation and Functionalization of C—H Bonds*; K. I. Goldberg, A. S. Goldman (Eds.), ACS Symposium Series 885, American Chemical Society, Washington DC, 2004, pp. 1–43.
11 A. R. Dick, M. S. Sanford, *Tetrahedron* 2006, *62*, 2439.
12 K. Godula, D. Sames, *Science* 2006, *312*, 67.
13 R. H. Crabtree, *J. Chem. Soc., Dalton Trans.* 2001, 2437.
14 P. Müller, C. Fruit, *Chem. Rev.* 2003, *103*, 2905.
15 P. Dauban, R. H. Dodd, *Synlett* 2003, 1571.
16 J. A. Halfen, *Curr. Org. Chem.* 2005, *9*, 657.
17 C. G. Espino, J. Du Bois, in *Modern Rhodium-Catalyzed Organic Reactions*; P. A. Evans (Ed.), Wiley-VCH, Weinheim, 2005, pp. 379–416.
18 T. Wirth (Ed), *Topics in Current Chemistry Vol. 224, Hypervalent Iodine Chemistry*; Springer-Verlag, Heidelberg, 2003.
19 T. Wirth, *Angew. Chem. Int. Ed.* 2005, *44*, 3656.
20 C. Greck, B. Drouillat, C. Thomassigny, *Eur. J. Org. Chem.* 2004, 1377.
21 J. M. Janey, *Angew. Chem. Int. Ed.* 2005, *44*, 4292.
22 M. Johannsen, K. A. Jørgensen, *Chem. Rev.* 1998, *98*, 1689.
23 F. Tiemann, *Ber. Dtsch. Chem. Ges.* 1891, *24*, 4162.
24 W. Lwowski (Ed.), *Nitrene*, Interscience, New York, 1970.
25 C. J. Moody, in *Comprehensive Organic Synthesis*; B. M. Trost, I. Fleming (Eds.), Pergamon Press, Oxford, 1992, *Vol. 7*, pp. 21–38.
26 K. H. Saunders, *J. Chem. Soc.* 1955, 3275.
27 G. Smolinsky, *J. Am. Chem. Soc.* 1960, *82*, 4717.
28 G. Smolinsky, *J. Org. Chem.* 1961, *26*, 4108.
29 W. Lwowski, T. W. Mattingly, *Tetrahedron Lett.* 1962, 277.
30 W. Lwowski, T. W. Mattingly, *J. Am. Chem. Soc.* 1965, *87*, 1947.
31 M. F. Sloan, W. B. Renfrow, D. S. Breslow, *Tetrahedron Lett.* 1964, 2905.
32 D. S. Breslow, M. F. Sloan, N. R. Newburg, W. B. Renfrow, *J. Am. Chem. Soc.* 1969, *91*, 2273.
33 A. G. Anastassiou, H. E. Simmons, F. D. Marsh, *J. Am. Chem. Soc.* 1965, *87*, 2296.
34 A. G. Anastassiou, H. E. Simmons, *J. Am. Chem. Soc.* 1967, *89*, 3177.
35 G. Smolinsky, B. I. Feuer, *J. Am. Chem. Soc.* 1964, *86*, 3085.
36 H. Kwart, A. A. Kahn, *J. Am Chem. Soc.* 1967, *89*, 1951.
37 D. S. Breslow, M. F. Sloan, *Tetrahedron Lett.* 1968, 5349.

38 D. Carr, T. P. Seden, R. W. Turner, *Tetrahedron Lett.* 1969, 477.

39 K. B. Sharpless, T. Hori, L. K. Truesdale, C. O. Dietrich, *J. Am. Chem. Soc.* 1976, *98*, 269.

40 D. H. R. Barton, R. S. Hay-Motherwell, W. B. Motherwell, *J. Chem. Soc., Perkin Trans. 1* 1983, 445.

41 R. A. Abramovitch, T. D. Bailey, T. Takaya, V. Uma, *J. Org. Chem.* 1974, *39*, 340.

42 Y. Yamada, T. Yamamoto, M. Okawara, *Chem. Lett.* 1975, 341.

43 R. Breslow, S. H. Gellman, *J. Chem. Soc., Chem. Commun.* 1982, 1400.

44 R. Breslow, S. H. Gellman, *J. Am. Chem. Soc.* 1983, *105*, 6728.

45 J. P. Mahy, G. Bedi, P. Battioni, D. Mansuy, *New J. Chem.* 1989, *13*, 651.

46 E. W. Svastits, J. H. Dawson, R. Breslow, S. H. Gellman, *J. Am. Chem. Soc.* 1985, *107*, 6427.

47 J. P. Mahy, G. Bedi, P. Battioni, D. Mansuy, *Tetrahedron Lett.* 1988, *29*, 1927.

48 D. A. Evans, M. M. Faul, M. T. Bilodeau, *J. Am. Chem. Soc.* 1994, *116*, 2742.

49 L. E. Overman, A. L. Tomasi, *J. Am. Chem. Soc.* 1998, *120*, 4039.

50 P. Dauban, R. H. Dodd, *Org. Lett.* 2000, *2*, 2327.

51 P. Müller, C. Baud, Y. Jacquier, *Tetrahedron* 1996, *52*, 1543.

52 I. Nägeli, C. Baud, G. Bernardinelli, Y. Jacquier, M. Moran, P. Müller, *Helv. Chim. Acta* 1997, *80*, 1087.

53 X.-Q. Yu, J.-S. Huang, X.-G. Zhou, C.-M. Che, *Org. Lett.* 2000, *2*, 2233.

54 C. G. Espino, J. Du Bois, *Angew. Chem. Int. Ed.* 2001, *40*, 598.

55 P. Dauban, L. Sanière, A. Tarrade, R. H. Dodd, *J. Am. Chem. Soc.* 2001, *123*, 7707.

56 J. Du Bois, *CHEMTRACTS-Org. Chem.* 2005, *18*, 1.

57 Y. Cui, C. He, *Angew. Chem. Int. Ed.* 2004, *43*, 4210.

58 C. G. Espino, P. M. Wehn, J. Chow, J. Du Bois, *J. Am. Chem. Soc.* 2001, *123*, 6935.

59 M. Okada, S. Iwashita, N. Koizumi, *Tetrahedron Lett.* 2001, *41*, 7047.

60 P. M. Wehn, J. Lee, J. Du Bois, *Org. Lett.* 2003, *5*, 4823.

61 R. E. Meléndez, W. D. Lubell, *Tetrahedron* 2003, *59*, 2581.

62 F. J. Durán, L. Leman, A. A. Ghini, G. Burton, P. Dauban, R. H. Dodd, *Org. Lett.* 2002, *4*, 2481.

63 J. J. Fleming, K. W. Fiori, J. Du Bois, *J. Am. Chem. Soc.* 2003, *125*, 2028.

64 K. W. Fiori, J. J. Fleming, J. Du Bois, *Angew. Chem. Int. Ed.* 2004, *43*, 4349.

65 J.-L. Liang, S.-X. Yuan, J.-S. Huang, W.-Y. Yu, C.-M. Che, *Angew. Chem. Int. Ed.* 2002, *41*, 3465.

66 J.-L. Liang, S.-X. Yuan, J.-S. Huang, C.-M. Che, *J. Org. Chem.* 2004, *69*, 3610.

67 J.-L. Zhang, J.-S. Huang, C.-M. Che, *Chem. Eur. J.* 2006, *12*, 3020.

68 C. Fruit, P. Müller, *Helv. Chim. Acta* 2004, *87*, 1607.

69 C. Fruit, P. Müller, *Tetrahedron: Asymmetry* 2004, *15*, 1019.

70 J. Zhang, P. W. H. Chan, C.-M. Che, *Tetrahedron Lett.* 2005, *46*, 5403.

71 P. Müller, C. Baud, Y. Jacquier, *Can. J. Chem.* 1998, *76*, 738.

72 J.-L. Liang, S.-X. Yuan, P. W. H. Chan, C.-M. Che, *Org. Lett.* 2002, *4*, 4507.

73 J.-L. Liang, S.-X. Yuan, P. W. H. Chan, C.-M. Che, *Tetrahedron Lett.* 2003, *44*, 5917.

74 C. G. Espino, K. W. Fiori, M. Kim, J. Du Bois, *J. Am. Chem. Soc.* 2004, *126*, 15378.

75 M. Kim, J. V. Mulcahy, C. G. Espino, J. Du Bois, *Org. Lett.* 2006, *8*, 1073.

76 A. Al Mourabit, P. Potier, *Eur. J. Org. Chem.* 2001, 237.

77 S.-M. Au, S.-B. Zhang, W.-H. Fung, W.-Y. Yu, C.-M. Che, K.-K. Cheung, *Chem. Commun.* 1998, 2677.

78 S.-M. Au, J.-S. Huang, W.-Y. Yu, W.-H. Fung, C.-M. Che, *J. Am. Chem. Soc.* 1999, *121*, 9120.

79 S. K.-Y. Leung, W.-M. Tsui, J.-S. Huang, C.-M. Che, J.-L. Liang, N. Zhu, *J. Am. Chem. Soc.* 2005, *127*, 16629.

80 X.-G. Zhou, X.-Q. Yu, J.-S. Huang, C.-M. Che, *Chem. Commun.* 1999, 2377.

81 S.-M. Au, J.-S. Huang, C.-M. Che, W.-Y. Yu, *J. Org. Chem.* 2000, *65*, 7858.

82 G. M. Loudon, A. S. Radhakrishna, M. R. Almond, J. K. Blodgett, R. H. Boutin, *J. Org. Chem.* 1984, *49*, 4272.

83 I. M. Lazbin, G. F. Koser, *J. Org. Chem.* 1986, *51*, 2669.
84 R. M. Moriarty, C. J. ChanyII , R. K. Vaid, O. Prakash, S. M. Tuladhar, *J. Org. Chem.* 1993, *58*, 2478.
85 J.-L. Liang, J.-S. Huang, X.-Q. Yu, N. Zhu, C.-M. Che, *Chem. Eur. J.* 2002, *8*, 1563.
86 Y. Kohmura, T. Katsuki, *Tetrahedron Lett.* 2001, *42*, 3339.
87 M. Yamawaki, H. Tsutsui, S. Kitagaki, M. Anada, S. Hashimoto, *Tetrahedron Lett.* 2002, *43*, 9561.
88 M. M. Díaz-Requejo, P. J. Pérez, *J. Organomet. Chem.* 2006, *690*, 5441.
89 M. M. Díaz-Requejo, T. R. Belderraín, M. C. Nicasio, S. Trofimenko, P. J. Pérez, *J. Am. Chem. Soc.* 2003, *125*, 12078.
90 M. R. Fructos, S. Trofimenko, M. M. Díaz-Requejo, P. J. Pérez, *J. Am. Chem. Soc.* 2006, *128*, 11784.
91 C. Liang, F. Robert-Peillard, C. Fruit, P. Müller, R. H. Dodd, P. Dauban, *Angew. Chem. Int. Ed.* 2006, *45*, 4641.
92 E. S. Levchenko, N. Y. Derkach, A. V. Kirsanov, *Zh. Obshch. Khim.* 1962, *32*, 1208.
93 E. S. Levchenko, L. N. Markovskii, Yu. G. Shermolovich, *Zh. Org. Khim.* 2000, *36*, 167.
94 S. Tsushima, Y. Yamada, T. Onami, K. Oshima, M. O. Chaney, N. D. Jones, J. K. Swartzendruber, *Bull. Chem. Soc. Jpn.* 1989, *62*, 1167.
95 D. Leca, A. Toussaint, C. Mareau, L. Fensterbank, E. Lacôte, M. Malacria, *Org. Lett.* 2004, *6*, 3573.
96 P. H. Di Chenna, F. Robert-Peillard, P. Dauban, R. H. Dodd, *Org. Lett.* 2004, *6*, 4503.
97 C. Fruit, F. Robert-Peillard, G. Bernardinelli, P. Müller, R. H. Dodd, P. Dauban, *Tetrahedron: Asymmetry* 2005, *16*, 3484.
98 R. E. White, *Inorg. Chem.* 1987, *26*, 3916.
99 T. Katsuki, *Chem. Lett.* 2005, *34*, 1304.
100 S. Cenini, S. Tollari, A. Penoni, C. Cereda, *J. Mol. Catal. A.: Chem.* 1999, *137*, 135.
101 F. Ragaini, A. Penoni, E. Gallo, S. Tollari, C. Li Gotti, M. Lapadula, E. Mangioni, S. Cenini, *Chem. Eur. J.* 2003, *9*, 249.
102 K. Omura, M. Murakami, T. Uchida, R. Irie, T. Katsuki, *Chem. Lett.* 2003, *32*, 354.
103 T. Uchida, Y. Tamura, M. Ohba, T. Katsuki, *Tetrahedron Lett.* 2003, *44*, 7965.
104 K. Omura, T. Uchida, R. Irie, T. Katsuki, *Chem. Commun.* 2004, 2060.
105 T. Bach, C. Körber, *J. Org. Chem.* 2000, *65*, 2358.
106 H. Takada, Y. Nishibayashi, K. Ohe, S. Uemura, C. P. Baird, T. J. Sparey, P. C. Taylor, *J. Org. Chem.* 1997, *62*, 6512.
107 D. P. Albone, P. S. Aujla, P. C. Taylor, S. Challenger, A. M. Derrick, *J. Org. Chem.* 1998, *63*, 9569.
108 D. P. Albone, S. Challenger, A. M. Derrick, S. M. Fillery, J. L. Irwin, C. M. Parsons, H. Takada, P. C. Taylor, D. J. Wilson, *Org. Biomol. Chem.* 2005, *3*, 107.
109 R. Vyas, B. M. Chanda, A. V. Bedekar, *Tetrahedron Lett.* 1998, *39*, 4715.
110 R. Vyas, G.-Y. Gao, J. D. Harden, X. P. Zhang, *Org. Lett.* 2004, *6*, 1907.
111 B. M. Chanda, R. Vyas, A. V. Bedekar, *J. Org. Chem.* 2001, *66*, 30.
112 J. U. Jeong, B. Tao, I. Sagasser, H. Henniges, K. B. Sharpless, *J. Am. Chem. Soc.* 1998, *120*, 6844.
113 S. I. Ali, M. D. Nikalje, A. Sudalai, *Org. Lett.* 1999, *1*, 705.
114 S. Minakata, D. Kano, Y. Oderaotoshi, M. Komatsu, *Angew. Chem. Int. Ed.* 2004, *43*, 79.
115 G. D. Kishore Kumar, S. Baskaran, *Chem. Commun.* 2004, 1026.
116 T. Baumann, M. Bächle, S. Bräse, *Org. Lett.* 2006, *8*, 3797.
117 W. Adam, K. J. Roschmann, C. R. Saha-Möller, *Eur. J. Org. Chem.* 2000, 557.
118 J.-L. Liang, X.-Q. Yu, C.-M. Che, *Chem. Commun.* 2002, 124.
119 J. Du Bois, J. Hong, E. M. Carreira, M. W. Day, *J. Am. Chem. Soc.* 1996, *118*, 915.
120 W. Lwowski, T. J. Maricich, *J. Am. Chem. Soc.* 1965, *87*, 3630.
121 S. Fioravanti, G. Luna, L. Pellacani, P. A. Tardella, *Tetrahedron* 1997, *53*, 4779.
122 S. Fioravanti, L. Pellacani, S. Tabanella, P. A. Tardella, *Tetrahedron* 1998, *54*, 14105.

123 S. Fioravanti, A. Morreale, L. Pellacani, P. A. Tardella, *Eur. J. Org. Chem.* 2003, 4549.

124 S. Fioravanti, D. Colantoni, L. Pellacani, P. A. Tardella, *J. Org. Chem.* 2005, *70*, 3296.

125 S. Fioravanti, A. Morreale, L. Pellacani, P. A. Tardella, *Tetrahedron Lett.* 2003, *44*, 3031.

126 H. Lebel, K. Huard, S. Lectard, *J. Am. Chem. Soc.* 2005, *127*, 14198.

127 M. B. Andrus, J. C. Lashley, *Tetrahedron* 2002, *58*, 845.

128 Y. Kohmura, K.-I. Kawasaki, T. Katsuki, *Synlett* 1997, 1456.

129 J. S. Clark, C. Roche, *Chem. Commun.* 2005, 5175.

130 C. W. Hamilton, D. S. Laitar, J. P. Sadighi, *Chem. Commun.* 2004, 1628.

131 L. He, P. W. H. Chan, W.-M. Tsui, W.-Y. Yu, C.-M. Che, *Org. Lett.* 2004, *6*, 2405.

132 A. R. Dick, K. L. Hull, M. S. Sanford, *J. Am. Chem. Soc.* 2004, *126*, 2300.

133 L. V. Desai, K. L. Hull, M. S. Sanford, *J. Am. Chem. Soc.* 2004, *126*, 9542.

134 D. Kalyani, M. S. Sanford, *Org. Lett.* 2005, *7*, 4149.

135 H.-Y. Thu, W.-Y. Yu, C.-M. Che, *J. Am. Chem. Soc.* 2006, *128*, 9048.

136 G. Dyker, *Angew. Chem. Int. Ed.* 1999, *38*, 1698.

137 F. Kakiuchi, N. Chatani, *Adv. Synth. Catal.* 2003, *345*, 1077.

138 W. C. P. Tsang, N. Zheng, S. L. Buchwald, *J. Am. Chem. Soc.* 2005, *127*, 14560.

139 T. Uemura, S. Imoto, N. Chatani, *Chem. Lett.* 2006, *35*, 842.

140 X. Chen, X.-S. Hao, C. E. Goodhue, J.-Q. Yu, *J. Am. Chem. Soc.* 2006, *128*, 6790.

141 H. M. L. Davies, M. S. Long, *Angew. Chem. Int. Ed.* 2005, *44*, 3518.

142 B. M. Trost, J. L. Gunzner, O. Dirat, Y. H. Rhee, *J. Am. Chem. Soc.* 2002, *124*, 10396.

143 H. Huang, J. S. Panek, *Org. Lett.* 2003, *5*, 1991.

144 K. A. Parker, W. Chang, *Org. Lett.* 2003, *5*, 3891.

145 K. A. Parker, W. Chang, *Org. Lett.* 2005, *7*, 1785.

146 T. Yakura, Y. Yoshimoto, C. Ishida, S. Mabuchi, *Synlett* 2006, 930.

147 A. Hinman, J. Du Bois, *J. Am. Chem. Soc.* 2003, *125*, 11510.

148 H.-S. Byun, L. He, R. Bittman, *Tetrahedron* 2000, *56*, 7051.

149 P. M. Wehn, J. Du Bois, *J. Am. Chem. Soc.* 2002, *124*, 12950.

150 M. Rezaei, T. M. Harris, C. J. Rizzo, *Tetrahedron Lett.* 2003, *44*, 7513.

151 P. M. Wehn, J. Du Bois, *Org. Lett.* 2005, *7*, 4685.

152 S. Toumieux, P. Compain, O. R. Martin, *Tetrahedron Lett.* 2005, *46*, 4731.

153 S. Toumieux, P. Compain, O. R. Martin, M. Selkti, *Org. Lett.* 2006, *8*, 4493.

154 R. V. Stevens, *Acc. Chem. Res.* 1984, *17*, 289.

155 J. J. Fleming, J. Du Bois, *J. Am. Chem. Soc.* 2006, *128*, 3926.

156 J. Yang, R. Weinberg, R. Breslow, *Chem. Commun.* 2000, 531.

157 J. Li, P. W. H. Chan, C.-M. Che, *Org. Lett.* 2005, *7*, 5801.

158 R. D. Richardson, T. Wirth, *Angew. Chem. Int. Ed.* 2006, *45*, 4402.

3
Nitroalkenes as Amination Tools

Roberto Ballini, Enrico Marcantoni, and Marino Petrini

3.1
Introduction

The outstanding electron-withdrawing power exerted by the nitro group allows conjugate nitroalkenes to be included in a profusion of synthetic procedures involving electrophilic substrates. As α,β-unsaturated compounds, conjugate nitroalkenes act as efficient Michael acceptors and dienophiles in many reactions directed towards the formation of new carbon-carbon and carbon-heteroatom bonds [1–3]. Since nitroalkanes are finally obtained in these reactions, a fundamental aspect that contributes to keeping nitroalkenes involved at the very heart in many practical processes stems from the ability of the nitro group to be converted into other useful functionalities [4, 5]. Among these transformations, reduction of the nitro group occupies a prominent position as it provides an effective route to primary amines. The whole process, encompassing both C—C bond formation and reduction, can be made diastereo- or enantioselective through the use of modern techniques pertaining to asymmetric synthesis, finally affording optically active amino derivatives. This aspect is particularly important when account is taken of the fact that the amino group is present in a plethora of pharmacologically active substances even more seriously required in their enantiopure forms for clinical purpose. This chapter focuses on some recent synthetic approaches that make use of nitroalkenes as pivotal reagents for the introduction of an amino group into the molecular framework and covers the procedures that have appeared in the literature since the early 1990s. The discussion only includes those methods in which the amino moiety is introduced by reduction of the nitro group.

3.2
General Strategies for the Synthesis of Nitroalkenes

Most of the available synthetic methods for the preparation of nitroalkenes fall into two main categories: a) dehydration of β-nitro alcohols obtained from nitroaldol reactions, and b) direct electrophilic substitution of nitro groups onto alkenes by use of various nitrating agents (Scheme 3.1).

Amino Group Chemistry. From Synthesis to the Life Sciences. Edited by Alfredo Ricci

ISBN: 978-3-527-31741-7

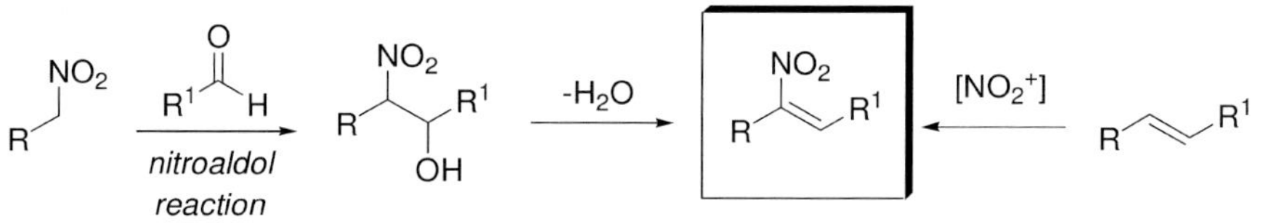

Scheme 3.1 General strategies for the preparation of nitroalkenes.

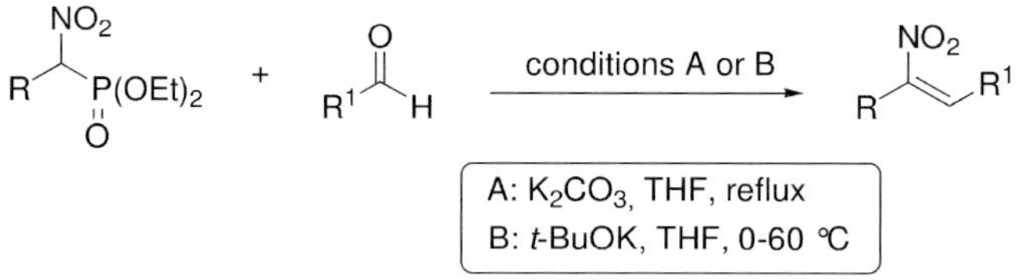

Scheme 3.2 Synthesis of nitroalkenes by Horner–Wadsworth–Emmons reactions.

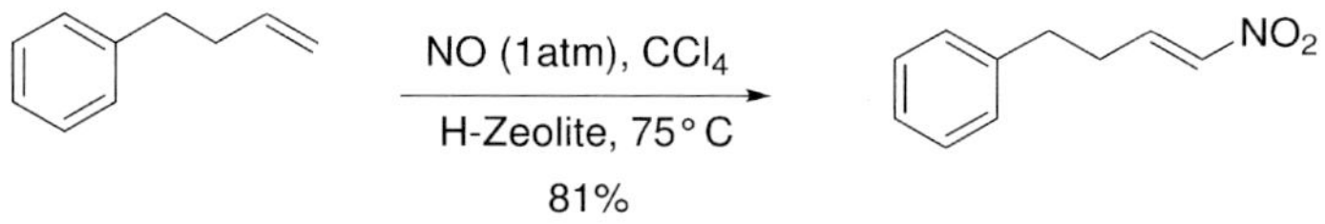

Scheme 3.3 Synthesis of a nitroalkene by direct nitration.

The nitroaldol reaction, also known as the Henry reaction, provides a rapid route to various β-nitro alcohols through treatment of nitroalkanes with aldehydes under a range of different reaction conditions that also include heterogeneous catalysis, solventless procedures, and microwave activation [6, 7]. The dehydration step to afford the nitroalkenes can be achieved by suitable functionalization of the hydroxy group followed by a base-induced elimination of the resulting derivative [8, 9], or alternatively, this operation can also be carried out in a "one-pot" fashion, allowing isolation of the nitroalkene from its remote precursors [10]. Arylaldehydes often directly give nitroalkenes in base-promoted reactions with nitroalkanes as a result of a Knoevenagel process. Nitroalkenes can be also formed by exploitation of a modified Horner–Wadsworth–Emmons reaction between a 2-nitrophosphonate and an aldehyde (Scheme 3.2) [11, 12].

Nitration of alkenes would represent a straightforward procedure to prepare nitroalkenes but is rarely used because of its general lack of selectivity and low yields. Nitric oxide is particularly effective in converting terminal alkenes into 1-nitroalkenes under various reaction conditions (Scheme 3.3) [13–15]. Nitrite and nitrate salts of alkali metals can be profitably used to convert alkenes and alkynes into nitroalkenes [16, 17].

An interesting procedure, especially useful for the preparation of 2,2-disubstituted 1-nitroalkenes, is based on the conjugate addition of complex zinc cuprates to nitroalkenes (Scheme 3.4) [18].

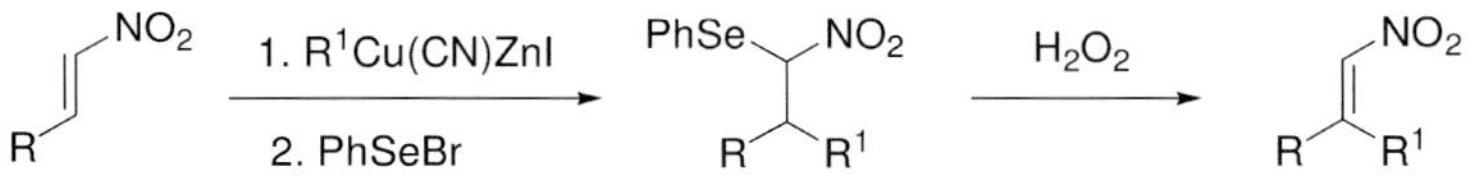

Scheme 3.4 Synthesis of 2,2-disubstituted 1-nitroalkenes.

The intermediate nitronate anions formed upon organometallic addition are trapped by treatment with PhSeBr and the resulting adducts are eliminated under oxidative conditions, giving 2,2-disubstituted 1-nitroalkenes. The reversibility of the Henry reaction makes the utilization of ketones for the direct preparation of 2,2-disubstituted 1-nitroalkenes troublesome. However, trapping of the unstable nitroalkene intermediate with a sulfide allows a clean and efficient two-step synthesis of 2,2-disubstituted 1-nitroalkenes by subsequent oxidative elimination [19].

3.3 Synthesis of Alkylamines

3.3.1 Monoamines

The conjugate addition of a carbon nucleophile to a nitroalkene, followed by reduction of the nitro group, represents an efficient route to a certain number of primary alkylamines. Nucleophiles bearing chiral auxiliary groups are able to exert a certain degree of diastereo- and enantioselection in reactions with various nitroalkenes, finally affording optically active amines. This strategy is illustrated by the reaction between the enolate anion of a chiral γ-lactam **1** and a nitroolefin, resulting – after reduction, protection of the amino group, and removal of the chiral auxiliary – in an α-substituted-γ-lactam **2** (Scheme 3.5) [20].

Alkylzinc reagents are particularly effective in catalytic enantioselective additions to nitroalkene derivatives [21]. Chiral dipeptide phosphines **3** require the presence of copper(I) salts to produce high levels of enantioselection. Catalytic reduction of the resulting nitroalkanes provides the corresponding amines in enantioenriched form (Scheme 3.6) [22].

The astonishing increase in the number of asymmetric transformations promoted by purely organic molecules has recently signaled a renaissance of organocatalysis, which can now replace organometallic complexes in almost all their former utilizations. Addition of trisubstituted carbon nucleophiles derived from unsymmetrical 2-substituted 1,3-dicarbonyl compounds generates two stereocenters in the reaction with β-substituted nitroalkenes (Scheme 3.7) [23]. Enhanced diastereo- and enantiocontrol in this reaction can be accomplished by use of a cinchona alkaloid **4**, although low temperatures are necessary.

According to frontier orbital theory, the electrophilic character of nitroalkenes is associated with low-energy LUMOs, which make these compounds particularly suitable for pericyclic reactions. The Diels–Alder reaction between

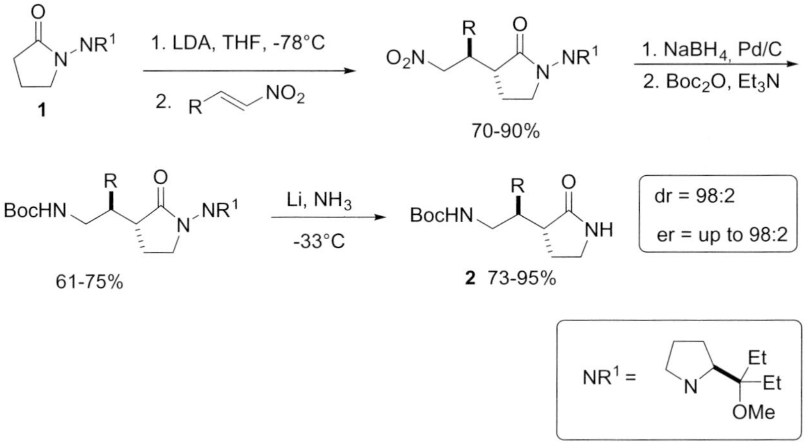

Scheme 3.5 Diastereoselective addition of chiral γ-lactam enolates to nitroalkenes.

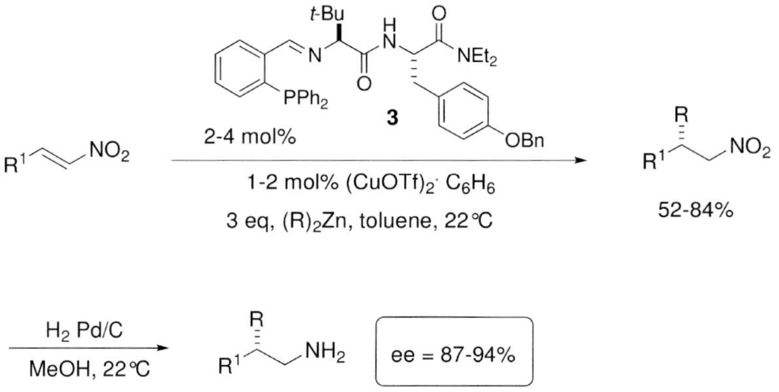

Scheme 3.6 Catalytic enantioselective additions of alkylzinc reagents to nitroalkenes.

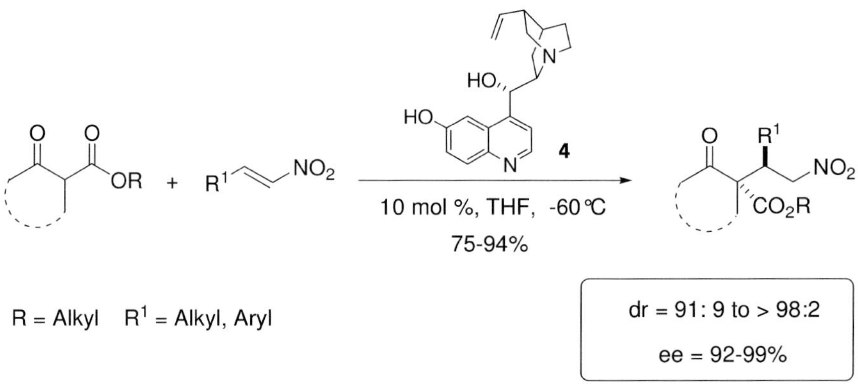

Scheme 3.7 Organocatalytic enantioselective additions of 2-substituted 1,3-dicarbonyl compounds to nitroalkenes.

3-methyl-1-nitrobut-1-ene and cyclopentadiene mainly affords the *exo* cycloadduct, which on catalytic hydrogenation affords a bicyclic amine particularly active on imidazoline I receptors (Scheme 3.8) [24].

A Diels–Alder reaction is also employed for the assembly of the 2-aminocyclohexenone core of (−)-calicheamicinone, the aglycone of the antitumor agent calicheamicin γ_1^I (Scheme 3.9) [25].

A derivative of 3-nitropropenoic acid esterified with (−)-8-phenylmenthol derivative **6** is treated with acyl ketene acetal **5**, resulting in the [4+2] cycloadduct **7** with surprisingly high diastereoselectivity. Reduction of ketoester **7** with $NaBH_4$ also

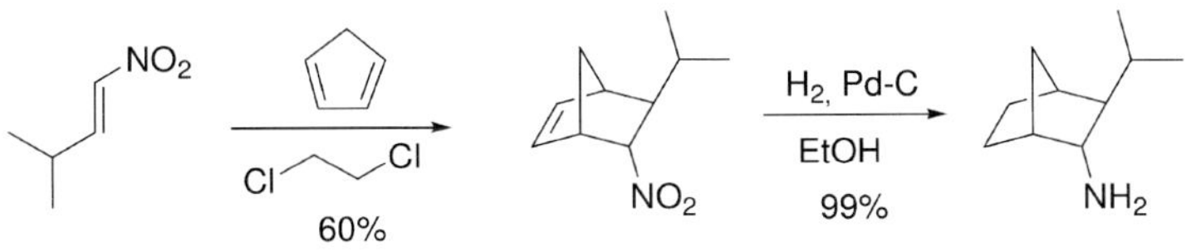

Scheme 3.8 Diels–Alder reaction between 3-methyl-1-nitrobut-1-ene and cyclopentadiene.

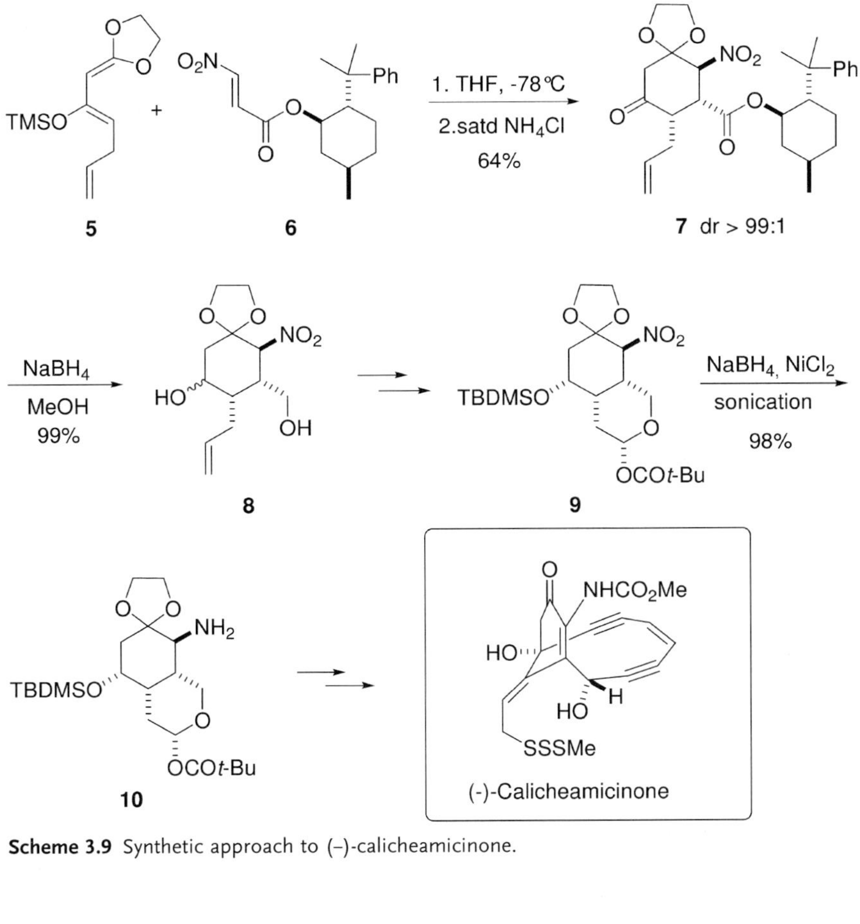

Scheme 3.9 Synthetic approach to (−)-calicheamicinone.

results in the cleavage of the chiral auxiliary and gives diol **8** with poor stereoselectivity. After oxidative cleavage of the double bond and protection of the resulting lactol, the nitro group in compound **9** is selectively reduced to the primary amine **10** in high yield by use of nickel boride, generated by treatment of $NaBH_4$ with $NiCl_2$. (−)-Calicheamicinone, the structure of which was confirmed by X-ray spectroscopy, is obtained from this amino derivative after a certain number of steps.

3.3.2
Amino Acid Derivatives

Synthetic approaches to the preparation of α-amino acids through the use of nitroalkenes as substrates are rather scarce. Lewis acids are able to promote the addition of isopropyl radicals to β-substituted α,β-unsaturated α-nitro esters at low temperatures. The presence of Bu_3SnH for hydride trapping of the intermediate radical also causes the formation of reduction products, though this can be minimized by use of zinc Lewis acids. Chiral ligands such as bisoxazolines can produce moderate levels of enantioselection in the obtained α-nitro esters, which are converted into the corresponding α-amino acid derivatives by simple catalytic hydrogenation (Scheme 3.10) [26, 27].

More traditionally, β-substituted α,β-unsaturated α-nitro esters **11** can add organolithium and Grignard reagents to give the corresponding α-nitro esters **12**, which are amenable to further functionalization at the reactive 2-position (Scheme 3.11) [28]. Quaternization at C-2 can be accomplished by means of palladium-catalyzed allylic alkylations, arylation with triphenylbismuth dichloride, and phosphine-promoted Michael additions. Chemoselective reduction of the nitro groups in some of these derivatives is best effected with metals such as tin or zinc in acetic acid, which also provides protection of the resulting amino group as the corresponding acetamide in compounds **13** and **14**.

Nucleophilic addition of stabilized carbanions onto 3-bromo-1-nitropropene results in a preliminary conjugate addition followed by an intramolecular ring-closure of the intermediate nitronate anion **15** to give cyclopropyl derivatives with

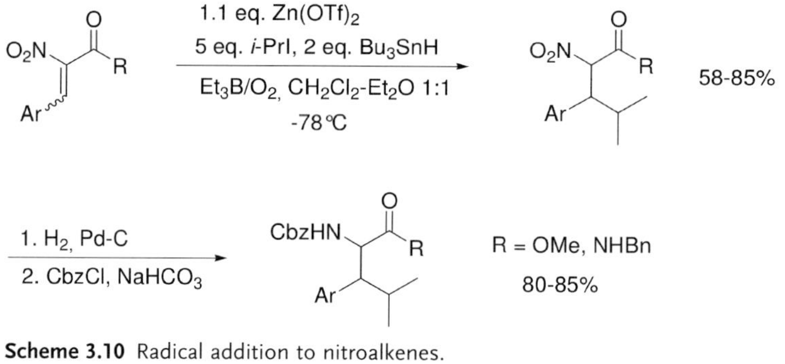

Scheme 3.10 Radical addition to nitroalkenes.

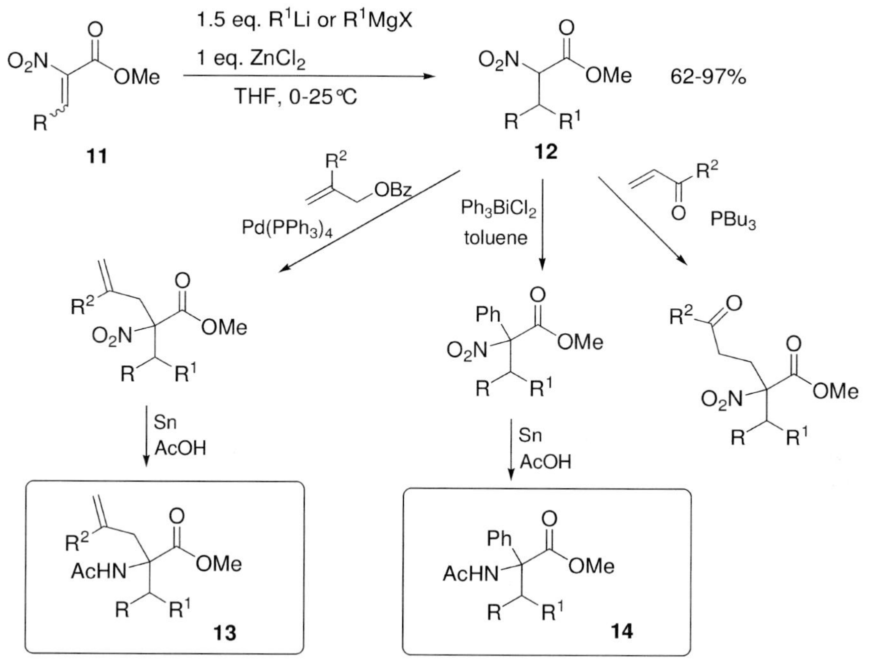

Scheme 3.11 Organometallic additions to nitroalkenes and synthetic utilization of the obtained 2-nitro esters.

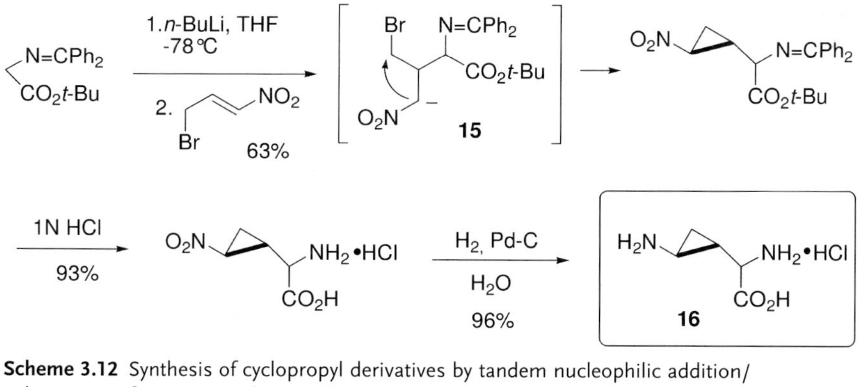

Scheme 3.12 Synthesis of cyclopropyl derivatives by tandem nucleophilic addition/substitution of an ester enolate to 3-bromo-1-nitropropene.

outstanding *E* stereoselectivity (Scheme 3.12) [29]. Acid hydrolysis and reduction of the nitro group affords 2-aminocyclopropylglycine **16**, a member of an interesting class of amino acid derivatives present in plants and microorganisms.

The structural attributes of nitroalkenes make them ideal substrates for the synthesis of β-amino acids and related derivatives. These unusual amino acids, although existing in nature, form β-peptides, which are rather resistant to degradation by various peptidases. Furthermore, β-peptides are endowed with stable tertiary structures that can be profitably used for several studies involving protein structures. The chiral nitroalkene **17**, bearing a stereocenter in close proximity to the unsaturated system, is able to provide the corresponding Michael adducts **18** and **19** with satisfactory diastereoselectivity (Scheme 3.13) [30]. As usual, reduction of the nitro group produces the amino moiety, which requires suitable protection before the next synthetic operations. The carboxylic groups in the target compounds **20** and **21** are generated by chemoselective hydrolysis of the acetonide group followed by oxidative cleavage of the diol.

Chiral catalysis is more efficient in producing optically active compounds through conjugate additions to nitroalkenes. A set of organozinc reagents can be used to create new carbon-carbon bonds enantioselectively in reactions involving nitroalkene **22** with the aid of minimum amounts of a BINOL-based chiral catalyst **23** that is usually employed in the presence of copper salts (Scheme 3.14) [31]. After reduction of the nitro group and protection of the resulting primary amine, the acetal can be simply hydrolyzed to the parent amino aldehyde under acid conditions or it can be directly oxidized to the corresponding acid **24** with H_5IO_6. The utilization of acetal groups as precursors of either carbonyl or carboxy systems in

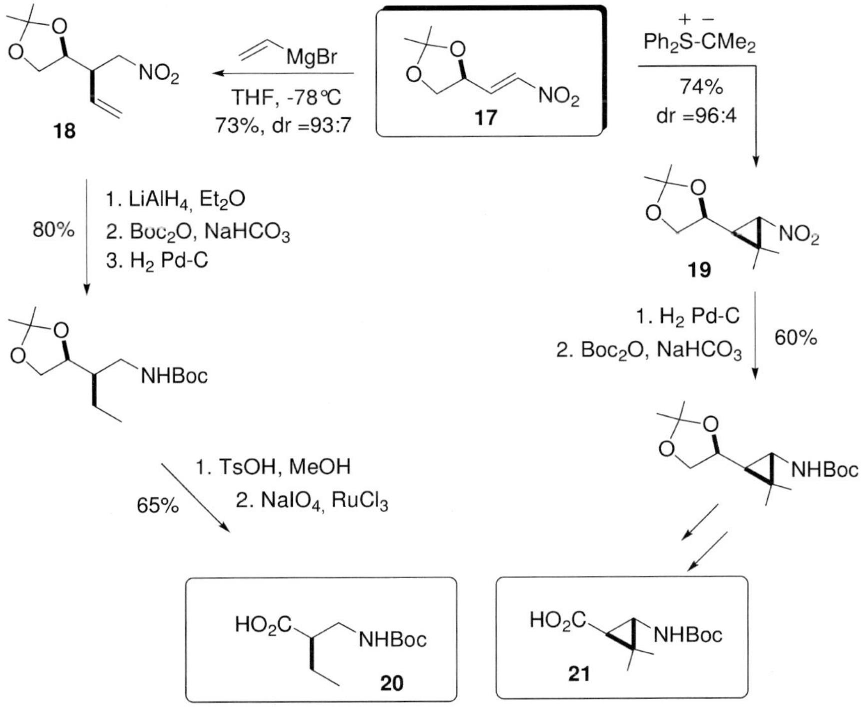

Scheme 3.13 Synthesis of β-amino acid derivatives from optically active nitroalkene **17**.

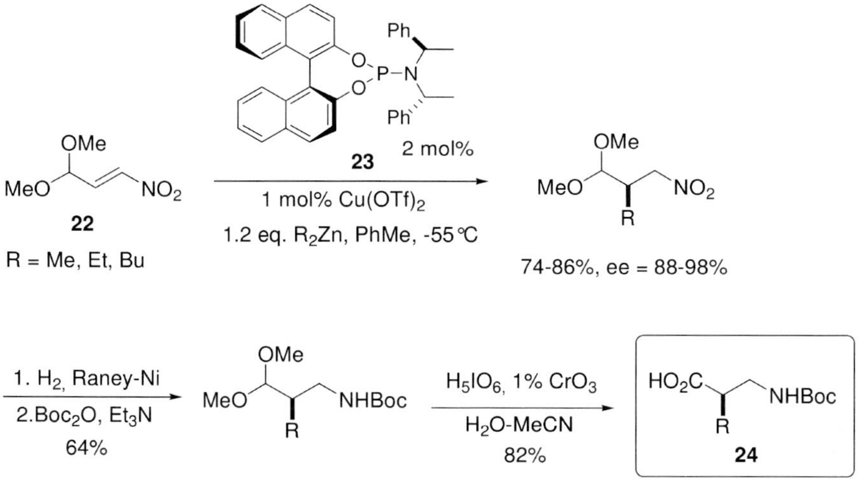

Scheme 3.14 Catalytic enantioselective addition of alkylzinc reagents to nitroalkene **22**.

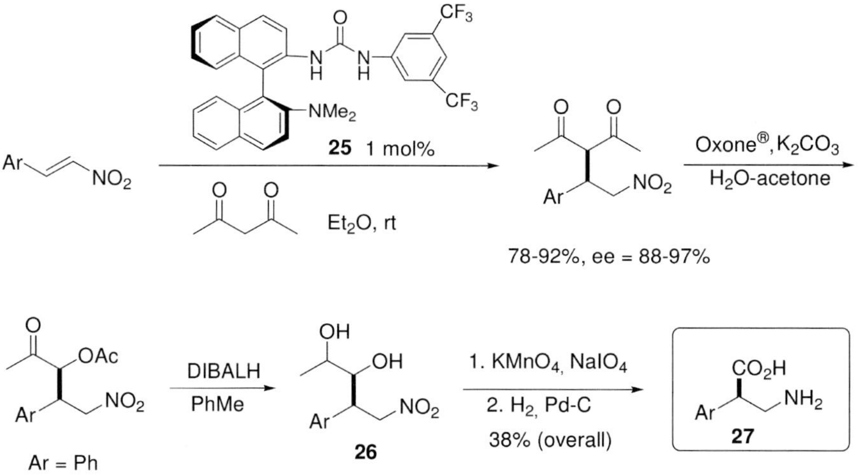

Scheme 3.15 Catalytic enantioselective addition of acetyl acetone to β-nitrostyrene.

these reactions could be avoided by direct use of nitroalkenes bearing ester groups [32, 33], although the *ee*s of the obtained adducts were slightly lower than those obtained by use of an acetal moiety.

A complementary strategy for the preparation of β-amino acids from simple nitrostyrenes involves the conjugate addition of carbon nucleophiles acting as masked carboxylic groups (Scheme 3.15) [34].

The organocatalyzed addition of acetylacetone to nitrostyrenes in the presence of chiral catalyst **25** occurs in satisfactory yields and enantioselectivities even at room temperature. Transformation of the diketone framework into a carboxylic group involves a preliminary Baeyer–Villiger process and a selective reduction of the keto group to produce diol **26**. Oxidative cleavage of the diol and reduction of the nitro group finally affords the expected β-amino acid **27**.

The Diels–Alder reaction between ethyl β-nitroacrylate and furan follows the known *endo* rule, affording the expected bicyclo derivative (Scheme 3.16) [35, 36]. Reduction and *N*-Boc protection affords a bridged amino ester **28**, a central intermediate for the preparation of a diasteromeric series of polyhydroxylated cyclohexyl-β-amino acids. Ring-cleavage of the bridged amino ester **28** under strongly basic conditions gives the cyclohexadiene derivative **29**, which can be selectively osmylated to provide the 4,5-*cis* compound **30**. Conversely, epoxidation of cyclohexadiene **29**, followed by nucleophilic ring-opening and double bond reduction, produces the 4,5-*trans* isomer **31**. Finally, direct osmylation on the unsaturated bridged amino ester gives the 3,4,5-*cis-cis* diastereomer **32** after conventional synthetic manipulations.

γ-Aminobutyric acid (GABA) is an inhibitory neurotransmitter acting in the central nervous systems of mammalians. Baclofen, a lipophilic analogue of GABA, is currently in use as an antispastic agent and it has been demonstrated that its pharmacological activity resides exclusively in the *R* enantiomer. The stereogenic

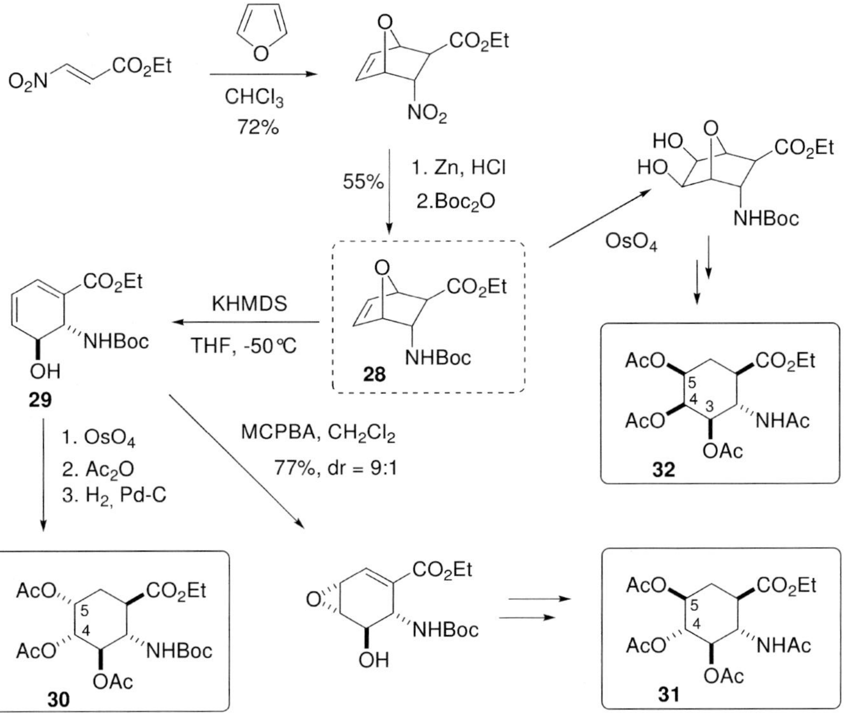

Scheme 3.16 Synthesis of polyhydroxylated cyclohexyl-β-amino acids.

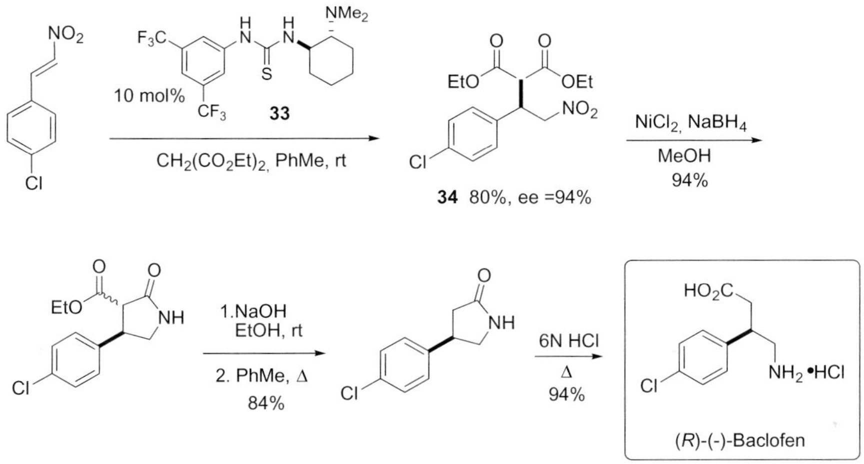

Scheme 3.17 Synthesis of the GABA analogue (*R*)-(–)-baclofen.

center that characterizes this compound is generated by an organocatalytic enantioselective process involving *p*-chloronitrostyrene, diethyl malonate, and the chiral thiourea **33** (Scheme 3.17) [37].

Chiral thioureas are particularly effective in promoting the addition of stabilized carbanions to nitroalkenes, which are activated toward the reaction by interaction between the oxygens of the nitro group and the NH of the thiourea. A subsequent reduction of the nitro group in the adduct **34** with nickel boride occurs concomitantly with lactamization and after decarboxylation the obtained pyrrolidinone is further hydrolyzed to afford (–)-baclofen hydrochloride salt. In a related synthesis, (–)-baclofen is prepared through the reaction between the same nitroalkene and an enantiopure Fisher-type amino carbene [38].

3.3.3 Amino Alcohols

The constant interest in the stereoselective synthesis of 1,2-amino alcohols is mainly justified by their widespread occurrence in many bioactive natural products. Optically active vicinal amino alcohols are also profitable starting materials for the synthesis of many chiral catalysts and auxiliaries that are used in several asymmetric processes [39]. Asymmetric nitroaldol reactions represent an effective procedure for the preparation of chiral β-nitro alcohols, which can readily be converted into 1,2-amino alcohols by simple reduction. A complementary strategy to the synthesis of vicinal amino alcohols consists of the conjugate addition of oxygen nucleophiles to nitroalkenes and reduction of the obtained β-oxygenated derivatives. This oxy-Michael procedure can be made enantioselective by addition of an optically active alkoxide anion capable of discriminating between the two enantiotopic faces of the nitroalkene. This synthetic operation can be accomplished by (–)-*N*-formylnorephedrine **35**, which is able to react, in the presence of sodium hydride, with nitroalkenes to give β-nitro ethers **36** with outstanding diastereoselectivity (Scheme 3.18) [40]. Reduction of the nitro group and protection of the obtained primary amine has to precede the cleavage of the benzyl ether with sodium in liquid ammonia to afford the enantioenriched *N*-Boc amino alcohol **37**.

A related approach employs chiral lactol **38** as a source of the alkoxide anion in reactions with various nitroalkenes (Scheme 3.19) [41].

The utilization of KHMDS, in the presence of a crown ether, is essential in order to ensure consistent levels of stereoselectivity. After reduction of the nitro group and protection of the resulting amine, the lactol moiety is cleaved by use of a polymer-bound acid resin, which avoids any reduction in the enantiomeric purity

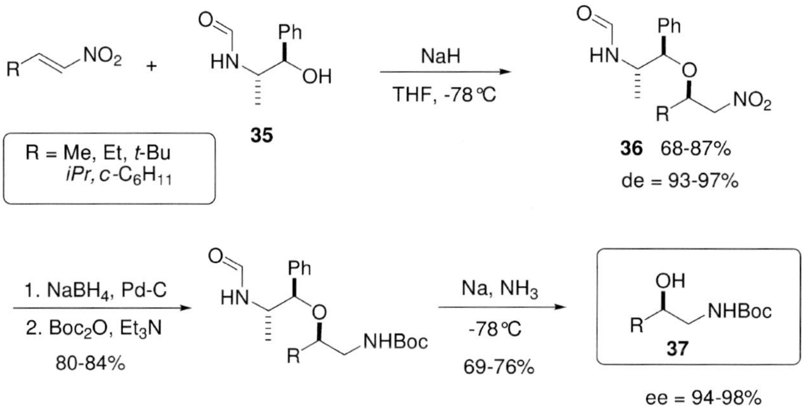

Scheme 3.18 Conjugate additions of optically active oxygen nucleophiles to nitroalkenes.

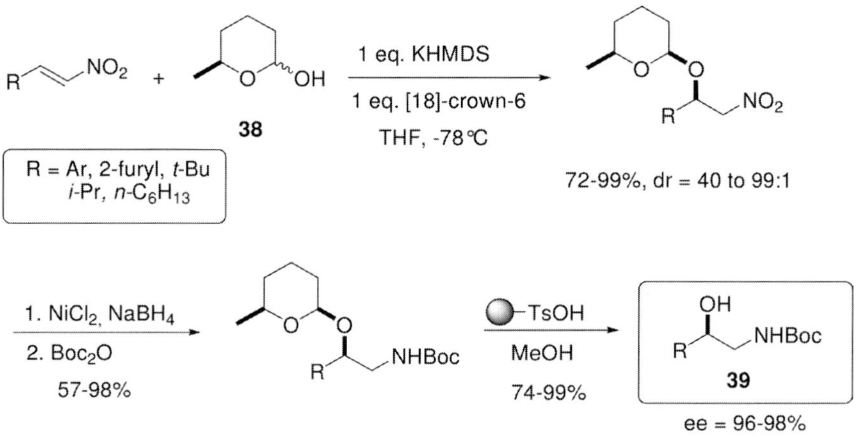

Scheme 3.19 Conjugate additions of a chiral lactol to nitroalkenes.

of the obtained *N*-Boc amino alcohol **39**. Application of these procedures to α,β-disubstituted nitroalkenes results in modest control of the newly created α-stereocenter and excellent control at the β-position, with a moderate preference for the *syn* stereoisomer [42, 43].

The efficiency and feasibility of the above synthetic strategy is demonstrated by its application to the preparation of several β-aryl ethanolamine compounds that display important pharmacological activity (Scheme 3.20) [44, 45].

The oxy-Michael products **40** obtained through reactions between nitroalkenes and allylic alcohols can be cyclized to the corresponding tetrahydrofuran derivatives in the presence of $CuCl_2$ (Scheme 3.21) [46]. The ring-closure occurs with modest diastereoselectivity by a mechanism in which the radical generated α to the nitro group attacks the unsaturated system and the formed resulting exocyclic radical is trapped by $CuCl_2$. Reduction of the nitro moiety to the amino group

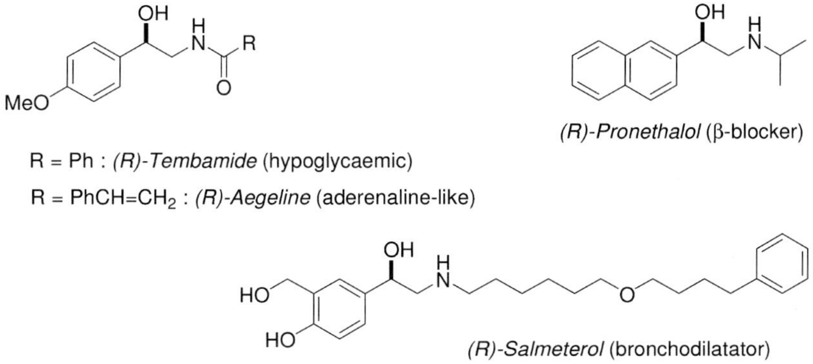

Scheme 3.20 Optically active 2-amino alcohol derivatives with pharmacological activity.

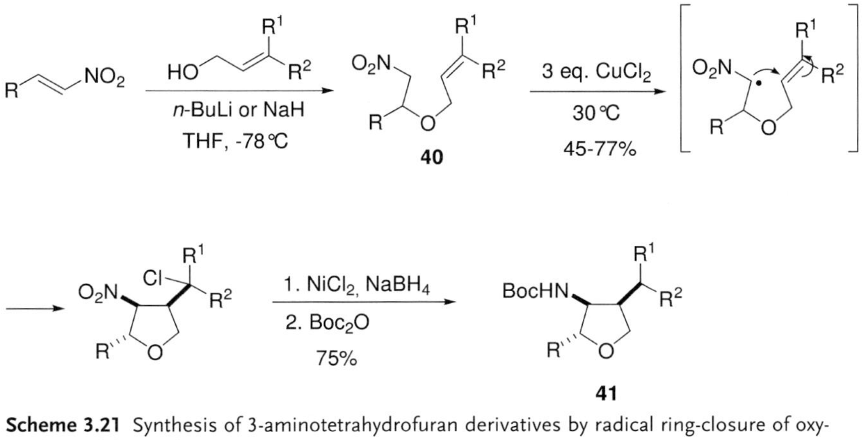

Scheme 3.21 Synthesis of 3-aminotetrahydrofuran derivatives by radical ring-closure of oxy-Michael adducts of nitroalkenes.

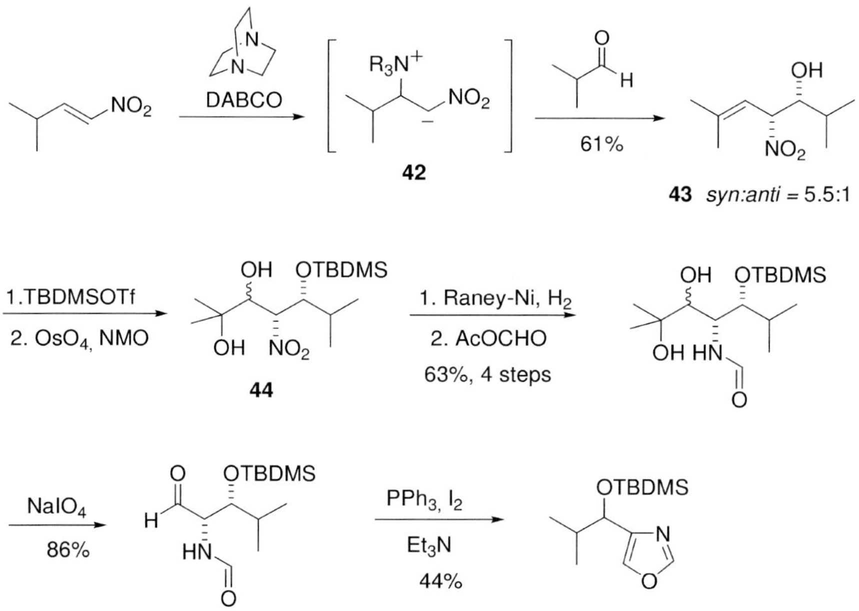

42

43 *syn:anti* = 5.5:1

44

45

46

Scheme 3.22 Baylis-Hillman-like reactions of nitroalkenes with aldehydes.

usually involves a concomitant dechlorination, resulting in 3-amino tetrahydrofurans **41**. 4-Chlorobut-2-yn-1-ol can be used in a similar synthetic operation involving a tandem conjugate addition/ring-closure with a completely ionic mechanism providing nitroallenyl tetrahydrofurans [47].

A unusual application of nitroalkenes can be found in a Baylis–Hillman-like reaction in which a tertiary amine such as 1,4-diazabicyclo[2.2.2]octane (DABCO) is able to add to the unsaturated system to generate a stabilized zwitterionic system **42** (Scheme 3.22) [48].

This intermediate reacts with isobutyraldehyde to give an unsaturated nitro alcohol **43**, with a certain preference for the *syn* stereoisomer. Protection of the hydroxy group and osmylation of the double bond affords compound **44**, which upon reduction of the nitro group and oxidative cleavage of the diol moiety affords amino aldehyde **45**. This can be converted into oxazole **46**, a synthetic unit present in many pharmacologically active drugs.

3.3.4
Diamino Derivatives

Reactions between nitrogen nucleophiles and nitroalkenes provide a rapid route to 2-aminonitroalkanes, which, after reduction of the nitro group, supply vicinal diamino derivatives [49]. A suitable choice of the nitrogen nucleophile allows a differentially protected diamino compound to be obtained. Alternatively, reduction and deprotection of the newly introduced amino appendage can be accomplished in the same synthetic operation. Michael addition of *O*-ethylhydroxylamine to nitroalkenes produces the corresponding 2-hydroxylamino nitroalkanes, which can be reduced at both nitrogenated groups in a "one-pot" procedure to afford the free 1,2-diamino compounds (Scheme 3.23) [50]. The procedure works satisfactorily both for linear and for cyclic nitroalkenes, with the only exception of nitrocyclohexene, which gives only a modest yield of 1,2-cyclohexanediamine.

Additions of chiral amines to cyclic nitroalkenes usually occur with high degrees of stereoselectivity [51], though the obtained optically active nitramines are often unstable compounds that decompose or suffer partial racemization during the reduction step. This drawback can be overcome by use of a large excess of SmI_2 as reducing agent, ensuring efficient reduction and preservation of the stereocenters' integrities (Scheme 3.24) [52].

To obtain primary diamino derivatives it is of extreme importance that the chiral ammonia equivalent used in the conjugate addition be easily removable, thus liberating the free amino group. Chiral hydrazines obtained from natural amino acids or sugar derivatives are good reagents for these reactions, which usually proceed with satisfactory diastereoselectivities, although long reaction times (6–25 days) are required to complete the process (Scheme 3.25) [53].

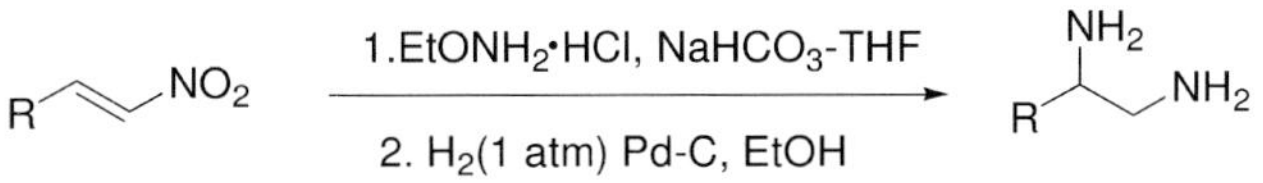

R	Ph	4-MeOPh	$PhCH_2$	$Ph(CH_2)_2$	n-C_8H_{17}	c-C_6H_{11}
yield (%)	88	64	76	88	65	69

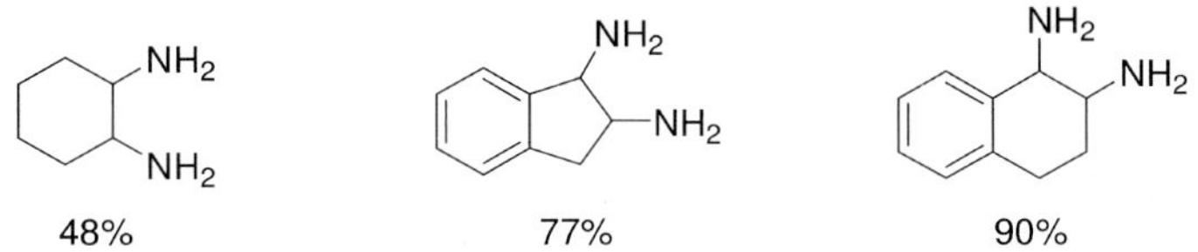

48% 77% 90%

Scheme 3.23 Synthesis of 1,2-diamino derivatives.

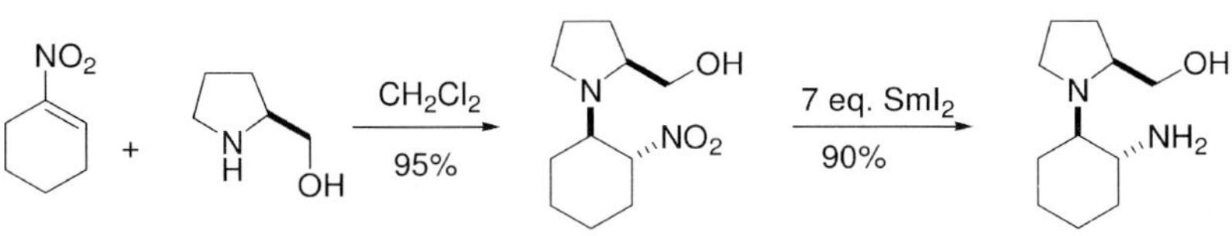

dr > 99:1 dr > 99:1

Scheme 3.24 Diastereoselective conjugate addition of chiral prolinol to 1-nitrocyclohexene.

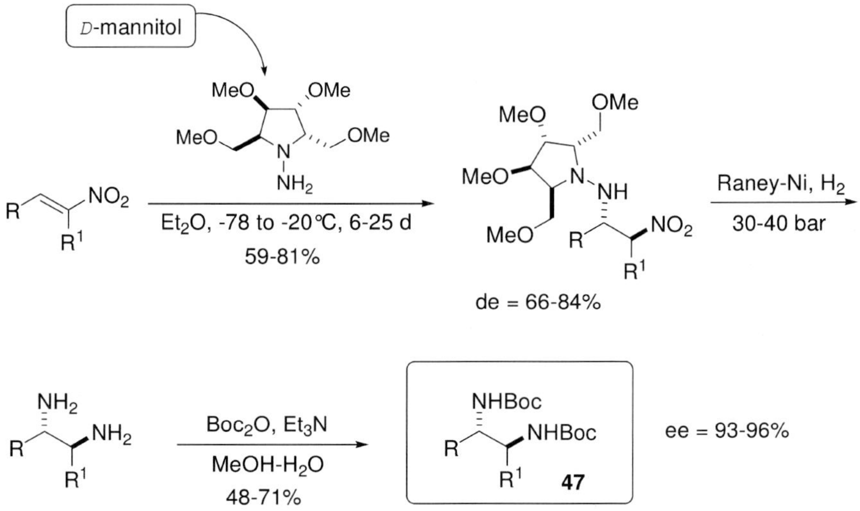

Scheme 3.25 Conjugate additions of a sugar-derived hydrazine to nitroalkenes.

The diastereomeric ratio can be greatly improved after a single crystallization of the solid product obtained, while catalytic hydrogenation is usually able to effect both reduction of the nitro group and removal of the chiral appendage of the hydrazine unit. For practical reasons these diamino compounds are frequently isolated as carbamoyl derivatives **47**.

Chiral oxazolidin-2-ones are among the most commonly employed auxiliaries in asymmetric synthesis involving the generation of metal enolates, but these chiral heterocycles can also serve as valuable sources of nitrogen nucleophiles in reactions with nitroalkenes (Scheme 3.26) [54, 55].

The anion of oxazolidin-2-one **48**, generated with *t*-BuOK, is able to exert a considerable degree of facial discrimination on the nitroalkene, giving the corresponding adduct with outstanding diastereoselectivity. After reduction of the nitro group by catalytic transfer hydrogenation with ammonium formate, cleavage of the heterocyclic ring can be achieved in a single step with lithium metal in liquid ammonia. This procedure offers poorly reproducible results, however, thus making preferable a preliminary hydrolytic cleavage of the ring under strongly acidic conditions followed by catalytic hydrogenation of the intermediate benzylic amine **49**. Despite the harsh conditions used, the final 1,2-diamines **50**, isolated as hydrochloride salts, still retain excellent *ee* values. Cleavage of the oxazolidin-2-one ring for the synthesis of trifluorodiamino derivatives **51** is carried out by an exchange reaction with ethylenediamine followed by hydrogenolysis of the obtained benzylamine [56]. Electrochemically generated anions of chiral oxazolidin-2-ones are also effective in the addition with various nitroalkenes, showing comparable results in terms of yield and diastereoselectivity [57].

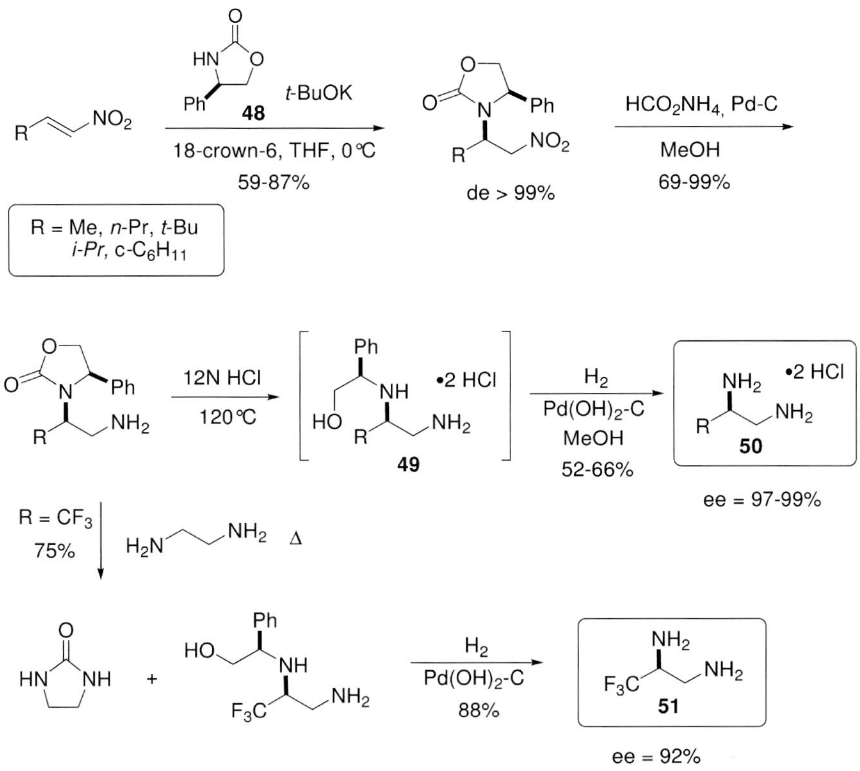

Scheme 3.26 Synthesis of chiral 1,2-diamines with use of 2-oxazolidin-2-one **48** as nucleophilic nitrogen donor.

In peptides, substitution of an ordinary [CONH] group with a stereogenic [CH(CF_3)NH] unit generates a new class of peptidomimetics with predictable conformations in solution, and these find useful applications in bioavailability studies. Such compounds can be prepared by kinetically controlled aza-Michael addition of natural α-amino acid esters to (*E*)-3,3,3-trifluoro-1-nitropropene (Scheme 3.27) [58].

The efficiency of the process is affected by the natures of the solvent and the base employed, resulting in the preferential formation of **52** as the *syn* diastereomer. A further elaboration of the resulting adduct involves the catalytic hydrogenation of the nitro group and subsequent coupling of the resulting amine with a phenylalanine unit, providing ψ[CH(CF_3)NH]Gly tripeptides **53** in good overall yields.

1,2-Diamino sugars are compounds of considerable practical interest since they find application as drug carriers. Their preparation starts from nitrosugars, which, as common nitro alcohols, can be dehydrated to the corresponding nitroalkenes **54** by known methods (Scheme 3.28) [59].

Conjugate addition of primary and secondary amines, followed by reduction of the nitro groups, gives 2,3-diamino derivatives that are purified after acetylation.

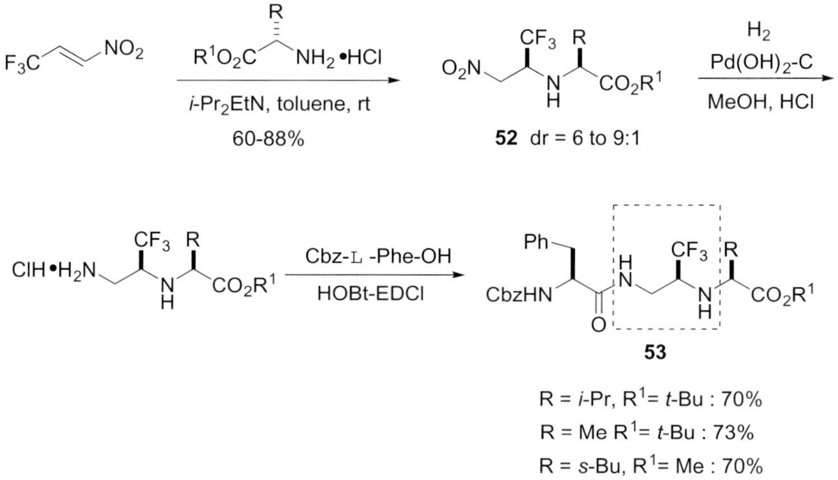

Scheme 3.27 Synthesis of trifluoromethylated peptidomimetics.

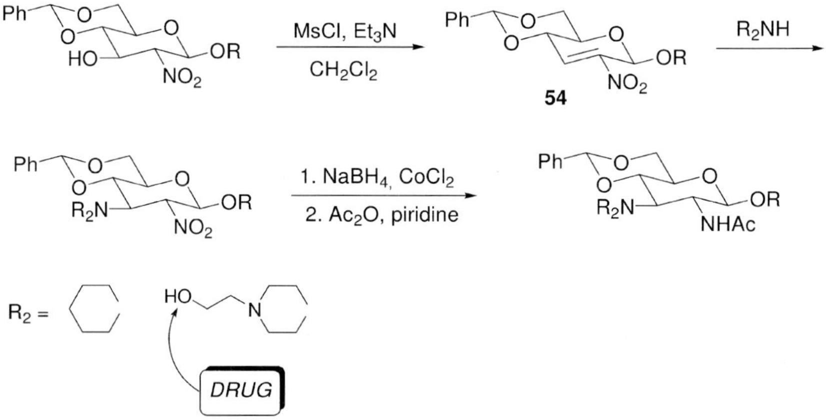

Scheme 3.28 Synthesis of aminosugars acting as drug carriers.

Anticancer drugs such as chlorambucil are usually linked to a hydroxyethyl appendage in the introduced secondary amine. Sugar-derived nitroalkenes are also involved in the synthesis of purine and pyrimidine nucleosides of D-galactosamine [60].

An interesting synthetic approach used for the preparation of 1,3-diamino derivatives is based on the conjugate addition of α-alkylamino carbanions to nitroalkenes. Such a nucleophilic reagent is conveniently prepared by preliminary lithiation of a stannyl derivative, followed by suitable transmetalation to give the desired organometallic compound **55** (Scheme 3.29) [61].

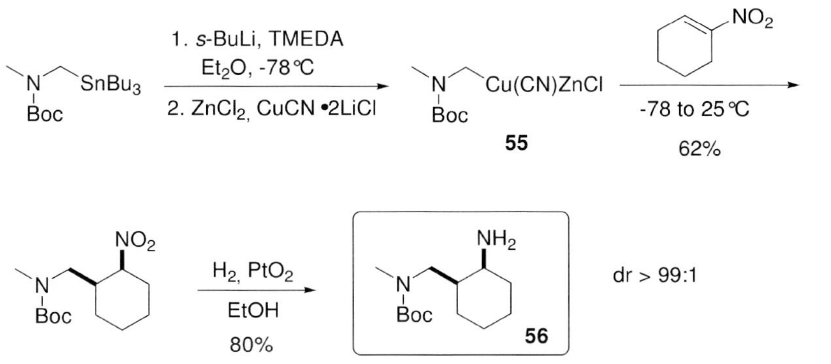

Scheme 3.29 Preparation of 1,3-diamino derivatives through conjugate additions of α-alkylamino carbanions to nitroalkenes.

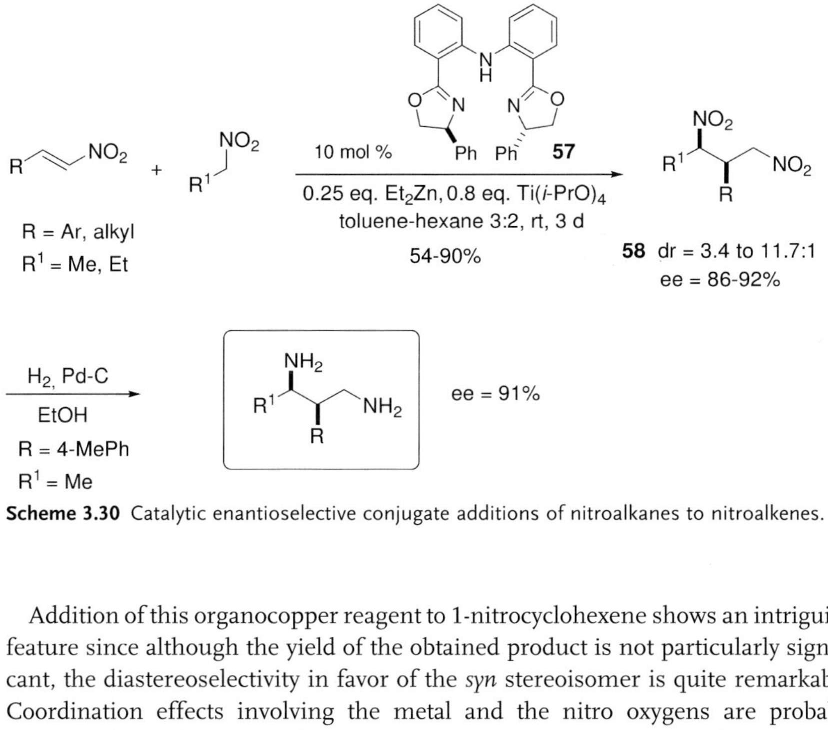

Scheme 3.30 Catalytic enantioselective conjugate additions of nitroalkanes to nitroalkenes.

Addition of this organocopper reagent to 1-nitrocyclohexene shows an intriguing feature since although the yield of the obtained product is not particularly significant, the diastereoselectivity in favor of the *syn* stereoisomer is quite remarkable. Coordination effects involving the metal and the nitro oxygens are probably responsible for the observed unusual stereoselectivity. A routine reduction of the nitro group completes the synthesis of the 1,3-diamino compound **56**.

Nitroalkanes, as well as many other methylene active compounds, are formidable nucleophiles in conjugate addition reactions [62]. Addition of simple nitroalkanes to nitroolefins in the presence of a chiral tridentate bisoxazoline **57** and diethylzinc affords 1,3-dinitro derivatives **58** with satisfactory diastereo- and enantioselectivities (Scheme 3.30) [63].

Despite the activation of diethylzinc provided by Ti(*i*-PrO)$_4$, the process requires long reaction times, and secondary nitroalkanes are not suitable for the addition. Although reduction of the obtained 1,3-dinitro compounds is illustrated for a single example, this procedure represents a potentially useful entry point to a whole series of optically active 1,3-diamino derivatives.

3.4
Pyrrolidine Derivatives

3.4.1
Pyrrolidinones

A general strategy for the preparation of pentaatomic nitrogen heterocycles involves the preliminary conjugate addition of ester enolates to nitroalkenes, followed by reduction of the nitro group. The obtained γ-amino ester spontaneously cyclizes to the parent pyrrolidin-2-one, which may be further reduced to afford a pyrrolidine ring. Both metal enolates and ketene acetals are reactive toward a wide array of nitroalkenes. In particular, silyl ketene acetals require activation by a Lewis acid that can strongly affect the diastereoselectivity of the process. The reaction between nitrostyrene and the TBDMS-protected propene acetal **59** is promoted by methylaluminium bis (2,6-di-*t*-butyl-4-methyl phenoxide) (MAD) and involves an open-chain transition state that gives rise to the corresponding *syn*-nitro ester **60** in a stereoconvergent fashion (Scheme 3.31) [64].

Reduction of the nitro group by catalytic transfer hydrogenation gives a 3,4-disubstituted pyrolidin-2-one **61** with complete retention of the original relative stereochemistry.

The metalated chiral bislactim ether **62** act as masked α-amino enolates in reactions with electron-poor olefins such as nitroalkenes. Best results are obtained with

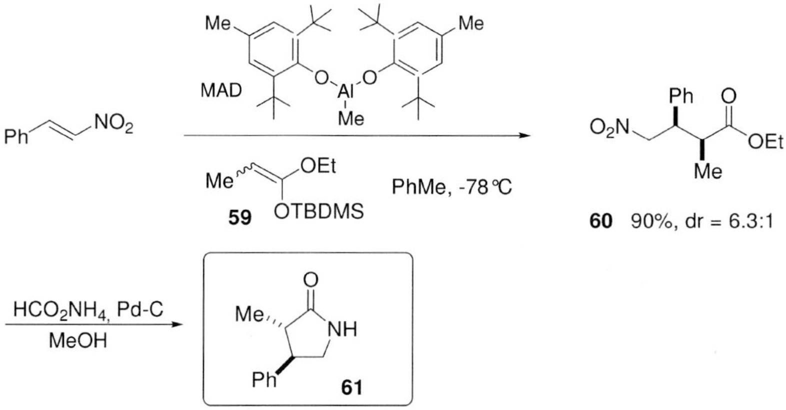

Scheme 3.31 MAD-catalyzed conjugate addition of a silyl ketene acetal to β-nitrostyrene.

titanium derivatives, which give the corresponding adducts differing only in the configurations at C-1′ (Scheme 3.32) [65].

Acid hydrolysis releases the free α-amino acid group and subsequent reduction of the nitro group allows the formation of the 3-aminopyrrolidin-2-one system.

Lithium enolates of cyclic esters can be employed in a similar strategy for the preparation of pyrrolidinones bearing a free hydroxy group (Scheme 3.33). Chiral 1,3-dioxolan-4-one enolate **63** reacts with nitroalkenes, ultimately affording optically active 3-hydroxypyrolidinones **64** [66], while the same process carried out on γ-lactone enolates **65** gives 3-hydroxyethylpyrrolidinones **66** [67].

Conjugate additions of enolates to nitroalkenes result in nitronate anions, which are usually quenched with a proton source. An alternative way to reach the carbon

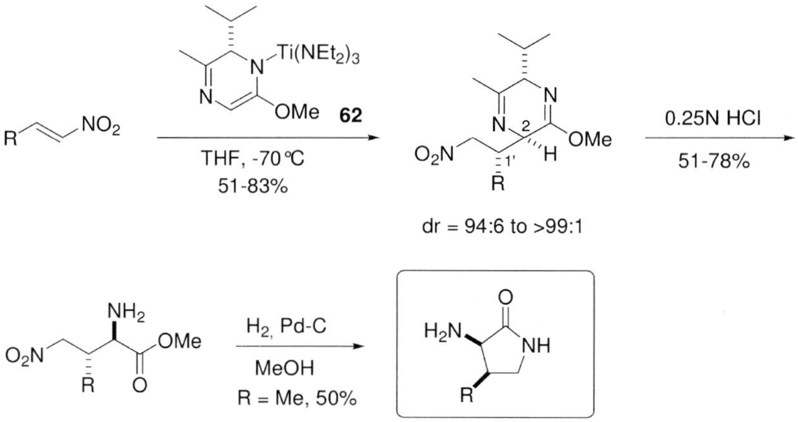

Scheme 3.32 Diastereoselective synthesis of 4-substituted 3-aminopyrrolidin-2-ones.

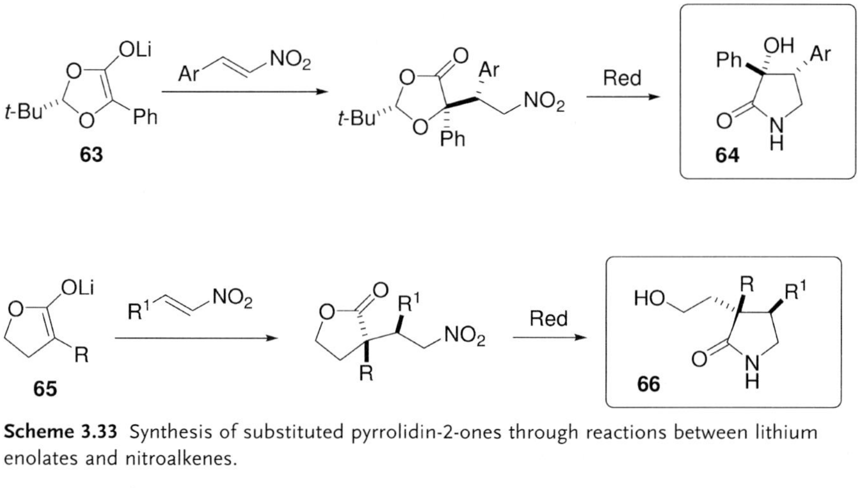

Scheme 3.33 Synthesis of substituted pyrrolidin-2-ones through reactions between lithium enolates and nitroalkenes.

backbone of the final molecule is to add a reactive carbon electrophile such as an aldehyde, exploiting a tandem Michael-nitroaldol reaction. The lithium enolates of cycloalkanecarboxylate esters **67** react with nitroethylene and the resulting nitronate is then trapped with formaldehyde (Scheme 3.34) [68]. Reduction and consequent ring-closure affords spirobicyclic lactams **68** of different ring sizes.

As previously stated, optically active oxazolidin-2-ones are very popular chiral auxiliaries in asymmetric synthesis. Their *N*-acyl metal enolates react with high diastereofacial selectivity with a number of functionalized nitroalkenes [69]. A practical application of this strategy is described for the enantioselective synthesis of the antidepressant drug (*R*)-rolipram (Scheme 3.35) [70].

The reaction between the sodium enolate of (*S*)-4-benzyl-3-acetyloxazolidin-2-one and the nitrostyrene derivative **69** affords the corresponding adduct with good diastereoselectivity, and this compound is further purified by crystallization. Catalytic hydrogenation ensures the debenzylation of the phenolic oxygen and the reduction of the nitro group, resulting in the formation of the lactam **70** with release of the chiral auxiliary. *O*-Alkylation of the free hydroxy group completes the synthesis of (*R*)-rolipram. Use of other systems such as (+)-camphor methyl ketone enolates gives results comparable to those obtained with oxazolidin-2-ones

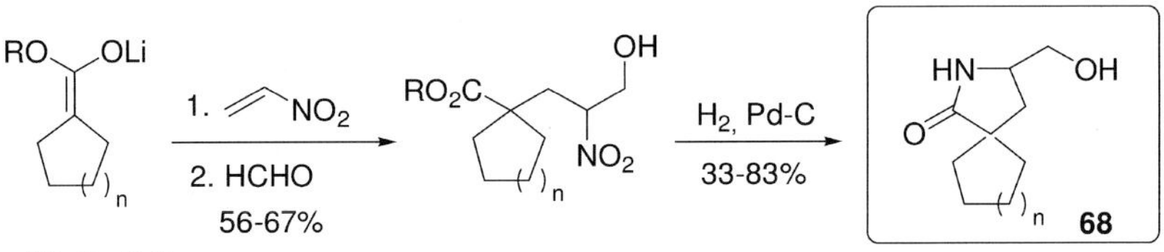

Scheme 3.34 Two-step synthesis of spirobicyclic lactams.

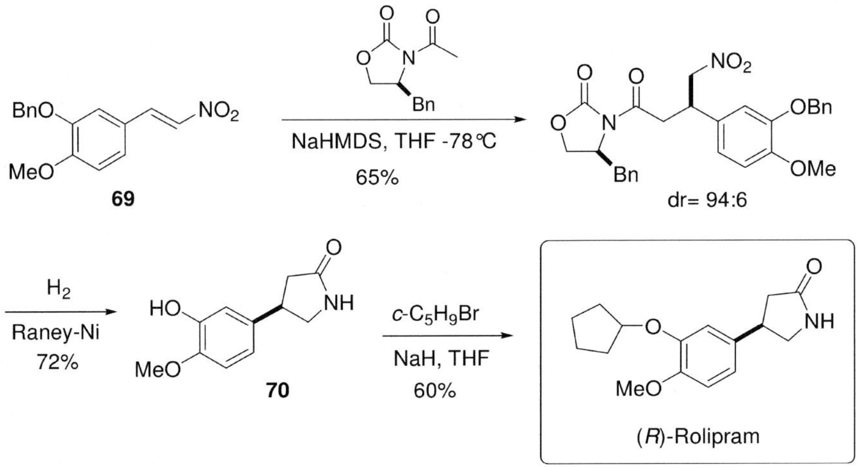

Scheme 3.35 Total synthesis of the antidepressant drug (*R*)-rolipram.

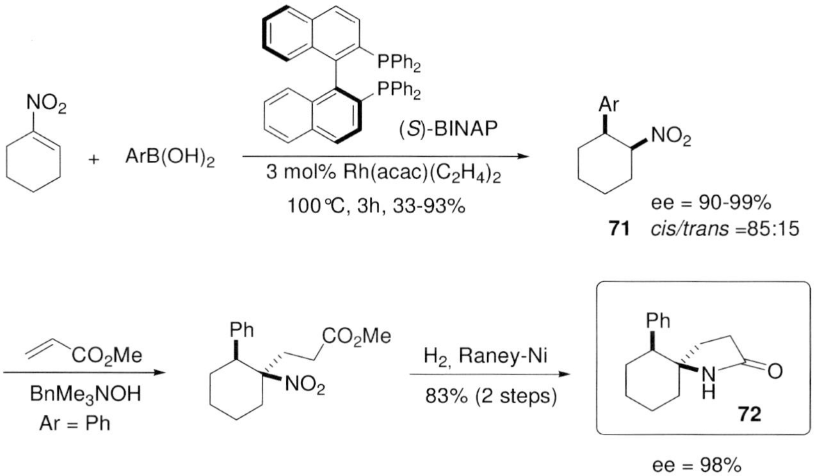

Scheme 3.36 Catalytic enantioselective conjugate additions of arylboronic acids to nitroalkenes.

in the reaction with nitroalkenes, although cleavage of the chiral auxiliary requires oxidative conditions [71].

Organoboronic acids are particularly effective in rhodium-catalyzed conjugate additions that involve the transfer of alkenyl or aryl groups. These reagents are fairly stable to oxygen and moisture, permitting the reactions to be run in aqueous solution. Rhodium complexes such as $Rh(acac)(C_2H_4)_2$ are usually associated with chiral BINAP ligands and require high temperatures (90–100 °C) to afford the desired addition products **71**. Conjugate addition of the aryl moiety to nitrocyclohexene occurs with high enantiofacial discrimination but moderate *cis* stereoselectivity, because protonation of the intermediate nitronate anion is favored on the opposite side with respect to the aryl group (Scheme 3.36) [72]. Treatment of nitrocyclohexane **71** with methyl acrylate affords the corresponding adduct with outstanding diastereoselectivity, and this adduct affords the spirobicyclo pyrrolidinone **72** in almost enantiopure form upon reduction and lactamization.

3.4.2
Pyrrolidines

All the synthetic procedures described for the preparation of pyrrolidinones could in principle be used to access pyrrolidines by further reduction of the lactam moiety [73]. The key step in the synthesis of Sch 50971, an agonist of the histamine H_3 receptor, involves the diastereoselective addition of a chiral *N*-propanoyloxazolidin-2-one titanium enolate to the imidazolyl nitroolefin **73** (Scheme 3.37) [74]. After reduction and cyclization to the corresponding pyrrolidinone **74**, the lactam group is further reduced with $LiAlH_4$ to afford Sch 50971.

Most of the reducing agents needed to transform γ-lactams into pyrrolidines (e.g., $LiAlH_4$, borane, etc.) are poorly chemoselective, so alternative strategies to

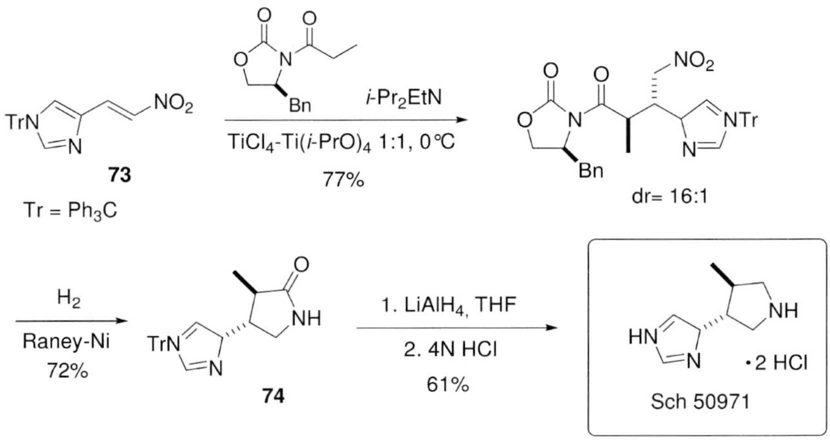

Scheme 3.37 Synthesis of the agonist of the histamine H3 receptor Sch 50971.

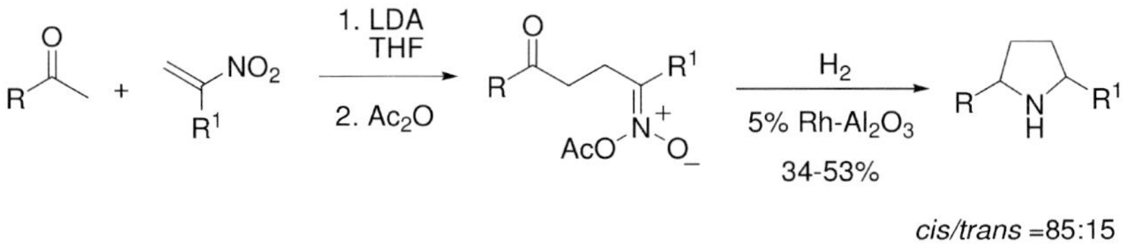

Scheme 3.38 Reactions between ketone enolates and nitroalkenes for the preparation of 2,5-disubstituted pyrrolidines.

produce these saturated heterocycles are often preferable. Reactions between ketone enolates and nitroalkenes give the corresponding nitronate anions, which are trapped as acetyl derivatives (Scheme 3.38) [75]. Catalytic hydrogenation of the acyl nitronates involves a series of transformations including reduction to the amino group, pyrroline formation by intramolecular condensation, and finally C=N reduction to the pyrrolidine ring.

Kainic acids exhibit interesting neurophysiological activity and their stereoselective synthesis has been the subject of considerable investigation. Treatment of the potassium enolate of α-ketoglutaric acid dimethyl ester with a nitrostyrene derivative produces the expected adduct **75** with favorable stereochemistry, while the lithium enolate fails in the same process because of its instability (Scheme 3.39) [76].

The hydroxylamine formed upon reduction of nitro group with ammonium formate reacts with the keto group, resulting after acid treatment in a cyclic nitrone **76**. Unfortunately, reduction of the nitrone into 1-pyrrolidinol **77** favors the unwanted 2,3-*cis* stereoisomer, though this can easily be separated from its epimer. Reduction of the pyrrolidin-1-ol and Cbz protection of the resulting pyrrolidine, however, are carried out on the epimeric mixture and finally the "wrong" stereo-

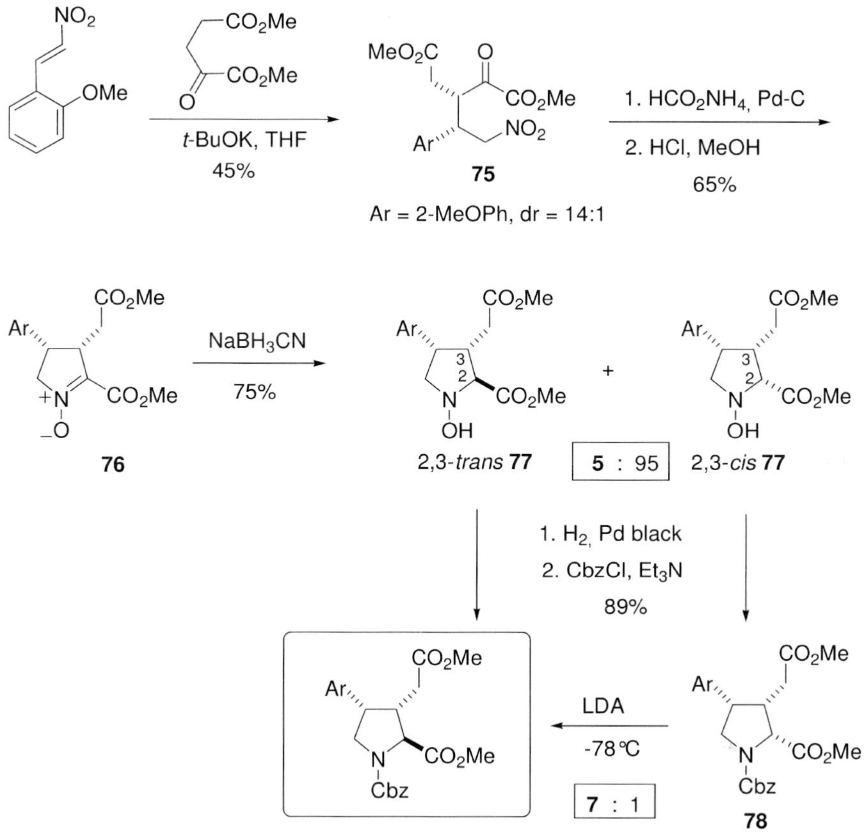

Scheme 3.39 Synthesis of kainic acid derivatives.

isomer **78** can be successfully equilibrated to the desired product under basic conditions.

Proline and its derivatives can be used as organocatalysts in Michael additions of aldehydes and ketones to nitroalkenes (Scheme 3.40). It is curious to observe that in these procedures chiral pyrrolidines are also used as a key to the preparation of other optically active pyrrolidines. The catalyst probably acts by forming an intermediate chiral enamine with the carbonyl derivative, which is the actual reagent in the reaction with the nitroalkene.

The chemical yields of these additions, as well as the observed *syn* diastereoselectivity, are generally satisfactory, while the enantioselectivity strongly depends on the nature of the catalyst employed. L-Proline gives unsatisfactory results in reactions between ketones and nitroalkenes [77], while (*S*)-2-(morpholinomethyl)-pyrrolidine (**79**) displays better levels of enantioselectivity for the same process with aldehydes [78, 79]. As usual, reduction of the adduct directly affords the substituted pyrrolidine, which is protected at nitrogen before isolation.

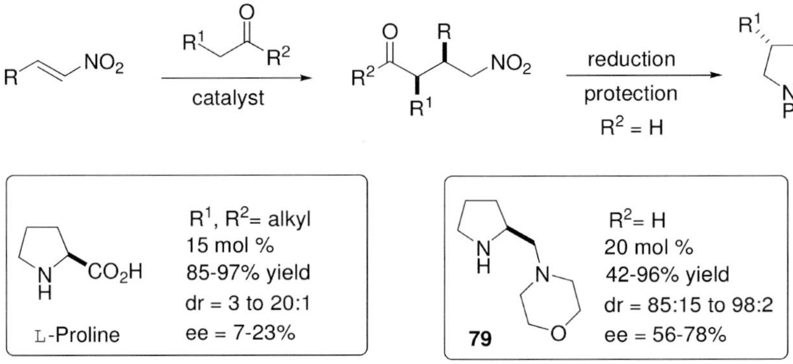

Scheme 3.40 Enantioselective synthesis of 3,4-disubstituted pyrrolidines.

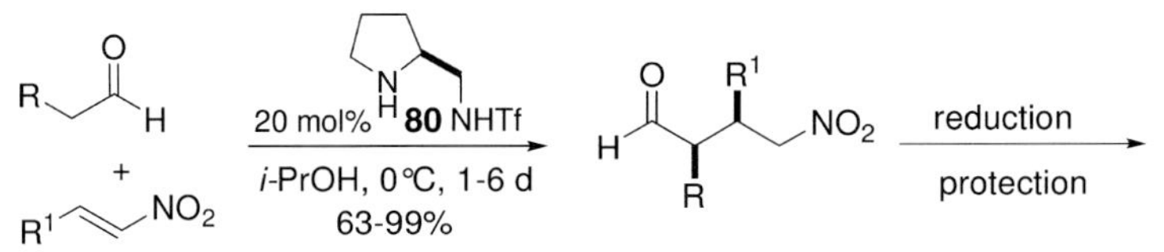

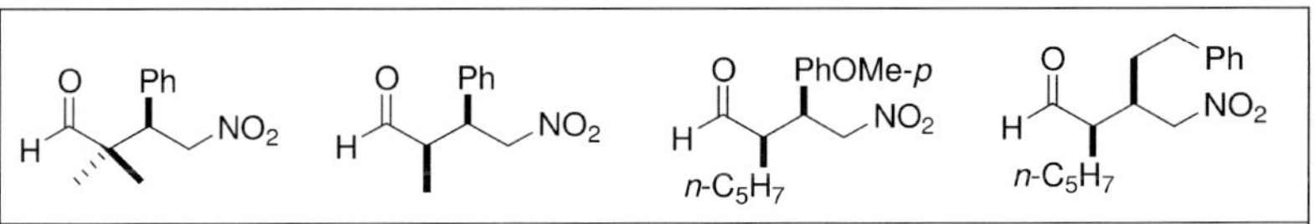

dr	-	12:1	20:1	50:1
ee%	90	97	99	22

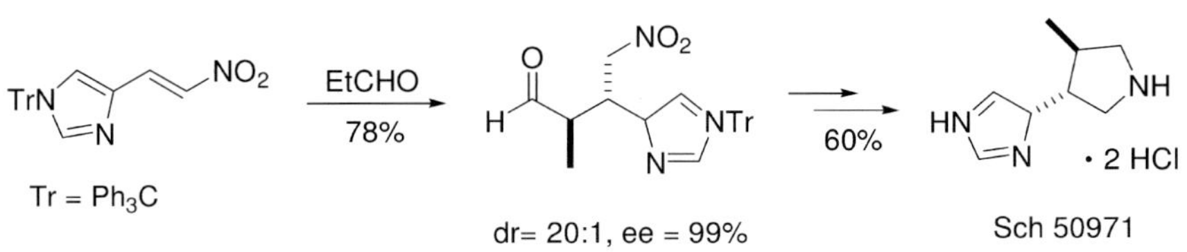

Scheme 3.41 Organocatalytic enantioselective conjugate additions of aldehydes to nitroalkenes.

Slight structural modifications of the organocatalyst often give rise to consistent improvements in the stereoselectivity of the process. Pyrrolidine **80** is remarkably efficient in promoting the conjugate addition of aldehydes to aryl nitroalkenes while being poorly effective with alkyl nitroalkenes (Scheme 3.41) [80].

The synthesis of Sch 50971 by this procedure is considerably more efficient than that previously portrayed in Scheme 3.37. Application of this method to cycloalkanones gives excellent results (*dr* = 30 to 50 : 1; *ee* = 88–99%), but only modest levels of stereoselectivity (*dr* = 3 to 50 : 1; *ee* = 53–99%) are obtained with acyclic ketones.

Readily enolizable compounds such as β-ketoesters react with nitroalkenes in the presence of catalytic amounts of chiral bisoxazoline-magnesium triflate complex and *N*-methylmorpholine (NMM) as co-catalyst (Scheme 3.42) [81].

The β-hydrogen residue present in the adduct is too acidic to allow configurational stability at the corresponding carbon, but selectivity at the other stereocenter is good. This procedure is amenable to scaling-up to 13 moles and can be employed for the synthesis of phosphodiesterase type 4 IC86518 [82], (*R*)-rolipram, and endothelin-A antagonist ABT-546 [83]. The last compound is prepared by coupling a suitable β-ketoester with a nitroalkene to afford the expected adduct **81** with 88%

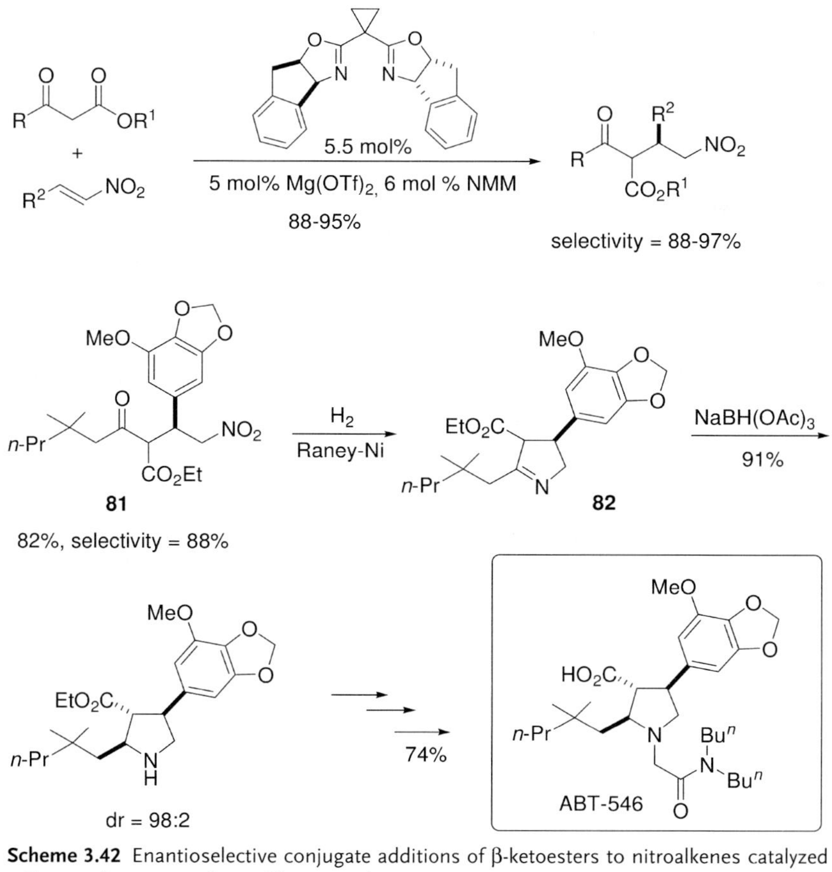

Scheme 3.42 Enantioselective conjugate additions of β-ketoesters to nitroalkenes catalyzed by a bisoxazoline-magnesium triflate complex.

selectivity. Reduction of the nitro group then gives pyrroline **82**, together with lesser amounts of the parent nitrone. Stereoselective reduction of this mixture is best effected with sodium triacetoxyborohydride under acid conditions, giving a pyrrolidine system that is converted into ABT-546 by simple synthetic manipulations.

As electron-poor olefins, nitroalkenes are formidable dienophiles in Diels–Alder reactions producing six-member ring systems. However, a closer inspection of the structural features of nitroalkenes soon reveals that these compounds can be also regarded as heterodienes that can be used in inverse electron demand [4+2] cycloadditions with electron-rich alkenes (Scheme 3.43) [84].

These Lewis acid-catalyzed cycloadditions of nitroalkenes with alkyl enol ethers occur with high regioselectivity and produce cyclic nitronates **83** that upon catalytic hydrogenation are converted into substituted pyrrolidines [85]. These transformations probably involve preliminary cleavage of the N—O bonds with subsequent reduction to the hydroxylamines and formation of aldehydes from their hemiacetals. Intramolecular reactions of these two groups give nitrones **84**, which are further reduced to the pyrrolidines.

The stereochemical outcomes of cycloadditions of (*E*)-nitroalkenes with enol ether **85**, bearing a chiral appendage on the oxygen, strongly depend on the natures of the Lewis acids used to promote the reactions (Scheme 3.44).

Titanium catalysts generally promote an *endo* approach of the alkene to the *re* face of the nitroolefin, while an opposite preference (*exo* approach to the *si* face of the nitroalkene) is observed with aluminium catalysts [86]. Reduction of the obtained cyclic nitronates affords optically active 3-substituted pyrrolidines with variable degrees of enantioselection, which are generally more satisfactory for aluminium-catalyzed reactions operating in the *exo* mode. The whole process represents an interesting example of *stereodivergent synthesis*.

This procedure can be also adopted for the synthesis of 3,4-disubstituted pyrrolidines through the use of chiral enol ether **86**, which affords the corresponding cyclic nitronate with high regio- and stereoselectivity (Scheme 3.45).

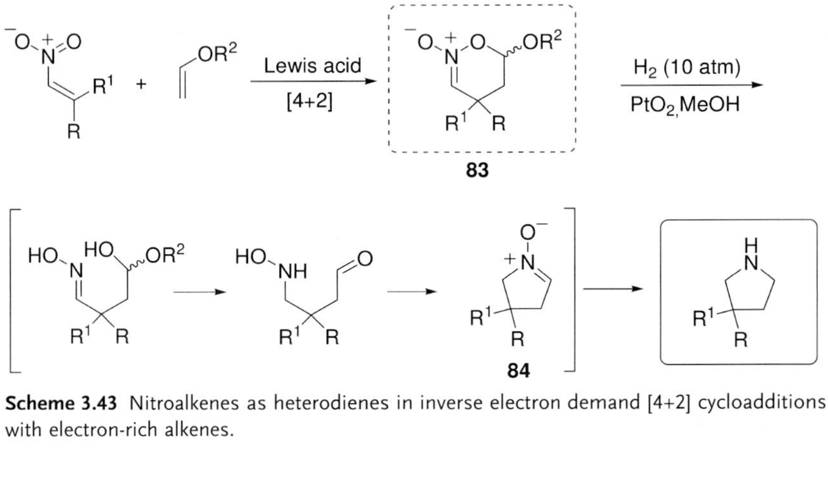

Scheme 3.43 Nitroalkenes as heterodienes in inverse electron demand [4+2] cycloadditions with electron-rich alkenes.

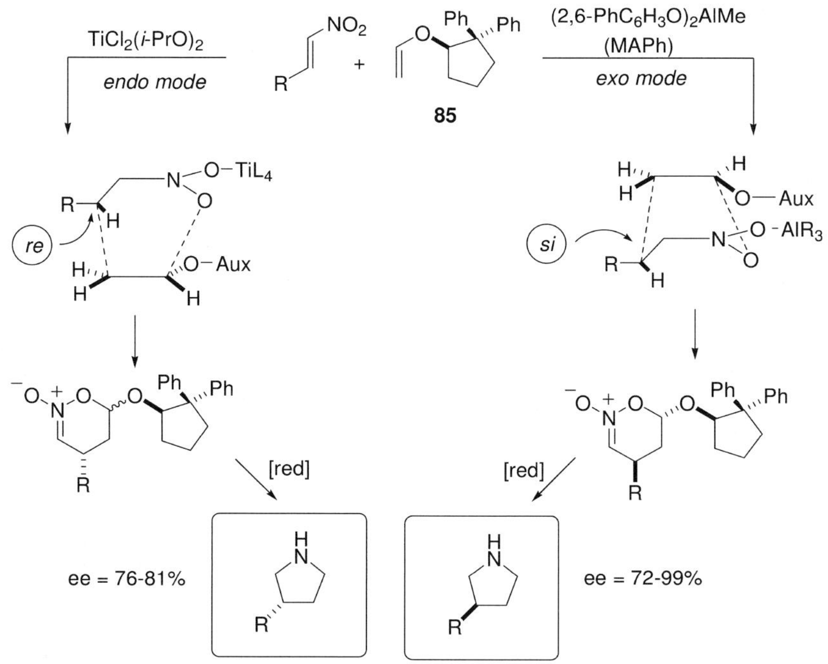

Scheme 3.44 Stereodivergent synthesis of 3-substituted pyrrolidines.

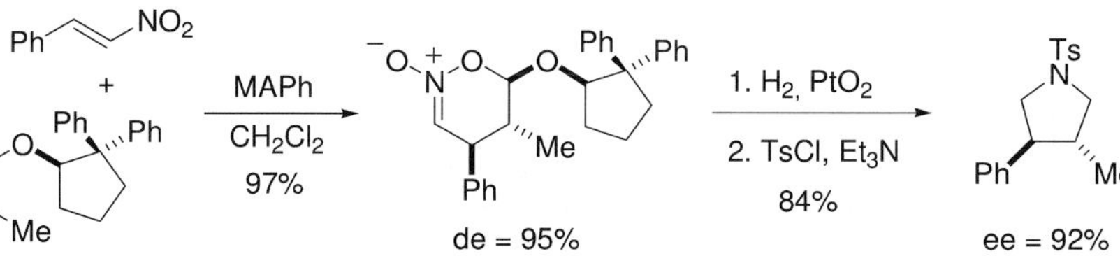

Scheme 3.45 Cycloaddition between a stereodefined chiral enol ether and β-nitrostyrene.

None of the previously used Lewis acids is effective when 2-acyloxyvinyl ethers are used in cycloaddition reactions with nitroalkenes. Only $SnCl_4$ is able to promote reactions of 2,2-alkylarylnitroalkenes with vinyl ether **87**, by an *exo* approach that finally affords 3-hydroxy-4,4-arylalkylpyrrolidines **88** (Scheme 3.46) [87]. A reversal in the selectivity is observed with use of 2-cyclohexyl-1-nitroethene, which prefers to react with the same enol ether in *endo* fashion. This opposite selectivity can be explained in terms of a lower positive charge density on the nitrogen atom in nitrostyrenes with respect to 2-alkyl-1-nitroalkenes, which causes a reduced Coulombic interaction with the oxygen of the enol ether.

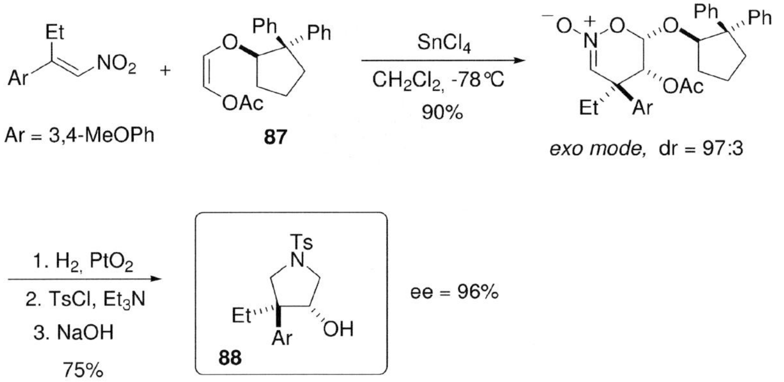

Scheme 3.46 Reactions between 2,2-alkylarylnitroalkenes and vinyl ether **87** following an *exo* approach.

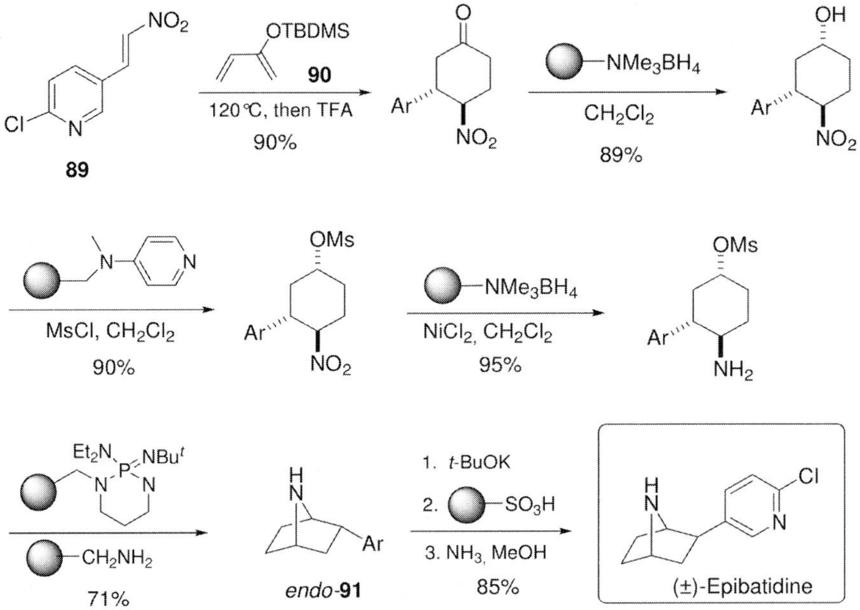

Scheme 3.47 Total synthesis of racemic epibatidine by use of polymer-bound reagents.

A simple Diels–Alder reaction between nitroalkene **89** and *O*-silylated-2-oxadiene **90** constitutes the first step toward the total synthesis of the analgesic alkaloid epibatidine (Scheme 3.47) [88].

The main feature of the proposed synthesis resides in the utilization, for the next steps, of polymer-supported (P-S) reagents, which avoids any chromatographic purification of the intermediates obtained. The stereoselective reduction of the obtained nitroketone is thus accomplished with (P-S) borohydride, while the subsequent mesylation of the resulting hydroxy group is promoted by (P-S) aminomethylpyridine. A (P-S) borohydride in the presence of $NiCl_2$ produces a reduction of the nitro group, while a (P-S) phosphazene base assists the key trans-annular cyclization more efficiently than the usual thermal conditions. Since the obtained bicyclic derivative **91** is the wrong (*endo*) isomer, it must be epimerized at the α-pyridyl proton to give racemic epibatidine.

The assembling of the nitrocyclohexanone unit can be also achieved by an enantioselective Michael addition in the presence of a chiral thiourea organocatalyst, followed by a further conjugate addition of the resulting intermediate **92** (Scheme 3.48) [89, 90].

Allylic ester cleavage and decarboxylation leaves a nitrocycloalkanone that is stereoselectively reduced to cycloalkanol **93** with L-Selectride®. The correct stereochemistry of the nitro group is achieved through a preliminary elimination of the methoxy group and a subsequent diastereoselective reduction of the formed nitrocyclohexene with sodium cyanoborohydride. The obtained nitrocyclohexanol **94** is transformed into (–)-epibatidine by simple procedures.

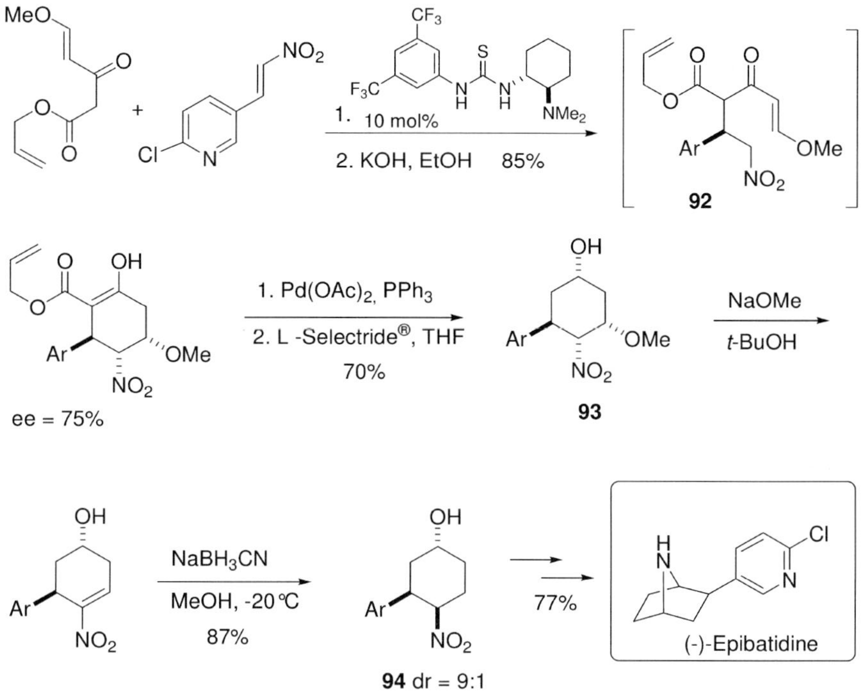

Scheme 3.48 Enantioselective total synthesis of (–)-epibatidine.

3.5
Piperidines and Piperazines

Formation of the piperidine nucleus would require a preliminary addition of a homoenolate anion to a nitroalkene in order to form a 1,5-difunctional derivative. Since the availability of synthetic equivalents of homoenolate anions is not so wide as for enolates, the number of procedures for the preparation of piperidines by use of nitroalkenes is quite small. Configurationally stable chiral aza-homoenolate anions are accessible by lithiation of *N*-Boc allylic amines in the presence of (–)-sparteine (**95**; Scheme 3.49) [91, 92].

Treatment of these anions with nitroalkenes provides the corresponding nitroenamines **96** with satisfactory *syn* selectivities and enantiomeric ratios. Hydrolysis of the enamino groups and oxidation of the intermediate aldehydes gives carboxylic acids, which after reduction of the nitro groups and esterification afford a piperidin-2-one systems **97**. Finally, reduction of the lactam groups and protection of the free secondary amines results in the formation of 3,4-disubstituted piperidines.

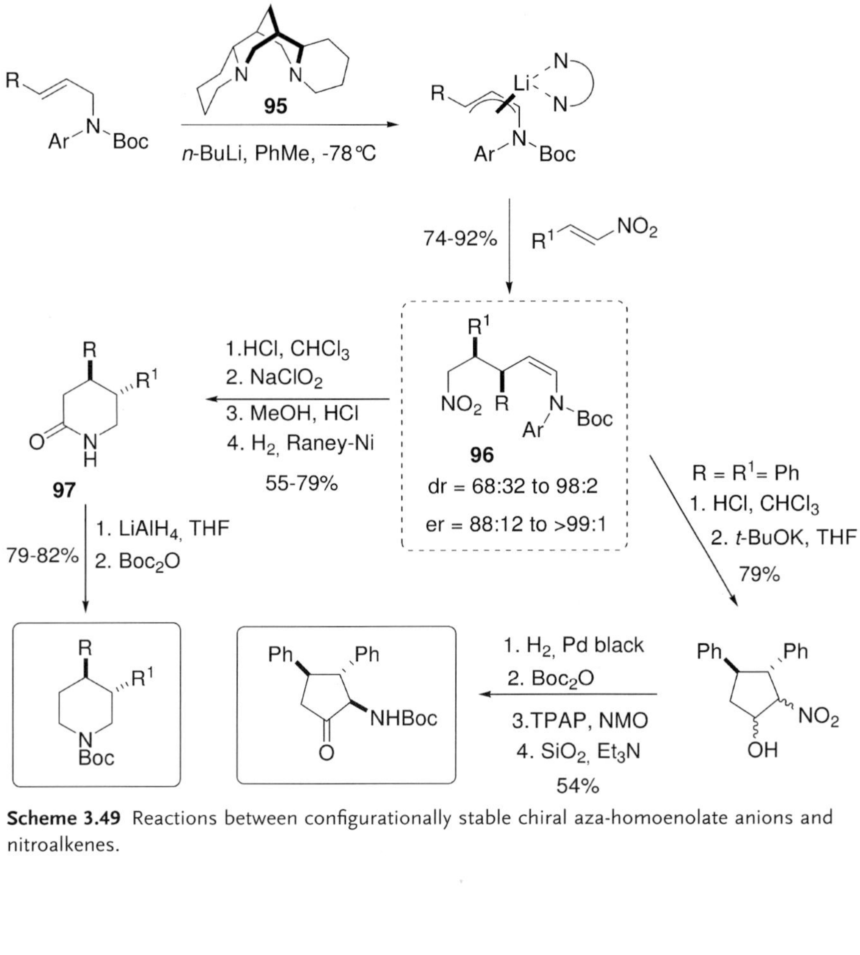

Scheme 3.49 Reactions between configurationally stable chiral aza-homoenolate anions and nitroalkenes.

Alternatively, the nitro aldehydes obtained by acidic cleavage of **96** can undergo intramolecular nitroaldol reactions under basic conditions. Reduction/protection of the nitro groups and oxidation of the alcoholic functions provides a strategy for the preparation of chiral 2-aminocyclopentan-1-ones [93].

The unavailability of the other enantiomer of (–)-sparteine makes this procedure unsuitable for the preparation of both enantiomeric forms of chiral 3,4-disubstituted piperidines. This drawback can be circumvented by an indirect method, as described for the synthesis of the antidepressant agent (+)-femoxetine (Scheme 3.50).

The chiral piperidin-2-one **98**, obtained by treatment of a lithiated allylamine with nitroethylene, is converted into the corresponding enolate and then diastereoselectively carboxylated. Reduction of the ester and lactam groups, followed by protection exchange, affords alcohol **99**, which upon mesylation and etherification gives the expected product with the opposite configuration to that obtained from piperidin-2-one **97**.

Preparation of piperazines usually entails alkylation of 1,3-diamino derivatives with 1,2-dihaloalkanes [94]. Optically active 2-piperazinones can be obtained in a two-step procedure starting from natural amino acid esters (Scheme 3.51) [95]. Because of the known instability of nitroethylene, this reactive alkene can be

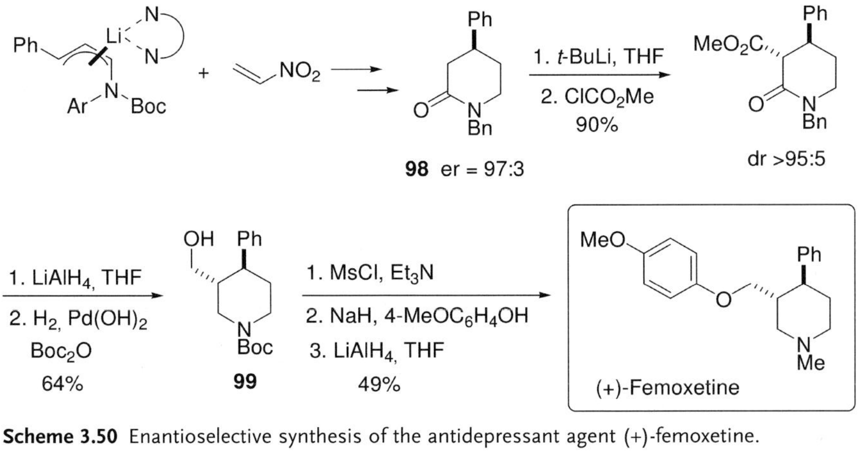

Scheme 3.50 Enantioselective synthesis of the antidepressant agent (+)-femoxetine.

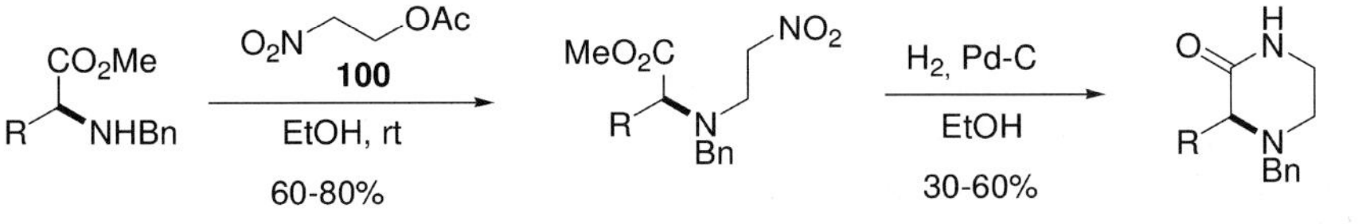

Scheme 3.51 Synthesis of optically active 2-piperazinones through reactions between amino acid esters and **100** as a nitroethylene equivalent.

profitably substituted by 2-acetoxy-1-nitroethane (**100**), which produces nitroethylene and undergoes reaction with the amino acid ester under the reaction conditions. Upon reduction of the nitro group the intermediate primary amine spontaneously cyclizes to the corresponding piperazinone.

3.6
Pyrrolizidines and Related Derivatives

The glycosidase inhibitory activity displayed by several alkaloids featuring pyrrolizidine units has stimulated a search for efficient synthetic procedures directed towards their preparation [96]. The reactivity of nitroalkenes as heterodienes in inverse-electron-demand Diels–Alder reactions is profitably used for the construction of the pyrrolidine ring but can be also adapted to the synthesis of more complex polycyclic derivatives incorporating nitrogen. As previously described, [4+2] cycloadditions between nitroalkenes and chiral enol ethers generate cyclic nitronates, which upon reductive cleavage give pyrrolidine systems. These cyclic nitronates, however, are reactive 1,3-dipoles for [3+2] cycloadditions with various alkenes such as dimethyl maleate, producing the corresponding bicyclo derivatives (Scheme 3.52) [97, 98].

The 1,3-dipolar cycloadditions generally occur with *exo* selectivities, favored by steric effects and particularly high with (*Z*)-disubstituted dipolarophiles [99]. Reduction of the N—O bonds and cyclization of the resulting intermediates gives pyrrolizidinones, which can be converted into dihydroxylated necine bases through further synthetic transformations. Regiochemical control in [3+2] cycloadditions of nitronates may be difficult when using oxygenated dipolarophiles that are necessary for suitable introduction of hydroxy groups at C-1 in the pyrrolizidine ring. Trialkylsilyl substituents are known to undergo Tamao–Fleming oxidation, thus acting as surrogates for hydroxy groups in several synthetic processes. Five of the six stereocenters required for the synthesis of the pyrrolizidine alkaloids (+)-7-epiaustraline [100, 101] and (+)-causarine [102, 103] are created in a tandem [4+2]/[3+2] cycloaddition involving nitroalkene **101**, a chiral enol ether, and a (*Z*)-vinylsilane **102** (TDSO = dimethylthexylsilyloxy) (Scheme 3.53).

The substituents at C-4 in the cyclic nitronates play fundamental roles in determining the correct facial approach of the dipolarophile. Stereoselective reduction of the carbonyl groups and mesylation must precede the cleavage of the bicyclic rings to afford the pyrrolizidine skeletons, and stereospecific peracid oxidation of the carbon-silicon bonds introduces a further hydroxy group in the molecular structure of each final product.

The absolute stereochemistry at C-7 in the final pyrrolizidine nucleus can be controlled by the nature of the *O*-substituent in the nitroalkene. Use of silyloxy nitroalkene **103** allows for the generation of a stereocenter with the opposite configuration (to that obtained with nitroalkene **101**) at C-7a in cycloadduct **104** (Scheme 3.54) [104].

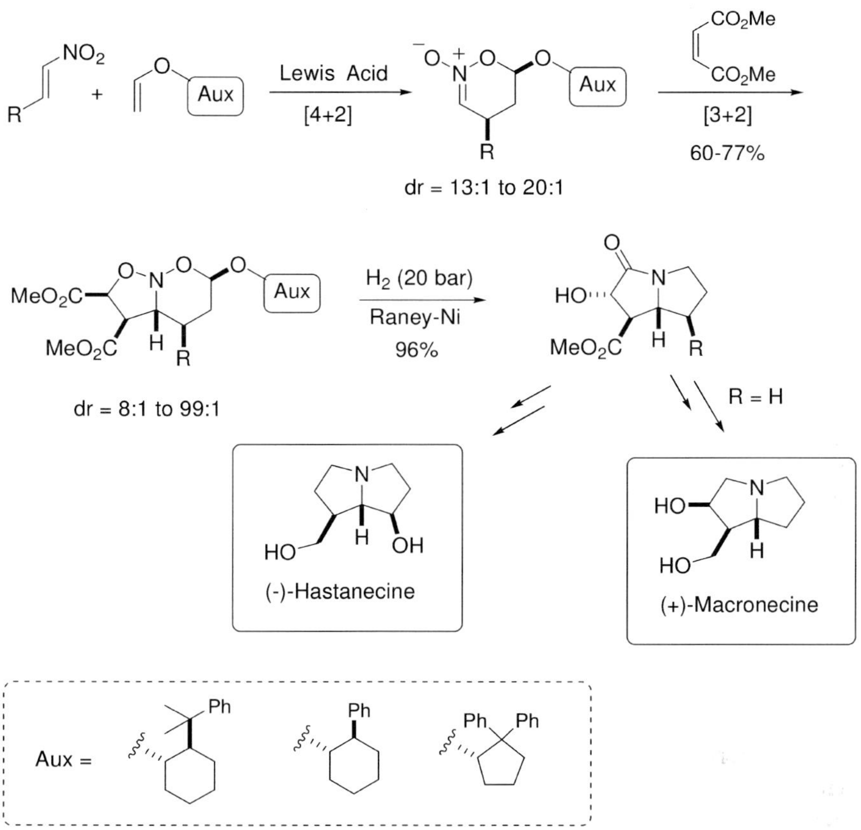

Scheme 3.52 Synthesis of optically active pyrrolizidine alkaloids by sequential [4+2]/[3+2] cycloadditions.

Furthermore, use of a butadiene-1-yl group linked to silicon allows the exploitation of a tandem *inter*molecular [4+2]/*intra*molecular [3+2] cycloaddition that forces the second process to occur with exclusive *endo* selectivity. Oxidation of the carbon-silicon bond in one of the final stages of the synthesis permits the correct stereochemistry between C-1 and C-7 to be introduced, resulting in the preparation of (+)-1-epiaustraline. A related approach that involves dihydroxylation of the vinyl group in the tricyclic intermediate **104** before reduction provides an efficient route to some indolizidine alkaloids of the castanospermine family [105].

Fumarate-derived nitroalkenes can be also involved in tandem *inter*molecular [4+2]/*intra*molecular [3+2] cycloadditions to afford tricyclic intermediates with satisfactory *exo* selectivity (Scheme 3.55) [106, 107].

The presence of the lactone carbonyl in cycloadduct **105** is problematic for the subsequent reductive rearrangement to the pyrrolizidine nucleus and it must be reduced beforehand to the corresponding lactol. Unfortunately, the utilization of

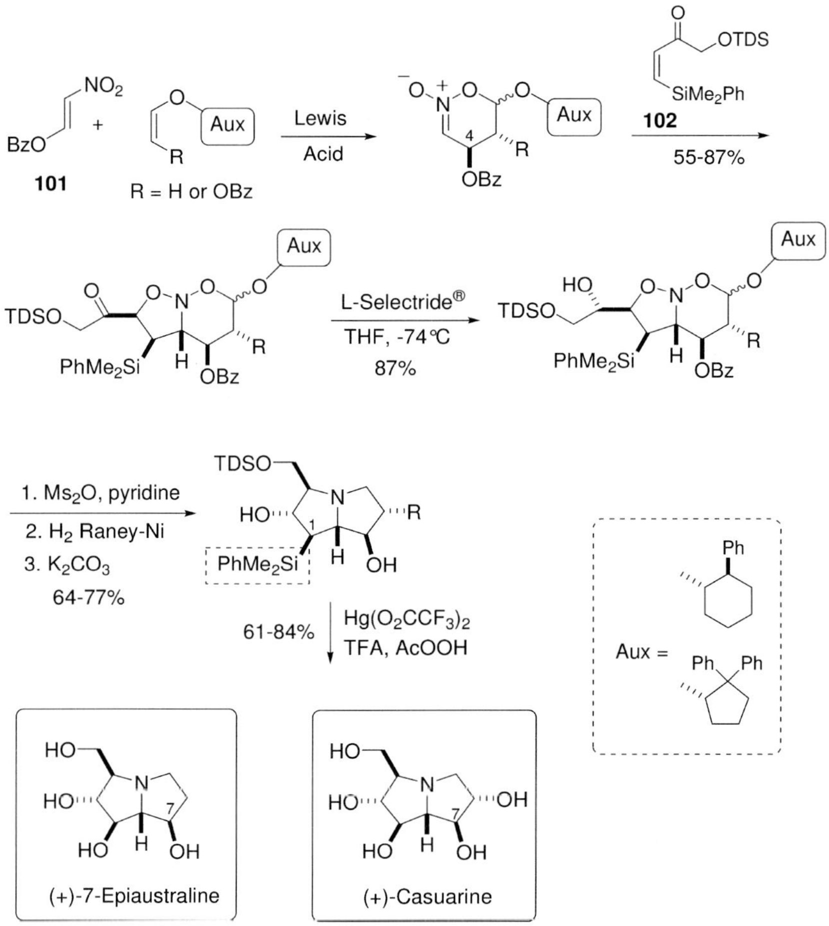

Scheme 3.53 A joint synthetic approach to the synthesis of (+)-7-epiaustraline and (+)-casuarine.

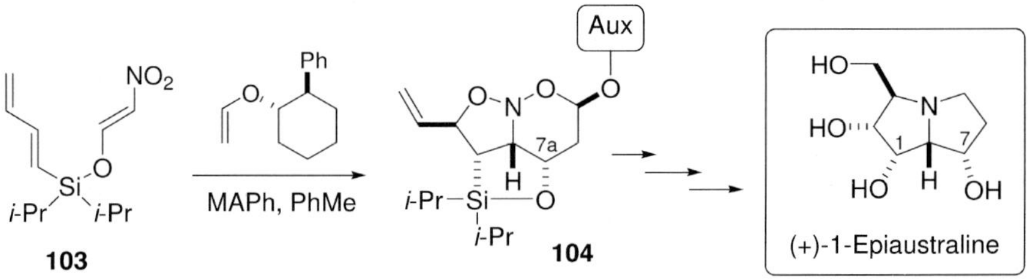

Scheme 3.54 *endo*-selective tandem intermolecular [4+2]/intramolecular [3+2] cycloaddition.

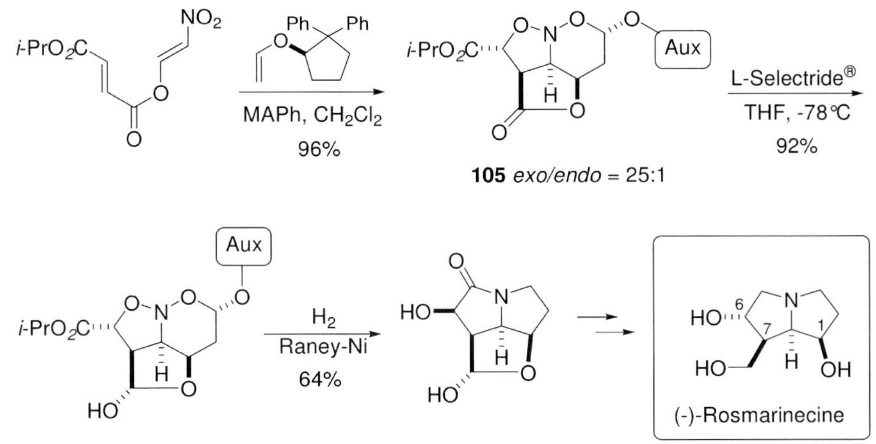

Scheme 3.55 *exo*-selective tandem intermolecular [4+2]/intramolecular [3+2] cycloaddition of a fumarate-derived nitroalkene.

the fumarate nitroalkene, necessary because of the instability of the stereoisomeric maleate, brings about the wrong stereochemistry at C-6 in the final target molecule. The correct configuration at this stereocenter is obtained through a Mitsunobu reaction, while other simple functional group transformations allow the total synthesis of (–)-rosmarinecine.

Barton radical deoxygenation at C-6 in the same intermediate can be used for the preparation of the 6-deoxy analogue of (–)-rosmarinecine – namely (–)-platynecine [108] – and a related procedure involving reactions between fumarate nitroalkenes and chiral 2-acetoxyvinyl ethers can furthermore be applied to the synthesis of (+)-crotanecine [109]. As previously discussed for the synthesis of pyrrolidines, the *exo/endo* selectivities in such intermolecular [4+2] cycloadditions are strongly affected by the natures of the Lewis acids used to promote the process [110]. Therefore, for reactions between nitroalkenes bearing enoate groups and chiral enol ethers, it is possible to select the appropriate stereoisomers simply by moving from titanium- to aluminium-based systems (Scheme 3.56) [111–113]. Reductive cleavage of each of the cycloadducts affords a pair of tricyclic derivatives that have an enantiomeric relationship.

The reactions between regioisomeric 2-nitroalkenes **106** and chiral enol ethers in the same tandem process afford spiro tricyclic nitroso acetals of variable ring size, depending on the number of methylene tethers present in the nitroalkene. The stereochemical outcome of each process is determined by the length of the tether and the configuration at the anomeric center (Scheme 3.57) [114]. Furthermore, better stereoselectivities are usually observed with nitroalkenes bearing enoate groups with the (*E*) configuration. Reduction of the nitroso acetals affords spirocyclic pyrrolizidinone derivatives as the final products.

Nitrocyclopentene derivatives are also profitable substrates in [4+2] cycloadditions with enol ethers. Only one diastereomer of the obtained couple of bicyclic

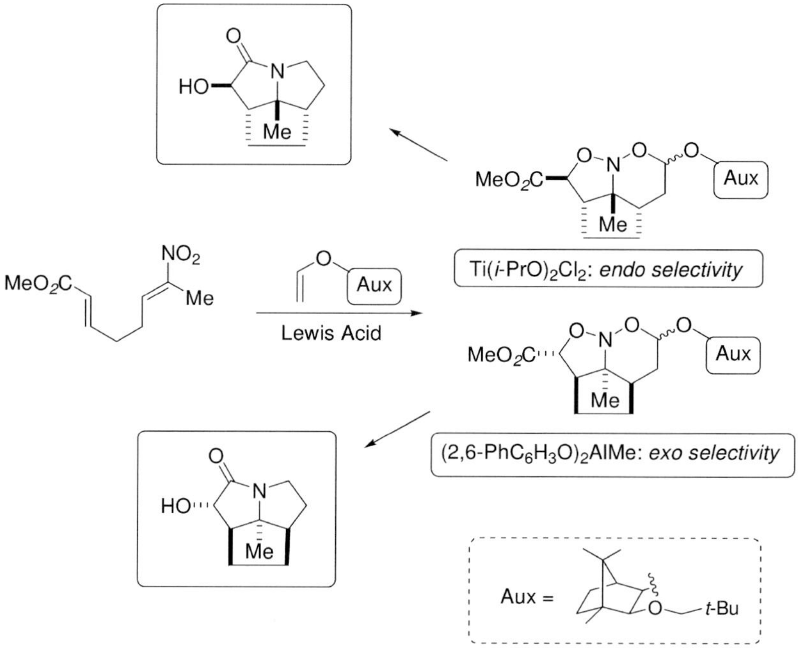

Scheme 3.56 Effect of the nature of the Lewis acid on the observed *exo/endo* selectivity.

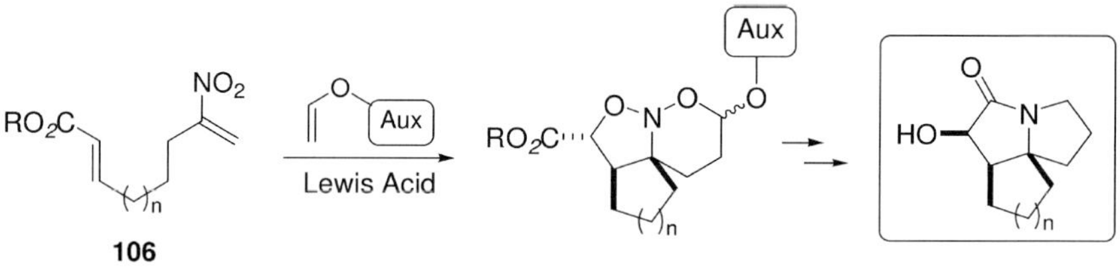

106

Scheme 3.57 Synthesis of spirocyclic pyrrolizidinone derivatives.

nitronates undergoes a spontaneous [3+2] reaction to afford nitroso acetal **107** (Scheme 3.58) [115]. Reductive rearrangement and functional group transformations on this derivative give 1-azafenestrane featuring the all *cis* configuration at the ring fusions.

The inclusion of a further unsaturation in the dienophile, as is the case in 2-alkoxy-1,4-dienes, allows tandem [4+2]/[3+2] cycloadditions in which bridged nitroso acetals **108** are formed after the intermediate cyclic nitronates have been heated at reflux for prolonged periods (Scheme 3.59) [116, 117]. Compounds **108**

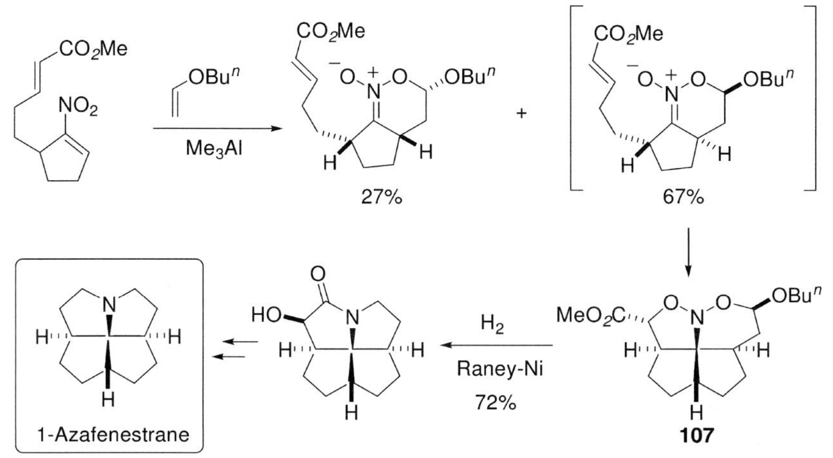

Scheme 3.58 Synthetic approach to the synthesis of 1-azafenestrane.

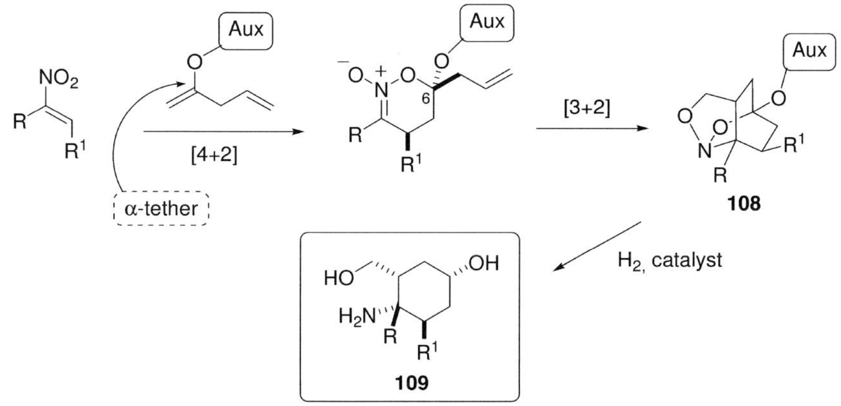

Scheme 3.59 Synthesis of aminocarbasugar derivatives.

are the products of α-tethered processes and upon reduction give amino cyclohexanols **109** closely reminiscent of aminocarbasugar compounds.

The utilization of 1-alkoxy-1,4-dienes for the same type of reaction gives 5-allylnitronates, cyclization of which occurs in a β-tethered mode to afford differently shaped bridged nitroso acetals **110** (Scheme 3.60) [118–120]. Conventional reductive cleavage of these compounds affords densely functionalized aminocyclopentanol units that are present in several pharmacologically active compounds such as glycosidase inhibitors and carbacyclic nucleosides.

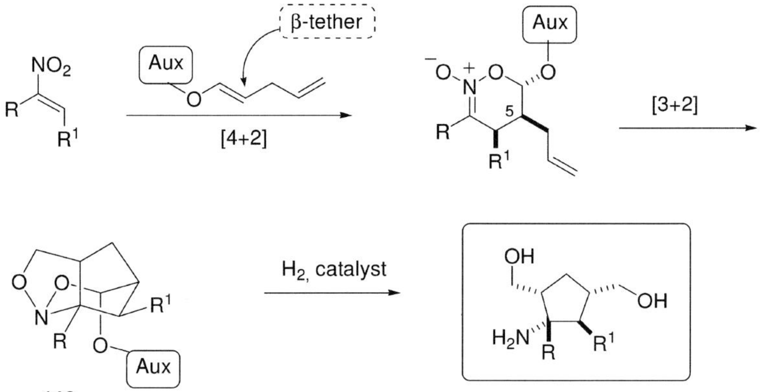

110

Scheme 3.60 Tandem [4+2]/[3+2] cycloadditions between 1-alkoxy-1,4-dienes and nitroalkenes.

3.7
Arene-Fused Nitrogen Heterocycles

3.7.1
Pyrroloindole Derivatives

A number of clinically important alkaloids used for treatment of glaucoma and myasthenia gravis feature the pyrroloindole nucleus. Most of them also contain chiral quaternary centers that can be generated through reactions between lithium enolates of 3-substituted 2-oxindoles and chiral nitroenamines (Scheme 3.61) [121, 122].

Complete reduction of the nitroalkene and methoxycarbonylation of the resulting primary amine affords an intermediate **111** that, on further reduction with $LiAlH_4$, gives (–)-esermethole, which can in turn easily be converted into its analogue (–)-physostigmine. A related procedure is also suitable for the preparation of racemic debromoflustramine B from 3-nitrovinylindole derivatives [123].

3.7.2
Carbolines and their Tryptamine Precursors

Carboline alkaloids are tricyclic derivatives usually found in plants and exhibit a wide range of biological activities. Most of the synthetic routes for the preparation of carbolines involve ring-closure of their tryptamine precursors in Pictet–Spengler reactions. These tryptamine systems can be prepared by nitroaldol reactions/eliminations starting from indole carbaldehydes and nitroalkanes (Scheme 3.62) [124, 125]. The obtained nitroalkenes can be completely reduced to the tryptamines by complex hydrides.

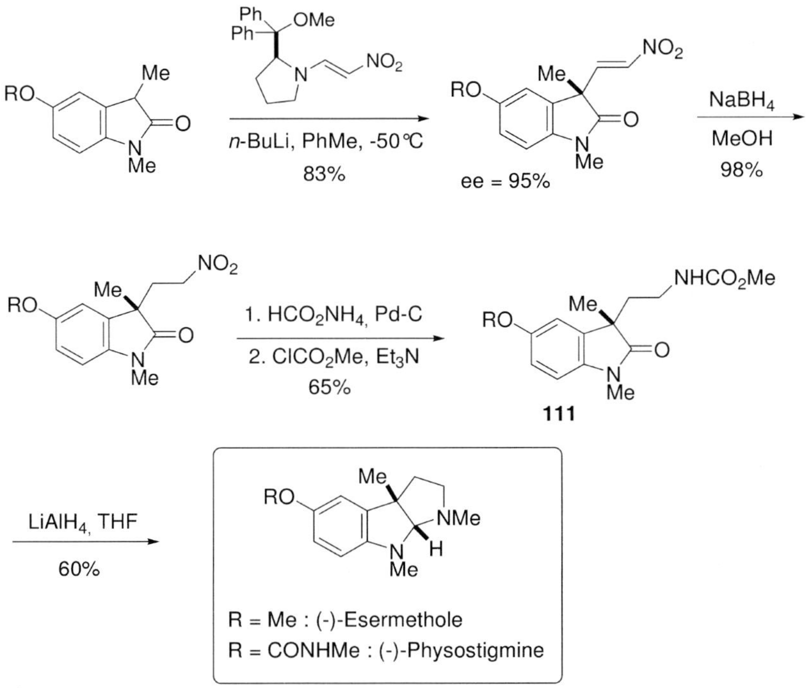

Scheme 3.61 Synthesis of (–)-esermethole and (–)-physostigmine drugs featuring pyrroloindole skeletons.

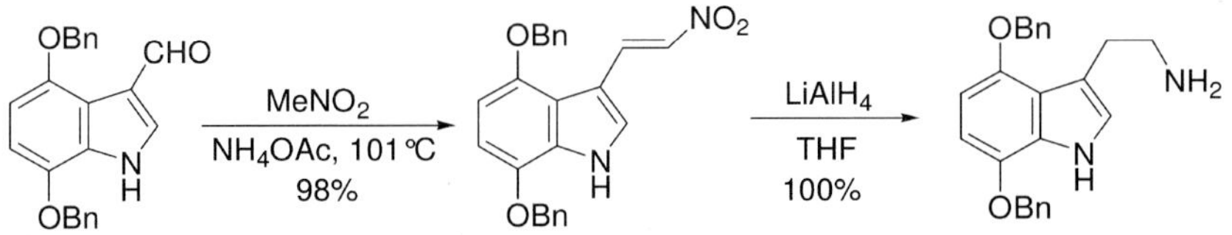

Scheme 3.62 Reduction of indolyl nitroalkenes for the preparation of tryptamines.

Acid promoted Friedel–Crafts reactions between indoles and nitroalkenes represent by far the most commonly employed procedure for the introduction of a nitroethyl framework at the 3-position of the indole nucleus. Both Brønsted and Lewis acid derivatives can be used to facilitate the nitroalkene addition to afford 3-indolyl derivatives that by reduction of the nitro group give the corresponding tryptamines **112** (Scheme 3.63).

The subsequent ring-closures to generate the tetrahydrocarboline skeletons can be carried out by treatment with aqueous formaldehyde under neutral conditions [126]. Alternatively, dihydrocarbolines can be obtained from the tryptamines by use of formyl derivatives such as glyoxylic acid or ethyl formate, which undergo decarboxylation or deethoxylation from the intermediate Pictet–Spengler products

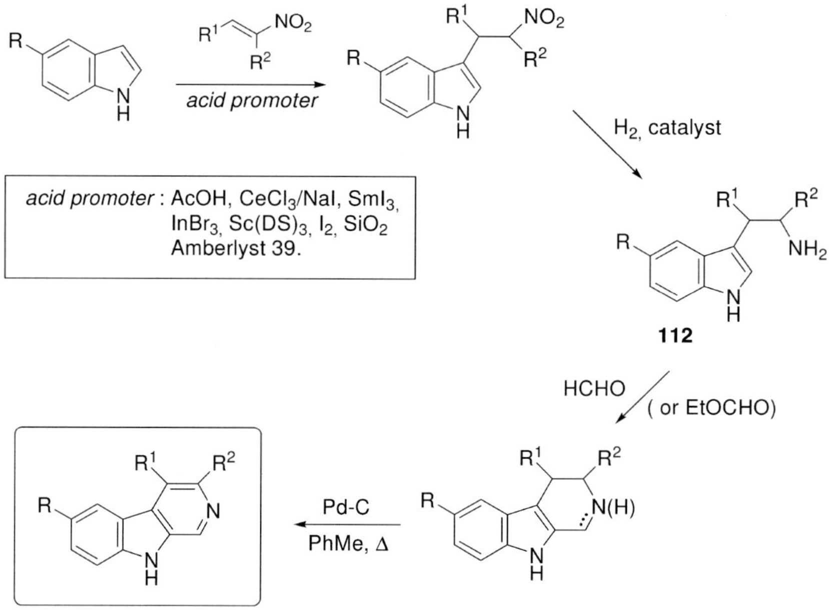

Scheme 3.63 Synthesis of carbolines through Friedel–Crafts reactions between indoles and nitroalkenes.

[127–129]. Final aromatization to the carbolines is achieved by dehydrogenation under catalytic conditions.

Enantioselective additions of indoles to nitroalkenes have been reported only very recently, through the use of different catalytic systems. Modest levels of enantioselection are observed with chiral salen-AlCl (*ee* = 28–48%) or 1,2-bis-sulfonamides (*ee* = 11–63%) [130, 131]. Chiral thiourea organocatalysts are more effective in promoting the same process to afford 3-indolyl derivatives **113** with satisfactory *ee*s (Scheme 3.64) [132].

The catalyst is believed to activate the nitroalkene through hydrogen bonding, while the free hydroxy group interacts with the NH of the indole, favoring nucleophilic attack on the *si* face of the nitroalkene. One of the adducts prepared by this procedure has been converted into the corresponding tryptamine by catalytic transfer hydrogenation and then transformed into a tetrahydro-β-carboline by a Pictet–Spengler reaction with benzaldehyde.

A complementary synthesis of the same adducts can be carried out through the generation of an indolyl anion by a suitable base, followed by the addition of the nitroolefin. This procedure is less widely exploited than the acid-catalyzed route and generally involves the utilization of Grignard reagents as basic promoters. Treatment of indole with a chiral nitroalkene in the presence of ethylmagnesium bromide affords the expected adduct **114**, and this is then converted after a few synthetic steps into the pharmacologically active alkaloid (−)-brevicolline (Scheme 3.65) [133].

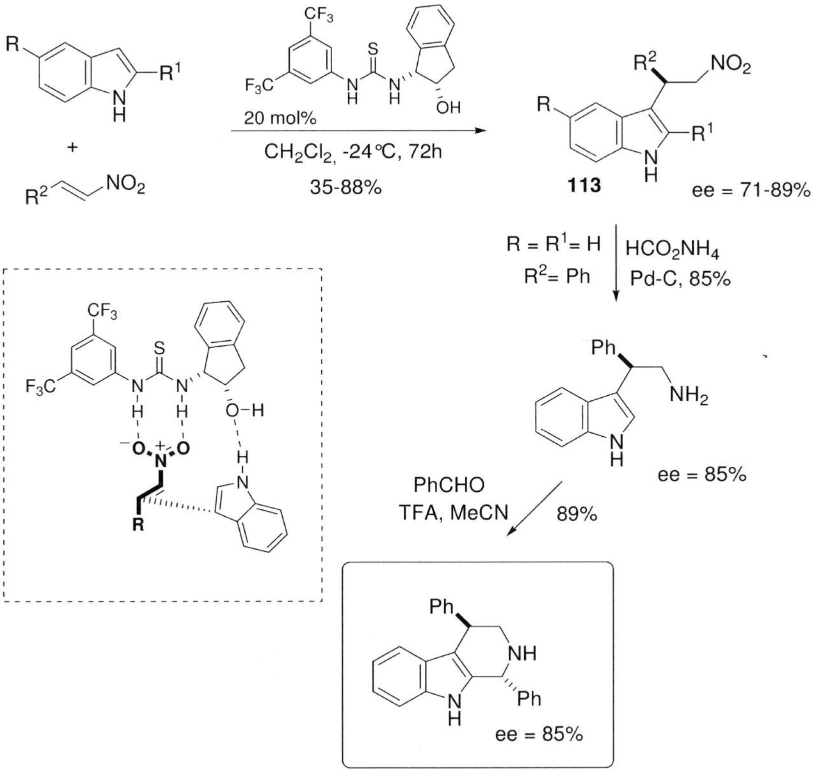

Scheme 3.64 Catalytic enantioselective Friedel–Crafts reactions between indoles and nitroalkenes promoted by chiral thiourea.

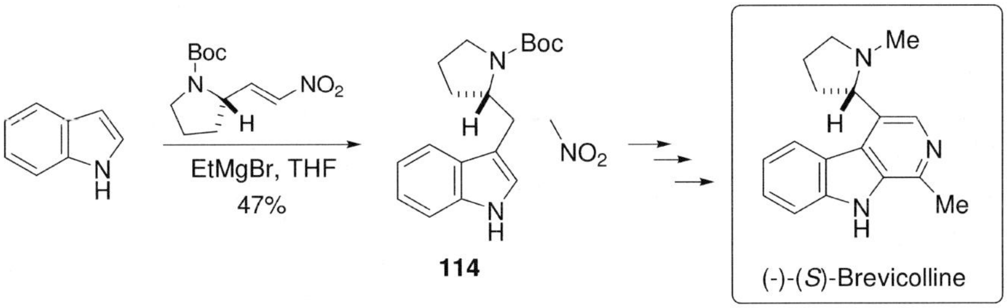

Scheme 3.65 Synthetic approach to optically active (–)-(*S*)-brevicolline.

3.7.3
Arene-Fused Piperidine Compounds

Tetrahydroisoquinoline derivatives can be prepared from nitroalkenes by different synthetic approaches. The high reactivity displayed by nitroalkenes toward addition by organometallic reagents can be suitably used to introduce different frameworks in the piperidine ring. The reaction between a nitrostyrene derivative and 4-methoxyphenylmagnesium bromide gives the corresponding adduct, which is reduced to the primary amine by treatment with $LiAlH_4$ (Scheme 3.66) [134]. An intramolecular Friedel–Crafts reaction induced by addition of formaldehyde occurs regioselectively on the dimethoxyaryl group and generates the piperidine ring, resulting in the synthesis of the natural alkaloid cherylline.

A direct procedure to build up nitrocyclohexene units involved the Diels–Alder reaction between a nitrostyrene and butadiene (Scheme 3.67) [135].

Chemoselective reduction of the nitro group in the presence of olefins can be achieved by use of aluminium amalgam. Protection of the amino group is instrumental for the subsequent dihydroxylation of the double bond but also serves for the subsequent ring closure to the piperidinone ring by exploitation of an electrophilic substitution promoted by trifluoromethanesulfonic anhydride (Tf_2O). The obtained hydroxylated heterocyclic derivative is a deoxy analogue of the anticancer drug pancratistatin.

3-Nitro-dihydronaphthalenes are often used as building blocks for ring construction to provide more complex structural entities. Organolithium reagents add to these nitroalkenes, but the process is not always efficient in terms of chemical yield [136, 137]. A remarkable result is obtained on treatment of the nitroalkene **115** with an aryl organolithium derivative prepared in the presence of the chiral ligand **116** (Scheme 3.68) [138, 139].

Although the initial diastereomeric ratio is unremarkable (*trans*/*cis* = 7 : 3), this value can be greatly improved by equilibration under basic conditions and, after *O*-deprotection and a single crystallization, the optically pure nitro alcohol **117** is obtained. Reduction of the nitro group and ring-closure under acidic conditions

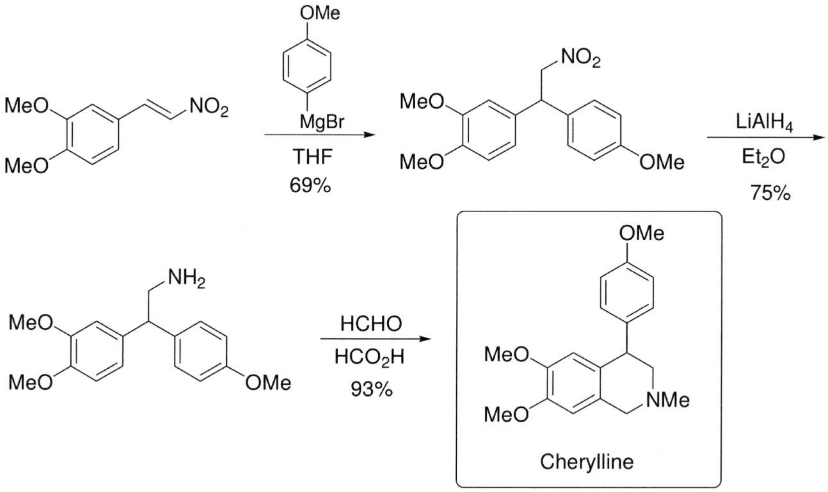

Scheme 3.66 Synthesis of the tetrahydroisoquinoline alkaloid cherylline.

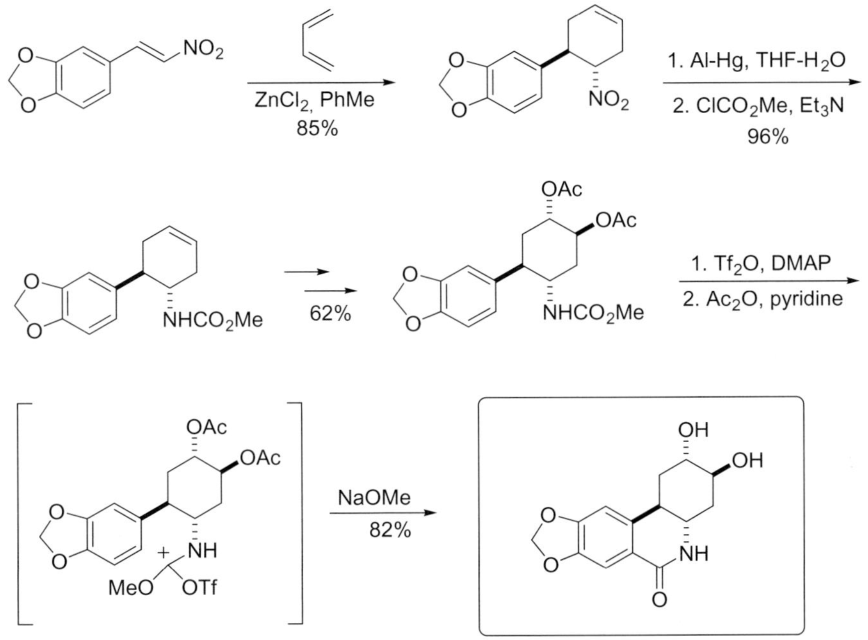

Scheme 3.67 Synthesis of a deoxy analogue of the anticancer drug pancratistatin.

affords a tetracyclic derivative that, after demethylation of the phenolic oxygens, affords the dopamine D1 full agonist (+)-dihydrexidine.

The superior electrophilic character displayed by nitroalkenes in comparison with other α,β-unsaturated compounds permits chemoselective addition of organometallic reagents when enoate systems are also present in the substrate. This approach forms the basis for some syntheses of the natural alkaloids lycoranes, promising drug candidates for treatment of Alzheimer's disease. Treatment of the nitroalkene **118** with the arylithium reagent **119** results in the exclusive attack at the nitroolefin moiety (Scheme 3.69) [140].

The subsequent intramolecular Michael addition to the resulting adduct **120** occurs efficiently but with poor diastereoselectivity, affording a nitrocyclohexanyl derivative **121**. Reduction of the nitro group, followed by lactamization and hydride reduction, generates perhydroindole **122**, and methoxycarbonylation of the pyrrolidine nitrogen provides the requisite substituent for a Pictet-type ring-closure to provide the lactam system, which is then reduced to β-lycorane. The stereoisomeric α-lycorane can be obtained from the minor isomer of the cyclohexanyl nitroester **121**, while utilization of *t*-butyl 2-(2-nitro-2-cyclohexenyl) acetate in the reaction with the same organolithium derivative provides a route to γ-lycorane [141].

Nitroallyl alkanoates act as multiple coupling reagents since their reactions with nucleophiles result in conjugate addition/eliminations that produce regioisomeric nitroalkenes suitable for second conjugate additions. In particular, treatment of

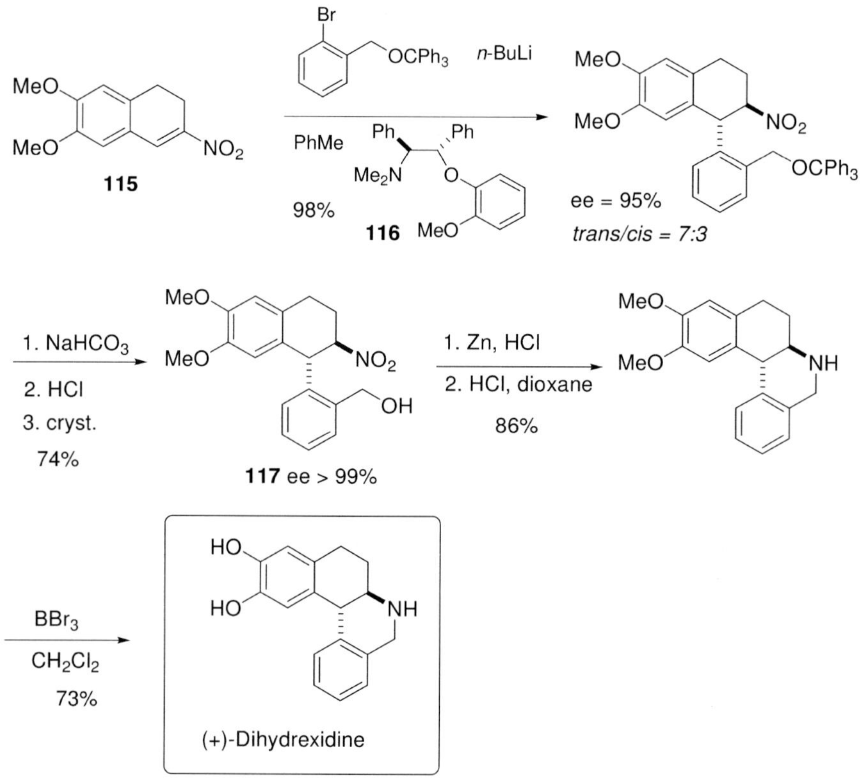

Scheme 3.68 Enantioselective synthesis of dopamine D1 full agonist (+)-dihydrexidine.

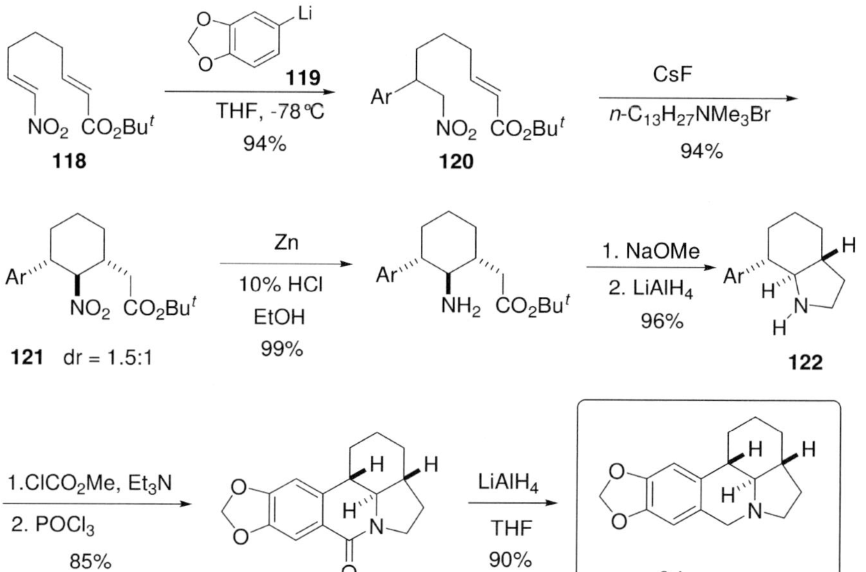

Scheme 3.69 Chemoselective conjugate addition of an aryllithium to a nitroalkene as a key step to β-lycorane.

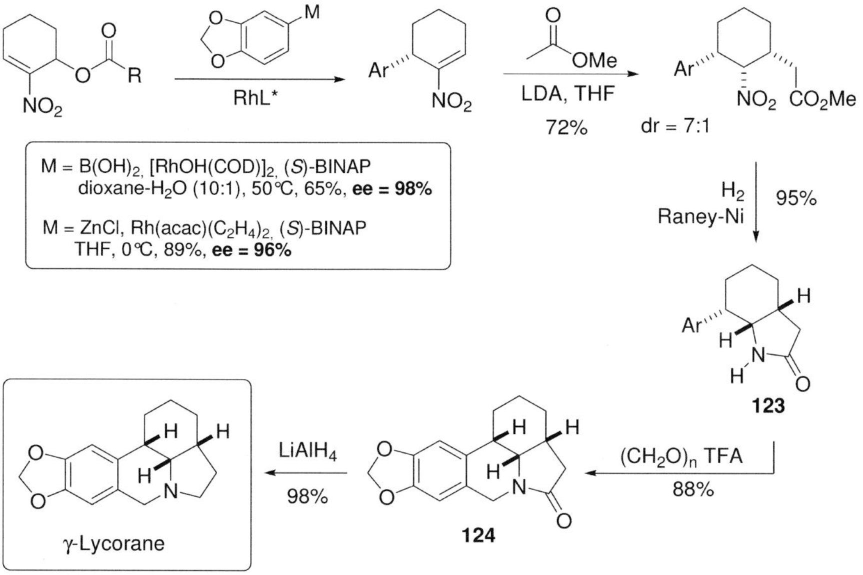

Scheme 3.70 Enantioselective conjugate addition of arylmetals to a nitroalkene as key step to chiral γ-lycorane.

2-nitro-2-cyclohexen-1-ol esters with arylboronic acids or arylzinc reagents in the presence of a rhodium complex and a chiral ligand generates substituted nitrocyclohexenes with a considerable degree of enantioselection (Scheme 3.70) [142, 143].

The resulting nitrocyclohexene derivative reacts with the lithium enolate of methyl acetate to give an all-*cis* adduct preferentially, and this affords a perhydroindolone system **123** upon reduction of the nitro group and ring-closure. The last cycle of the γ-lycorane skeleton is installed by treatment of **123** with paraformaldehyde and trifluoroacetic acid; reduction of the amido group in compound **124** then completes the synthesis of the desired alkaloid.

Cephalotaxine is a naturally occurring alkaloid that, while biologically inactive, displays potent antileukemic activity after esterification with some hydroxy acid derivatives. A Pd(II)-promoted [3+2] methylenecyclopentane anellation of nitrostyrene **125** with acetate **126** allows the first carbacyclic unit to be built up (Scheme 3.71) [144].

Interestingly, the subsequent Michael addition of the cyclic nitro compound to methyl acrylate occurs with high diastereoselectivity to afford a single isomer that is reduced at the nitro group and then spirolactamized. After some synthetic manipulations involving lactam reduction, double bond oxidative cleavage, and nitrogen derivatization, the obtained sulfoxide **127** undergoes a Pummerer rearrangement. The intermediate carbocation **128** reacts with the aromatic ring to form the benzazepinone ring, while functional group adjustment finally affords the target compound.

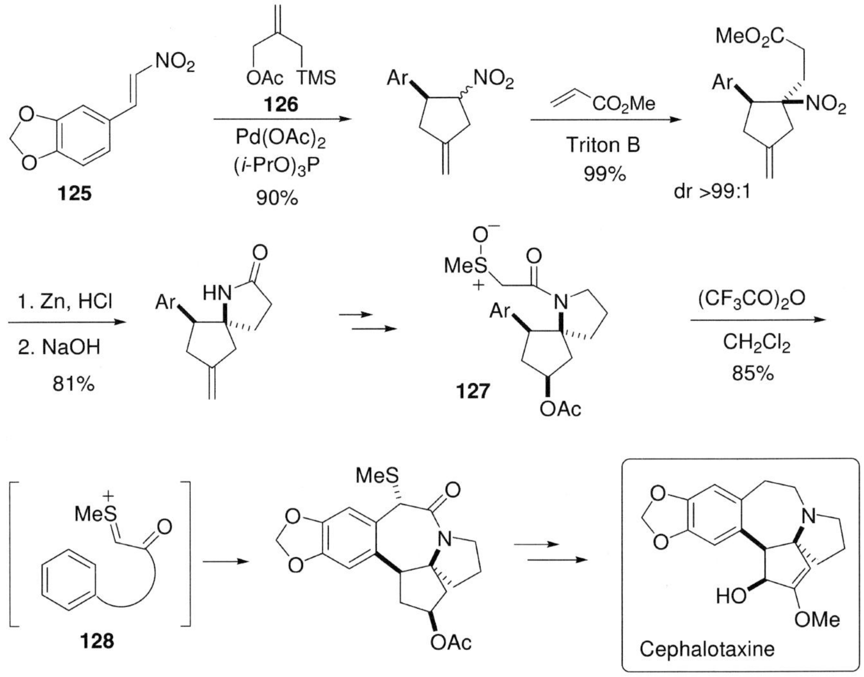

Scheme 3.71 Synthesis of the alkaloid cephalotaxine.

3.8 Other Polycyclic Derivatives

The azabornane skeleton is present in several substance-P antagonists and a general strategy for its synthesis involves the intramolecular ring-closure of suitable pyrrolidine derivatives. Tandem double conjugate additions of 4-benzylaminoenoate **129** with nitroalkenes provide a straightforward entry to 3-nitropyrrolidines **130**, featuring a 3,4-*cis* relationship (Scheme 3.72) [145].

As usual, the desired 3,4-*trans* isomers can easily be obtained by equilibration of **129** under basic conditions. Reduction of the nitro and ester groups by treatment with different reagents produces amino alcohols that are subsequently *N*-Boc-protected. Upon mesylation of the hydroxy groups, concurrent nucleophilic displacements occur, generating the ammonium salts of azabornane systems, which on hydrogenolytic debenzylation afford the free amino derivatives.

A Diels–Alder reaction between nitroalkenyl ester **131** and Danishefsky's diene represents the first step in the synthesis of the octahydroquinoline system **132**, which can be profitably employed for the preparation of 9-azasteroids (Scheme 3.73) [146]. Ketalization of the carbonyl group precedes the reduction of the nitro group, which, because of steric hindrance, cannot be reduced directly to the amine but is converted into the parent hydroxylamine by treatment with aluminium

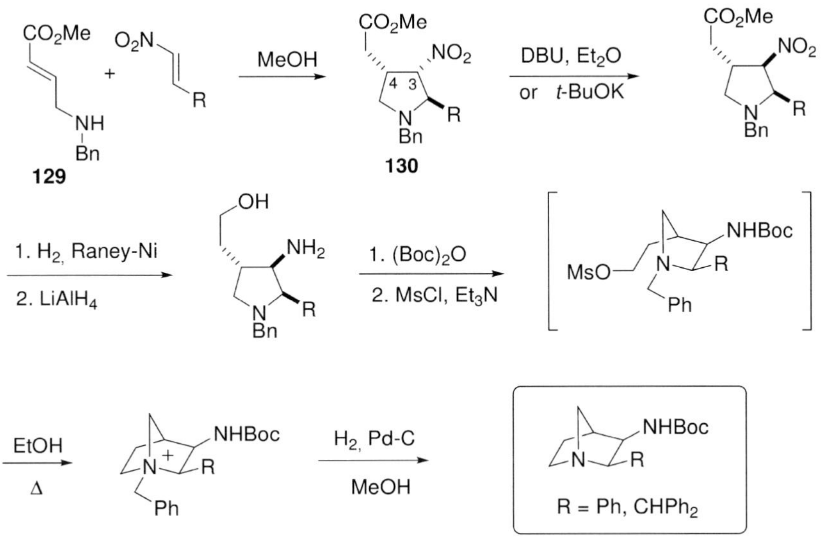

Scheme 3.72 Tandem double conjugate additions of 4-benzylaminoenoate with nitroalkenes.

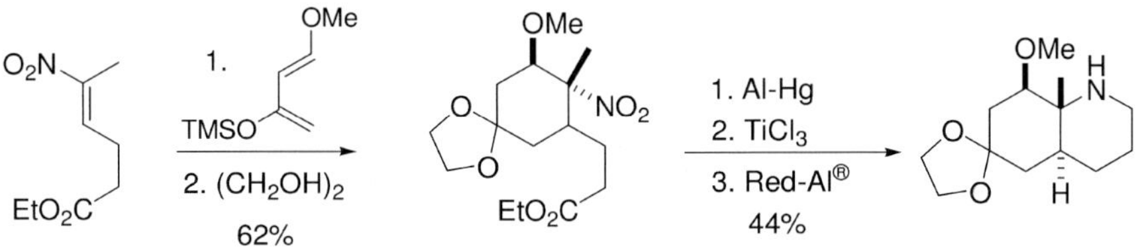

Scheme 3.73 Synthesis of octahydroquinoline **132**.

amalgam and then further reduced with $TiCl_3$. After a spontaneous cyclization, the lactam is reduced to the octahydroquinoline **132** under standard conditions.

The 2-azabicylo[3.3.1]nonane substructure **133**, featuring an exocyclic double bond, is a typical structural feature of *Strychnos* alkaloids, a well known class of physiologically active compounds. The planned synthetic route requires an aminocyclohexenyl unit that is obtained by a Diels–Alder cycloaddition between nitrostyrene and butadiene, followed by reduction of the nitro group and tosylation of the resulting amine (Scheme 3.74) [147]. Nitrogen alkylation with 2,3-dibromopropene provides the substrate for the subsequent Heck reaction, resulting in the formation of the desired bridged structure.

Formation of pyrrolizidine rings by means of *inter*[4+2]/*intra*[3+2] cycloadditions usually requires the presence in the starting nitroalkene of an enoate group, which is responsible for the lactam formation during the reductive cleavage of the cyclic nitroso acetal. When this function is replaced by a terminal enone protected at the carbonyl as a ketal, a tricyclic nitroso acetal **134** is formed under these conditions (Scheme 3.75). Catalytic hydrogenation of compound **134** results in the formation

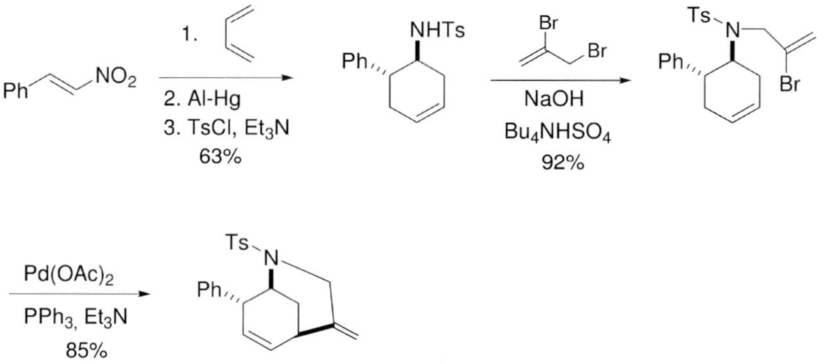

133

Scheme 3.74 Synthetic approach to the bridged substructure of *Strychnos* alkaloids.

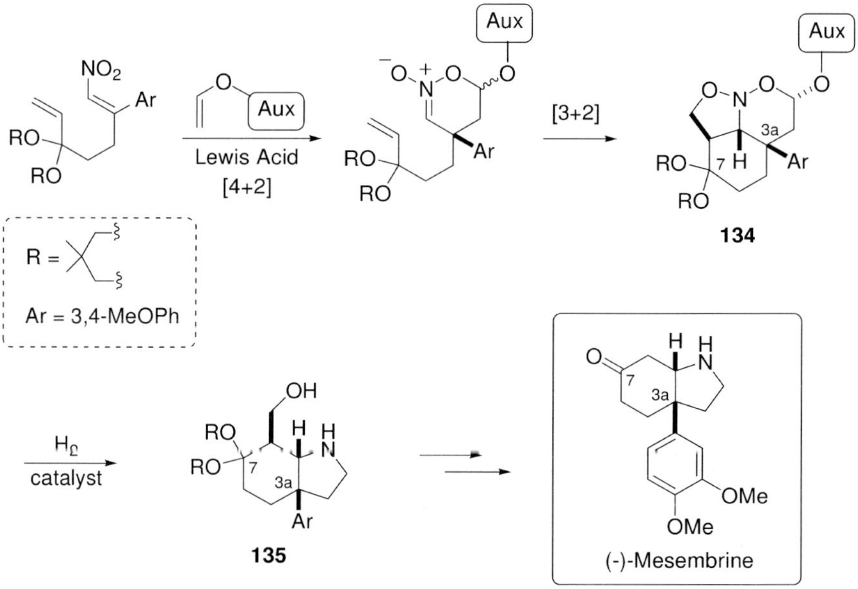

134

135

Scheme 3.75 Enantioselective synthesis of (–)-mesembrine.

of a octahydroindolone system **135**, present in many compounds of practical interest, such as the *Sceletium* alkaloid (–)-mesembrine [148].

As a further option among various tandem cycloadditions involving nitroalkenes as the diene elements, the *intra*[4+2]/*intra*[3+2] tandem cycloaddition has been recently introduced for the synthesis of polycyclic derivatives (Scheme 3.76) [149, 150].

Nitroalkene units are linked through suitable alkyl chains to isolated alkene and terminal enoate systems. The *intra*[3+2] cycloadditions occur in separate steps, upon heating of the reaction mixtures containing intermediates **136**, providing the

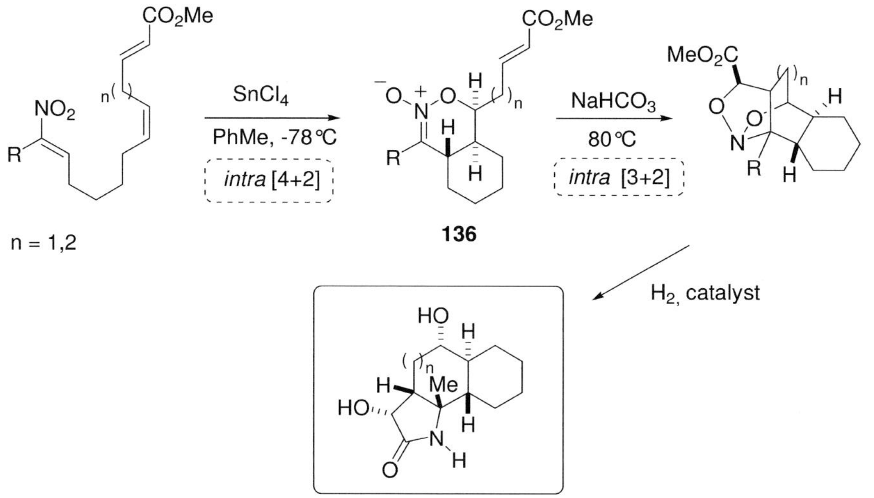

Scheme 3.76 $SnCl_4$-promoted intramolecular[4+2]/[3+2] tandem cycloaddition.

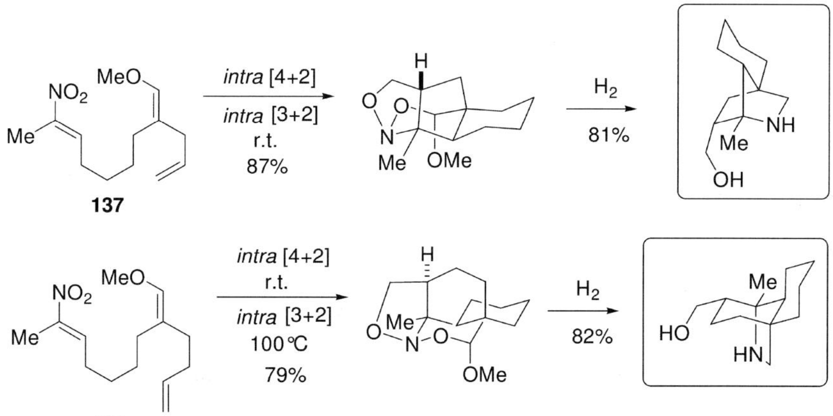

Scheme 3.77 Synthesis of bridged nitrogen heterocycles by intra[4+2]/intra[3+2] tandem cycloaddition.

expected nitroso acetals. Final reduction of the nitroso acetals affords tricyclic lactams with central rings of different sizes depending on the lengths of the alkyl tethers between the dienophile and the dipolarophile components.

The same process can be also carried out with the related substrates **137** and **138**, with different arrangements of the linkages between the nitroalkene components and the other reactive alkene counterparts (Scheme 3.77).

Formation of the cyclic nitroso acetal in the *intra*[3+2] process occurs rapidly at room temperature for the allyl tethered nitroalkene **137**, while heating at 100 °C is required for the homoallylic analogue **138**. Reductive cleavage of the nitroso acetals

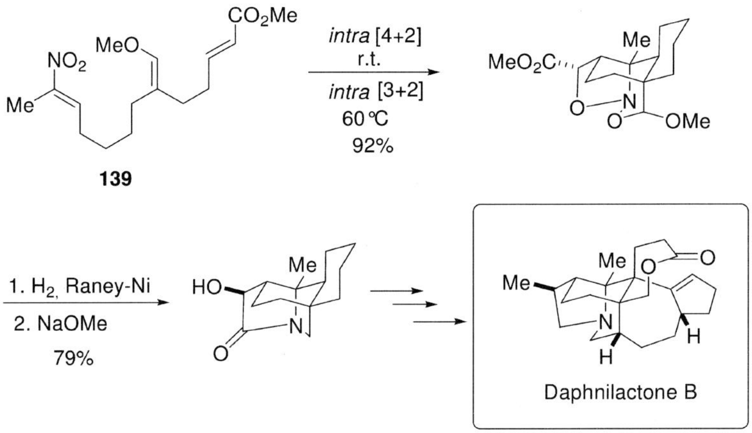

Scheme 3.78 Synthetic route to daphnilactone B.

provides architecturally complex bridged frameworks that often constitute the cores of interesting polycyclic compounds.

A model study geared toward the total synthesis of the *Daphniphyllum* alkaloid daphnilactone B employs this strategy to build up four of the six rings of the target compound starting from nitroalkene **139** (Scheme 3.78) [151].

3.9 Conclusion

Nitroalkenes are firmly established as pivotal intermediates in many synthetic processes resulting in the preparation of amino derivatives. As reactive carbon electrophiles, nitroalkenes are deeply involved in conjugate addition reactions with a certain number of carbon and heteronucleophiles. The electron-poor characters of their double bonds also make nitroalkenes very efficient dienophiles in cycloaddition reactions geared towards the formation of six-membered ring carbacycles. On the other hand, nitroalkenes can also be regarded as heterodienes with reduced electronic density suitable for inverse-electron-demand [4+2] cycloadditions with electron-rich alkenes. The highly ordered transition states that often typify these processes are of fundamental importance in the preparation of stereodefined compounds that nowadays represents a major goal of advanced research in organic synthesis. The ease of conversion of the nitro into an amino group creates a direct connection between nitroalkenes and their reduction products. Finally, the presence of the amino group in a plethora of pharmaceutically active molecules promises many exciting and important developments in the chemistry of nitroalkenes in the near future.

References

1 Berner, O. M.; Tedeschi, L.; Enders, D. *Eur. J. Org. Chem.* 2002, 1877.
2 Perekalin, V. V.; Lipina, E. S.; Berestovitskaya, V. M.; Efremov, D. A. *Nitroalkenes: Conjugated Nitro Compounds*, Wiley, Chichester, 1994.
3 Barrett, A. G. M. *Chem. Soc. Rev.* 1991, *20*, 95.
4 Ono, N. *The Nitro Group in Organic Synthesis*, Wiley-VCH, New York, 2001.
5 The Chemistry of Amino, Nitroso, Nitro and Related Groups Patai, S., Ed., Wiley, Chichester, 1996.
6 Rosini, G. In *Comprehensive Organic Synthesis*, Trost, B. M. Ed., Pergamon, Oxford, 1991, Vol.*2*, p.321.
7 Luzzio, F. A. *Tetrahedron* 2001, *57*, 915.
8 Melton, J.; Mc Murry, J. E. *J. Org. Chem.* 1975, *40*, 2138.
9 Knochel, P.; Seebach, D. *Synthesis* 1982, 1017.
10 Ballini, R.; Castagnani, R.; Petrini, M. *J. Org. Chem.* 1992, *57*, 2160.
11 Franklin, A. S. *Synlett* 2000, 1154.
12 Fujii, M. *Chem. Lett.* 1992, 933.
13 Sreekumar, R.; Padmakumar, R.; Rugmini, P. *Tetrahedron Lett.* 1998, *39*, 2695.
14 Mukaiyama, T.; Hata, E.; Yamada, T. *Chem. Lett.* 1995, 505.
15 Li, R.; Liu, Z.; Zhou, Y.; Wu, L. *Synlett* 2006, 1367.
16 Campos, P. J.; García, B.; Rodríguez, M. A. *Tetrahedron Lett.* 2000, *41*, 979.
17 Yasubov, M. S.; Perederina, I. A.; Filimonov, V. D.; Park, T.-H.; Chi, K.-W. *Synth. Commun.* 1998, *28*, 833.
18 Denmark, S. E.; Marcin, L. R. *J. Org. Chem.* 1993, *58*, 3850.
19 Lin, W.-W.; Jang, Y.-J.; Wang, Y.; Liu, J.-T.; Hu, S. R.; Wang, L.-Y.; Yao, C.-F. *J. Org. Chem.* 2001, *66*, 1984.
20 Enders, D.; Teschner, P.; Raabe, G. *Synlett* 2000, 637.
21 Schäfer, H.; Seebach, D. *Tetrahedron* 1995, *51*, 2305.
22 Mampreian, D. M.; Hoveyda, A. H. *Org. Lett.* 2004, *6*, 2829.
23 Li, H.; Wang, Y.; Tang, L.; Wu, F.; Liu, X.; Guo, C.; Foxman, B. M.; Deng, L. *Angew. Chem., Int. Ed.* 2005, *44*, 105.
24 Munk, S. A.; Lai, R. K.; Burke, J. E.; Arasasingham, P. N.; Kharlamb, A. B.; Manlapaz, C. A.; Padillo, E. U.; Wijono, M. K.; Hasson, D. W.; Wheeler, L. A.; Garst, M. E. *J. Med. Chem.* 1996, *39*, 1193.
25 Clive, D. L. J.; Bo, Y.; Selvakumar, N.; McDonald, R.; Santarsiero, B. D. *Tetrahedron* 1999, *55*, 3277.
26 Srikanth, G. S. C.; Castle, S. L. *Org. Lett.* 2004, *6*, 449.
27 He, L.; Srikanth, G. S. C.; Castle, S. L. *J. Org. Chem.* 2005, *70*, 8140.
28 Fornicola, R. S; Oblinger, E.; Montgomery, J. *J. Org. Chem.* 1998, *63*, 3528.
29 Zindel, J.; De Meijere, A. *Synthesis* 1994, 190.
30 Hübner, J.; Liebscher, J.; Pätzel, M. *Tetrahedron* 2002, *58*, 10485.
31 Duursma, A.; Minnaard, A. J.; Feringa, B. *J. Am. Chem. Soc.* 2003, *125*, 3700.
32 Rimkus, A.; Sewald, N. *Org. Lett.* 2003, *5*, 79.
33 Rimkus, A.; Sewald, N. *Synthesis* 2004, 135.
34 Wang, J.; Li, H.; Duan, W.; Zu, L.; Wang, W. *Org. Lett.* 2005, *7*, 4713.
35 Masesane, I. B.; Steel, P. G. *Synlett* 2003, 735.
36 Masesane, I. B.; Steel, P. G. *Tetrahedron Lett.* 2004, *45*, 5007.
37 Okino, T.; Hoashi, Y.; Furukawa, T.; Xu, X.; Takemoto, Y. *J. Am. Chem. Soc.* 2005, *127*, 119.
38 Licandro, E.; Maiorana, S.; Baldoli, C.; Capella, L.; Perdicchia, D. *Tetrahedron: Asymmetry* 2000, *11*, 975.
39 Ager, D. J.; Prakash, I.; Schaad, D. R. *Chem. Rev.* 1996, *96*, 835.
40 Enders, D.; Haertwig, A.; Raabe, G.; Runsink, J. *Angew. Chem., Int. Ed.* 1996, *35*, 2388.
41 Adderley, N. J.; Buchanan, D. J.; Dixon, D. J.; Lainé, D. I. *Angew. Chem., Int. Ed.* 2003, *42*, 4241.

42 Enders, D.; Haertwig, A.; Raabe, G.; Runsink, J. *Eur. J. Org. Chem.* 1998, 177.
43 Buchanan, D. J.; Dixon, D. J.; Hernandez-Juan, F. A. *Org. Biomol. Chem.* 2004, *2*, 2932.
44 Buchanan, D. J.; Dixon, D. J.; Scott, M. S.; Lainé, D. I. *Tetrahedron: Asymmetry* 2004, *15*, 195.
45 Buchanan, D. J.; Dixon, D. J.; Looker, B. E. *Synlett* 2005, 1948.
46 Jahn, U.; Rudakov, D. *Synlett* 2004, 1207.
47 Dumez, E.; Faure, R.; Dulcère, J.-P. *Eur. J. Org. Chem.* 2001, 2577.
48 Barrett, A. G. M.; Kohrt, J. T. *Synlett* 1995, 415.
49 Lucet, D.; Le Gall, C.; Mioskowski, C. *Angew. Chem., Int. Ed.* 1998, *37*, 2580.
50 Imakawa, K.; Hata, E.; Yamada, T.; Mukaiyama, T. *Chem. Lett.* 1996, 291.
51 Morris, M. L; Sturgess, M. A. *Tetrahedron Lett.* 1993, *34*, 4746.
52 Sturgess, M. A.; Yarberry, D. J. *Tetrahedron Lett.* 1993, *34*, 4746.
53 Enders, D.; Wiedemann, J. *Synthesis* 1996, 1443.
54 Lucet, D.; Toupet, L.; Le Gall, C.; Mioskowski, C. *J. Org. Chem.* 1997, *62*, 2682.
55 Lucet, D.; Sabelle, S.; Kosterlitz, O.; Le Gall, C.; Mioskowski, C. *Eur. J. Org. Chem.* 1999, 2583.
56 Turconi, J.; Lebeau, L.; Paris, J.-M.; Mioskowski, C. *Tetrahedron* 2006, *62*, 8109.
57 Feroci, M.; Inesi, A.; Palombi, L.; Rossi, L. *Tetrahedron: Asymmetry* 2001, *12*, 2331.
58 Molteni, M.; Volonterio, A.; Zanda, M. *Org. Lett.* 2003, *5*, 3887.
59 Vega-Pérez, J. M.; Candela, J. I.; Blanco, E.; Iglesias-Guerra, F. *Tetrahedron* 1999, *55*, 9641.
60 Winterfeld, G. A.; Das, J.; Schmidt, R. R. *Eur. J. Org. Chem.* 2000, 3047.
61 Strekowski, L.; Gulevich, Y.; Van Aken, K.; Wilson, D. W.; Fox, K. R. *Tetrahedron Lett.* 1995, *36*, 225.
62 Ballini, R.; Bosica, G.; Fiorini, D.; Palmieri, A.; Petrini, M. *Chem. Rev.* 2005, *105*, 933.
63 Du, D.-M.; Xu, J.; Zhang, S.-W.Lu, S.-F. *J. Am. Chem. Soc.* 2006, *128*, 7418.
64 Tucker, J. A.; Clayton, T. L.; Mordas, D. M. *J. Org. Chem.* 1997, *62*, 4370.
65 Busch, K.; Groth, U. M.; Kühnle, Schöllkopf, U. *Tetrahedron* 1992, *48*, 5607.
66 Blay, G.; Fernández, I.; Monje, B.; Pedro, J. R. *Tetrahedron* 2004, *60*, 165.
67 Forzato, C.; Nitti, P.; Pitacco, G.; Valentin, E.; Morganti, S.; Rizzato, E.; Spinelli, D.; Dell'Erba, C.; Petrillo, G.; Tavani, C. *Tetrahedron* 2004, *60*, 11011.
68 Posner, G. H.; Crouch, R. D. *Tetrahedron* 1990, *46*, 7509.
69 Brenner, M.; Seebach, D. *Helv. Chim. Acta* 1999, *82*, 2365.
70 Mulzer, J.; Zuhse, R.; Schmiechen, R. *Angew. Chem., Int. Ed.* 1992, *31*, 870.
71 Palomo, C.; Aizpurua, J. M.; Oiarbide, M.; García, J. M.; González, A.; Odriozola, I.; Linden, A. *Tetrahedron Lett.* 2001, *42*, 4829.
72 Hayashi, T.; Senda, T.; Ogasawara, M. *J. Am. Chem. Soc.* 2000, *122*, 10716.
73 Hoshiko, T.; Ishihara, H.; Shino, M.; Mori, N. *Chem. Pharm. Bull.* 1993, *41*, 633.
74 Aslanian, R.; Lee, G.; Iyer, R. V.; Shih, N.-Y.; Piwinski, J. J.; Draper, R. W.; McPhail, A. T. *Tetrahedron: Asymmetry* 2000, *11*, 3867.
75 Miyashita, M.; Awen, B. Z. E.; Yoshikoshi, A. *Chem. Lett.* 1990, 239.
76 Maeda, H.; Kraus, G. A. *J. Org. Chem.* 1997, *62*, 2314.
77 List, B.; Pojarliev, P.; Martin, H. J. *Org. Lett.* 2001, *3*, 2423.
78 Betancort, J. M.; Barbas, III, C. F. *Org. Lett.* 2001, *3*, 3737.
79 Betancort, J. M.; Sakthivel, K.; Thayumanavn, R.; Tanaka, F.; Barbas, III, C. F. *Synthesis* 2004, 1509.
80 Wang, J.; Li, H.; Lou, B.; Zu, L.; Guo, H.; Wang, W. *Chem. Eur. J.* 2006, *12*, 4321.
81 Ji, J.; Barnes, D. M; Zhang, J.; King, S. A.; Wittenberger, S. J.; Morton, H. E. *J. Am. Chem. Soc.* 1999, *121*, 10215.
82 Nichols, P.J.; DeMattei, J. A.; Barnett, B. R.; LeFur, N. A.; Chuang, T.-H.; Piscopo, A. D.; Koch, K. *Org. Lett.* 2006, *8*, 1495.
83 Barnes, D. M; Ji, J.; Fickes, M. G.; Fitzgerald, M. A.; King, S. A.; Morton, H. E.; Plagge, F. A.; Preskill, M.; Wagaw, S.

H.; Wittenberger, S. J.; Zhang, J. *J. Am. Chem. Soc.* 2002, *124*, 13097.
84 Denmark, S. E.; Thorarensen, A. *Chem. Rev.* 1996, *96*, 137.
85 Denmark, S. E.; Marcin, L. R. *J. Org. Chem.* 1993, *58*, 3857.
86 Denmark, S. E.; Marcin, L. R. *J. Org. Chem.* 1995, *60*, 3221.
87 Denmark, S. E.; Schnute, M. E. *J. Org. Chem.* 1994, *59*, 4576.
88 Habermann, J.; Ley, S. V.; Scott, J. S. *J. Chem. Soc., Perkin Trans. 1* 1999, 1253.
89 Hoashi, Y.; Yabuta, T.; Takemoto, Y. *Tetrahedron Lett.* 2004, *45*, 9185.
90 Hoashi, Y.; Yabuta, T.; Yuan, P.; Miyabe, H.; Takemoto, Y. *Tetrahedron* 2006, *62*, 365.
91 Johnson, T. A.; Curtis, M. D.; Beak, P. *J. Am. Chem. Soc.* 2001, *123*, 1004.
92 Johnson, T. A.; Jang, D. O.; Slafer, B. W.; Curtis, M. D.; Beak, P. *J. Am. Chem. Soc.* 2002, *124*, 11689.
93 Johnson, T. A.; Curtis, M. D.; Beak, P. *Org. Lett.* 2002, *4*, 2747.
94 Mouhtaram, M.; Jung, L.; Stambach, J. F. *Tetrahedron* 1993, *49*, 1391.
95 Pollini, G. P.; Baricordi, N.; Benetti, S.; De Risi, C.; Zanirato, V. *Tetrahedron Lett.* 2005, *46*, 3699.
96 Liddel, J. R. *Nat. Prod. Rep.* 2002, *19*, 773.
97 Denmark, S. E.; Thorarensen, A. *J. Org. Chem.* 1994, *59*, 5672.
98 Denmark, S. E.; Hurd, A. R. *J. Org. Chem.* 1998, *63*, 3045.
99 Denmark, S. E.; Seierstad, M.; Herbert, B. *J. Org. Chem.* 1999, *64*, 884.
100 Denmark, S. E.; Herbert, B. *J. Am. Chem. Soc.* 1998, *120*, 7357.
101 Denmark, S. E.; Herbert, B. *J. Org. Chem.* 2000, *65*, 2887.
102 Denmark, S. E.; Hurd, A. R. *Org. Lett.* 1999, *1*, 1311.
103 Denmark, S. E.; Hurd, A. R. *J. Org. Chem.* 2000, *65*, 2875.
104 Denmark, S. E.; Cottell, J. J. *J. Org. Chem.* 2001, *66*, 4276.
105 Denmark, S. E.; Martinborough, E. A. *J. Am. Chem. Soc.* 1999, *121*, 3046.
106 Denmark, S. E.; Thorarensen, A.; Middleton, D. S. *J. Org. Chem.* 1995, *60*, 3574.
107 Denmark, S. E.; Thorarensen, A.; Middleton, D. S. *J. Am. Chem. Soc.* 1996, *118*, 8266.
108 Denmark, S. E.; Parker, Jr., D. L.; Dixon, J. A. *J. Org. Chem.* 1997, *62*, 435.
109 Denmark, S. E.; Thorarensen, A. *J. Am. Chem. Soc.* 1997, *119*, 125.
110 Denmark, S. E.; Seierstad, M. *J. Org. Chem.* 1999, *64*, 1610.
111 Denmark, S. E.; Senanayake, C. W. B. *J. Org. Chem.* 1993, *58*, 1853.
112 Denmark, S. E.; Schnute, M. E.; Senanayake, C. W. B. *J. Org. Chem.* 1993, *58*, 1859.
113 Denmark, S. E.; Schnute, M. E.; Marcin, L. R.; Thorarensen, A. *J. Org. Chem.* 1995, *60*, 3205.
114 Denmark, S. E.; Middleton, D. S. *J. Org. Chem.* 1998, *63*, 1604.
115 Denmark, S. E.; Montgomery, J. I.; Kramps, L. A. *J. Am. Chem. Soc.* 2006, *128*, 11620.
116 Denmark, S. E.; Guagnano, V.; Dixon, J. A.; Stolle, A. *J. Org. Chem.* 1997, *62*, 4610.
117 Denmark, S. E.; Juhl, M. *Helv. Chim. Acta* 2002, *85*, 3712.
118 Denmark, S. E.; Dixon, J. A. *J. Org. Chem.* 1997, *62*, 7086.
119 Denmark, S. E.; Dixon, J. A. *J. Org. Chem.* 1998, *63*, 6167.
120 Denmark, S. E.; Dixon, J. A. *J. Org. Chem.* 1998, *63*, 6178.
121 Node, M.; Itoh, A.; Masaki, Y.; Fuji, K. *Heterocycles* 1991, *32*, 1705.
122 Fuji, K.; Kawabata, T.; Ohmori, T.; Shang, M.; Node, M. *Heterocycles* 1998, *47*, 951.
123 Somei, M.; Yamada, F.; Izumi, T.; Nakajou, M. *Heterocycles* 1997, *45*, 2327.
124 Knöller, H.-J.; Hartmann, K. *Synlett* 1993, 755.
125 Fujii, A.; Fujima, Y.; Harada, H.; Ikunaka, M.; Inoue, T.; Kato, S.; Matsuyama, K. *Tetrahedron: Asymmetry* 2001, *12*, 3235.
126 Bartoli, G.; Bosco, M.; Giuli, S.; Giuliani, A.; Lucarelli, L.; Marcantoni, E.; Sambri, L.; Torregiani, E. *J. Org. Chem.* 2005, *70*, 1941.
127 Hollinshead, S. P.; Trudell, M. L.; Skolnick, P.; Cook, J. M. *J. Med. Chem.* 1990, *33*, 1062.

128 Cox, E. D.; Diaz-Arauzo, H.; Huang, Q.; Reddy, M. S.; Ma, C.; Harris, B.; McKernan, R.; Skolnick, P.; Cook, J. P. *J. Med. Chem.* 1998, *41*, 2537.

129 Busacca, C. A.; Dong, Y. *Synth. Commun.* 2000, *30*, 501.

130 Bandini, M.; Garelli, A.; Rovinetti, M.; Tommasi, S., Umani-Ronchi, A. *Chirality* 2005, *17*, 522.

131 Zhuang, W.; Hazell, R. G.; Jørgensen, K. A. *Org. Biomol. Chem.* 2005, *3*, 2566.

132 Herrera, R. P.; Sgarzani, V.; Bernardi, L.; Ricci, A. *Angew. Chem., Int. Ed.* 2005, *44*, 6576.

133 Mahboobi, S., Wiegrebe, W.; Popp, A. *J. Nat. Prod.* 1999, *62*, 577.

134 Ruchirawat, S.; Tontoolarug, S.; Sahakitpichan, P. *Heterocycles* 2001, *55*, 635.

135 McNulty, J.; Mao, J.; Gibe, R.; Mo, R.; Wolf, S.; Pettit, G. R.; Herald, D. L.; Boyd, M. R. *Bioorg. Med. Chem. Lett.* 2001, *11*, 159.

136 Michaelides, M. R.; Hong, Y.; DiDomenico, S.; Bayburt, E. K.; Asin, K. E.; Britton, D. R.; Lin, C. W.; Shiosaki, K. *J. Med. Chem.* 1997, *40*, 1585.

137 Gu, Y. G.; Bayburt, E. K.; Michaelides, M. R.; Lin, C. W.; Shiosaki, K. *Bioorg. Med. Chem. Lett.* 1999, *9*, 1341.

138 Yamashita, M.; Yamada, K.; Tomioka, K. *J. Am. Chem. Soc.* 2004, *126*, 1954.

139 Yamashita, M.; Yamada, K.; Tomioka, K. *Tetrahedron* 2004, *60*, 4237.

140 Yasuhara, T.; Nishimura, K.; Yamashita, M.; Fukuyama, N.; Yamada, K.; Muraoka, O.; Tomioka, K. *Org. Lett.* 2003, *5*, 1123.

141 Yasuhara, T.; Osafune, E.; Nishimura, K.; Yamashita, M.; Yamada, K.; Muraoka, O.; Tomioka, K. *Tetrahedron Lett.* 2004, *45*, 3043.

142 Dong, L.; Xu, Y.-J.; Cun, L.-F.; Cui, X.; Mi, A.-Q.; Jiang, Y.-Z.; Gong, L.-Z. *Org. Lett.* 2005, *7*, 4285.

143 Dong, L.; Xu, Y.-J.; Yuan, W.-C.; Cui, X.; Cun, L.-F.; Gong, L.-Z. *Eur. J. Org. Chem.* 2006, 4093.

144 Ikeda, M.; Okano, M.; Kosaka, K.; Kido, M.; Ishibashi, H. *Pharm. Chem. Bull.* 1993, *41*, 276.

145 O'Neill, B. T.; Thadeio, P. F.; Bundesmann, M. W.; Elder, A. M.; McLean, S.; Bryce, D. K. *Tetrahedron* 1997, *53*, 11121.

146 Stanetty, P.; Kremslehner, M.; Mihovilovich, M. D. *Tetrahedron Lett.* 2000, *41*, 1717.

147 Rawal, V. H.; Michoud, C. *Tetrahedron Lett.* 1991, *32*, 1695.

148 Denmark, S. E.; Marcin, L. R. *J. Org. Chem.* 1997, *62*, 1675.

149 Denmark, S. E.; Gomez, L. *Org. Lett.* 2001, *3*, 2709.

150 Denmark, S. E.; Gomez, L. *J. Org. Chem.* 2003, *68*, 8015.

151 Denmark, S. E.; Baiazitov, R. Y. *J. Org. Chem.* 2006, *71*, 593.

4
Isocyanide-Based Multicomponent Reactions (IMCRs) as a Valuable Tool with which to Synthesize Nitrogen-Containing Compounds

Alexander Doemling

Abstract

IMCR is an emerging technique, based on the extraordinary reactivity of isocyanides as starting materials, used to synthesize organic compounds in an efficient and resource-saving way. Thanks to its exquisite reactivity and its huge functional group tolerability in both primary and secondary transformations, virtually hundreds of organic chemical scaffolds are amenable. The immense chemical space available through IMCR will provide the basis for new discoveries both in material sciences and in life sciences.

4.1
Introduction

Historically, a fairly large part of today's knowledge relating to the reactivity of isocyanides in the context of multicomponent reactions (MCRs) has been contributed by only a few scientists, including the Italian Passerini, the Estonian Ugi, and the Dutch van Leusen. Interestingly, despite the fact that isocyanides have been known for more than 150 years, their chemistry was investigated only sporadically over the first 100 year period. Only with the advent of a general and efficient production method for isocyanides from primary amines, developed by Ugi, was the chemistry of isocyanides investigated in depth (Scheme 4.1) [1]. In addition to the previous poor synthetic accessibility of the compounds, many scientists also seemed to be deterred by the unusual and unpleasant odors of isocyanides.

From a structural standpoint, numerous X-ray structure analyses favor the first mesomeric description, due to the linearity and the NC distances of isocyanides of ~116 pm in **1**, which are comparable to typical triple bonds in nitriles R—CN (114 pm in acetonitrile). From a mechanistic standpoint, the reactivities of isocyanides in many ways reflect those of the isolobal carbenes.

Usually in isocyanide-based MCRs (IMCRs) the isocyanide is an explicit part of the product scaffold, but there are certain reactions in which the isocyanide con-

Amino Group Chemistry. From Synthesis to the Life Sciences. Edited by Alfredo Ricci

ISBN: 978-3-527-31741-7

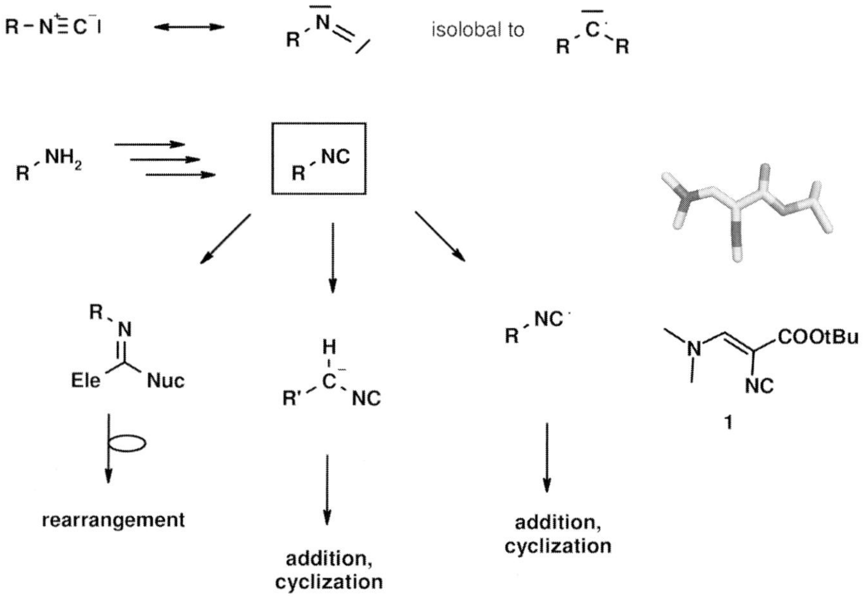

Scheme 4.1 Isocyanides can be viewed in valence bond theory as linear charged zwitterions with a positive nitrogen and a negatively charged carbon or as a bent carbene-type neutral species. The main reactivity pathways of isocyanides include α-addition (insertion) and subsequent rearrangement, including cyclization reactions.

tributes to the final product only the isocyano-*C* and the residual atoms including the nitrogen are cleaved off. In general, therefore, IMCRs are ideally suited for the synthesis of diverse *N*-containing scaffolds in organic chemistry.

Isocyanide reactivity is represented by three distinct patterns, by far the most important of which is the α-addition of electrophiles and nucleophiles at the isocyanide carbon. This unusual reactivity of an atom in a functional group in adding both a nucleophile and an electrophile provides the basis for most of the other secondary reactions. In this respect isocyanides behave similarly to carbenes (Scheme 4.2). In contrast, ordinary functional groups normally react at different atom centers with electrophiles and to nucleophiles.

The second type of reactivity is a product of the highly acidic character of the hydrogen attached to the carbon adjacent to the isocyano functional group, while finally, thanks to the energetically accessible SUMO orbital, isocyanides easily undergo radical reactions. Among the numerous examples of isocyanide radical reactions appeared in literature documenting the usefulness of isocyanide-based radical reactions, the elegant work of Curran et al. on the rapid assembly of the anti cancer agent camptothecin and derivatives [2] is noteworthy.

Thanks to the eminently powerful and selective reactivity of the isocyano group, most reactions are easily performed at room temperature or below. Moreover, the

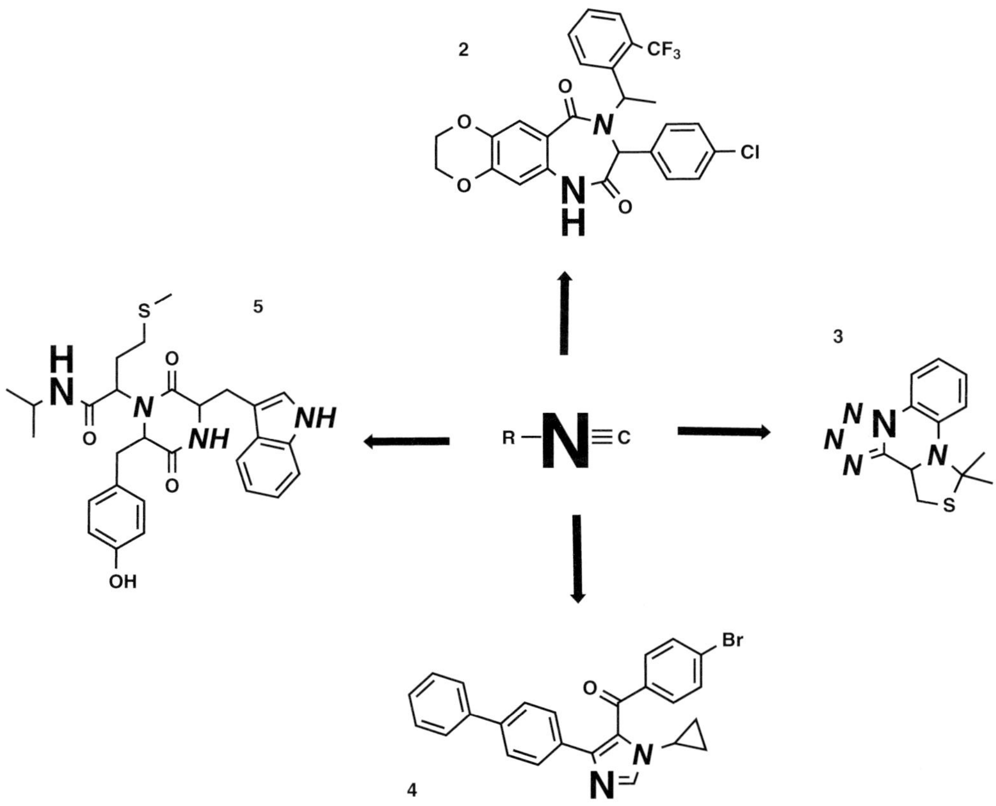

Scheme 4.2 The variety of a few representative structures **2–5** accessible by isocyanide-based MCR. The nitrogen atoms in black are derived from the isocyanide, whereas the others are introduced by other starting *materials*.

reactions of isocyanides are often compatible with a wide range of functional groups, making the products originating directly from an IMCR ideal substrates for further secondary reactions. In most reactions the isocyanide moiety is directly involved in the assembly of the product, so isocyanide-based reaction products are almost inevitably nitrogen-containing organic molecules (Scheme 4.2).

Why are IMCRs such a popular and fruitful group of reactions in organic chemistry? As a result of the immense diversity of possible reaction products, IMCRs are becoming increasingly popular in the organic chemical community, and more and more organic scaffolds based on IMCR and possessing interesting new properties are being described [3]. Several reviews on the chemistry of isocyanides and their MCRs have appeared recently [4], the following review highlights the *N*-contributing aspects of isocyanide chemistry to the product families involved. Because of the exponential number of publications in this area of organic chemistry and the limited space available here, this chapter is confined to a personal compendium of recent and classical IMCR contributions.

4.2 The Ugi Reaction

The Ugi reaction was described for the first time in 1959 [5]. In a short period of several months the main variants of today's Ugi reactions were discovered and described by Ugi et al. The name "Ugi reaction" is misleading, however, since it is not a single reaction, but a family of reactions. In fact, the most general definition of Ugi reactions is "the α-addition of a nucleophile and a Schiff base or enamine (electrophile) onto an isocyanide carbon and a subsequent rearrangement". Depending mainly on the nature of the nucleophile and to a minor extent on the nature of the amine component, several primary reaction products are formed (Scheme 4.3). A generally accepted simplified mechanism is depicted below. After the formation of the Schiff base, the electrophilicity of the imine is enhanced by the protonation of the Schiff base nitrogen or by complexation with a Lewis acid. The isocyanide then adds to form an iminium ion; these intermediate

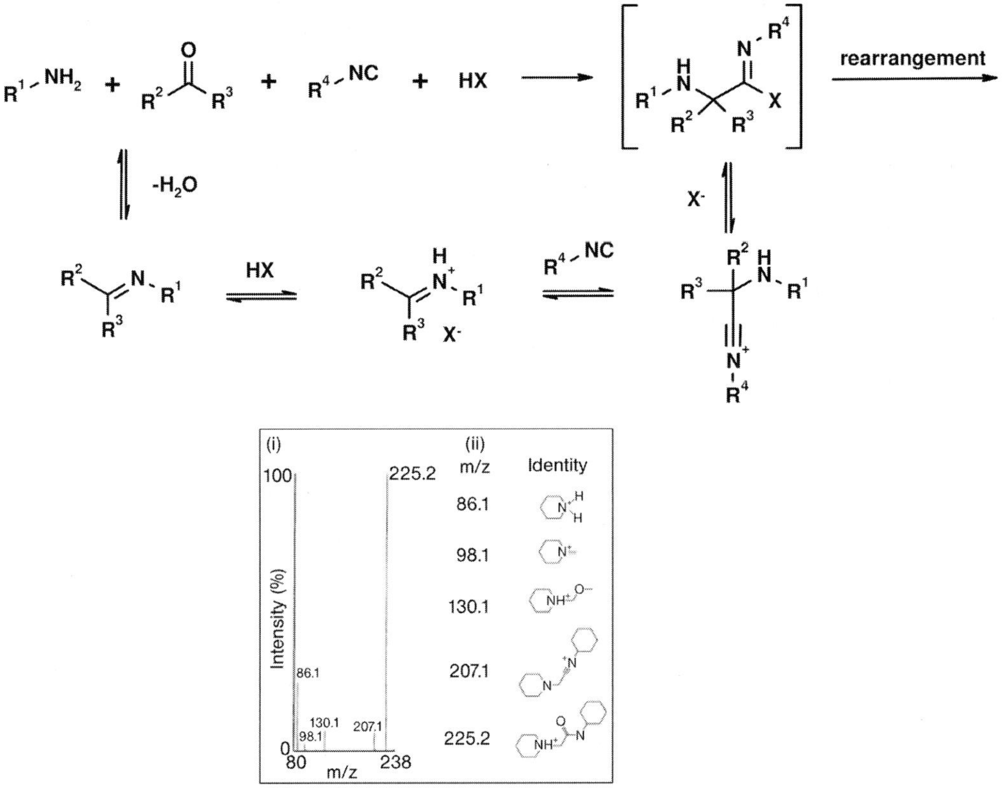

Scheme 4.3 General mechanism for the Ugi reaction. The intermediate nitrilium ion can easily be observed by mass spectroscopy (box).

iminium ions (nitrilium ions) have regularly been detected by mass spectroscopy during the reactions [6].

These primary Ugi scaffolds are α-acylaminocarbonamide, α-aminocarbonamide, hydantoin-4-imide, 2-thio-hydantoin-4-imide, β-lactam, urethane, thioamide, selenoamide, amidine, diacylimide, and tetrazole [7]. Because of the many varieties of intramolecular reactions and the plethora of secondary reactions, however, the number of accessible scaffolds by far exceeds the original family of Ugi reaction products. Intramolecular reactions can afford a variety of heterocycles, but Ugi reactions all have in common – independently of the particular scaffold – that a potentially very large chemical space is accessible. This very large chemical space is a direct consequence of the underlying combinatorial mathematics (e.g., reaction of 1000 educts of each of the four groups of starting materials in the classical α-acylaminocarbonamide scaffold yields 1×10^{12} products [7].

The classical Ugi reaction involves the four starting materials – carboxylic acid, primary amine, aldehyde or ketone, and isocyanide – in the formation of the α-acylaminocarbonamide scaffold (Scheme 4.4). The assumed intermediate α-adduct is shown in brackets. A crucial substructure of this intermediate is an *N*-hetero carboxylic acid anhydride (left box), resembling a carboxylic acid anhydride, which accounts for its high intramolecular acylating power during the Mumm rearrangement. Almost independently of the natures of the specific components, the reaction products are formed in moderate to often excellent yields. Several attempts have been made to deduce reactivity scales for the four educt classes and their dependence on their electronic and steric natures. The difficulty in the use of such general reactivity rows is that the reactivities of the educts are interdependent, especially in MCRs.

The classical four-component Ugi condensation (U-4CR) has been utilized numerous times in the assembly of polyfunctionalized N-containing organic molecules such as **6** [8], but also in drug discovery as demonstrated (Scheme 4.5) by the syntheses of the oxytocin agonist **7** [9] and of the growth hormone secretagogue (GHS) **8** [10].

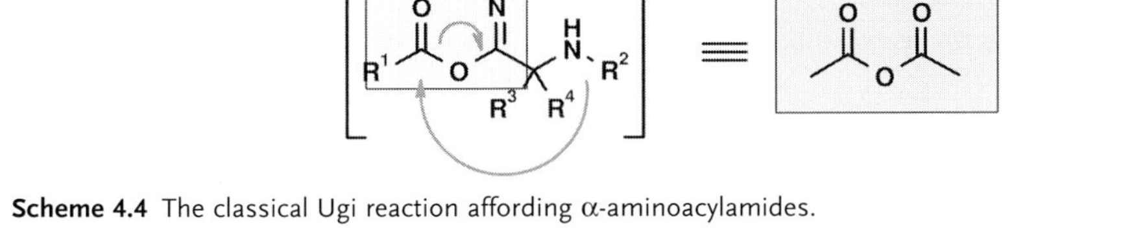

Scheme 4.4 The classical Ugi reaction affording α-aminoacylamides.

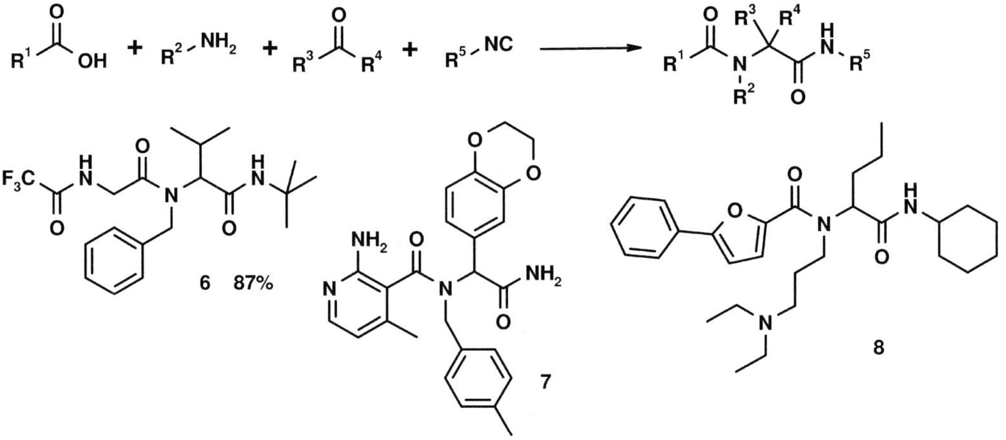

Scheme 4.5 Core key units and target compounds synthesized by the Ugi four-component condensation (U-4CR)

Recently the reactivities of each of 10 different isocyanides, carboxylic acids, and aldehydes, plus one resin-bound primary amine, were systematically investigated by an elegant MS method [11]. However, it should be noted that the reactivities of the same reactants in liquid-phase synthesis could potentially differ to a sizeable extent. Several reactants that obviously reacted badly or not at all in the SPOS have been reported to react smoothly under liquid-phase conditions. Factors that need to be considered in the case of combinatorial libraries of Ugi-type IMCRs include the often poor solubilities of carboxylic acids. Even when the apparent solubilities of the reactants are low, however, reaction can take place rapidly thanks to a solubility equilibrium coupled with subsequent fast reaction of the educts. Other effects that certainly play a role are reactions at the surface of a starting material. It still sounds counterintuitive, however, that the Ugi reaction is reported to run effectively in water or water/organic solvent mixtures [12, 13]. Part of the Ugi mechanism – the formation of the Schiff base, a dehydration reaction – proceeds smoothly in water. An advantage of the reaction in water could be the potential precipitation of the hydrophobic Ugi product, which considerably favors the purification. Unfortunately, however, there is no standard way to perform Ugi reactions. The identification of the best procedure with which to perform a particular reaction is mostly a task of optimization, and can potentially result in a considerable improvement in the yield. This is certainly a disadvantage with regard to the production of combinatorial libraries, in which one would like to use one universal procedure that should work best for all reactions. There is therefore obviously a trade-off between speed and quality, although this holds generally for all reactions and is not a particular issue in Ugi chemistry. In the case of combinatorial libraries involving Ugi reactions this is more obvious, however, since a badly reacting starting material potentially results in the drop-off of one or even several reaction plates.

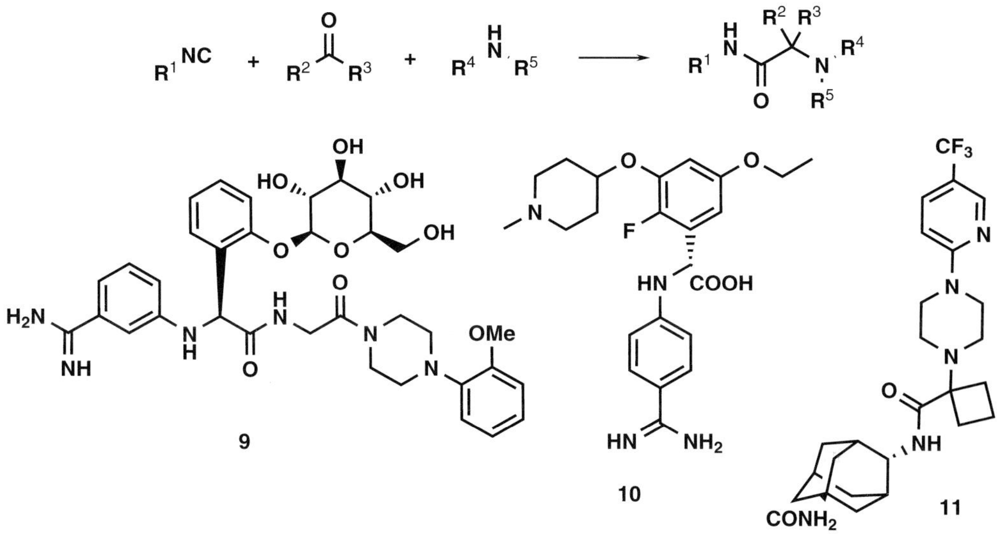

Scheme 4.6 U-3CR of isocyanides, oxo components and 1° or 2° amines.

In general the Ugi-4CR backbone incorporates a tertiary amide. In solution it is commonly possible to distinguish the two rotamers around the tertiary amide bond, which often makes interpretation of the ^{1}H and ^{13}C NMR spectra cumbersome.

The absence of the carboxylic acid – in a U-3CR (Scheme 4.6) – results in the production of an α-aminocarbonamide scaffold. This scaffold is of greater potential pharmacological importance, since (unlike the U-4CR scaffold) it has been reported to produce benign pharmacokinetic/pharmacodynamic (PKPD) properties, including oral bioavailability and water solubility. The reaction has been performed in a variety of experimental settings. Of importance for the conversion and yield are the catalyst utilized and the solvent system; acetic acid, BF_3, or $InCl_3$ have thus been reported to be beneficial in protic solvents and even in water.

Several representative biologically important molecules obtained by U-3CRs of isocyanides, such as the orally bioavailable FXa inhibitor **9** [14], the FVIIa inhibitor **10** [15], and the recently described 11-β-HSD-1 inhibitor **11** [16], are shown in Scheme 4.6.

Variation of the acid component by use of hydrogen azide can provide α-amino tetrazoles (Scheme 4.7). In earlier procedures HN_3 generated from the corresponding salts was used, but nowadays the use of $TMSN_3$ is highly recommended for safety reasons and gives comparable yields. In this reaction a total of five nitrogen atoms are introduced into the product: one originating from the amine, one from the isocyanide, and three from the azide. This versatile reaction can be performed with primary and secondary amines as well as with aldehydes and ketones, yielding the corresponding tetrazoles. The reaction regularly gives good to excellent

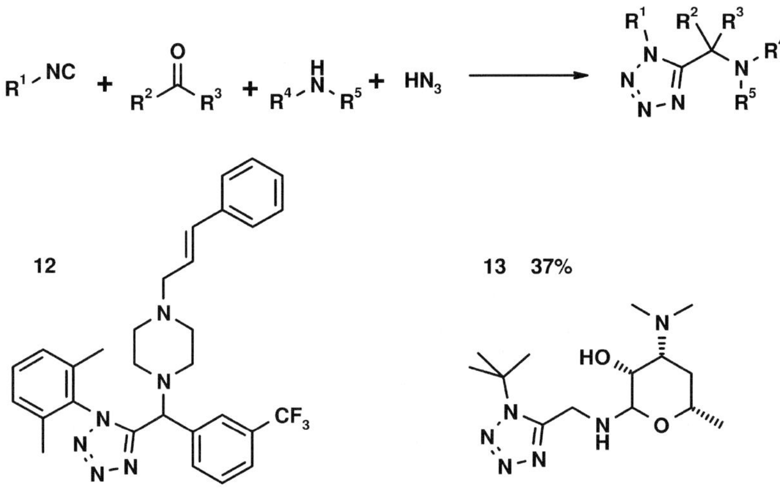

Scheme 4.7 U-4CR affording tetrazoles with three points of diversity.

yields. Use of a suitable "convertible" isocyanide (e.g., 1-(isocyanomethyl)-1,2,3-beuzotriazole (BetMIC)) can provide α-amino acid isosteres α-amino tetrazoles in good yield [17]. Recently, a series of potent bioavailable potent MCH1 receptor antagonists such as **12** capable of traversing the blood/brain barrier has been reported [18]. In another publication, tetrazole derivative **13** incorporating the erythromycin sugar moiety desosamine for the synthesis of new antibiotics was described [19]. Because of the formation of a secondary or tertiary amine this scaffold is particularly interesting in medicinal chemistry. Often the compounds show very good water solubility.

In situ generation of hydrogen cyanate and thiocyanate and reaction with a Schiff base and an isocyanide results in hydantoin-4-imides **15** and 2-thiohydantoin-4-imides **14** (Scheme 4.8) [20, 21]. Reportedly, when an aldehyde is used as the carbonyl component the synthesis takes place smoothly to give hydantoin-4-imides, which are formed only exceptionally when starting from ketones. In contrast, 2-thiohydantoin-4-imides are obtained in high yields from ketones, but not from aldehydes. A solid-supported transformation using an isocyanide on Wang resin has been described (**16**) [22]. Despite the attractive backbone structure and the great variability, this reaction has not in the past received the attention it deserves.

Monomethyl carbonate formed in situ from carbon dioxide in methanol serves as an acid component in Ugi reactions, e.g. with primary amines, aldehydes or ketones, and isocyanides, forming urethanes **17** (Scheme 4.9) [23, 24]. The scope and limitations of this reaction have been investigated in detail more recently [25]; it turned out that the reaction could be used to introduce common protecting groups such as the Fmoc or Alloc groups smoothly. Both carbon sulfide oxide and carbon disulfide react in a more complex way to form mixtures containing ureas and carbamates in rather low yields.

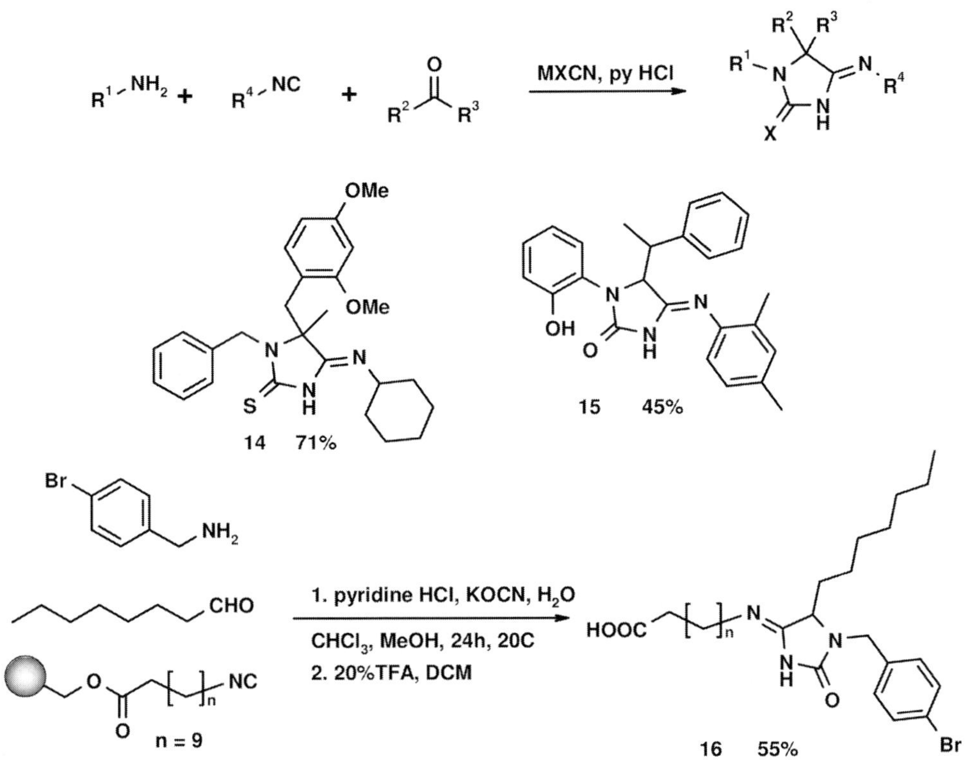

Scheme 4.8 General reaction scheme of the U-4CR affording 2-thiohydantoin-4-imides (**14**) and 2-hydantoin-4-imides (**15**), together with some examples. A solid-phase variation has also been performed.

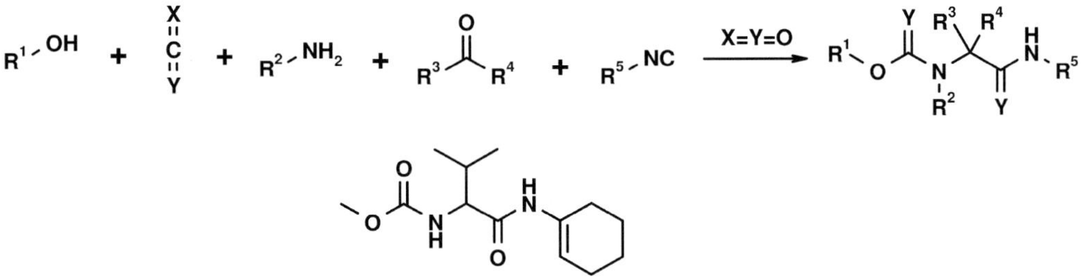

17 80%

Scheme 4.9 CO_2 in Ugi-type IMCRs yields carbamates such as **17**.

The quinolinium ion can be understood as an analogue of the iminium ion in terms of its reactivity with isocyanides and carboxylic acids to form *N*-alkyl-3-acyl-4-hydroquinoline-4-carbonamide [26]. Recently the reaction has been broadened in scope through the use of dihydropyridines (Scheme 4.10) [27, 28, 29].

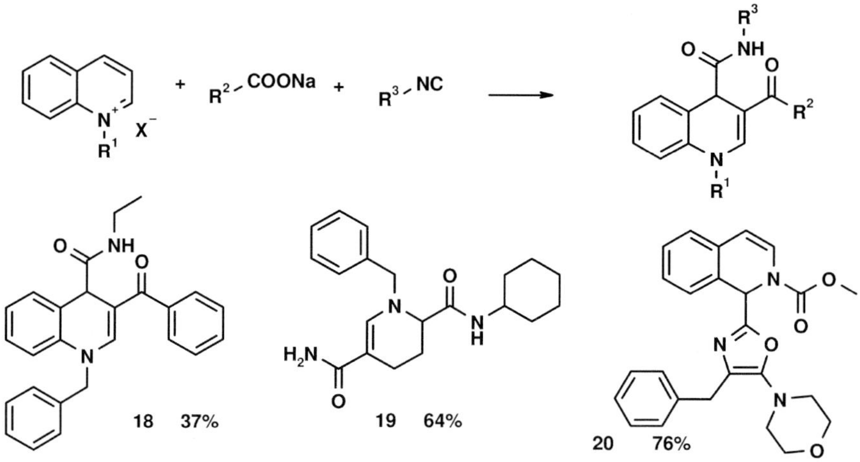

Scheme 4.10 Quinolinium and dihydropyridines in U-MCRs.

4.2.1 Intramolecular Ugi Reactions Involving Two Functional Groups

One way to achieve cyclic products based on the Ugi reaction is to use bifunctional starting materials. All potential combinations of bifunctional starting materials in the Ugi reaction are depicted in Scheme 4.11.

Most but not all combinations have been achieved so far, as is demonstrated by the totally unprecedented reaction of keto isocyanides in Ugi-type reactions.

β-Amino acids incorporate both the necessary carboxylic acid function and the amine function for the Ugi reaction. Ugi discovered that β-amino acids would react smoothly to yield the corresponding β-lactams in a reaction believed to proceed via a seven-membered α-adduct. The secondary amine of the intermediate is intramolecularly transacylated to form the more ring-strained β-lactam ring system (Scheme 4.12).

Because of its different mechanism this reaction offers a complementary approach to β-lactams, giving easy access to otherwise inaccessible compounds such as the nucleobase chimeras **21** [30], β-lactam steroid **22** [31], and thiazole β-lactams **23** [32].

The reaction has been used by many groups to build up bicyclic β-lactam antibiotics [33]. Thus, to the best of my knowledge, the synthesis of 6-penicilinic acid amide as performed by Ugi is the shortest and most convergent synthesis performed so far (Scheme 4.13) [33]. Interestingly, the synthesis involves two MCRs: the Asinger-4CR to form the thiazolidine imine **28** and a subsequent Ugi reaction yielding **30**. Subsequent functional group transformations yield 6-APA.

N-Monoalkylated β-amino acids have recently been used for the synthesis of chiral β-aminoacids [34]. Moreover, through screening of libraries of Ugi-derived β-lactams the discovery of a new type of noncompetitive HIV protease inhibitors has been described [35].

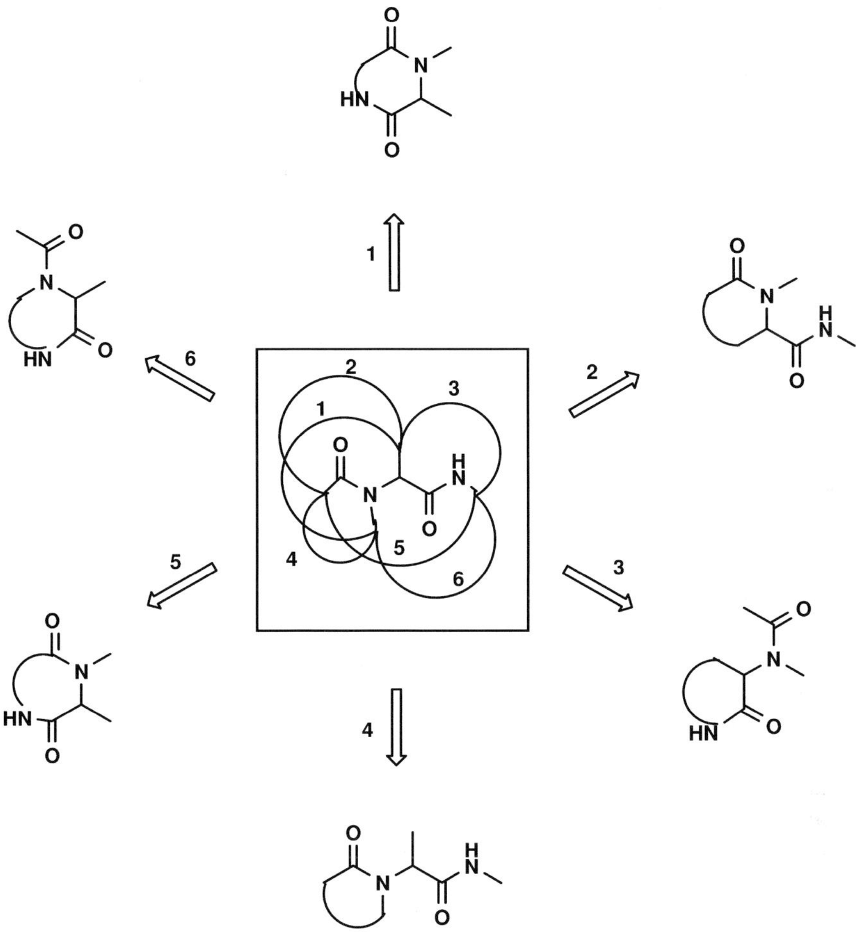

Scheme 4.11 All six possible intramolecular variations of Ugi IMCRs based on bifunctional starting materials.

α-Amino acids do not give the expected α-lactams but react differently. The intermediate six-membered α-adduct does not intramolecularly transacylate but rather acylates nucleophiles in an intermolecular fashion. Depending on the nature of the side chain of the α-amino acid, intramolecular acylation is possible as well (Scheme 4.14). Morpholine-2-one **35** is accessible through the use of hydroxyacetaldehyde as an aldehyde component. Interestingly, morpholine-2-ones derived from libraries of this particular Ugi reaction have recently been described as potent, selective, and novel T-type Ca^{2+} channel blockers [36].

γ-Amino and δ-amino acids, together with aldehydes and isocyanides, have not so far been successfully transformed into meaningful products. However, dipeptides unprotected at their C and N termini have been described (Scheme 4.15) as

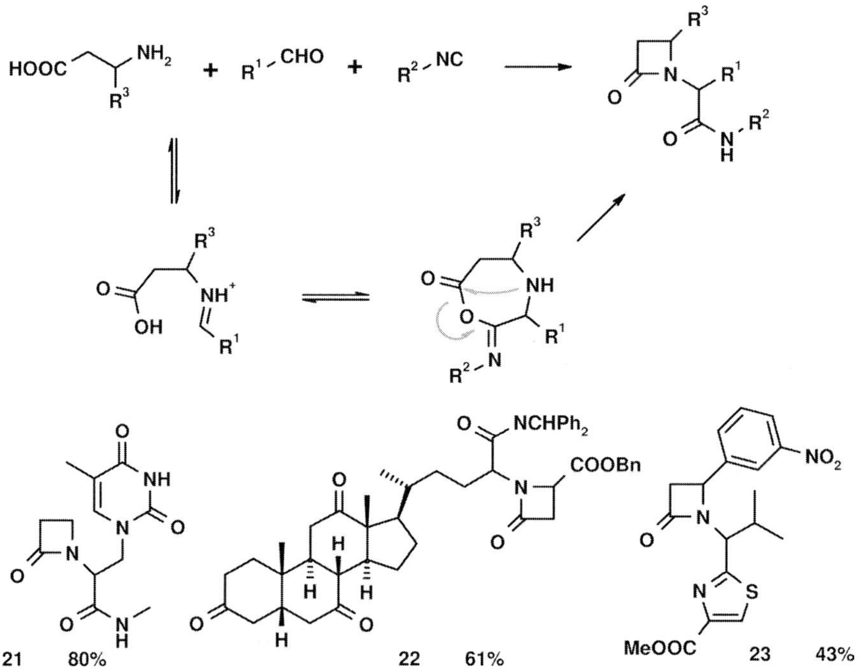

Scheme 4.12 β-Lactam rings from bifunctional β-amino acids by the U-3CR and the proposed reaction mechanism.

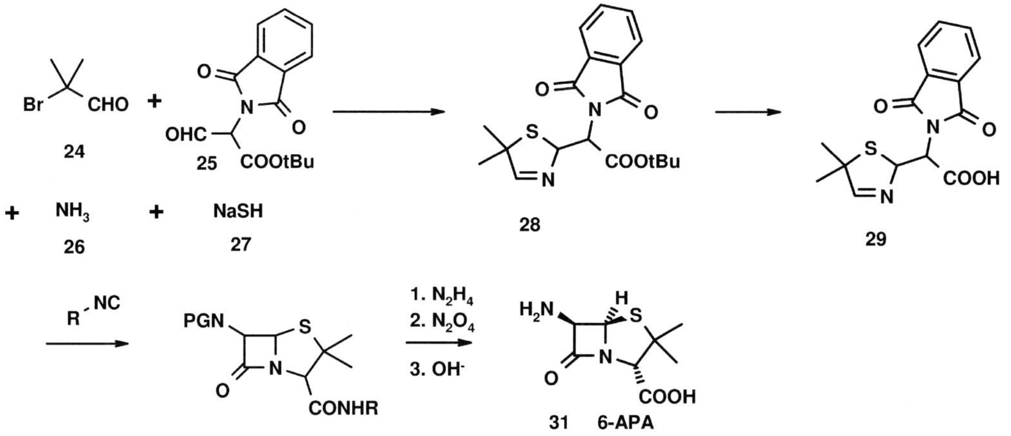

Scheme 4.13 Probably the shortest synthetic route to 5-APA derivatives, involving an Asinger 4-CR and a subsequent Ugi 3-CR.

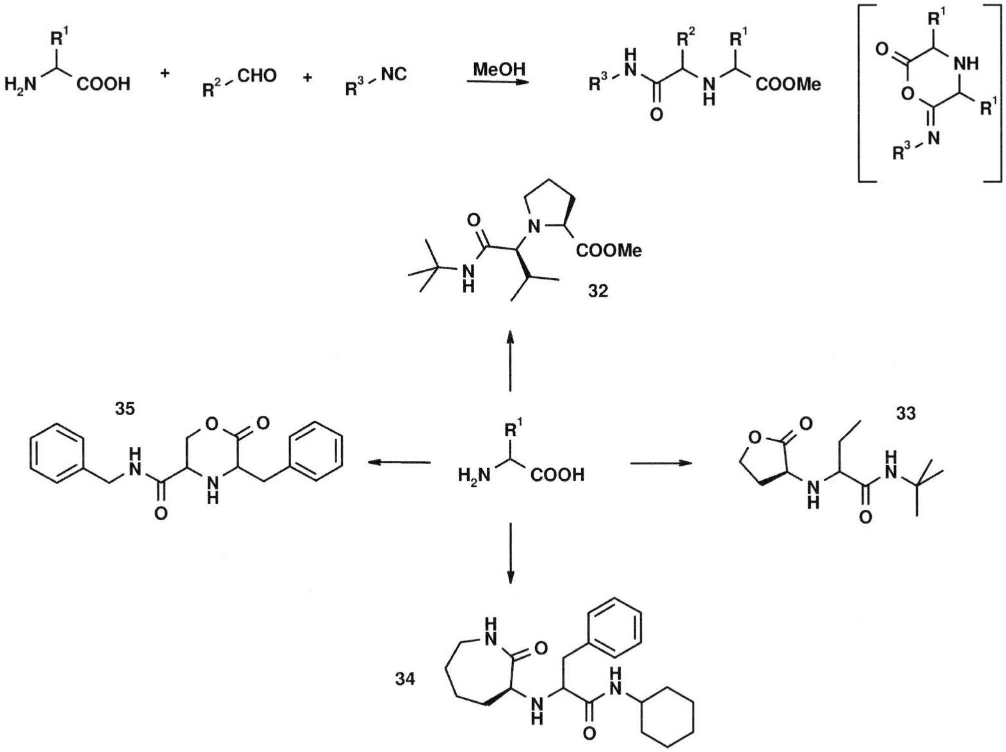

Scheme 4.14 The use of bifunctional α-amino acids in Ugi reactions. The intermediate six-membered α-adduct is shown in brackets.

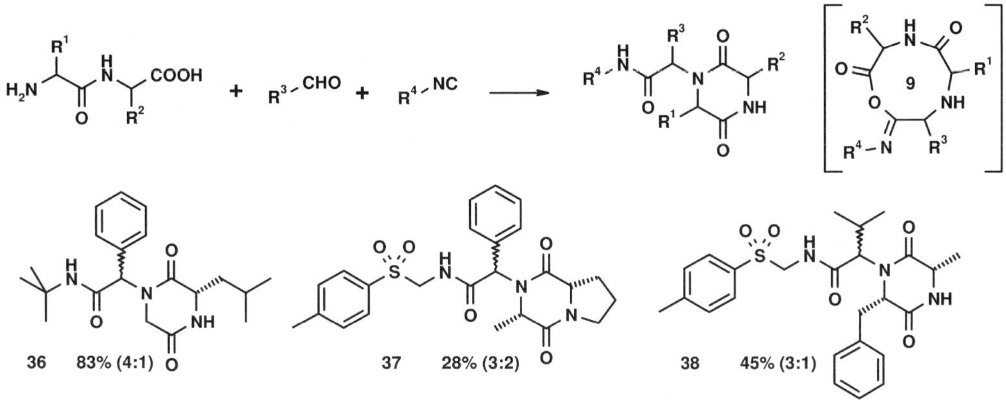

Scheme 4.15 Diketopiperazine synthesis with stereocontrol over two of the three stereocenters. The intermediate nine-membered α-adduct is shown in brackets.

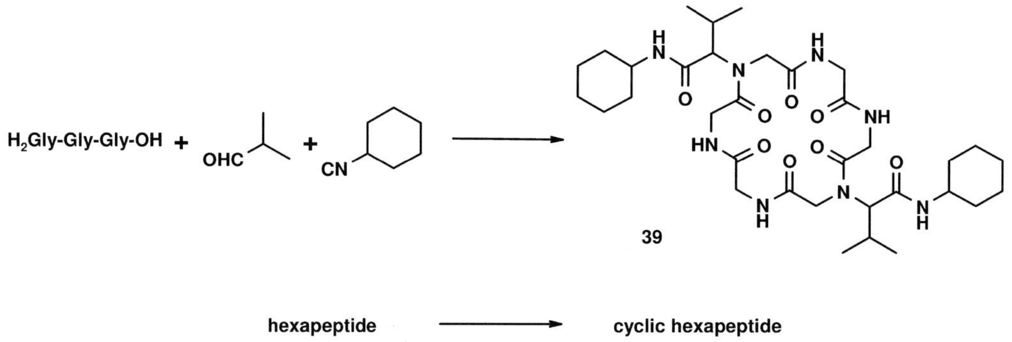

Scheme 4.16 Cyclic hexapeptides can be smoothly formed by U-4CRs with bifunctional peptides.

useful starting materials for preparation of libraries of diketo piperazines such as **36–38** [37].

Unprotected tripeptides do not react to form cyclic tripeptides but rather dimerize and form symmetric cyclic hexapeptides. Accordingly, the corresponding linear hexapeptides also cyclize to form the cyclic hexapeptides (Scheme 4.16) [38]. The formed α-adduct intermediate must be assumed to be a 20-membered large ring. The potential for the formation of cyclic peptide libraries has been certainly not yet fully exploited but rather only scratched at the surface.

The potential of bifunctional starting materials to form macrocyclic Ugi products has recently been investigated in depth (Scheme 4.17). In a series of papers, Wessjohann et al. published the syntheses of macrocycles of different size and shape from bisisocyanides or other bifunctional building blocks, involving either natural product fragments or purely synthetic components [39]. Potential applications and developments as specific sensors or macromolecular devices are envisioned.

Keto carboxylic acids have often been used as bifunctional starting materials in Ugi reactions (Scheme 4.18). From these, it is possible to obtain lactams of different ring size, depending on the distance between the keto and the carboxylic acid groups in the starting material [40, 41, 42]. Several groups have reported on the performance of the Ugi reaction under aqueous conditions. The advantage is that the product often precipitates during the reaction and can easily be filtered off and washed to provide clean products. The performance of Ugi reactions under aqueous reactions has been claimed to be general and to represent an efficient reaction procedure, especially in terms of purification. Pirrung et al. reported on a β-lactam synthesis with β-keto acids, primary amines, and isocyanides [43]. Use of heterocyclic keto carboxylic acids provides interesting heterocyclic lactams.

Interestingly, when molecules bearing isocyanide and carboxylic acid functionalities (**47**) are treated with Schiff bases (**48**), intermediate α-adduct oxazolidinones (**49**) can be isolated. Compound **49** can be further treated with a nucleophile (e.g., a primary or secondary amine) to yield bis-amide **50** (Scheme 4.19).

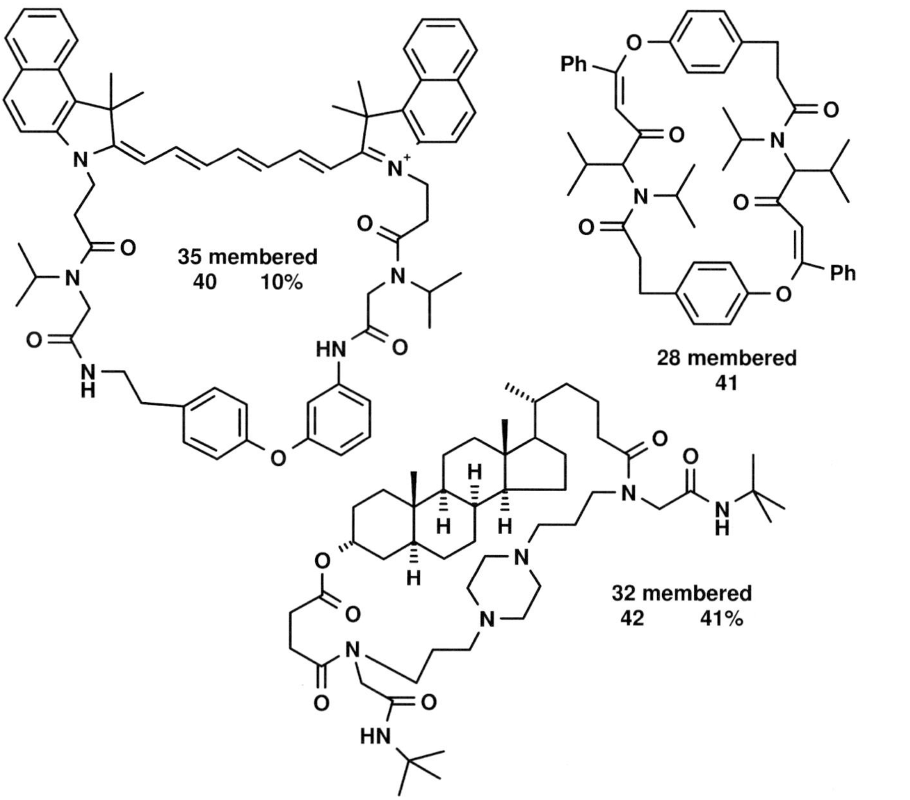

Scheme 4.17 Macrocyclic Ugi products formed by Ugi macrocyclization.

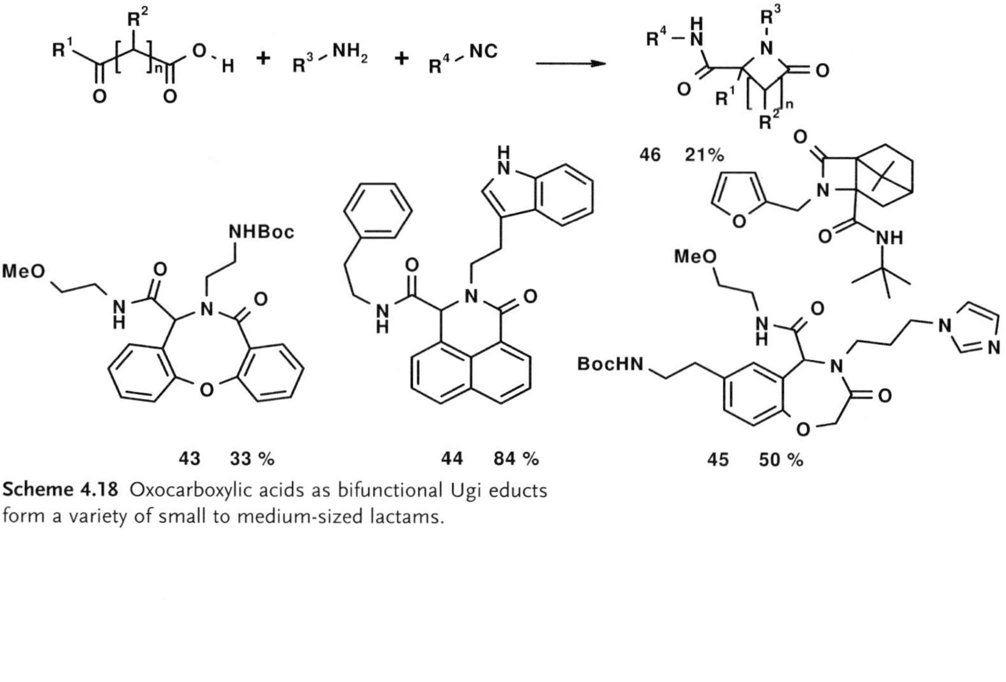

Scheme 4.18 Oxocarboxylic acids as bifunctional Ugi educts form a variety of small to medium-sized lactams.

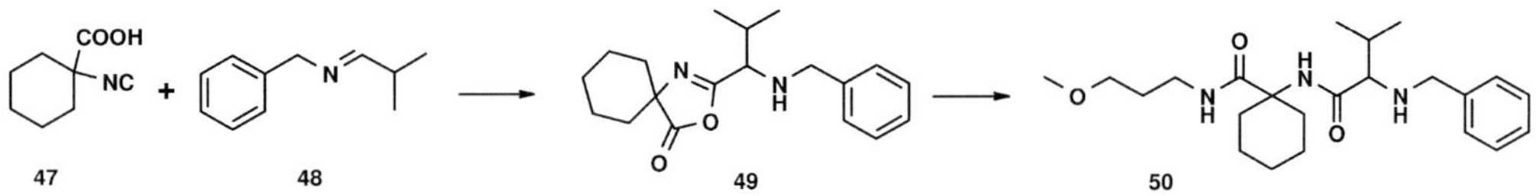

47 48 49 50

Scheme 4.19 Reactions of bifunctional molecules containing both carboxy and isocyano moieties with Schiff bases.

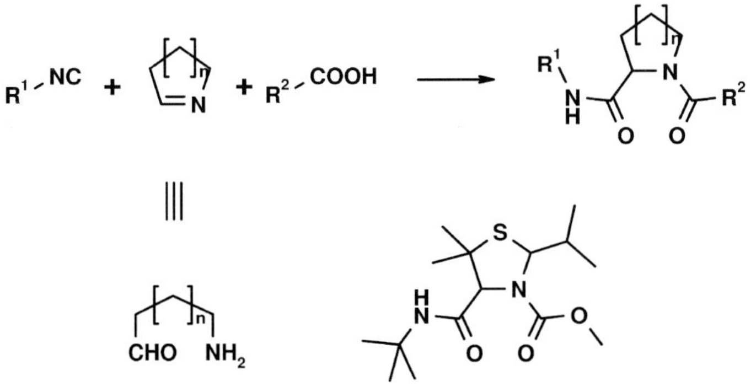

51 43%

Scheme 4.20 Cyclic Schiff bases derived formally from α,ω-amino aldehydes.

Cyclic Schiff bases are derived from α,ω-amino ketones or amino aldehydes, and preformed cyclic imines are excellent substrates in Ugi reactions (Scheme 4.20). The use of Asinger heterocycles has been widely described in the literature, and the Asinger 4-CR is an excellent route to a variety of heterocyclic S- or O-containing Schiff bases. Not surprisingly, the products can typically be isolated in good yields, and the corresponding products are clearly *cis*-amide, proline-type peptidomimetics with potential interesting biological activities. Compound **51** was prepared similarly, through treatment of seven distinct starting materials in one-pot style by making use of in situ Asinger and Ugi reactions [44].

One of the few Ugi variations – according to the general definition of hr Ugi reaction – that was never performed by Ugi, but which has only recently been discovered, is the Groebcke reaction of five- or six-membered amidines with aldehydes and ketones to form bicyclic nitrogen-containing ring systems. This reaction (Scheme 4.21) was published at the same time by several groups [45, 46, 47]. The exocyclic part of the amidine functions as an amine, whereas the cyclic nitrogen acts as an acid component, while the reaction must be catalyzed by a Brönsted or a Lewis acid. Several catalysts have been described, including acetic acid, *p*-toluenesulfonic acid, $HClO_4$, $Sc(OTf)_3$, BF_3, and $ZnCl_2$, as well as silica sulfuric acid, which seems to be particularly advantageous as it can easily be filtered off [48]. Because of the similarity of the products obtained by this procedure to several drugs on the market, this Ugi variation is nowadays extremely popular and has appeared many times in the patent literature.

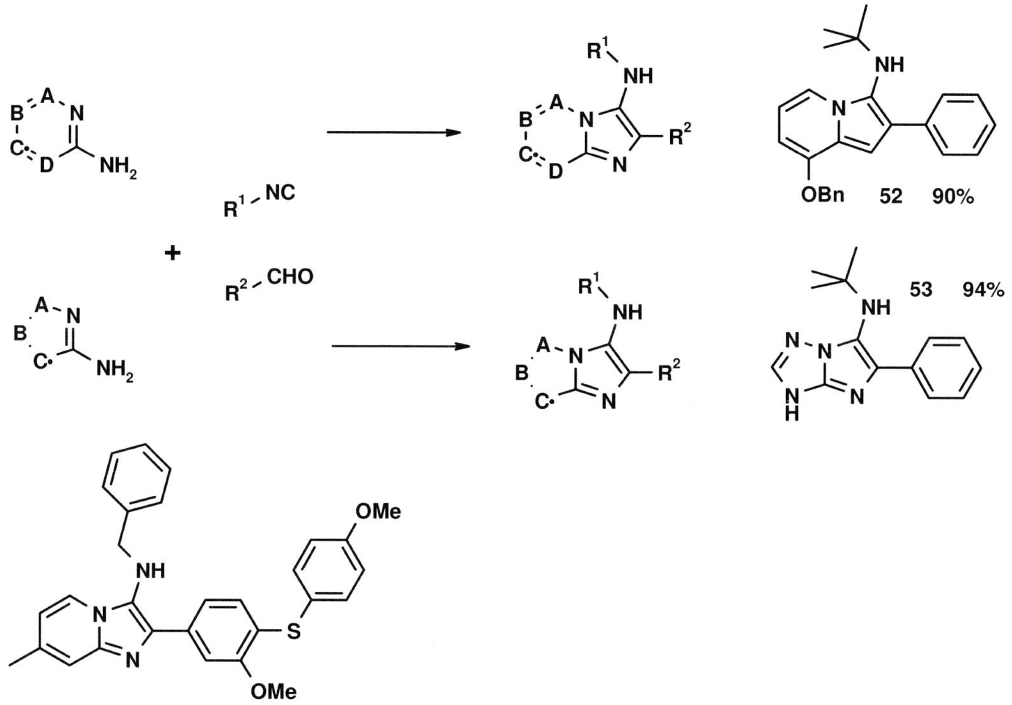

Scheme 4.21 3-CRs of heteroaromatic five- or six-membered amidines yielding bicyclic nitrogen heterocycles.

The reaction has been advantageously performed under solvent-free conditions with use of Montmorillonite clay [49], under microwave conditions [50], and also on solid support [51]. The introduction of formaldehyde apparently gives rise to problems in this reaction, but the use of glyoxylic acid and MP-glyoxylate as formaldehyde equivalents successfully produces 2-unsubstituted 3-aminoimidazo heterocycles through decarboxylation [52].

Use of an α,β-unsaturated isocyanide substituted with an β-leaving group in conjunction with a thiocarboxylic acid has recently been shown to afford in moderate to good yields of substituted thiazoles in one-pot fashion (Scheme 4.22). The discovery of this new scaffold represents a viable process for the design of a new MCR. Initially, with use of thioacetic acid in a classical U-4CR reaction the exclusive formation of an α-aminoacyl thioamide was observed. Surprisingly, no trace of the other expected α-aminothioacylamide regioisomer was detected. It is known that thiazoles can be formed by a variety of Hantzsch-type processes from thioamides. All of these, however, use primary thioamides. The question therefore arose of whether the surprising selective outcome of the above reaction could be advantageously used for a new thiazole synthesis. Indeed, the use of a multifunctional isocyanide – one of the 1-isocyano-2,2-dimethylacetic acid esters first

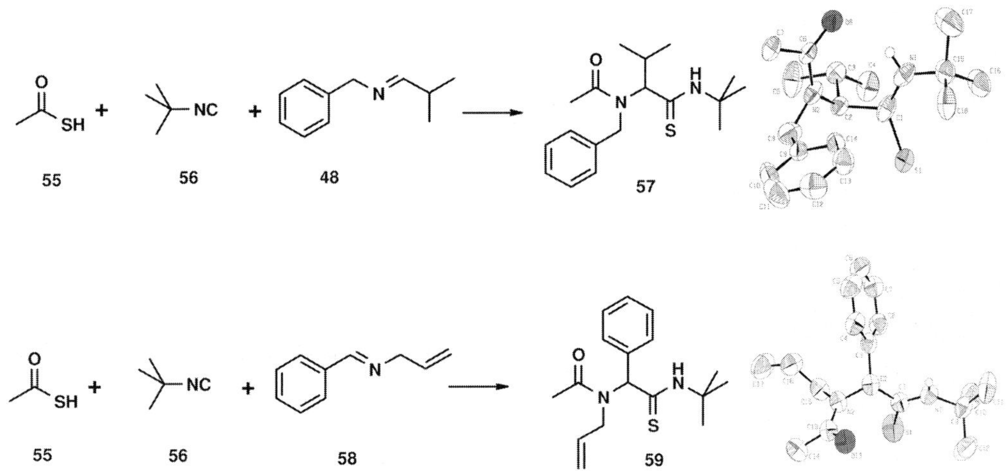

Scheme 4.22 A highly regioselective U-4CR with thioacetic acid, which formed the basis for the discovery of a new thiazole MCR synthesis.

described by Schöllkopf – together with other components as primary amines, aldehydes or ketones, and thiocarboxylic acids resulted in the formation of 2-methylaminoacylthiazoles. The reaction is highly compatible with many functional groups and can be used for the synthesis of thousands of thiazoles. Moreover, solid-phase variations, with either the Schöllkopf isocyanide or the primary amine resin-bound, have also been described. The Ugi product, reminiscent of a isocyanide-derived secondary amide, is transformed into a thiazole, thus yielding different ADMET properties and rendering the structures more Pfizer rule-like (Scheme 4.23).

The combinatorial chemistry of 3-(dimethylamino)-2-isocyanoacrylic acid methyl ester is highly diversity-oriented, reflecting the high degree of molecule functionalization, including the isocyanide and the ester moieties, the *N*,*N*-dimethylamino leaving group, and the α,β-unsaturation. Many scaffolds are easily accessible in large numbers and with great diversity, while several solid-phase variants of this isocyanide have also been described [53, 54, 55]. The potential of this multifunctional synthon is summarized in a recent review [56].

4.2.2
The Ugi Reaction and Secondary Transformations

A simple and efficient way of gaining access to a variety of new combinatorial scaffolds is to use an initial MCR followed by a secondary transformation such as a ring-closure. As can be seen in the increasing number of publications in recent years, this approach has become particularly popular and successful for investigation of an area devoted to new drug-like chemicals.

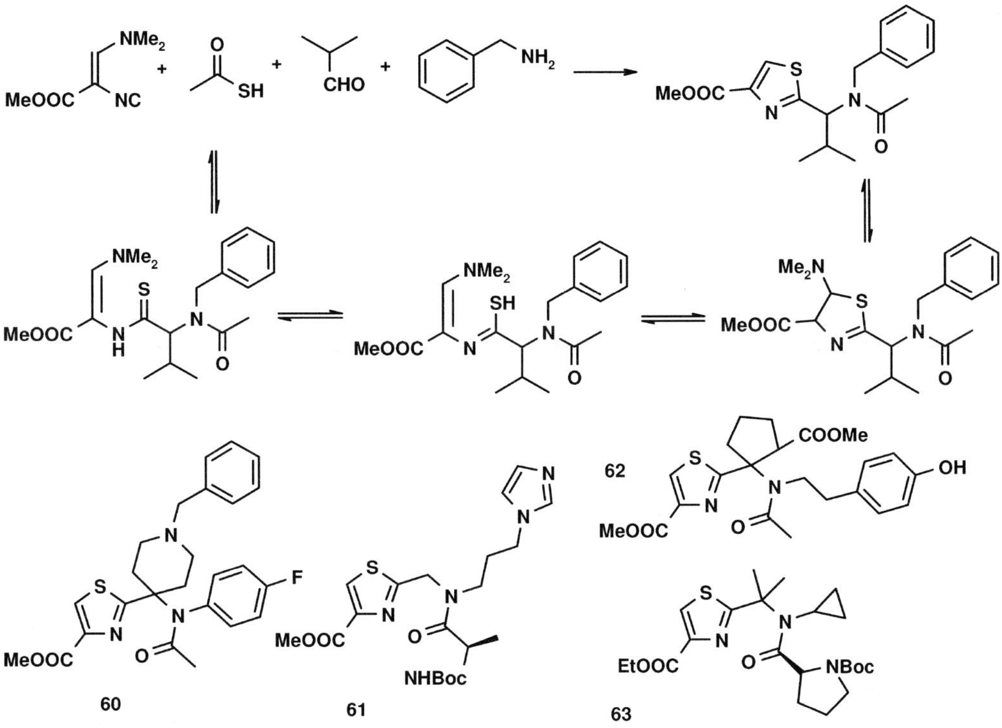

Scheme 4.23 Thiazoles are accessible through a new IMCR involving a U-4CR between a thiocarboxylic acid and the highly functionalized 3-(dimethylamino)-2-isocyanoacrylic acid methyl ester.

Hulme et al. introduced an especially efficient group of secondary transformations after Ugi reactions by using bifunctional mono-*N*-protected and secondary orthogonal bifunctional starting materials [57]. Thus, after the initial Ugi reaction has taken place by all the possible pathways the intermediate Ugi product is *N*-deprotected and subjected to a subsequent reaction, usually a heterocyclic ring formation (Scheme 4.24). Hulme coined the expression UDC (*U*gi-*D*eprotect-*C*yclization) for this very useful group of transformations

γ-Lactone, benzimidazole, benzodiazepine, diketopiperazine, ketopiperazine, morpholine, and quinazoline scaffolds are thus available in broad diversity (Scheme 4.25). All of these backbones have been reported to be amenable to high-throughput synthesis and tens of thousands of products have been synthesized, analyzed, and purified.

An interesting and unique feature of MCRs is that a particular backbone is often amenable by several synthetic routes, from differentially substituted starting materials [58]. The pharmacologically important benzodiazepine backbone, for example, can be made by IMCR-type approaches in over six different ways. Such diversity in approaching a basic structure is of great value, since each backbone, together

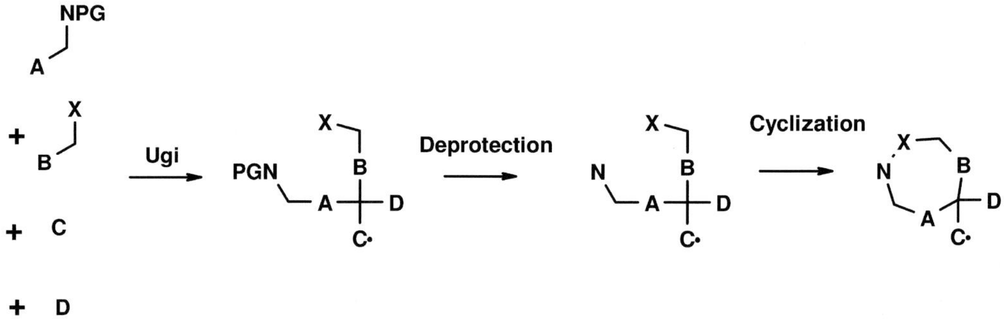

Scheme 4.24 General outline of a UDC scheme.

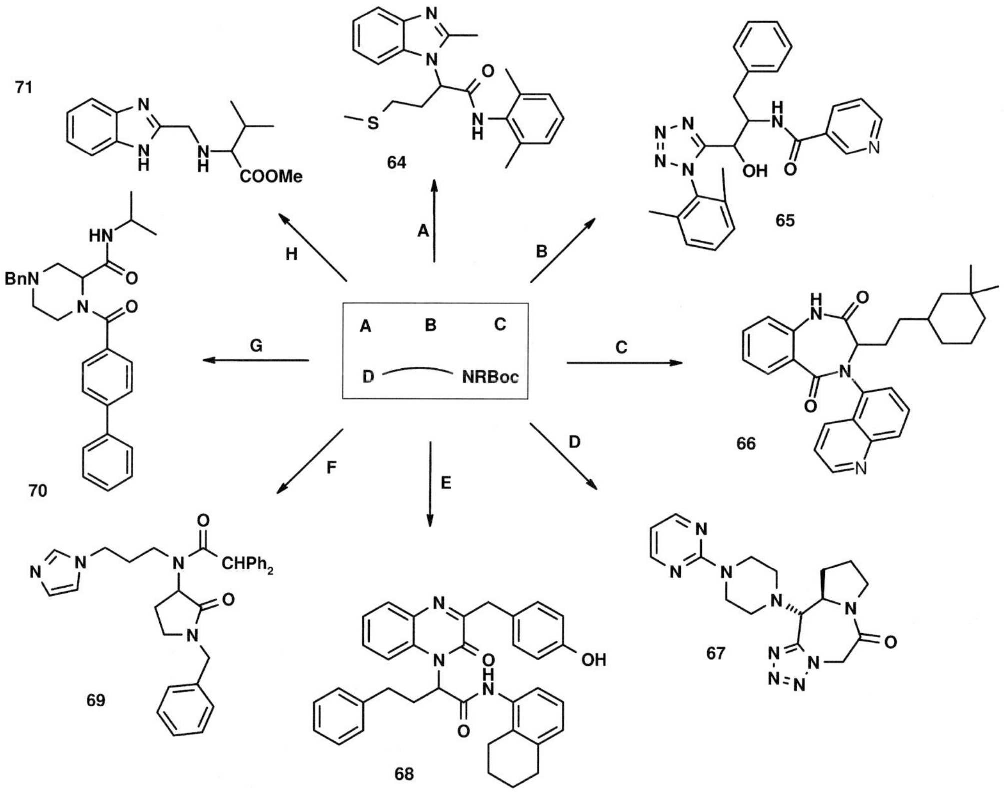

Scheme 4.25 UDC leading to a variety of useful scaffolds **64–71**.

with its corresponding educts, opens up a different chemical space, which also gives differential biological activities (Scheme 4.26).

The benzodiazepine scaffold is a privileged structure. According to the 1988 definition, when the term was introduced, a privileged structure is "a single

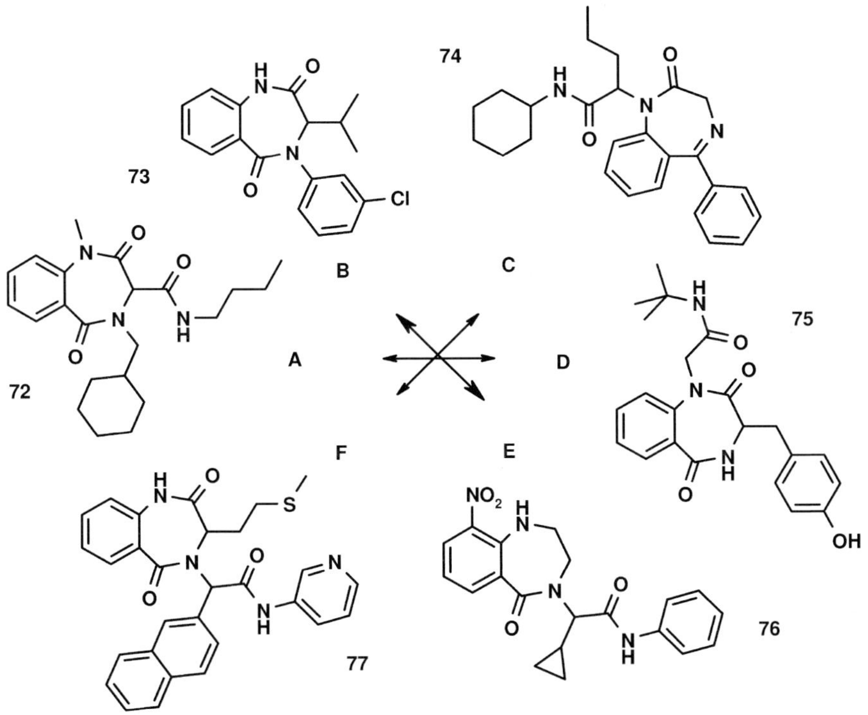

Scheme 4.26 Differential access to the benzodiazepine scaffold (compounds **72–77**) by IMCR. A) OHCCOOMe, RNC, RNH_2, anthranilic acids. B) Cyclohexenylisocyanide, *N*-Boc-anthranilic acid, amine. C) Isocyanide, aldehyde, aminobenzophenone, α-azido acid. D) *N*-Boc-amino acid, anthranilic acid ester, aldehyde, isocyanide. E) 2-Fluoro-5-nitrobenzoic acid, ethylenediamine, isocyanide, aldehyde. F) *o*-Nitrobenzoic acid, isocyanides, amino acid esters, aldehydes.

molecular framework able to provide ligands for diverse receptors" [59]. The benzodiazepine valium is an archetypal sedative and benzodiazepines are commonly used as peptidomimetics for β-turns [60]. Moreover, benzodiazepines as α-helix mimetics have also recently been described for the first time [61], while in 2005 an excellent application of a UDC backbone in drug discovery was reported by Parks et al. of Johnson & Johnson, dealing with the disruption of the protein-protein interaction antagonists between p53 and HDM2, an attractive approach for cancer therapy, effected by benzodiazepines (Scheme 4.27) [62]. Using UDC technology these workers were able to refine this hit by synthesizing more than 20 000 derivatives [63]. *N*-Protected anthranilic acids, primary amines, aldehydes, and a special isocyanide react to form an Ugi intermediate. This sequence is a UDC involving an initial Ugi-4CR of the "convertible" cyclohexenyl isocyanide, subsequent *N*-deprotection, and final in situ ring-closure under acidic conditions. The convertible cyclohexenyl isocyanide serves to introduce one of the

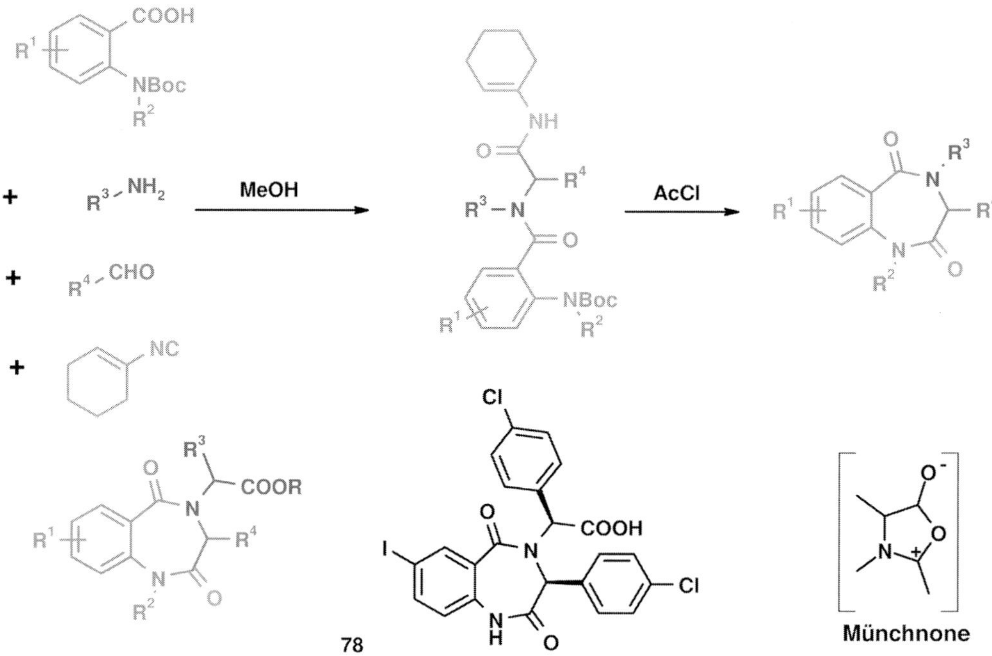

Scheme 4.27 UDC benzodiazepine as a scaffold of choice for the discovery of antagonists of protein interaction.

two carbonyl groups by forming the zwitterionic Münchnone intermediate upon acidification (Scheme 4.27).

Thanks to the availability of many different starting materials and the generality of the reaction a large library was synthetically accessible, and optimized benzodiazepines have been tested in xenograft models with highly promising results [64]. Currently a derivative is undergoing preclinical development. This example nicely underscores the potency of IMCR in terms of effectiveness and enabling: no other chemical methods are able to provide a 20 000 benzodiazepine library in a one-pot procedure from commercially available starting materials in a relatively short time.

Zhu and others investigated the chemistry of α-isocyanoacetamides derived from α-amino acids in the context of IMCR. The chemistry of α-isocyanoacetamides can be summarized as follows. The initial step involves a U-3CR of the α-isocyanoacetamides, amines, and aldehydes, with the intermediate Ugi product undergoing oxazole ring-formation under the usually employed (reflux) conditions. The obtained 5-aminooxazoles are electron-rich dienes and can undergo a variety of secondary addition reactions (Scheme 4.28). Interestingly, a high-yielding one-pot synthesis allows an easy supply of a large array of α-isocyanoacetamides, the starting materials for this reaction [65].

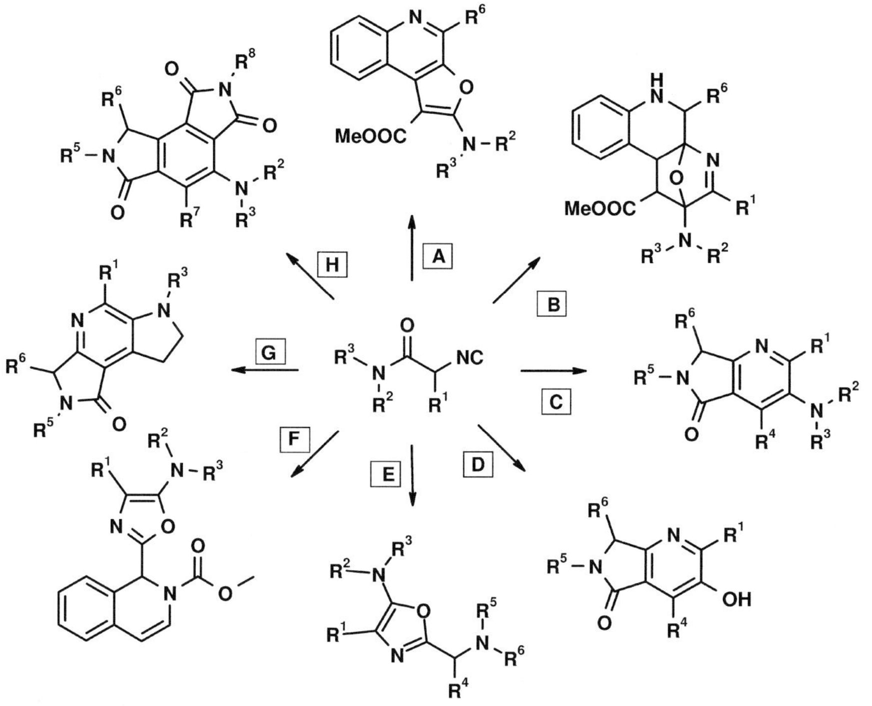

Scheme 4.28 Diversity in IMCRs of α-isocyanoacetamides. A) Furoquinolines. B) Tetrahydroquinolines. C) Pyrrolopyridine. D) Complementary pyrrolopyridine. E) 5-Aminooxazoles. F) 2-Oxazolo-tetrahydroisoquinolines. G) Dipyrrolopyridine. H) Hexasubstituted benzenes.

4.3
Passerini Reaction

Although discovered more than 30 years before the Ugi reaction, the Passerini reaction (P-3CR) never received the same attention as the former [66]. The two reactions might look quite similar simply on inspection of the stoichiometry; at a first approximation one might describe the Ugi reaction as a Passerini reaction with the addition of an amine. This, however, does not reflect the fundamental differences between the two reactions in terms of mechanism, variability, stereochemical induction, and applicability.

P-3CRs with isocyanides, oxo components, and carboxylic acids yield α-hydroxyacyl amides. The reactions are generally performed in aprotic lipophilic solvents between −20 °C and +60 °C, depending on the natures of the starting materials. The diversity of the Passerini reaction (Scheme 4.29) in terms of reacting acid components is rather small (the use of HN_3, for example has only rarely been

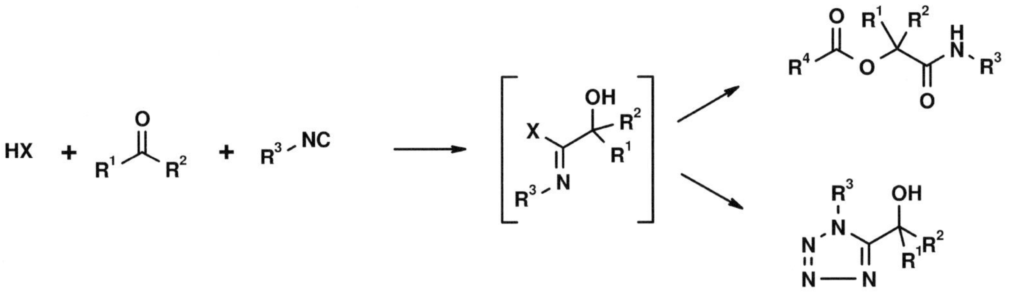

Scheme 4.29 A simplified mechanism of the Passerini reaction.

described). Moreover, unlike in the Ugi reaction, where this variant is an important application, the Passerini tetrazole variant is experimentally difficult to perform: HN_3 reacts in only a few cases and $Al(N_3)_3$ has been reported [67] as a more suitable N_3 source. Scaffold diversity clearly arises from secondary reactions of the Passerini reaction intermediates.

A detailed discussion of the stereochemical aspects of IMCRs is beyond the scope of this chapter, but great progress has been made, especially in the field of chiral induction in the Passerini reaction, and some recent developments are shown in Scheme 4.30. Despite several publications in this area, it has to be noted that the problem of stereoinduction in the Passerini reaction is not yet generally solved, the described procedures being valid only in particular cases.

It is interesting to speculate why Passerini-type MCRs seem to be stereochemically controllable, as shown by several publications in this field, whereas not one publication on catalytic stereoinduction during Ugi-type MCRs has appeared. This might be due to the different reaction mechanisms of these two reactions and to their very different reaction conditions. Whereas Ugi reactions are usually carried out in a protic solvent such as an alcohol or even water, Passerini-type MCRs are best performed in an aprotic apolar solvent (e.g., ether or an aromatic hydrocarbon).

The latter class of solvent systems is more easily amenable to selective Lewis acid complexation with the educts. Protic solvents, however, could easily suppress effective complexation with the starting materials of the Ugi reaction.

Many heterocyclic backbones involving a Passerini reaction and a secondary transformation have been described recently. The triply substituted oxazoles **92** have been reported to assemble during reactions between α-ketoaldehydes, carboxylic acids, and isocyanides and subsequent ammonia-mediated cyclisation [72], while ketocarboxylic acids react with isocyanides to yield [66] the expected lactones **95** and **97**. Haloketones yield α-epoxyamides **93** or azetidinones **94**, depending on the reaction conditions [73, 74], while 2-hydroxy-4-acylaminofurans **98** have been made by P-3CRs of α-ketoaldehydes and α-acidic acetic acids with subsequent Knoevenagel condensations [75]. Similarly, the combination of P-3CR and Wittig ring closure yields steroid-type multicyclic compounds **97** [76]. Finally, Schöllkopf isocyanide yields together with a thiocarboxylic acid 2-hydroxymethylthiazoles **99** in a Passerini-type reaction (Scheme 4.3.1) [77].

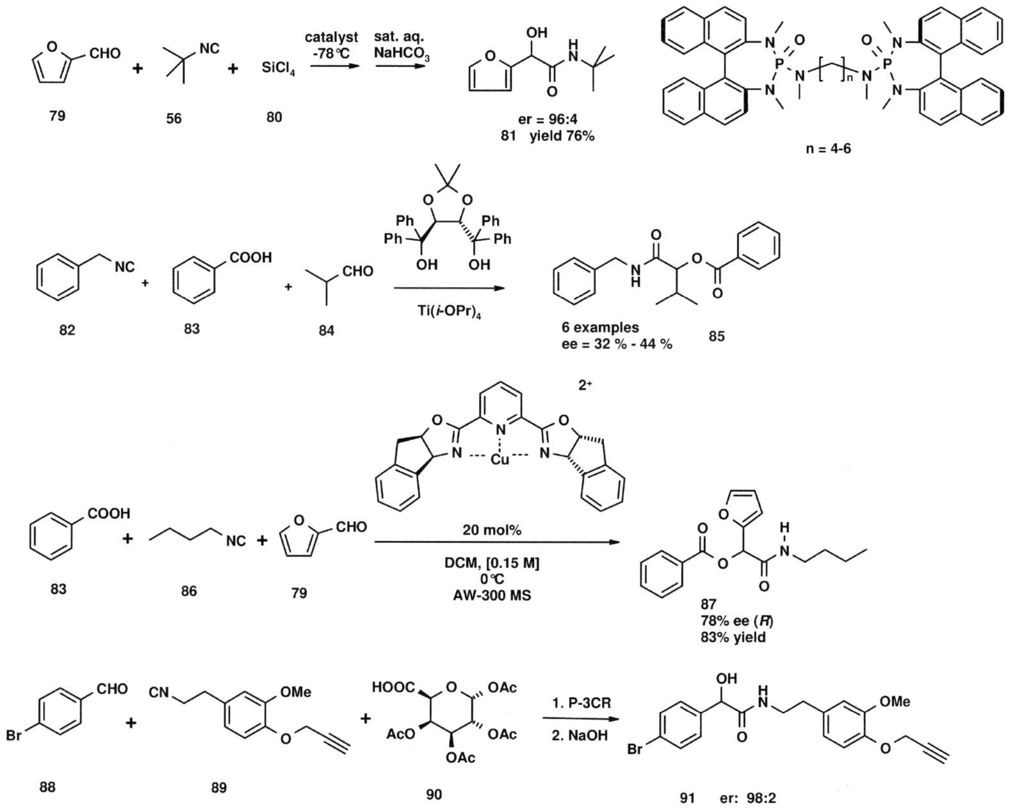

Scheme 4.30 Examples of enantio- and diastereoselective Passerini reactions. First line: Denmark's catalytic enantioselective P-2CR [68]. Second line: Dömling's enantioselective P-3CR using a chiral Lewis acid auxiliary [69]. Third line: Schreiber's enantioselective catalytic P-3CR. [70]. Fourth line: Solvay's diastereoselective P-3CR [71].

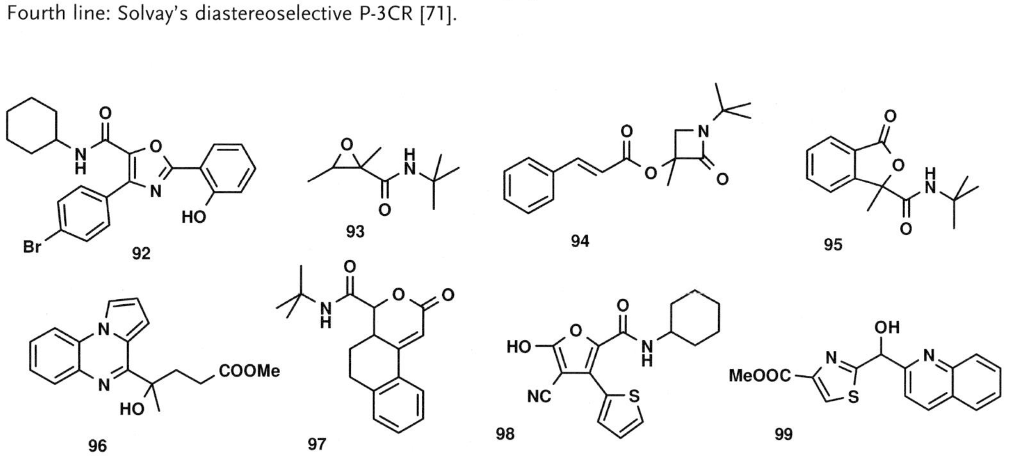

Scheme 4.31 Heterocycles by P-3CRs and secondary reactions.

At the same time, two groups published interesting sequences involving P-3CRs and finally affording the highly complex protease inhibitors **105–108** in only two or three steps (Scheme 4.32) [78, 79]. These P-3CRs involve the usage of amino acid-derived N-protected α-amino aldehydes; upon N-deprotection the acyl residue is N-transacylated. Thanks to the wide availability of chiral amino aldehydes, these reactions represent a convergent approaches towards highly substituted α-hydroxy or α-keto amides.

The abbreviation PADAM, standing for PAsserini Deprotection Migration, has recently been coined for this sequence. No other method affords such potentially valuable protease inhibitors in such a short sequence. A recent application of this reaction also points to its usefulness in natural product synthesis: Semple et al. performed a convergent synthesis of the protease inhibitor eurystatine using PADAM (Scheme 4.33) [80].

Macrocycles are amenable by a variety of methods that include IMCRs; the combination of an IMCR followed by a metathesis reaction to close the macrocycle has been particularly fruitful [81]. Reactions between cyclic carboxylic acid anhydrides and a variety of nucleophiles bearing terminal alkene moieties afford an array of carboxylic acids containing terminal alkene groups. Performing an IMCR with an educt containing an orthogonal terminal alkene yields an IMCR

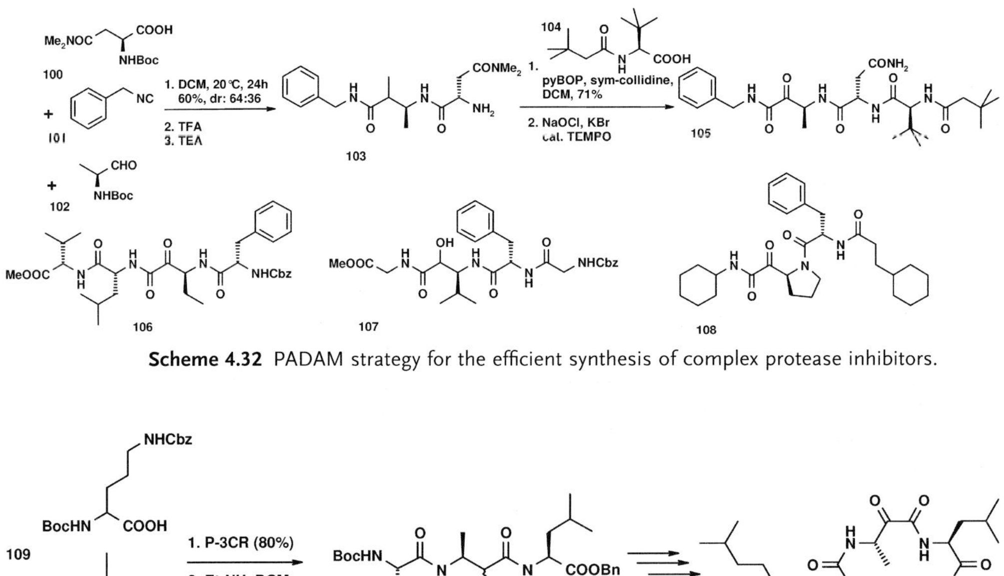

Scheme 4.32 PADAM strategy for the efficient synthesis of complex protease inhibitors.

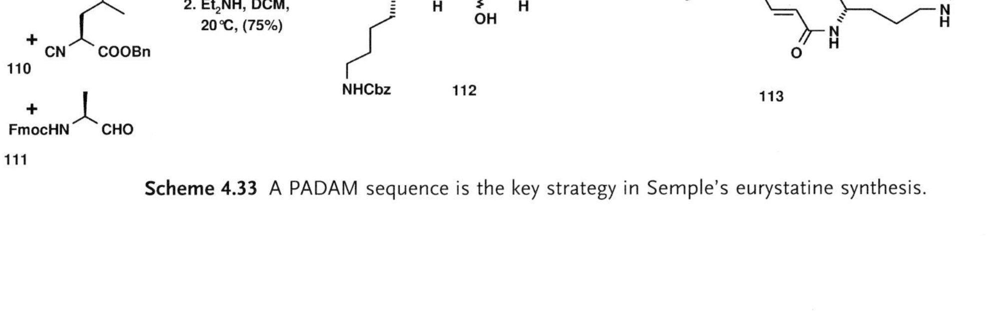

Scheme 4.33 A PADAM sequence is the key strategy in Semple's eurystatine synthesis.

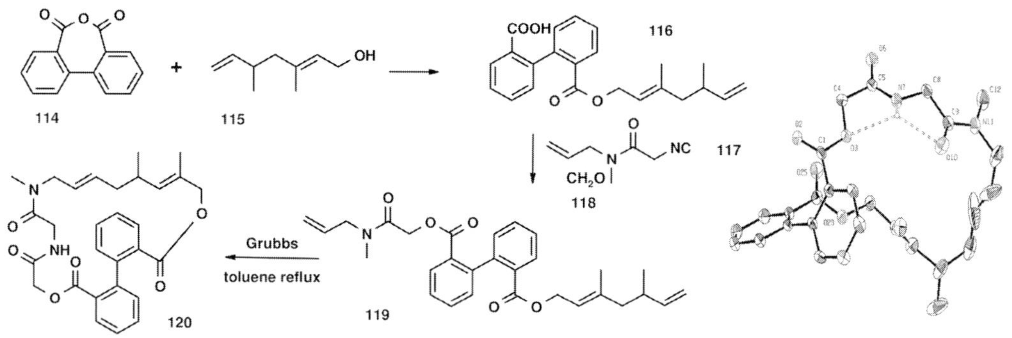

Scheme 4.34 A short sequence allowing access to a diverse conformational space of natural product-like macrocycles, involving an array synthesis of alkene carboxylic acids, subsequent IMCR, and final RCM. A prototypic macrocycle **120** is shown together with its 3D structure as determined by X-ray structure crystallography.

intermediate with a α,ω-diene relationship, which paves the ground for the final ring-closing metathesis (Scheme 4.34).

4.4 van Leusen Reaction

The Dutch chemist van Leusen discovered tosylmethylisocyanide (Tosmic) and subsequently investigated the reactivity and synthetic perspectives offered by this multifunctional reagent [82]. The chemistry of Tosmic is particularly diverse in relation to the other already known highly reactive types of isocyanides, as a result of its various pronounced reactivities, involving its α-acidic character, the reactivity of the isocyano function towards nucleophiles and electrophiles, and the leaving group ability of the sulfone (sulfinic acid abstraction). The chemistry of Tosmic thus involves Knoevenagel-type condensations, a connective C1 synthon, a ketone synthesis by alkylation and hydrolysis, and an Umpolung process, although only its MCR chemistry is reviewed here [83]. Synthetic aspects of Tosmic chemistry have recently been comprehensively summarized [84].

The van Leusen 3-CR of Tosmic or its α-substituted derivatives and aldehydes and primary amines yields 2,4,5-trisubstituted imidazoles (Scheme 4.35).

As shown in Scheme 4.35, the van Leusen 3-CR (vL 3-CR) mechanism involves the formation of a Schiff base, deprotonation of the Tosmic, nucleophilic addition onto the imine, secondary amine-induced isocyanide addition and ring-closure to the imidazolidine, and final sulfinic acid elimination with oxidative imidazole aromatization. It is a highly versatile MCR comparable to the Ugi reaction with respect to its functional group tolerance and the number of available starting materials, so very large libraries can potentially be produced ($>10^{12}$) and a very large chemical space can be exploited. Moreover the resulting product scaffold has

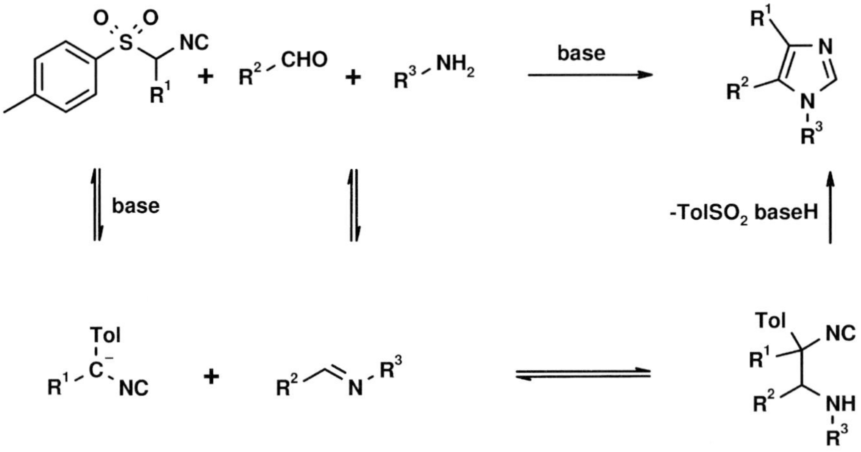

Scheme 4.35 The van Leusen 3-CR (vL 3-CR) mechanism.

potentially benign PKPD properties, with a large number of compounds obeying the Pfizer rules [85], so it represents very useful and widely applicable chemistry for the synthesis of compound libraries.

Tosmic and its derivatives are available by a number of processes, the most popular Tosmic synthesis involving another MCR, an α-aminoalkylation of an aldehyde with formamide and sulfinic acid, and dehydration of the resulting formamide by classical means [86]. Because of the α-acidic character of the Tosmic, however, alkylation procedures are also popular.

Tosmic had been used intensively in organic chemistry in the past, but the full potential of the van Leusen 3-CR was revealed by the seminal paper by Sisko et al. of SmithKline Beecham Pharmaceuticals, through the identification of previously unrecognized functional group compatibility [87]. Especially on the aldehyde and amine side, many functional groups are tolerated and do not interfere with the imidazole formation. Typical product examples are shown in Figure 4.1.

The aldehyde in the van Leusen 3-CR reaction displays broad tolerance towards a large variety of reactive functionalities. So far an additional formyl group such as that in glyoxal does not give rise to any interference and the same is the case with dicarbonyl compounds bearing aldehyde and ketone functions, hydroxy aldehydes, many heterocyclic systems, alkenes, and alkynes, to name just a few. In the same manner, carboxylic acids are tolerated in the amine component (i.e., chiral amino acids react with retention of the stereochemistry), but hydroxy, amino, disulfide, alkene, and alkyne functionalities can also be present. Ammonia undergoes the reaction to form 4,5-disubstituted imidazoles, whereas formaldehyde does not yield the expected 1,4-disubstituted imidazoles, although glyoxylic acid does indeed smoothly undergo the reaction to give 1,4-disubstituted imidazoles. The backbone of van Leusen imidazole can thus potentially be exploited in a wide variety of follow-up secondary reactions. Early examples involve alkene metathesis [88], the Heck reaction [89], C—H bond activation [90], or alkyne-azide cycloaddition reactions (Scheme 4.36) [91].

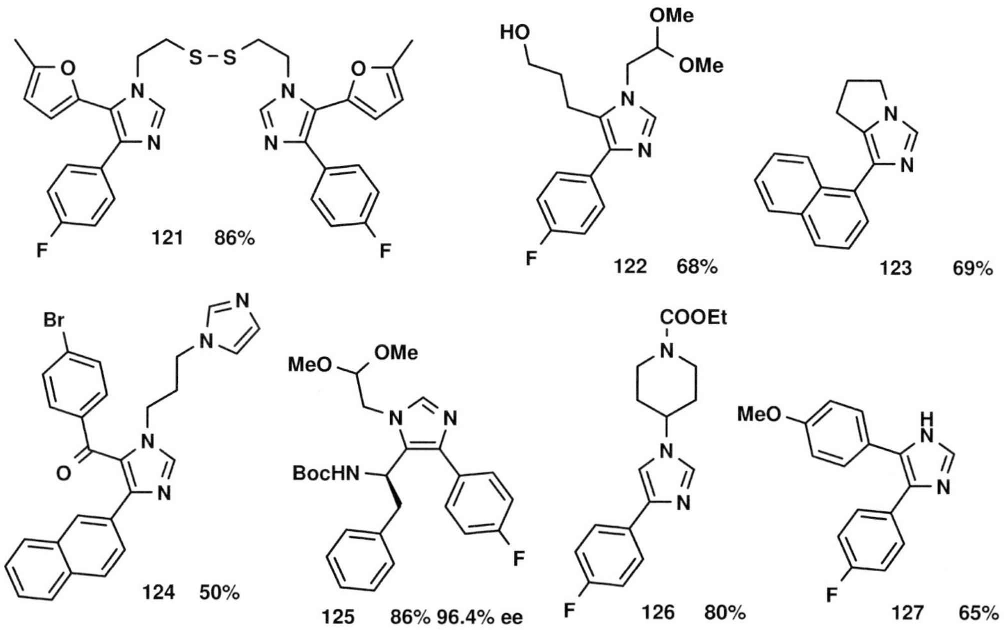

Figure 4.1 Examples of van Leusen 3-CR products with isolated yields underscoring the broad functional group tolerability.

Many more secondary transformations are obviously potentially viable, however, and will probably be exploited in the next future. In solid-phase variants of the van Leusen 3-CR, resin-bound Tosmic [92] and aldehyde [93] have been reported (Scheme 4.37). The van Leusen backbone is thus probably one of the most prolific primary MCR backbones in drug discovery.

Van Leusen products are ubiquitous in drug discovery, as emphasized by many patents and applications. Potent antimitotic van Leusen products inhibiting tubulin polymerization have been described recently [94]. Finally, it should be noted that a p38 inhibitor based on a van Leusen imidazole scaffold has undergone phase III clinical trials on hundreds of patients with arthritis. The corresponding Tosmic was prepared on a 500 kg scale and the subsequent vL-MCR was performed on a multikilogram scale (Scheme 4.38). This example shows that IMCRs can be scaled up and that the isocyanides can be safely handled even on a hundred kg scale [95].

4.5 Other IMCRs

That isocyanides form noncovalent adducts with activated carbon-carbon triple bonds (dimethyl acetylenedicarboxylate, DMAD), and that these can undergo subsequent addition reactions, was discovered as early as 1969 [96]. Although MCRs

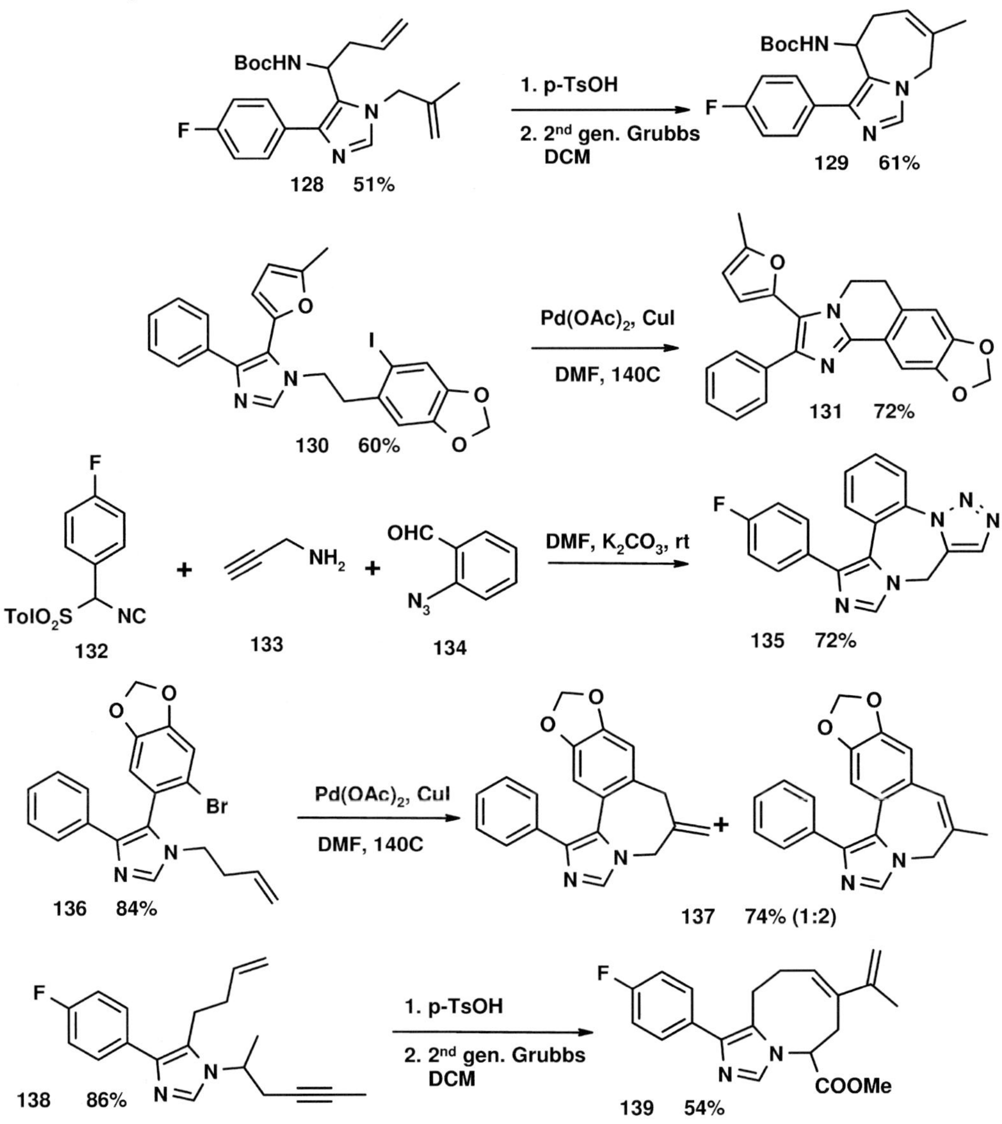

Scheme 4.36 Possible follow-up chemistry for van Leusen (vL) products. Two-step vL/metathesis sequences (**128**, **129**), vL/*C–H* activation ring-closure (**130**, **131**), vL/azido-alkyne ring-closure (**132–135**), vL/Heck (**136**, **137**), and vL/alkyne-alkene metathesis (**138**, **139**) have been carried out experimentally.

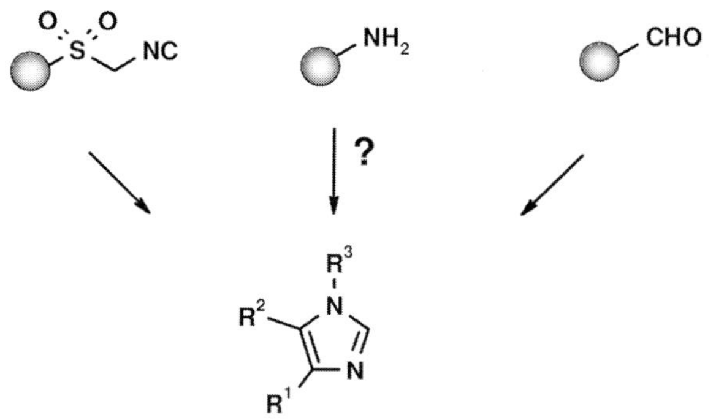

Scheme 4.37 Solid-phase variants of van Leusen's 3-CR employing resin-bound aldehyde and Tosmic, but not amine.

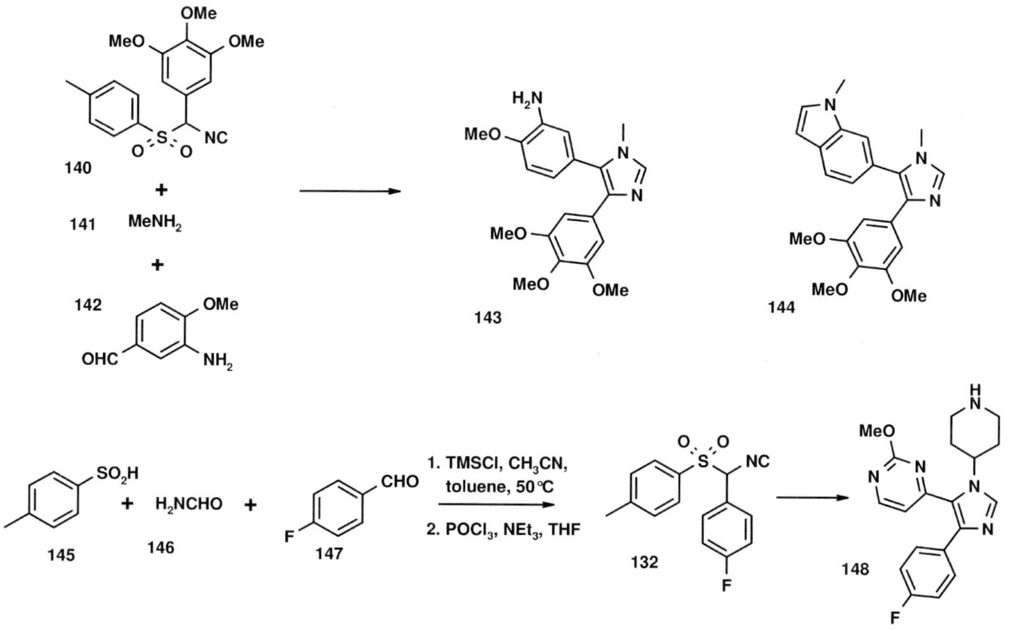

Scheme 4.38 Pharmaceutical applications of vL-IMCR:
a) synthesis of the antiproliferative tubulin binder **144**,
b) scalable synthesis of potent p38 kinase inhibitor **148**.

involving DMAD are in principle not apt to yield very large numbers of products, because the DMAD is almost invariably held constant, this class of IMCRs is nevertheless of special interest since a lot of different scaffolds are accessible from the intermediate DMAD isocyanide adduct (Scheme 4.39). Most investigation of the new IMCRs of DMAD has been by Saabhani's group in Iran.

DMAD is a versatile starting material for a variety of IMCRs. The crucial intermediate is the initial adduct of DMAD and isocyanide, the corresponding

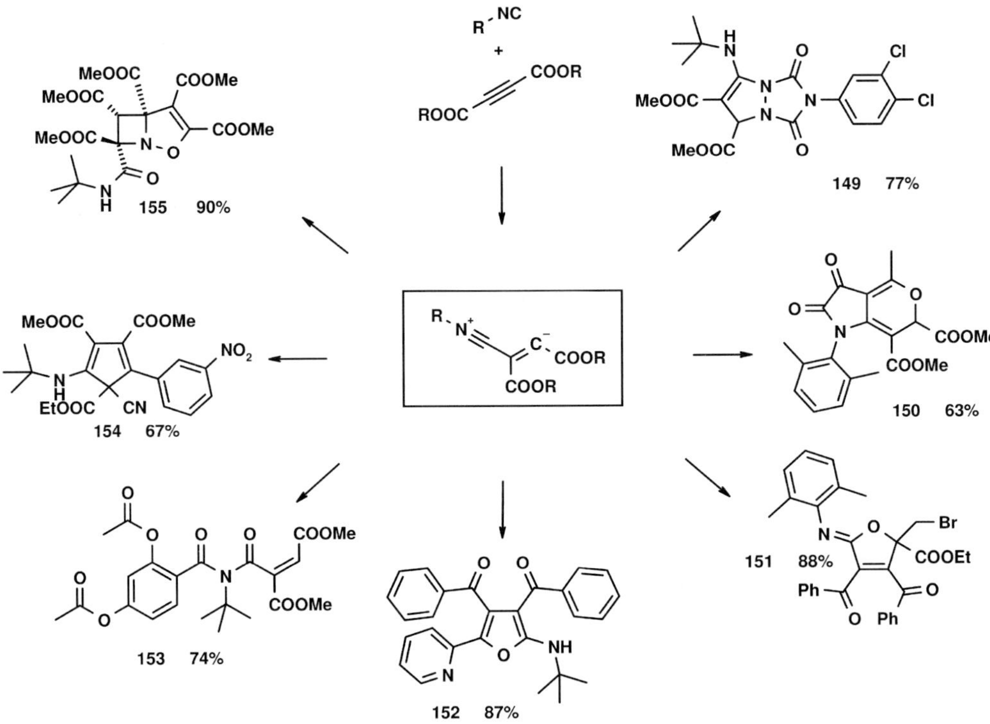

Scheme 4.39 Different scaffolds accessible from the intermediate DMAD isocyanide adduct.

zwitterion subsequently being able to undergo cycloaddition, ring-closure, and further additions. Substituted pyrrolotriazoles **149** [97], tetrahydropyranopyrroles **150** [98], iminodihydrofurans **151** [99], pyridylfurans **152** [100], butenedioate derivatives **153** [101], cyclopentadiene derivatives **154** [101], and azabicyclo[3.2.0]h eptenes **155** [102], among many others, can be achieved by this synthetic strategy.

4.6 Outlook

In this review a spotlight has been shone on a selection of the classical and new developments in IMCR chemistry. However it has become very clear from the numerous recently published applications that IMCRs represent a very potent synthetic technique well suited for further populating of the already large nitrogen-containing chemical space. No other single method allows such rapid and convergent synthesis of hundreds of nitrogen-containing scaffolds, and it is therefore safe to predict that IMCR will attract even more attention from the chemical and pharmaceutical communities. Future areas are likely to focus on new MCRs and their secondary transformations to access new scaffolds. Another area of high interest and largely unsolved so far deals with the stereoselective control of IMCRs. The appeal of the current MCR technology lies in its very mature state; that is, the corresponding procedure can be conveniently used in all kinds of discovery and chemistry projects. Moreover, the potential user of this technology can choose between a plethora of amenable scaffolds, displaying very different 2D structures and 3D pharmacophore and ADMET properties. Finally it remains to be said that the design of new MCR chemistry is a great intellectual challenge requiring a lot of theoretical knowledge of reactions of functional groups and orthogonal compatibility, as well as the thermodynamics and kinetics of organic reactions [104].

References

1 Ugi, I.; Meyr, R. *Angew. Chem.* 1958, *70*, 702.
2 Curran, D. P.; Ko, S.-B.; Josien, H. *Angew. Chem., Int. Ed. Engl.* 1996, *34*, 2683.
3 Dömling, A. *Chem. Rev.* 2006, *126*, 17.
4 Recent reviews a) Zhu, Bienayme, Multicomponent Reactions. Wiley-VCH 2005, Weinheim. b) Weber, L.; Illgen, K.; Almstetter, M. *Synlett* 1999, 366; c) Bienaymé, H.; Hulme, C.; Oddon, G.; Schmitt, P.; *Chem.-Eur. J.* 2000, *6*, 3321. d) Dömling, A.; Ugi, I. *Angew. Chem., Int. Ed.* 2000, *39*, 3168. e) Dömling, A. *Chem. Rev.* 2006, *126*, 17.
5 Ugi, I; Meyr, R.; Fetzer, U.; Steinbrückner, C. *Angew. Chem.* 1959, *71*: 386.
6 Mitchell, M. C.; Spikmans, V.; de Mello, A. J. *Analyst* 2001, *126*, 24.
7 Isonitrile Chemistry. Ugi, I. (Ed.), Academic Press 1971, New York.
8 Ugi, I.; Steinbrueckner, C. *Chem. Ber.* 1961, *94*, 2802.
9 Armour, D. R.; Bell, A. S.; Edwards, P. J.; Ellis, D.; Hepworth, D.; Lewis, M. L.; Smith, C. R. Preparation of heterocyclylcarboxamides as oxytocin inhibitors. WO 2004020414.
10 Shoda, M.; Harada, T.; Kogami, Y.; Tsujita, R.; Akashi, H.; Kouji, H.; Stahura, F. L.; Xue, L.; Bajorath, J. *J. Med. Chem.* 2004, *47*, 4286.
11 Portal, C.; Launay, D.; Merritt, A.; Bradley, M. J. *Comb. Chem.* 2005, *7*, 554.
12 Mironov, M. A.; Ivantsova, M. N.; Mokrushin, V. S. *Mol. Diversity* 2003, *6*, 193.
13 Pirrung, M. C.; Das Sarma, K. *J. Am. Chem. Soc.* 2004, *126*, 444.
14 Nerdinger, S.; Fuchs, T.; Illgen, K.; Eckl, R. Aryl amides that inhibit factor Xa activity. WO 2002068390.
15 Groebke Zbinden, K.; Banner, D. W.; Ackermann, J.; D'Arcy, A.; Kirchhofer, D.; Ji, Y.-H.; Tschopp, T. B.; Wallbaum, S.; Weber, L. *Bioorg. Med. Chem. Lett.* 2005, *15*, 817.
16 Sorensen, B.; Rohde, J.; Wang, J.; Fung, S.; Monzon, K.; Chion, W.; Pan, L.; Deng, X.; Stolarik, D.; Frevert, E. U.; Jacobson, P.; Link, J. T. *Bioorg. Med. Chem. Lett.* 2006, *23*, 5958.
17 Dömling; A.; Beck, B.; Magnin-Lachaux, M. *Tetrahedron Lett.* 2006, *47*, 4289.
18 Hulme, C. In *Multicomponent Reactions*; Zhu, J., Bienaymé, H., Eds.; Wiley: Weinheim, 2005; p 333.
19 Achatz, S.; Dömling, A. *Bioorg. Med. Chem. Lett.* 2006, *16*, 6360.
20 Ugi, I.; Offermann, K. *Chem. Ber.* 1964, *97*, 2276.
21 Ugi, I.; Rosendahl, F. K.; Bodesheim, F. *Justus Liebigs Ann. Chem.* 1963, *666*, 54.
22 Short, K. M.; Ching, B. W.; Mjalli, A. M. M. *Tetrahedron Lett.* 1996, *37*, 7489.
23 Ugi, I.; Steinbrückner, C. *Chem. Ber.* 1961, *94*, 2802.
24 Gross, H.; Gloede, J.; Keitel, I.; Dunath, D. *J. Pract. Chem.* 1968, *37*, 192.

25 Keating, T. A.; Armstrong, R. W. *J. Org. Chem.* 1998, *63*, 867.

26 Ugi, I.; Boettner, E. *Justus Liebigs Ann. Chem.* 1963, *670*, 74.

27 Masdeu, C.; Diaz, J. L.; Miguel, M.; Jiminez, O.; Lavilla, R. *Tetrahedron Lett.* 2004, *45*, 7907.

28 Diaz, J. L.; Miguel, M.; Lavilla, R. *J. Org. Chem.* 2004, *69*, 3550.

29 Tron, G. C.; Zhu, J. *Synlett* 2005, 532.

30 Dömling, A.; Kehagia, K.; Ugi, I. *Tetrahedron* 1995, *51*, 9519.

31 Dömling, A.; Starnecker, M.; Ugi, I. *Angew. Chem. Int. Ed. Engl.* 1995, *34*, 2238.

32 Kolb, J.; Beck, B.; Dömling, A. *Tetrahedron Lett.* 2002, *43*, 6897.

33 Ugi, I. *Angew. Chem.*, 1982, *94*, 826.

34 Basso, A.; Banfi, L.; Riva, R.; Guanti, G. *J. Org. Chem.* 2005, *70*, 575.

35 Sperka, T.; Pitlik, J.; Bagossi, P.; Tozer, J. *Bioorg. Med. Chem. Lett.* 2005, *15*, 3086.

36 Ku, I. W.; Cho, S.; Doddareddy, M. R.; Jang, M. S.; Keum, G.; Lee, J.-H.; Chung, B. Y.; Kim, Y.; Rhim, H.; Kang, S. B. *Bioorg. Med. Chem. Lett.* 2006, *16*, 5244.

37 Cho, S.; Keum, G.; Kang, S. B.; Han, S.-Y.; Kim, Y. *Mol. Div.* 2003, *6*, 283.

38 Failli, A.; Immer, H.; Götz, M. *Can. J. Chem.* 1979, *57*, 3257.

39 Wessjohann, L. A.; Ruijter, E. *Mol. Divers.* 2005, *9*, 159.

40 Gross, H.; Gloede, J.; Keitel, I.; Kunath, D. *J. prakt. Chem.*, 1968, *37*, 192.

41 Zhang, J.; Jacobson, A.; Rusche, J. R.; Herlihy, W. *J. Org. Chem.*, 1999, *64*, 1074.

42 Harriman, G. C. B. *Tetrahedron Lett.*, 1997, *38*, 5591.

43 Pirrung, M. C.; Das, S. K. *Synlett* 2004, 1425.

44 Dömling, A.; Ugi, I. *Angew. Chem. Intl. Ed. Engl.* 1993, *32*, 563.

45 Gröbke, K.; Weber, L.; Mehlin, F. *Synlett* 1998, 661.

46 Blackburn,C.; Guan, B.; Shiosaki, K.; Tsai, S. *Tetrahedron Lett.* 1998, *39*, 3635.

47 Bienaymé, H.; Bouzid, K. *Angew. Chem.* 1998, *110*, 2349; *Angew. Chem. Int. Ed.* 1998, *39*, 2234.

48 Shaabani, A.; Soleimani, E.; Malekim A. *Monatsh. Chem.* 2007, *138*, 73.

49 Varma, R. S.; Kumar, D. *Tetrahedron Lett.* 1999, *40*, 7665.

50 Ireland, S. M.; Tye, H.; Whittaker, M. *Tetrahedron Lett.* 2003, *44*, 4369.

51 Chen, J. J.; Golebiowski, A.; McClenaghan, J.; Klopfenstein, S. R.; West, L. *Tetrahedron Lett.* 2001, *42*, 2269.

52 Lyon, M. A.; Kercher, T. S. *Org. Lett.* 2004, *6*, 4989.

53 Henkel, B.; Sax, M.; Dömling, A. *Synlett* 2003, 2410.

54 Henkel, B. *Tetrahedron Lett.* 2004, *45*, 2219.

55 De Luca, L.; Giacomelli, G.; Porcheddu, A. *J. Comb. Chem.* 2005, *7*, 905.

56 Dömling, A.; Illgen, K. *Synthesis* 2005, *11*, 662.

57 Hulme, C.; Nixey, T. *Curr. Opin. Drug Discovery Dev.* 2003, *6*, 921.

58 Dömling, A. *Org. Chem. Highlights* 2005, July 5. URL: http://www.organic-chemistry.org/Highlights/2005/05July.shtm

59 Evans, B. E.; Rittle, K. E.; Bock, M. G.; Dipardo, R. M.; Freidinger, R. M.; Whitter, W. L.; Lundell, G. F.; Veber, D. F.; Anderson, P. S.; Chang, R. S. L.; Lotti, V. J.; Cerino, D. J.; Chen, T. B.; Kling, P. J.; Kunkel, K. A.; Springer, J. P.; Hirshfield, J. *J. Med. Chem.* 1988, *31*, 2235.

60 Hata, M.; Marshall, G. R. *Drug Discov. Des.* 2006, *20*, 321.

61 Cumming, M. D.; Schubert, C.; Parks, D. J.; Calvo, R. R.; LaFrance, L. V.; Lattanze, J.; Milkiewicz, K. L.; Lu, T. *Biol. Drug. Des.* 2006, *67*, 201.

62 Parks, D. J. et al. *Bioorg. Med. Chem. Lett.* 2005, *15*, 765.

63 Grasberger, B. L. et al. *J. Med. Chem.* 2005, *48*, 909.

64 Koblish, H. K. et al. *Mol. Cancer Ther.* 2006, *5*, 160.

65 Dömling, A.; Beck, B.; Fuchs, T.; Yazbak, A. *J. Comb. Chem*, 2006, *8*, 872.

66 Passerini, M. *Gazz. Chim. Ital.* 1923, *53*, 331.

67 Ugi, I.; Meyr, R. *Chem. Ber.* 1961, *94*, 2229.

68 Denmark, S. E.; Fan, Y. J. *Am. Chem. Soc.* 2003, *125*, 7825.

69 Kusebauch, U.; Beck, B.; Messer, K.; Herdtweck, E.; Dömling, A. *Org. Lett.* 2003, *5*, 4021.
70 Andreana, P. R.; Liu, C. C.; Schreiber, S. L. *Org. Lett.* 2004, *6*, 4231.
71 Frey, R.; Galbraith, S. G.; Guelfi, S.; Lamberth, C.; Zeller, M. *Synlett* 2003, 1536.
72 Bossio, R.; Marcaccini, S.; Pepino, R. *Liebigs Ann. Chem.* 1991, 1107.
73 Sebti, S., Foucaud, A. *Synthesis* 1983, 546.
74 Bossio, R.; Marcos, C. F.; Marcaccini, S.; Pepino, R. *Tetrahedron Lett.* 1997, *38*, 2519.
75 Bossio, R.; Marcaccini, S.; Pepino, R.; Torroba, T. *Synthesis* 1993, 783.
76 Beck, B.; Picard, A.; Herdtweck, E.; Dömling, A. *Org. Lett.* 2004, *6*, 39.
77 Henkel, B.; Beck, B.; Westner, B.; Mejat, B.; Dömling, A. *Tetrahedron Lett.* 2003, *44*, 8947.
78 Semple, E.; Owens, T. D.; Nguyen, K.; Levy, O. E. *Org. Lett.* 2000, *2*, 2769.
79 Banfi, L.; Guanti, G.; Riva, R. *Chem. Commun.* 2000, 985.
80 Owens, T. D.; Araldi, G.-L.; Nutt, R. F.; Semple, J. E. *Tetrahedron Lett.* 2001, *42*, 6271.
81 Beck, B.; Larbig, G.; Magnin-Lachaux, M.; Picard, A.; Herdtweck, E.; Dömling, A. *Org. Lett.* 2003, *5*, 1047.
82 van Leusen, A. M.; Wildeman, J.; Oldenziel, O. H. *J. Org. Chem.* 1977, *42*, 1153.
83 Dömling, A. http://www.organic-chemistry.org/frames.htm?http://www.organic-chemistry.org/Highlights/2005/05May.shtm.
84 van Leusen, D.; Van Leusen, A. M. *Org. Reac.* 2001, *57*, 417.
85 The Pfizer rules are based on a statistical analysis of bioavailability in a large company compound library data set. According to these rules compounds with more than one of the following characteristics are likely to have poor oral absorption: More than 5 H-bond donors; molecular weight >500; c log P > 5; sum of Ns and Os (a rough measure of H-bond acceptors) >10.
86 Sisko, J.; Mellinger, M.; Sheldrake, P. W.; Baine, N. H. *Org. Synth.* 2001, *77*, 198.
87 Sisko, J.; Kassick, A. J.; Mellinger, M.; Filan, J. J.; Allen, A.; Olsen, M. A. *J. Org. Chem.* 2000, *65*, 1516.
88 Gracias, V.; Darczak, D.; Gasiecki, A. F.; Djuric, S. W. *Tetrahedron Lett.* 2005, *46*, 9053.
89 Beebe, X.; Gracias, V.; Djuric, S. W. *Tetrahedron Lett.* 2006, *47*, 3225.
90 Gracias, V.; Gasiecki, A. F.; Paganao, T. G.; Djuric, S. W. *Tetrahedron Lett.* 2006, *47*, 8873.
91 Gracias, V.; Gasiecki, A. F.; Djuric, S. W. *Tetrahedron Lett.* 2006, *47*, 9049.
92 Kularni, A. B.; Ganesan, A. *Tetrahedron Lett.* 1999, *40*, 5633.
93 Samanta, S. K.; Kylanlahti, I.; Yli-Kauhaluoma, J.; *Bioorg. Med. Chem. Lett.* 2005, *15*, 3717.
94 Wang, L.; Woods, K. W.; Li, Q.; Barr, K. J.; McCroskey, R. W.; Hannick, S. M.; Gherke, L.; Credo, R. B.; Hui, Y.-H.; Marsh, K.; Warner, R.; Lee, J. Y.; Zielinski-Mozng, N.; Frost, D.; Rosenberg, S. H.; Sham, H. L. *J. Med. Chem.*2002, *45*, 1697.
95 Sisko, J.; Mellinger, M. *Pure Appl. Chem.* 2002, *74*, 1349.
96 Winterfeldt, E.; Schumann, D.; Dillinger, H. J. *Chem. Ber.* 1969, *102*, 1656.
97 Adib, M.; Sayahi, M. H.; Mahmoodi, N.; Bijanzadeh, H. R. *Helv. Chem. Acta* 2006, *89*, 1176.
98 Yavari, I.; Esnaashari, M. *Synthesis* 2005, *7*, 1049.
99 Yavari, I.; Hossaini, Z.; Sabbaghan, M. *Mol. Div.* 2006, *10*, 479.
100 Adib, M.; Sayahi, M. H.; Koloogani, S. A.; Mirzaei, P. *Helv. Chim. Acta* 2006, *89*, 299.
101 Alizadeh, A.; Rostamnia, S.; Zhu, L.-G. *Tetrahedron Lett.* 2006, *62*, 5641.
102 Nair, V.; Menon, R. S.; Beneesh, P. B., Bindu, S. *Org. Lett.* 2004, *5*, 767.
103 Yavari, I.; Moradi, L. *Helv. Chim. Acta* 2006, *89*, 1942.
104 Dömling, A. *Curr. Opin. Chem. Biol.*, 2000, *4*, 318.

5
Direct Catalytic Asymmetric Mannich Reactions and Surroundings

Armando Córdova and Ramon Rios

5.1
Introduction

The amino group is one of the most important functionalities in organic synthesis and in nature, and in this context the Mannich reaction is a classic method for the preparation of β-amino carbonyl compounds and therefore a very important carbon-carbon bond-forming reaction in organic synthesis. Its versatility and potential to create both functional and structural diversity have long stimulated the creativity of chemists [1] and it has been successfully employed numerous times as a key step both in natural product synthesis and in medicinal chemistry [2]. The first asymmetric Mannich reactions were diastereoselective and involved the addition of preformed enolates and enamines to preformed imines with the use of stoichiometric amounts of chiral auxiliaries [3]. Only recently have the first successful examples of catalytic asymmetric additions of enolates to imines been reported [4–6]. However, the preparation and instability of the preformed enolates used can be a disadvantage in these stereoselective Mannich reactions. More recently, elegant indirect organocatalytic enantioselective Mannich-type reactions have been developed [7–8]. Organosilver complexes are also excellent catalysts in indirect one-pot, three-component asymmetric Mannich reactions [9], while the most effective and atom-economic asymmetric Mannich reaction would be a catalytic process that involved the same equivalents of unmodified carbonyl donor, amine, and acceptor aldehyde [Eq. (1)] [10].

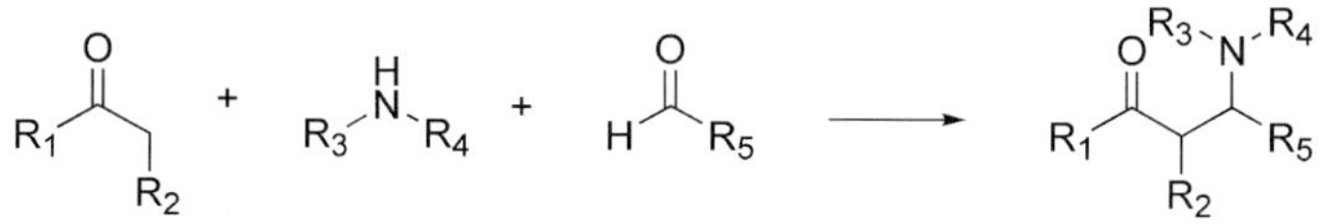

(1)

Recently, direct catalytic asymmetric Mannich-type reactions based on C—H activation of carbonyl compounds have been reported. The transformations are catalyzed both by organometallic complexes and by metal-free organic catalysts.

Amino Group Chemistry. From Synthesis to the Life Sciences. Edited by Alfredo Ricci

ISBN: 978-3-527-31741-7

5.2 Organometallic Catalysts

Shibasaki and co-workers have conducted extensive research into the use of heterobimetallic complexes as catalysts for asymmetric synthesis [11]. The reactions are catalyzed by heterobimetallic complexes that function both as Lewis acids and as Brønsted bases. Among them, $LaLi_3$tris(binaphthoxide) catalyst **1** (LLB) has proven to be an effective catalyst in direct asymmetric aldol reactions (Figure 5.1) [12]. On the basis of this research, Shibasaki et al. disclosed the first report of a direct catalytic asymmetric Mannich reaction [13].

In their initial one-pot, three-component experiment, propiophenone, $(CH_2O)n$, and pyrrolidine were allowed to react in the presence of a catalytic amount of LLB, affording the corresponding Mannich product with an *ee* of 64% (Scheme 5.1).

The yield of the Mannich product was only 16%, due to competing formation of $C_4H_8NCH_2NC_4H_8$. However, the chemoselectivity of the Mannich reaction can be significantly increased by in situ generation of the iminium ion through the use of aminomethyl ethers in combination with rare earth metal triflates and AlLibis(binaphthoxide) (ALB; **2**) as the catalyst (Figure 5.1) [Eq. (2)].

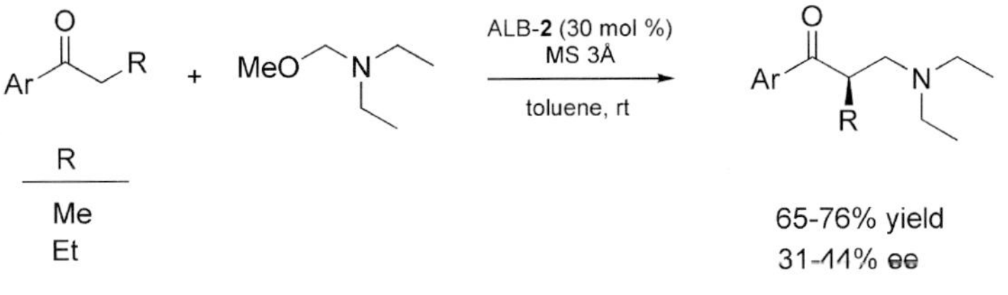

(2)

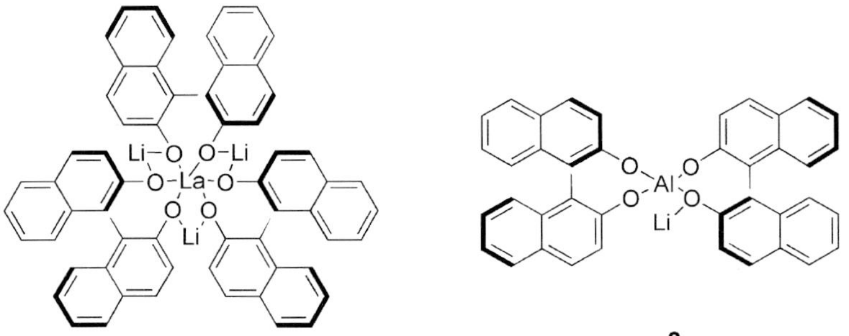

1

2

Figure 5.1 $LaLi_3$tris(binaphthoxide) catalyst **1** and AlLibis(binaphthoxide) (ALB) **2**.

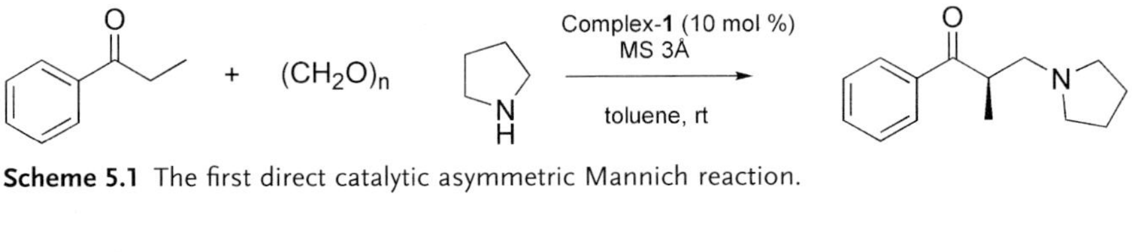

Scheme 5.1 The first direct catalytic asymmetric Mannich reaction.

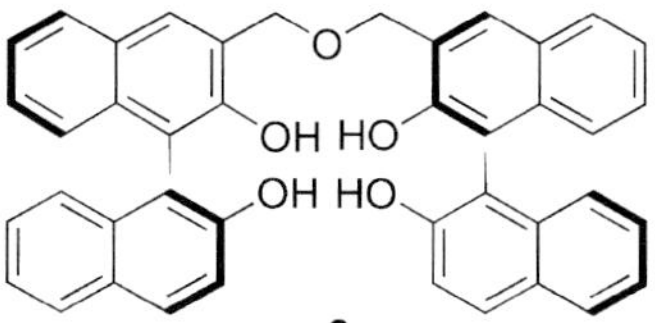

3

Figure 5.2 (*S*,*S*)-linked BINOL ligand **3**.

The cooperative complex of ALB (**2**) and $La(OTf)_3 \cdot nH_2O$ catalyzes direct asymmetric Mannich-type reactions with good selectivity, providing β-amino aryl ketones in good yields and with 31–44% *ees*.

Most recently, Shibasaki et al. reported that the Et_2Zn/linked-BINOL complex **3** (Figure 5.2) is an excellent catalyst for direct asymmetric Mannich-type reactions [14–15].

The Et_2Zn/linked-complex **3** was investigated as a catalyst because of its high selectivity in direct asymmetric *syn*-selective aldol reactions and Michael reactions with aryl hydroxyketones as the donors [16]. Mannich reactions between imines with various *N*-protecting groups, hydroxyaceto-2-methoxyphenone, and Et_2Zn-linked complex **3** showed *N*-diphenylphosphinoyl-protected (Dpp-protected) imines to be the most promising with regard to stereoselectivity. *N*-Dpp-protected imines are thus employed as acceptors and the corresponding Mannich adducts are isolated in high yields and with excellent enantioselectivities [Eq. (3)].

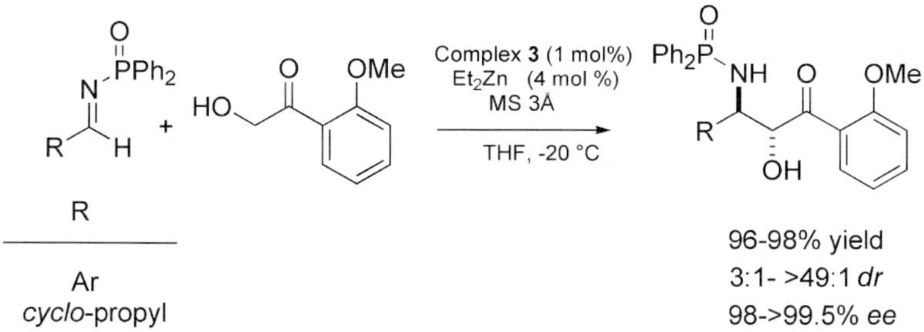

(3)

High *anti* diasteroselectivity is observed for several aromatic imines: in the case of *ortho*-substituted aromatic imines the two newly formed stereocenters are created with almost absolute stereocontrol. Aliphatic imines can also be used as substrates and the reaction can readily be performed on a gram scale with as little as 0.25 mol% of catalyst loading. Furthermore, the Mannich adducts are readily transformable into protected α-hydroxy-β-amino acids in high yield. The absolute stereochemistries of the Mannich adducts revealed that the Et_2Zn-linked complex **3** affords Mannich and aldol adducts with the same absolute configurations (2*R*), but that the diastereoselectivities of the amino alcohol derivatives are *anti*, unlike in the case of the *syn*-1,2-diol aldol products. The electrophiles thus approach the

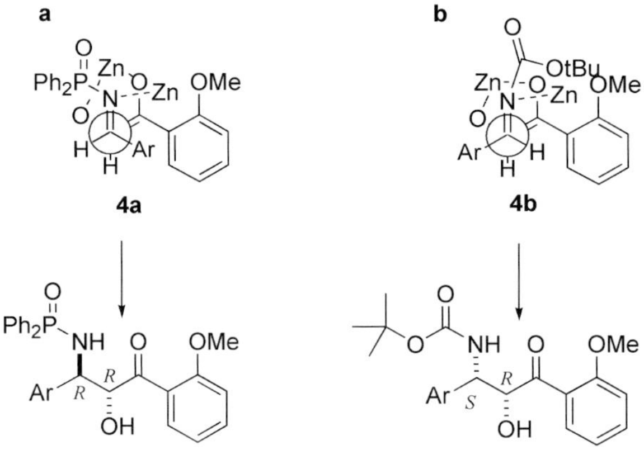

Figure 5.3 Plausible transition states **4a** and **4b**.

re-face of the Zn enolate in the Mannich reactions and the *si*-face in the aldol reactions. The *anti* selectivity is due to the bulky Dpp-group of the imine nitrogen; to avoid steric repulsion the Mannich-type reaction proceeds via the transition state **4a** (Figure 5.3a). Interestingly, a change in the protecting group on the preformed imines to a Boc group switched the electrophilic approach of the enolate derived from the Zn complex of **3** to the *si*-face (Figure 5.3b; transition state **4b**), thus furnishing Boc-protected *syn*-1,2-amino alcohols with excellent stereoselectivity [Eq. (4)] [15].

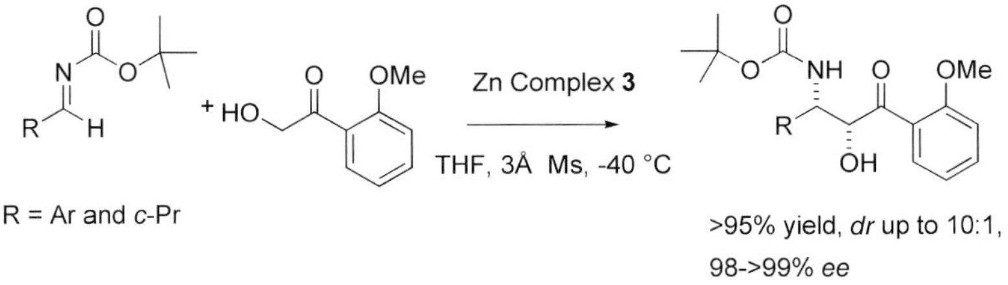

(4)

It has recently been shown that fine-tuned Zn/non-C_2-symmetric linked-BINOL complexes such as **3a–3d** (Figure 5.4) catalyze the direct Mannich reaction depicted in Eq. 3 with high *anti*- and enantioselectivity [17].

Trost et al. have discovered a novel design of dinuclear zinc catalysts that can catalyze diasteroselective and enantioselective direct aldol reactions [18]. The dinuclear zinc catalysts **5a–7a** are generated in situ by treatment of the appropriate ligands **5–7**, respectively, with 2 equiv. of diethylzinc in THF (Scheme 5.2).

Trost and co-workers have recently reported that these dinuclear zinc complexes catalyze Mannich reactions with unmodified aromatic hydroxy ketones as donors with excellent enantioselectivity [19]. Mannich-type reactions between *N*-*p*-methoxyphenyl-protected (PMP-protected) α-ethyl glyoxalate and hydroxyaceto-

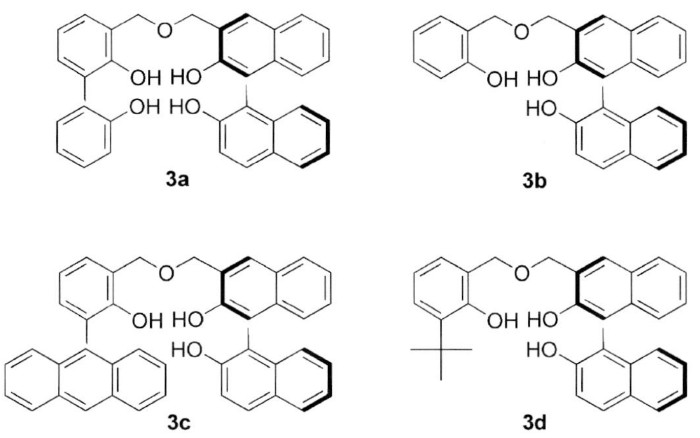

Figure 5.4 Non-C_2-symmetric (*S*,*S*)-linked BINOL complexes **3a–3d**.

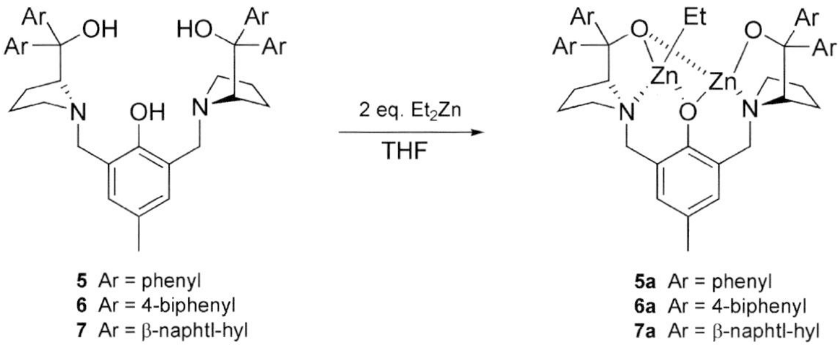

Scheme 5.2 Generation of dinuclear Zn catalysts **5a–7a**.

phenone in the presence of a catalytic amount of catalyst **5a** afforded the desired *N*-PMP-protected amino acid derivative in 76% yield with a *d.r.* of 7 : 1 and 95% *ee* [Eq. (5)].

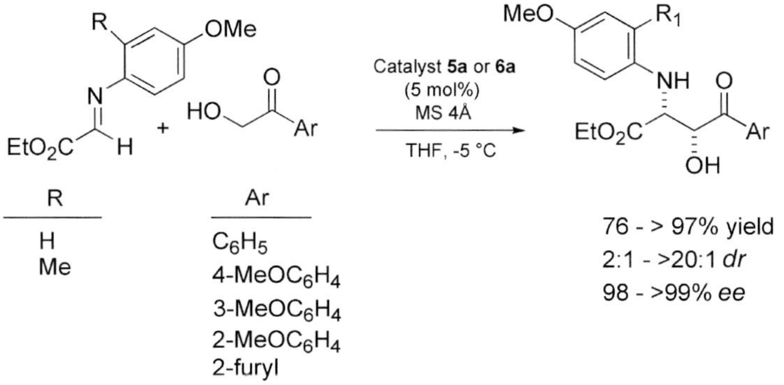

(5)

Imines derived from the more sterically demanding 2-methyl-4-methoxyaniline derivatives and ethyl glyoxylate provide even higher selectivity. The reaction also exhibits a significant ligand effect: when ligand **6** is used the yield and diastereoselectivity are further improved, so the biphenyl ligand **6** is used as the standard ligand for reactions with glyoxylate imines. The Zn_2-linked complex **6a** catalyzes Mannich reactions with more electron-rich aromatic hydroxy ketones and a single diastereoisomer is formed with *ortho*-substituted methoxy hydroxy ketone as the donor. The reactions catalyzed by the dinuclear zinc complex are also highly enantioselective for acceptor imines derived from aromatic aldehydes. In this case the β-naphthyl ligand **7** significantly increases the diastereoselectivity of the reaction.

The employment of acceptor imines derived from *ortho*-methoxyaniline also dramatically increases the diastereoselectivity of the reaction, which can explained in terms of a bidentate binding model with the *ortho*-substituted derivative, in which two-point binding of the imine – both through the nitrogen and through the methoxy group – helps to rigidify the dynamic nature of the imine-Lewis acid complex, preventing E/Z isomerization of the carbon-nitrogen double bond. Importantly, a nearly atom-economic process can be achieved, 1.1 equiv of ketone, and with no change in chemoselectivity after addition of "zincaphilic" additives such as Ph_3PS and Ph_3AsO. The 1,2-amino alcohol derivatives are valuable synthetic intermediates: a three-step procedure including Baeyer–Villiger oxidation and oxidative dearylation, for example, provides protected *syn* α-hydroxy-β-amino acids that are constituents of natural products such as taxol [2e]. NOE experiments confirmed the formation of *syn*-1,2-amino alcohols, which is the opposite of the case of the direct asymmetric aldol reaction catalyzed by the dinuclear zinc complex, in which vicinal diols with *anti* configurations are obtained [18].

Jørgensen and co-workers have developed direct asymmetric reactions that are catalyzed by chiral copper(II)bisoxazoline (BOX) complexes [20] and on the basis of this research have disclosed the first chiral copper(II)/(BOX) complex-catalyzed direct asymmetric Mannich reactions between activated carbonyl compounds and α-imino esters [21]. Screening of different C_2-symmetric ligands with copper(II) as the metal ion revealed that the $Cu(OTf)_2$/BOX complexes **8** and **9** (Figure 5.5) functioned as catalysts in the reaction between pyruvate and *N*-tosyl-protected α-imino ethyl glyoxylate. Different α-carbonyl esters can be used as nucleophiles, providing functionalized α-amino acid derivatives with excellent diastereoselectivity and enantioselectivity [Eq. (6)].

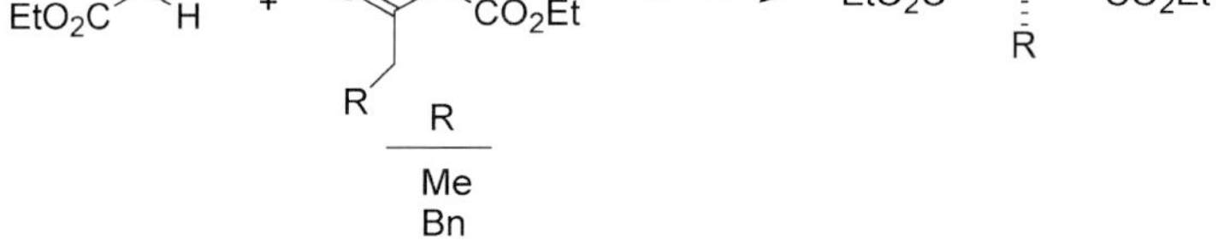

(6)

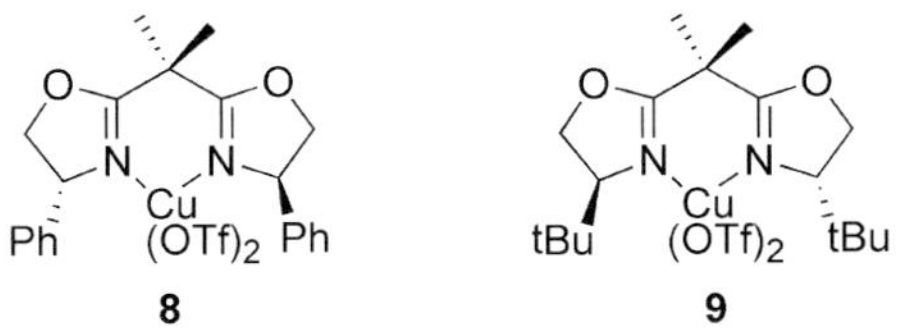

Figure 5.5 $Cu(OTf)_2$/BOX complex **8** and **9**.

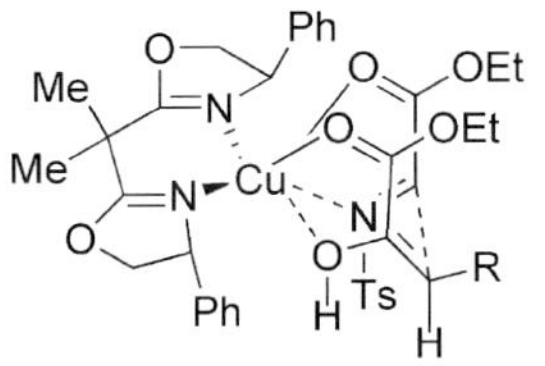

Figure 5.6 Plausible transition state.

The Mannich adducts are readily transformed into optically active α-amino-γ-lactones through a diastereoselective one-pot reduction and lactonization sequence and the tosyl group can be exchanged for a Boc group in a two-step procedure. The copper(II) ion is crucial for the success of this reaction [22]: it has the properties necessary both to generate the enol species in situ and, in combination with the C_2-symmetric ligand, also to coordinate it as the imine in a bidentate fashion. The reaction proceeds via a cyclohexane-like transition state with the *R* substituent of the enol in the less sterically crowded equatorial position, which is required in order to achieve the observed diastereoselectivity (Figure 5.6).

The chiral $Cu(OTf)_2$/BOX complexes **8** and **9** are also catalysts for asymmetric Mannich-type additions of unmodified malonates and β-ketoesters to activated *N*-tosyl-α-imino esters [23].

Shibasaki and Matsunaga have shown that indium-linked BINOL complexes act as catalysts in highly enantioselective direct catalytic asymmetric Mannich-type reactions of *N*-(2-hydroxyacetyl)pyrrole as an ester-equivalent donor [Eq. (7)] [24].

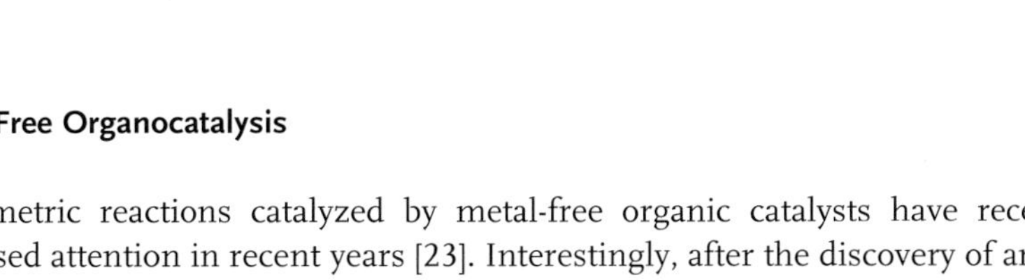

(7)

5.3 Metal-Free Organocatalysis

Asymmetric reactions catalyzed by metal-free organic catalysts have received increased attention in recent years [23]. Interestingly, after the discovery of amino acid-catalyzed stereoselective Robinson annulations in the early 1970s, there was no intensive research on this concept for other C—C bond-forming reactions for several decades, even though the reaction was (and still is) frequently used in the preparation of building blocks for the total synthesis of natural products [26–27]. It was not until almost three decades later that researchers demonstrated that amino acid derivatives function as catalysts for direct asymmetric intermolecular C—C bond-forming reactions [23]. The first direct organocatalytic asymmetric Mannich reaction was reported by List and co-workers [28], the design and support for such transformations being based on Kobayshi and co-workers' report of one-pot, three-component Mannich reactions and previous research on proline-catalyzed direct asymmetric aldol reactions [29–30]. They thus envisioned that a small chiral amine should be able to form enamines with unmodified ketones and mediate stereoselective additions to imines generated in situ in one-pot procedures. The proline-catalyzed one-pot, three-component reactions with acetone, *p*-anisidine, and aliphatic aldehydes proceed with excellent chemoselectivity, affording Mannich adducts in high yields and with 61%–99% *ee*s [Eq. (8)].

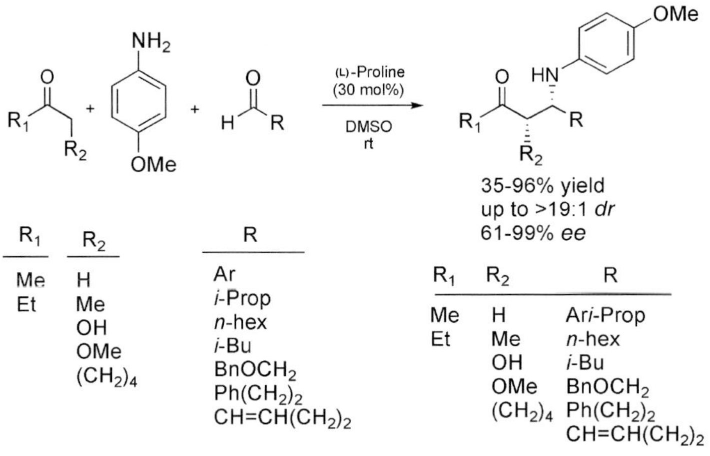

(8)

The L-proline-catalyzed Mannich reaction also takes place with other ketones as donors, aliphatic ketones providing two regioisomers with good enantioselectivities. The Mannich reaction is regiospecific with oxygen-substituted ketones, providing single regioisomers in good yields and with high *ee*s. In particular, the reaction with hydroxyacetone as the donor provides a new, highly chemo-, regio-, diastereo-, and enantioselective route to chiral 1,2-amino alcohols. The highest selectivity is observed when aromatic imines with electron-withdrawing groups are used as acceptors. The relative and absolute configurations were determined by X-ray analysis, confirming the formation of *syn*-1,2-amino alcohols. The amino alcohol products can also be converted into Boc-protected oxazolidinones in three steps, which revealed that L-proline affords β-amino ketones with *S* configurations.

Barbas and co-workers later demonstrated that amino acid derivatives other than proline can also catalyze direct asymmetric Mannich-type reactions with good enantioselectivities (Figure 5.7) [31].

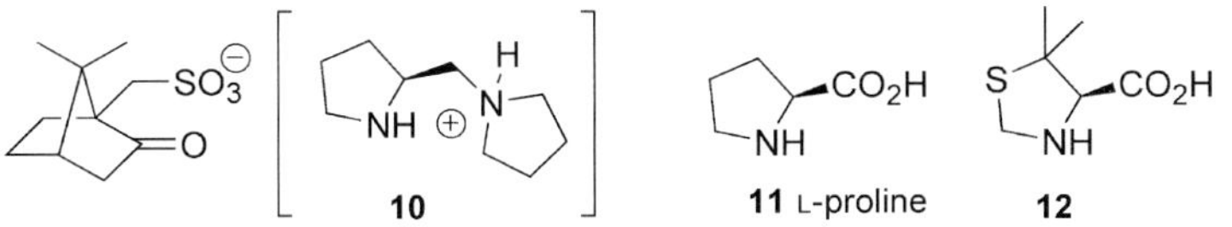

10 11 L-proline 12

Figure 5.7 Organic catalysts **10**, **11**, and **12**.

Their previous catalyst screening of aldol reactions and Robinson annulations had suggested the possibility of chiral amines being able to catalyze the Mannich reaction as well [32–33], and catalyst screening of Mannich-type reactions between *N*-OMP-protected aldimines and acetone revealed that the chiral diamine salt **10**, L-proline (**11**), and 5,5-dimethylthiazolidine-4-carboxylic acid (DMTC; **12**) functioned as catalysts for Mannich-type reactions, affording Mannich adducts in moderate yields and with 60–88% *ee*s. To extend the Mannich-type reactions to aliphatic imines, the DMTC-catalyzed reactions are performed as one-pot, three-component procedures (the *o*-anisidine component has to be exchanged for *p*-anisidine for the one-pot reactions to occur). These DMTC-catalyzed one-pot, three-component direct asymmetric Mannich reactions provide Mannich adducts in moderate yields and with 50–86% *ee*s.

Additions of nucleophiles to electrophilic glycine templates have served as excellent routes for the synthesis of α-amino acid derivatives [2c, 4–6]. In particular, imines derived from α-ethyl glyoxylate are excellent electrophiles for the stereoselective construction of optically active molecules [34]. This research and retrosynthetic analysis made us believe that amine-catalyzed asymmetric Mannich-type additions of unmodified ketones to glyoxylate-derived imines should be an attractive route for the synthesis of γ-keto-α-amino acid derivatives [35]. Initially, the L-proline-catalyzed direct asymmetric Mannich reaction between acetone and *N*-PMP-protected α-ethyl glyoxylate was examined in different solvents. This Mannich-type reaction was effective in all solvents tested and the corresponding amino acid derivative was isolated in excellent yield and enantioselectivity (*ee* >95%). Direct asymmetric Mannich-type additions with other ketones also afforded Mannich adducts in good yields and with excellent regio-, diastereo-, and enantioselectivities [Eq. (9)].

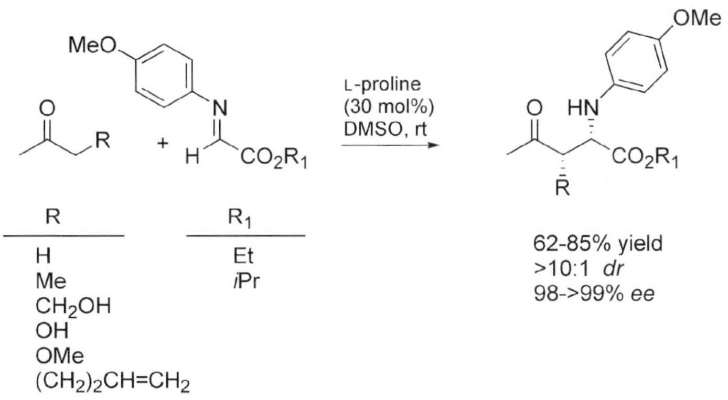

(9)

The reaction was regiospecific and C—C bond formation exclusively occurred on the most substituted side of the ketone donor. In contrast, the corresponding Mannich additions of non-oxygen-substituted ketones to other imines resulted in mixtures of regioisomers [28]. The amino acid derivatives can be further manipulated and the PMP group can be exchanged without racemization for a Boc protecting group in a one-pot dearylation/carbamate-formation procedure. The corresponding Boc-protected γ-keto-α-amino acid derivatives are useful building blocks in peptide chemistry and medicinal chemistry. Determination of the absolute and relative configurations showed that L-proline provided L-amino acids with *syn* relationships between the alkyl and amino groups.

The one-pot, three-component Mannich reaction can also be catalyzed with high enantioselectivity by simple acyclic amino acids and their derivatives [Eq. (10)] [36].

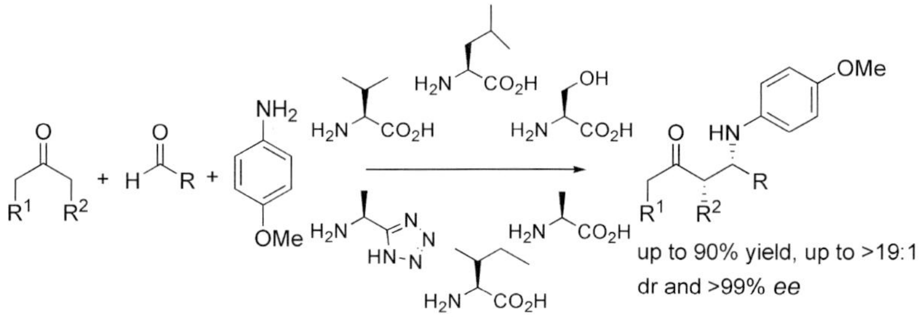

(10)

Proline and its derivatives also catalyze "classical" asymmetric Mannich reactions between aqueous formaldehyde, anilines, and ketones. This was the first successful direct catalytic α-hydroxymethylation of ketones, and the corresponding α-aminomethyl ketones were isolated in excellent yields with up to >99% *ees* (Scheme 5.3) [37]. The employment of α,β-unsaturated cyclic ketones gives access

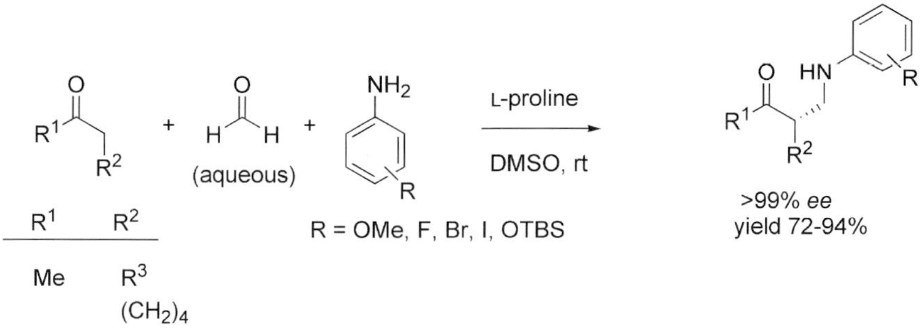

Scheme 5.3 Direct catalytic asymmetric "classical" Mannich reactions.

to aza-Diels–Alder products with up to >99% *ees* by a catalytic domino one-pot, three-component Mannich/Michael reaction pathway [38].

In addition, Ohsawa and co-workers have demonstrated that proline can catalyze direct catalytic Mannich reactions between protected 3,4-dihydro-β-carbolidine and ketones, yielding indole precursors in up to 92% *ees* [39].

The similarity in reaction mechanisms between proline- and 2-deoxyribose-5-phosphate aldolase-catalyzed direct asymmetric aldol reactions with acetaldehyde offered a precedent that a chiral amine might be able to catalyze stereoselective reactions through C—H activation of unmodified aldehydes, which might add to different electrophiles such as imines [40–41]. Indeed, proline is able to mediate direct catalytic asymmetric Mannich reactions with unmodified aldehydes as nucleophiles [42]. The first proline-catalyzed direct asymmetric Mannich-type reactions between aldehydes and *N*-PMP-protected α-ethyl glyoxylate proceeded with excellent chemo-, diastereo-, and enantioselectivities [Eq. (11)].

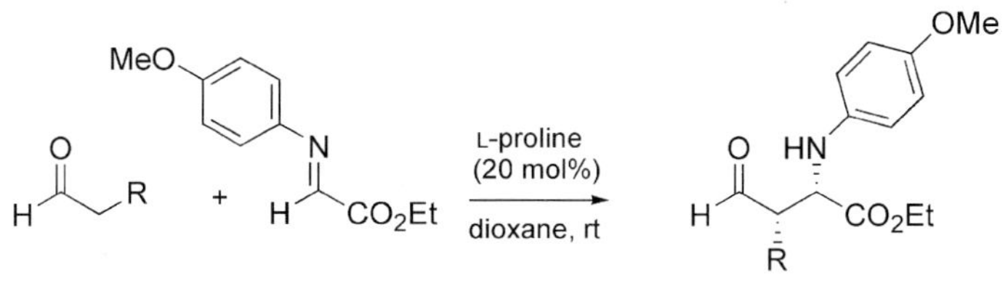

75-88% yield
3:1- >19:1 *dr*
93->99% *ee* (11)

The highest enantioselectivities were achieved in dioxane, THF, DMF, and NMP, affording the corresponding β-formyl-functionalized amino acid derivatives with high yields and stereoselectivities. The diastereoselectivity of the Mannich reaction increases with enhanced chain length of the donor: for aldehydes with more than six carbon atoms the two stereocenters are formed with almost absolute stereocontrol. The reaction is readily performed on a multigram scale, with only 2 equivalents of aldehyde being used, together with 10 mol% of catalyst. Mannich-type additions with unmodified aldehydes add a new dimension to the direct asymmetric Mannich reaction, since the aldehyde moiety allows further chemical manipulations and linkage to tandem reactions to furnish functional α-amino acid derivatives and β-lactams. The absolute configuration of the *iso*-valeraldehyde-derived lactam revealed that L-proline catalyzes the creation of L-amino acid derivatives with *syn* relative configurations. Expansion of the proline-catalyzed cross-Mannich-type reactions to other preformed imines is also possible [43–44]. Notably, performing the proline-catalyzed Mannich-type reactions at –20 °C in DMF or NMP (*N*-methylpyrrolidinone) circumvents the need for a syringe pump and yields the corresponding amino alcohols in high yields and with >99% *ees* in several cases [Eq. (12)] [43, 45].

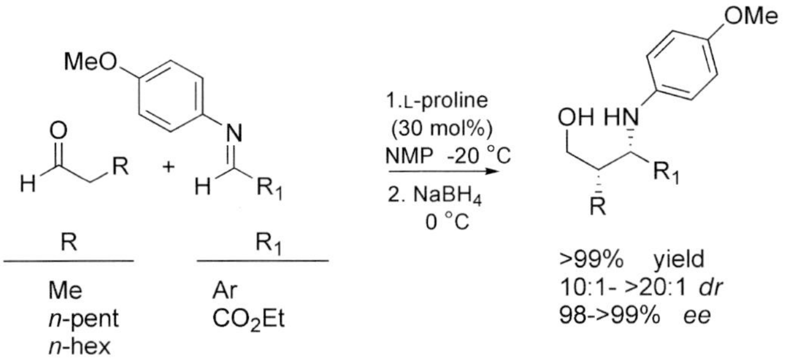

(12)

The reaction can be regarded as a regiospecific asymmetric synthesis of 3-amino-1-ols. Furthermore, proline is also able to catalyze one-pot, three-component direct catalytic enantioselective Mannich reactions with two unmodified aldehydes and anilines [Eq. (13)] [44].

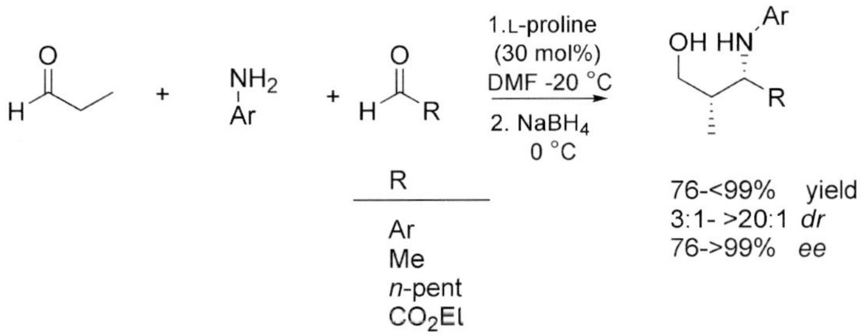

(13)

The corresponding β-amino aldehydes are reduced in situ and the corresponding amino alcohols are isolated in good yields with up to >99% *ee*s. The Mannich reactions proceed with excellent chemoselectivity and imine formation occurs with the acceptor aldehyde at a faster rate than C–C bond formation. Moreover, the one-pot, three-component direct asymmetric cross-Mannich reaction allows for aliphatic aldehydes to serve as acceptors. The absolute stereochemistry of the reaction was determined by synthesis, which revealed that L-proline provides *syn* β-amino aldehydes with (*S*) stereochemistry of the amino group. In addition, the proline-catalyzed direct asymmetric Mannich-type reaction has been connected to one-pot tandem cyanation and allylation reactions in THF and aqueous media to afford functional α-amino acid derivatives [43, 46], while the direct organocatalytic asymmetric α-aminomethylation of aldehydes is also possible [Eq. (14)] [47]. The reaction is catalyzed with high enantioselectivity by protected diarylprolinol derivatives with use of aminomethyl ethers as electrophiles. The transformation represents an efficient route to β^2-amino acids. In addition, amino methyl ethers are used as electrophiles in the proline-catalyzed enantioselective α-aminomethylation of cyclohexanones [48].

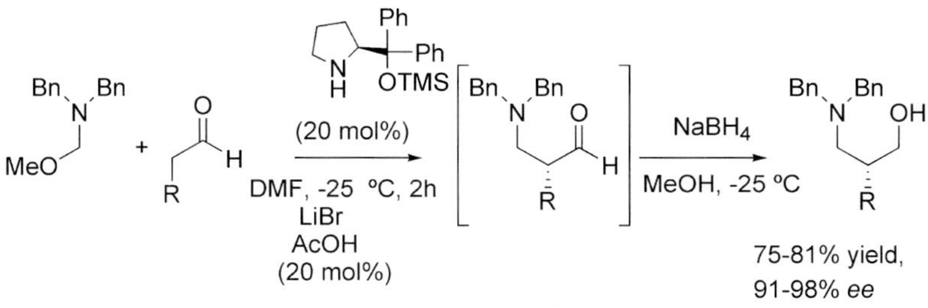

(14)

Enders and co-workers extended work on carbon-carbon bond-forming reactions using protected dihydroxy acetone derivatives as donors, resulting in the development of organocatalytic asymmetric syntheses of ketosugars [49]. The use of dihydroxyacetone equivalents has also been expanded to direct asymmetric Mannich reactions [Eq. (15)] [50].

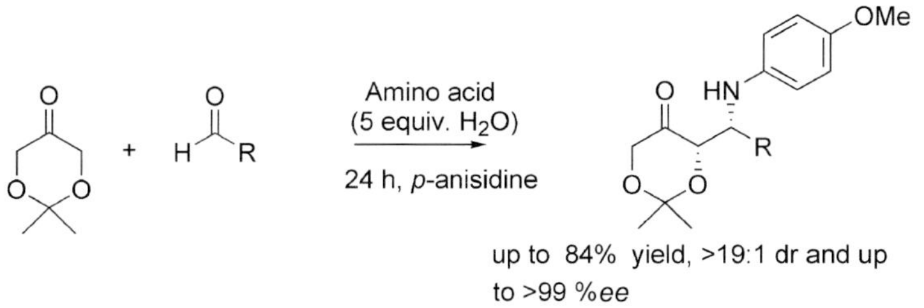

(15)

The reaction represents a simple route to aminosugars and polyhydroxylated amino acids. In addition, Enders and co-workers have also shown that proline catalyzes asymmetric additions of protected dihydroxy acetone to Boc-protected aldimines with high enantioselectivity. Amino sugars are also prepared by proline-catalyzed asymmetric self-Mannich reactions of protected α-oxyacetaldehydes, while the reaction is also linked in a one-pot process with Horner–Wittig–Emmons olefinations and subsequent catalytic diastereoselective dihydroxylation followed by cyclization to furnish iminosugars with up to 96% *ees* (Scheme 5.4) [51].

Proline and hydroxyproline are excellent catalysts for the addition of unmodified aldehydes to *N*-Boc-protected imines [52]. The reactions give the corresponding β-amino aldehydes in high yields and with 93 to >99% *ees*. In situ, oxidation gives access to the corresponding β-amino acids in one-pot procedures.

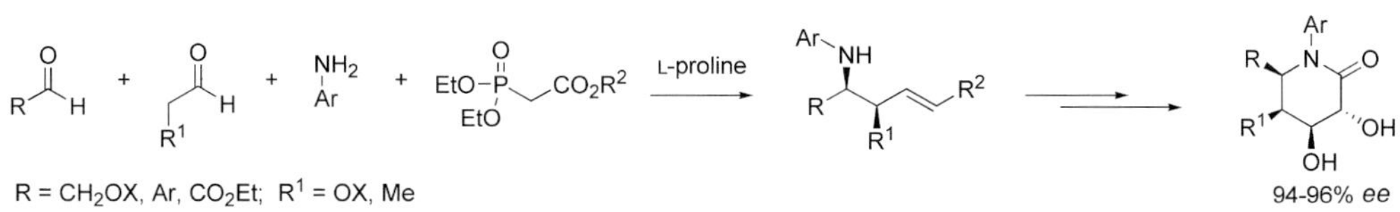

Scheme 5.4 Organocatalytic asymmetric synthesis of iminosugars.

The mechanism of these proline-catalyzed Mannich reactions is depicted in Scheme 5.5. The ketone or aldehyde donor reacts with proline to give an enamine, the preformed or in situ generated imine next reacts with the enamine to give, after hydrolysis, the enantiomerically enriched Mannich adduct, and the catalytic cycle can be repeated.

The stereochemical outcome of L-proline-catalyzed direct asymmetric Mannich reactions can be explained in terms of a *si*-facial attack on the imine, which has a *trans* configuration, by the *si*-face of the enamine, which has a *trans* configuration (Figure 5.8). The six-membered transition state is stabilized by hydrogen bonding between the nitrogen of the imine and the carboxylic group of the proline. A switch of the facial selectivity is disfavored due to steric repulsion between the PMP group of the imine and the pyrrolidine moiety of the enamine. This is the opposite of the case of the similar direct asymmetric aldol reaction, in which a *re*-facial attack occurs [29, 32, 40].

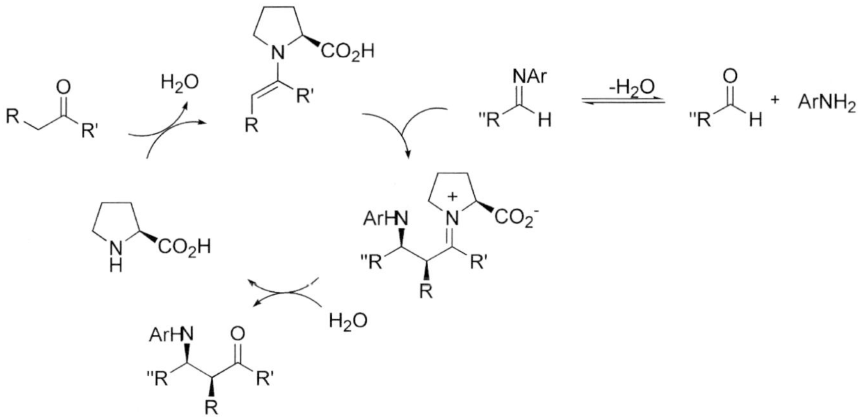

Scheme 5.5 The mechanism of the proline-catalyzed direct asymmetric Mannich reactions.

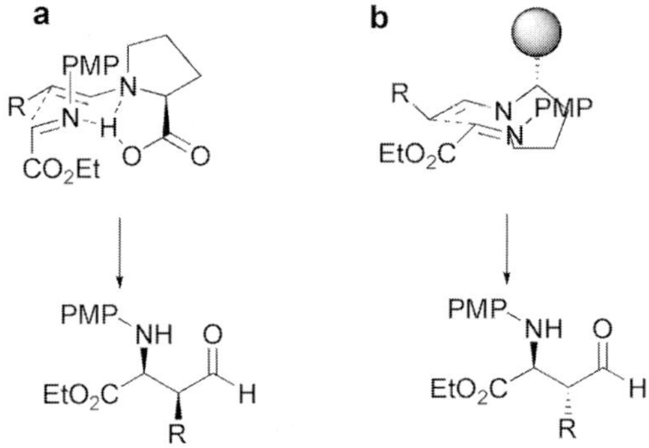

Figure 5.8 a) Transition state of the proline-catalyzed direct asymmetric Mannich reaction. b) Transition state of the chiral pyrrolidine-catalyzed direct asymmetric Mannich reaction.

From previous reports of *anti*-selective Mannich-type reactions with preformed chiral enamines or imines, there was a precedent that it would be theoretically possible for a chiral amine to catalyze a similar *anti*-selective direct asymmetric Mannich transformation [3h–i, 53]. Screening of chiral amines in the reaction between unmodified aldehydes and *N*-PMP-protected α-ethyl glyoxylate revealed that ether- and ester-functionalized proline-derived secondary amines were able to mediate the *anti*-selective transformation, albeit in lower yields than observed with proline [54]. (*S*)-2-Methoxymethylpyrrolidine (SMP), for example, is able to catalyze the direct asymmetric Mannich reaction with high *anti* selectivity to afford α-amino acid derivatives in moderate yield with 74–92% *ee*s. Notably, the employment of protected diarylprolinols as catalyst improved both the *anti* selectivity and the enantioselectivity of the reaction (up to >19:1 and 99% *ee*) [55]. The stereochemical outcome of the reaction was explained by the proposed transition state b in Figure 5.8. Thus, attack on the *si*-face of the imine with a *trans* configuration by the *re*-face of the chiral enamine gives the amino acid derivative with an *anti* configuration. Stabilization of the *trans* configuration and efficient shielding of the *si*-face of the chiral enamine can explain the high stereoselectivity of the reaction. In addition, plausible stabilization by Coulombic interactions between the amine atom of the imine and the δ^+ on the nitrogen of the pyrrolidine moiety of the chiral enamine, generated during the nucleophilic attack, contributes to the stabilization of the *si*-facial attack on the electrophile. The difference in facial selectivity between proline and its hydrophobic derivatives has also been observed in cross-Mannich reactions with ketimines by Jørgensen and co-workers [56]. An elegant approach to achieve *anti*-selective Mannich reactions is to employ an organic catalyst that forms an enamine with a *cis* configuration. Mauroka has shown that this is possible by using a designed organocatalyst to mediate the reaction between aldehydes and *N*-PMP-protected α-imino-glyoxylate (Figure 5.8a).

Barbas, Houk, and co-workers also utilized the same approach to develop an *anti*-selective chiral methyl-3-pyrrolidinecarboxylic acid catalyst for the Mannich reactions (vide infra; Figure 5.9b). Moreover, Barbas and Mauroka have also developed organocatalytic *anti*-selective direct Mannich reactions between ketones and *N*-PMP-protected α-imino-glyoxylates [57].

Another important approach to mediation of metal-free catalytic enantioselective Mannich-type reactions is through electrophilic activation of the preformed imines with chiral Brønsted acids [7, 8, 58]. Employing this strategy, Terada and co-workers performed chiral phosphoric acid-catalyzed direct asymmetric Mannich-type reactions between Boc-protected imines and acetoacetone, to furnish aryl β-amino ketones with up to 96% *ee*s. These are readily converted into α-amino acid derivatives (Scheme 5.6).

Akiyama independently utilized the same type of catalyst for highly enantioselective indirect Mannich-type reactions [59], also showing that chiral phosphoric acids are excellent catalysts for the aza-Diels–Alder reaction [60]. Rueping and co-workers reported the use of chiral phosphoric acids as excellent catalysts for the reaction between preformed imines and α,β-unsaturated cyclic ketones [61].

Jacobsen and co-workers reported thiourca-catalyzed nitro-Mannich (or aza-Henry) reactions with excellent yields and enantioselectivities (up to 97% *ee*s) [62].

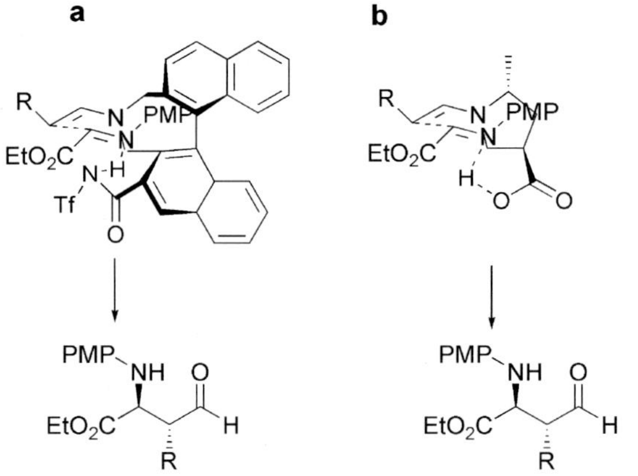

Figure 5.9 a) Transition state of the axially chiral amino sulfonamide-catalyzed direct asymmetric Mannich reaction. b) Transition state of the chiral 5-methyl-3-pyrrolidinecarboxylic acid-catalyzed direct asymmetric Mannich reaction.

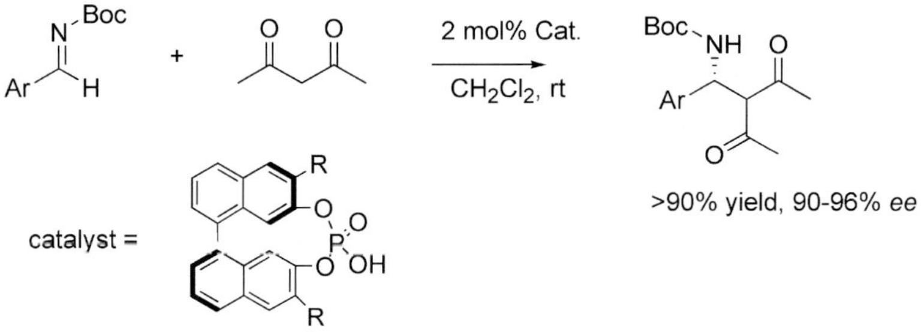

Scheme 5.6 Chiral phosphoric acid-catalyzed direct asymmetric Mannich reactions.

The thiourea catalyst promotes the stereoselective addition of a range of nitroalkanes to aromatic *N*-Boc imines with excellent yields and enantioselectivities [Eq. (16)]. These substrates also afforded good *syn* diastereoselectivities (7–16 : 1 *syn* : *anti*).

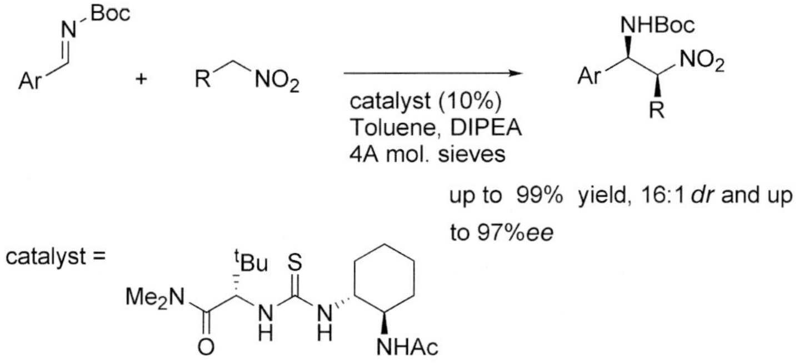

(16)

Ricci and co-workers described a phase-transfer-catalyzed (with *N*-benzylquininium chloride as a catalyst) nitro-Mannich (or aza-Henry) reaction with the aid of *N*-carbamoyl imines generated in situ from α-amido sulfones, with excellent yields and *ees* [63]. The chiral transfer catalyst acts in a dual fashion, first promoting the formation of the imine under mild reaction conditions and then activating the nucleophile for asymmetric addition [Eq. (17)].

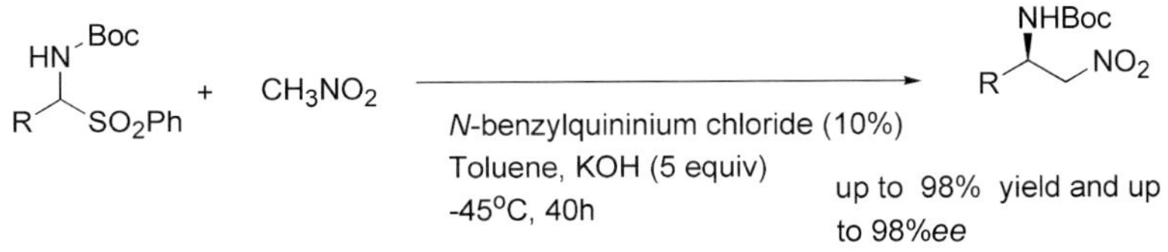

(17)

5.4 Conclusions

The direct asymmetric Mannich reaction is an important transformation for the synthesis of optically active chiral amines. As reported in this chapter, there are several catalytic methods for direct asymmetric Mannich-type reactions based on the employment either of metal catalysis or of organocatalysis. The versatility and complementation of the described catalytic methods make the direct Mannich-type reaction a versatile and fundamental transformation for the asymmetric construction of functional amines. This research field is rapidly expanding and each year significant improvements appear in the literature. From Robinson (1917) until now (2007), chemists have focused their efforts to improve this reaction for the synthesis of amine-containing compounds with exceptional levels of stereochemistry and atom economy, making this reaction one of the most powerful tools in organic synthesis.

Experimental

Typical Experimental Procedure for Direct Catalytic Asymmetric Mannich-Type Reactions with Aryl Hydroxyketones as Donors and with a Et_2Zn-linked BINOL Complex as the Catalyst [15]

In a typical experimental procedure, MS (3 Å, 100 mg) were activated in a test tube prior to use under reduced pressure at 160 °C for 3 h. After the system had cooled down to room temperature, it was placed under argon and a solution of (*S*,*S*)-linked-BINOL (0.005 mmol) in THF (0.167 mL) was added, followed by Et_2Zn (0.020 mmol). After the system had been stirred at −20 °C for 10 minutes, a solution of the aryl hydroxyketone (1.0 mmol) in THF (1.11 mL) and the imine (0.5 mmol) were added successively and the mixture was stirred at the same temperature. The stirring was continued for 9 h at this temperature and the reaction was quenched by addition of saturated aqueous NH_4Cl. The mixture was extracted with EtOAc, washed with brine, and dried over Na_2SO_4. The crude mixture was purified by column chromatography to afford the product.

Typical Experimental Procedure for a Highly Enantioselective Amino Acid-Catalyzed Route to Functionalized α-Amino Acids [35]

In a typical experimental procedure, *N*-PMP-protected α-imino ethyl glyoxylate (0.5 mmol) was dissolved in anhydrous dioxane and the corresponding aldehyde donor (0.75 mmol) was added, followed by L-proline (5 mol%). The total volume of the reaction mixture was 5 mL. After stirring for 2–24 h at room temperature, the mixture was worked up by addition of half-saturated ammonium chloride solution and extraction with ethyl acetate. The combined organic layers were dried over $MgSO_4$, filtered, and concentrated, and the crude residue was purified by column chromatography to afford the corresponding Mannich addition product.

References and Notes

1 The first example of the application of the Mannich reaction to natural product synthesis is attributed to Robinson in his synthesis of tropinone; R. Robinson, *J. Chem. Soc.* 1917, *762.*

2 For excellent reviews, see: a) E. F. Kleinmann, in *Comprehensive Organic Synthesis*; B. M. Trost, I. Flemming, Eds.; Pergamon Press: New York, 1991; Vol. 2, Chapter 4.1. b) M. Arend, B. Westerman, N. Risch, *Angew. Chem. Int. Ed.* 1998, *37,* 1044. c) S. Kobayashi, H. Ishitani, *Chem. Rev.* 1999, *99,* 1069. d) S. Denmark, O. J.-C. Nicaise, in *Comprehensive Asymmetric Catalysis*; E. N. Jacobsen, A. Pfaltz, H. Yamomoto, Eds.; Springer: Berlin, 1999, Vol. 2, 93. e) For examples, see: *Enantioselective Synthesis of* β*-Amino Acids*, E. Juaristi, Ed., Wiley-VCH, Weinheim, 1997.

3 a) D. Seebach, M. Hoffmann, *Eur. J. Org. Chem.* 1998, 1337. b) Y. Aoyagi, R. P. Jain, R. M. Williams, *J. Am. Chem. Soc.* 2001, *123,* 3472 and references therein. c) U. Schöllkopf, in *Topics in Current Chemistry*, F. L. Boschke, Ed., Springer Verlag, Berlin, 1983, Vol. 109, pp 45–85, g) D. A. Evans, F. Urpi, T. C. Somers, J. S. Clark, M. T. Bilodeau, *J. Am. Chem. Soc.* 1990, *112,* 8215. h) R. Kober, K. Papadopoulos, W. Miltz, D. Enders, W. Steglich, H. Reuter, H. Puff, *Tetrahedron* 1985, *42,* 1963. i) C. Palomo, M. Oiarbide, A. Landa, M. C. Gonzales-Rego, J. M. Garcia, A. Gonzales, J. M. Odriozola, M. Martin-Pastor, A. Linden, *J. Am. Chem. Soc.* 2002, *124,* 8637 and references therein.

4 a) H. Ishitani, M. Ueno, S. Kobayashi, *J. Am. Chem. Soc.* 1997, *119,* 7153. b) S. Kobayashi, T. Hamada, K. Manabe, *J. Am. Chem. Soc.* 2002, *124,* 5640. c) H. Ishitani, S. Ueno, S. Kobayashi, *J. Am. Chem. Soc.* 2000, *122,* 8180.

5 a) E. Hagiwara, A. Fujii, M. Sodeoka, *J. Am. Chem. Soc.* 1998, *120,* 2474. b) A. Fujii, E. Hagiwara, M. Sodeoka, *J. Am. Chem. Soc.* 1999, *121,* 545.

6 a) D. Ferraris, B. Young, T. Dudding, T. Lectka, *J. Am. Chem. Soc.* 1998, *120,* 4548. b) D. Ferraris, B. Young, C. Cox, T. Dudding, W. J. Drury, III, L. Ryzhkov, A. E. Taggi, T. Lectka, *J. Am. Chem. Soc.* 2002, *124,* 67. and references therein.

7 A. G. Wenzel, E. N. Jacobsen, *J. Am. Chem. Soc.* 2002, *124,* 12964.

8 T. Akiyama, J. Itoh, K. Yokota, K. Fuchibe, *Angew. Chem. Int. Ed.* 2004, *43,* 1566.

9 N. Josephsohn, M. L. Snapper, A. H. Hoveyda, *J. Am. Chem. Soc.* 2004, *126,* 3734.

10 B. M. Trost, *Science* 1991, *254,* 1471.

11 See: M. Shibasaki, H. Sasai, T. Arai, *Angew. Chem. Int. Ed. Engl.* 1997, *36,* 1236. and references therein.

12 a) N. Yoshikawa, Y. M. A. Yamada, J. Das, H. Sasai, M. Shibasaki, *J. Am. Chem. Soc.*

1999, *121*, 4168. c) N. Yoshikawa, N. Kumagai, S. Matsunaga, G. Moll, G. T. Ohshima, T. Suzuki, M. Shibasaki, *J. Am. Chem. Soc.* 2001, *123*, 2466. d) D. Sawada, M. Shibasaki, *Angew. Chem. Int. Ed.* 2000, *39*, 209.

13 S. Yamasaki, T. Iida, M. Shibasaki, *Tetrahedron Lett.* 1999, *40*, 307.

14 S. Matsunaga, N. Kumagai, N. Harada, S. Harada, M. Shibasaki, *J. Am. Chem. Soc.* 2003, *125*, 4712.

15 S. Matsunaga, T. Yoshida, H. Morimoto, N. Kumagai, M. Shibasaki, *J. Am. Chem. Soc.* 2004, *126*, 8777.

16 a) S. Matsunaga, J. Das, J. Roels, E. M. Vogel, N. Yamomoto, T. Iida, K. Yamaguchi, M. Shibasaki, *J. Am. Chem. Soc.* 2000, *122*, 2252. b) N. Kumagai, S. Matsunaga, T. Kinoshita, S. Harada, S. Okada, S. Sakamoto, K. Yamaguchi, M. Shibasaki, *J. Am. Chem. Soc.* 2003, *125*, 2169. and references therein. c) S. Harada, N. Kumagai, T. Kinoshita, S. Matsunaga, M. Shibasaki, *J. Am. Chem. Soc.* 2003, *125*, 2582. and references therein.

17 T. Yoshida, H. Morimoto, N. Kumagai, S. Matsunaga, M. Shibasaki, *Angew. Chem. Int. Ed.* 2005, *44*, 3470.

18 a) B. M. Trost, H. Ito, *J. Am. Chem. Soc.* 2000, *122*, 12003. b) B. M. Trost, H. Ito, E. Silcoff, *J. Am. Chem. Soc.* 2001, *123*, 3367. c) B. M. Trost, E. Silcoff, H. Ito, *Org. Lett.* 2001, *3*, 2497.

19 B. M. Trost, L. M. Terrell, *J. Am. Chem. Soc.* 2003, *125*, 338.

20 a) K. Juhl, N. Gathergood, K. A. Jørgensen, *Chem. Commun.* 2000, 2211. b) K. Juhl, K. A. Jørgensen, *J. Am. Chem. Soc.* 2002, *124*, 2420.

21 K. Juhl, N. Gathergood, K. A. Jørgensen, *Angew. Chem. Int. Ed.* 2001, *40*, 2995.

22 For reviews of C_2-bisoxazoline-Lewis acid complexes see: a) J. S. Johnson, D. A. Evans, *Acc. Chem. Res.* 2000, *33*, 325. b) K. A. Jørgensen, M. Johannsen, S. Yao, H. Audrain, J. Thorhauge, *Acc. Chem. Res.* 1999, *32*, 605.

23 M. Marigo, A. Kjaersgaard, K. Juhl, N. Gathergood, K. A. Jørgensen, *Chem. Eur. J.* 2003, *9*, 2395.

24 S. Harada, S. Handa, S. Matsunaga, M. Shibasaki, *Angew. Chem. Int. Ed.* 2005, *44*, 4365.

25 a) P. I. Dalko, L. Moisan, *Angew. Chem. Int. Ed.* 2001, *40*, 3726. b) B. List, *Tetrahedron* 2002, *58*, 5573. c) J. Gröger, J. Wilken, *Angew. Chem. Int. Ed.* 2001, *40*, 529. c) E. R. Jarvo, S. J. Miller, *Tetrahedron* 2002, *58*, 2481. d) R. O. Duthaler, *Angew. Chem. Int. Ed.* 2003, *42*, 975.

26 a) Z. G. Hajos, D. R. Parrish, Asymmetric Synthesis of Optically Active Polycyclic Organic Compounds. German Patent. D. E. 2102623, Jul 29, 1971. b) Z. G. Hajos, D. R. Parrish, *J. Org. Chem.* 1974, *39*, 1615. c) U. Eder, G. Sauer, R. Wiechert, Optically Active 1,5-Indanone and 1,6-Naphthalenedionene. German Patent DE 2014757, Oct 7, 1971. d) U. Eder, G. Sauer, R. Wiechert, *Angew. Chem. Int. Ed.* 1971, *10*, 496. e) C. Agami, *Bull. Soc. Chim. Fr.* 1988, 499.

27 For example, the total synthesis of taxol: S. J. Danishefsky, et al. *J. Am. Chem. Soc.* 1996, *118*, 2843.

28 a) B. List, *J. Am. Chem. Soc.* 2000, *122*, 9336. b) B. List, P. Porjalev, W. T. Biller, H. J. Martin, *J. Am. Chem. Soc.* 2002, *124*, 827.

29 a) B. List, R. A. Lerner, C. F. Barbas, III, *J. Am. Chem. Soc.* 2000, *122*, 2395. b) W. Notz, B. List, *J. Am. Chem. Soc.* 2000, *122*, 7386.

30 K. Manabe, S. Kobayashi, *Org. Lett.* 1999, *1*, 1965. and references therein.

31 W. Notz, K. Sakthivel, T. Bui, G. Zhong, Barbas, III, C. F. *Tetrahedron Lett.* 2001, *42*, 199.

32 K. Saktihvel, W. Notz, T. Bui, C. F. Barbas, III, *J. Am. Chem. Soc.* 2001, *123*, 5260.

33 T. Bui, C. F. Barbas, III, *Tetrahedron Lett.* 2000, *41*, 6951.

34 a) A. E. Taggi, A. M. Hafez, T. Lectka, *Acc. Chem. Res.* 2002, *36*, 10. b) S. Yao, M. Johannsen, R. G. Hazell, K. A. Jørgensen, *Angew. Chem. Int. Ed.* 1998, *37*, 3121.

35 A. Córdova, W. Notz, G. Zhong, J. M. Betancort, C. F. Barbas, III, *J. Am. Chem. Soc.* 2002, *124*, 1842.

36 I. Ibrahem, W. Zou, M. Engqvist, Y. Xu, A. Córdova, *Chem. Eur. J.* 2005, *11*, 7024.

37 a) I. Ibrahem, J. Casas, A. Córdova, *Angew. Chem. Int. Ed.* 2004, *43*, 6528. b) I. Ibrahem, W. Zao, J. Casas, H. Sundén, A. Córdova, *Tetrahedron* 2005, *62*, 357.

38 H. Sundén, I. Ibrahem, L. Eriksson, A. Córdova, *Angew. Chem. Int. Ed.* 2005, *44*, 4877.

39 T. Itoh, M. Yokoya, K. Miyauchi, K. Nagata, A. Ohsawa, *Org. Lett.* 2003, *5*, 4301.

40 a) A. Córdova, W. Notz, C. F. Barbas, III, *J. Org. Chem.* 2002, *67*, 301. b) A. Bøgevig, N. Kumaragurubaran, K. A. Jørgensen, *Chem. Commun.* 2002, 620. c) A. B. Northrup, D. W. C. MacMillan, *J. Am. Chem. Soc.* 2002, *124*, 6798. d) N. S. Chowdari, D. B. Ramachary, A. Córdova, C. F. Barbas, III, *Tetrahedron Lett.* 2002, *43*, 9591. e) A. Córdova, W. Notz, C. F. Barbas, III, *Chem. Commun.* 2002, *67*, 3034. f) A. Bøgevig, K. Juhl, N. Kumaragurubaran, W. Zhuang, K. A. Jørgensen, *Angew. Chem. Int. Ed.* 2002, *41*, 1790. g) B. List, *J. Am. Chem. Soc.* 2002, *124*, 5656.

41 T. D. Machajewski, C.-H. Wong, *Angew. Chem. Int. Ed.* 2000, *39*, 1352. and references therein.

42 A. Córdova, S. Watanabe, F. Tanaka, W. Notz, C. F. Barbas, III, *J. Am. Chem. Soc.* 2002, *124*, 1866.

43 a) A. Córdova, 225th ACS National Meeting, New Orleans, LA. Amine-Catalyzed Direct Asymmetric Mannich-type Reactions: Enantioselective Synthesis of Amino Acid and Amino Alcohol Derivatives. b) A. Córdova, *Synlett.* 2003. 1651. c) A. Córdova, *Chem. Eur. J.* 2004, *10*, 1987. c) I. Ibrahem, A. Córdova, *Tetrahedron Lett.* 2005, *46*, 2839.

44 The Mannich adducts are reduced to the more stabile γ-amino alcohols prior to isolation and *ee* determination, due to their lower risk of decomposition, epimerization and racemization.

45 a) H. Hayashi, W. Tsuboi, I. Ashime, T. Urushima, M. Shoji, K. Sakai, *Angew. Chem. Int. Ed.* 2003, *42*, 3677. b) Y. Hayashi, T. Urushima, M. Shoji, T. Uchimaru, I. Shiina, *Adv. Synth. Catal.* 2005, *347*, 1595.

46 a) S.-I. Watanabe, A. Córdova, F. Tanaka, C. F. Barbas, III, *Org. Lett.* 2002, *4*, 4519. b) A. Córdova, C. F. Barbas, III, *Tetrahedron Lett.* 2003, *44*, 1923.

47 a) Y. Chi, S. H. Gellman, *J. Am. Chem. Soc.* 2006, *128*, 6804. b) I. Ibrahem, G.-L. Zhao, A. Córdova, *Chem. Eur. J.* 2007, *13*, 683.

48 I. Ibrahem, P. Dziedzic, A. Córdova, *Synthesis* 2006, 4060.

49 a) D. Enders, C. Grondal, M. Vrettou, G. Raabe, *Angew. Chem. Int. Ed.* 2005, *44*, 4079. b) I. Ibrahem, A. Córdova, *Tetrahedron Lett.* 2005, *46*, 3363. c) I. Ibrahem, W.-W. Zou, Y. Xu, A. Córdova, *Adv. Synth. Catal.* 2006, *348*, 211.

50 B. Westermann, C. Neuhaus, *Angew. Chem. Int. Ed.* 2005, *44*, 4077.

51 W.-W. Liao, I. Ibrahem, A. Córdova, *Chem. Commun.* 2006, 674.

52 a) J. Vesely, R. Rios, I. Ibrahem, A. Córdova, *Tetrahedron Lett.* 2007, *48*, 421. b) J. W. Yang, M. Stadler, B. List, *Angew. Chem. Int. Ed.* 2007, *46*, 609.

53 V. Vinkovic, V. Sunjic, *Tetrahedron* 1997, *53*, 689.

54 A. Córdova, C. F. Barbas, III, *Tetrahedron Lett.* 2002, *43*, 7749.

55 a) I. Ibrahem, A. Córdova, *Chem. Commun.* 2006, 1760. b) J. Franzén, M. Marigo, D. Fielenbach, T. C. Wabniz, A. Kjaergaard, K. A. Jørgensen, *J. Am. Chem. Soc.* 2005, *127*, 18296.

56 W. Zhuang, S. Saaby, K. A. Jørgensen, *Angew. Chem. Int. Ed.* 2004, *43*, 4476.

57 a) S. Mitsumori, H. Zhang, P. H.-Y. Cheong, K. N. Houk, F. Tanaka, C. F. Barbas, III, *J. Am. Chem. Soc.* 2006, *128*, 1040. b) H. Zhang, M. Misfud, F. Tanaka, C. F. Barbas, III, *J. Am. Chem. Soc.* 2006, *128*, 9630. c) T. Kano, Y. Yamaguchi, O. Tokuda, K. Maruoka, *J. Am. Chem. Soc.* 2005, *127*, 16408. d) T. Kano, Y. Hato, K. Maruoka, *Tetrahedron Lett.* 2006, *47*, 8467.

58 D. Uraguchi, M. Terada, *J. Am. Chem. Soc.* 2004, *126*, 5356.

59 T. Akiyama, J. Itoh, K. Yokota, K. Fuchibe, *Angew. Chem. Int. Ed.* 2004, *43*, 1566.

60 a) T. Akiyama, H. Morita, K. Fuchibe, *J. Am. Chem. Soc.* 2006, *128*, 13070. b) J. Itoh, K. Fuchibe, T. Akiyama, *Angew. Chem. Int. Ed.* 2006, *45*, 4796.

61 a) M. Rueping, C. Azap, *Angew. Chem. Int. Ed.* 2006, *45*, 7832. b) see also: H. Liu, L.-F. Cun, A.-Q. Mi, Y.-Z. Jiang, L.-Z. Gong, *Org. Lett.* 2006, *8*, 6023. **DOI**: 10.1021/ol0624991t.

62 T. P. Yoon, E. N. Jacobsen, *Angew. Chem. Int. Ed.* 2005, *44*, 466.

63 F. Fini, V., Agarzani, D. Pettersen, R. P. Herrera, L. Bernardi, A. Ricci, *Angew. Chem. Int. Ed.* 2005, *44*, 7975.

6
Amino-Based Building Blocks for the Construction of Biomolecules

André Mann

6.1
Introduction

For the construction of molecules of life, nature uses a variety of building blocks such as amino acids, nucleic acids, or sugars. Their combinations and arrangements in aqueous medium by enzymes produce complex compounds responsible for biological events. Interestingly, examination of nature's favorite chemistry reveals a striking preference for generating carbon-heteroatom bonds over carbon-carbon bonds. Inspired by these natural processes, organic chemists have over the years identified and constructed an arsenal of small building blocks that enable the construction of elaborated target compounds [1]. Amines and their derivatives are widespread functional groups found in natural products, pharmaceuticals, and biologically important fine chemicals. This chapter focuses on the preparation of biomolecules from building blocks bearing amino residues. A building block can be defined in terms of two complementary statements: (i) a scaffold for the introduction of different structural elements that can generate weak interactions (ionic or hydrophobic interactions, hydrogen bonding) with a given biological target, and (ii) a chemical entity, carrying chemical information, that reacts with a suitable partner in a single reaction, unmasking a new reaction center available for further transformations. Monoprotected amino acids are the archetypes of building blocks for the construction of biopolymers, but a literature survey reveals the existence of a large number of elementary amino building blocks, some of them nowadays commercially available. In the last decade, the need for new building blocks has been satisfied by the considerable development of parallel and combinatorial syntheses. These molecules appear to play a key role in a number of seminal processes exemplified by the Schotten–Baumann and Suzuki reactions and by reductive amination [2, 3]. Some of them are illustrated in this chapter mainly because they are useful for the construction of important biomolecules or for the creation of molecular complexity. The data presented here are merely based on an arbitrary choice by the author and each building block is treated in a general frame giving its preparation and/or its use in recent applications. Throughout the chapter only recent literature reports are considered.

Amino Group Chemistry. From Synthesis to the Life Sciences. Edited by Alfredo Ricci

ISBN: 978-3-527-31741-7

6.2
Propargylamines (PLAs)

Propargylamines **1** (PLAs), typical amino building blocks, offer great synthetic potential thanks to the presence of five diversity sites (R^1–R^5). The commonest preparations of PLAs are based on the Mannich reaction. The reactivity of the triple bonds can be exploited in, among others, Sonogashira and Pauson–Khand reactions, in Diels–Alder and dipolar cycloadditions, and – because of the nitrogen nucleophilicity – in Schotten–Baumann couplings or for reductive aminations.

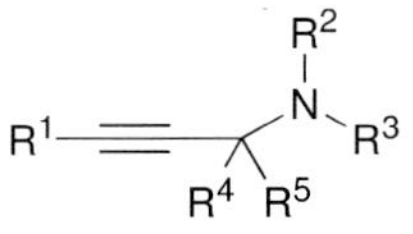

Propargylamines **1** (PLA)

PLAs are both biologically relevant and useful amino derivatives. Their substructures are present as such in several important biomolecules such as oxotremorine, a standard ligand for the muscarinic receptors, in I-MAO inhibitors such as deprenyl, clorgyline, and ladostigil, and in inhibitors of catecholamine metabolism. Notably, ladostigil is in clinical trials as an anti-Alzheimer's and neuroprotective drug [4], while a bis-PLA HIV-inhibitor is a promising non-nucleoside reverse transcriptase inhibitor [5]. An embedded PLA moiety can also be seen in the complex structure of dynemicin, a promising anticancer drug belonging to the ene-yne family [6] (Figure 6.1)

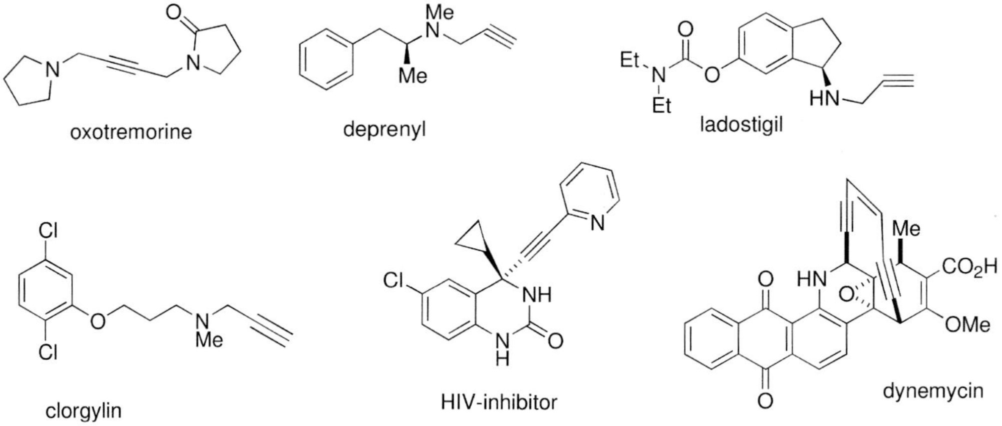

Figure 6.1 Examples of biomolecules containing propargylamine (PLA) subunits.

6.2.1
Synthesis of PLAs

The most attractive synthetic route to PLAs (**1**) is the classical Mannich reaction. The three-component, one-pot procedure (amine, aldehyde, and alkyne) appears to be the reaction of choice if a tertiary amino alkyne is desired, but the two-step approach involving formation of an imine and subsequent nucleophilic addition of a suitable alkyne is preferred if a secondary amino alkyne is wanted. Initially restricted to aryl acetylenes and to secondary amines, the reaction has been extended to less activated alkynes through the use of catalysts; even primary aminoalkynes **2** are accessible if suitable protection on the amine partners is employed. Several improvements have recently made access to chiral analogues possible through the use either of chiral ligands or of chiral auxiliaries; indeed, the preparation of this class of amines is a field actively under investigation. Some procedures for the preparation of PLAs are described here.

A procedure based on the addition of trimethylsilylacetylene to aldimines in the presence of an iridium(I) complex at room temperature has been reported (Scheme 6.1) to give high yields. Moreover the reaction can be run even under solventless conditions [7]

An efficient three-component coupling of aldehyde, alkyne, and amine performed with the aid of microwave irradiation in water with Cu(I) as the catalyst has also been reported (Scheme 6.2) for the synthesis of PLAs. The reaction is fast (5–20 min) and high-yielding, and the method is applicable to a wide range of substrates. Similar results have been obtained with other catalysts such as Cu(I) in the presence of Ru(III), Au(I) or Ag(I) [8, 9].

Several teams have developed three-component asymmetric versions of this reaction. With Cu(I) catalysis low levels of conversion of the starting materials into PLAs were observed in this reaction, but the addition of nitrogen-based ligands

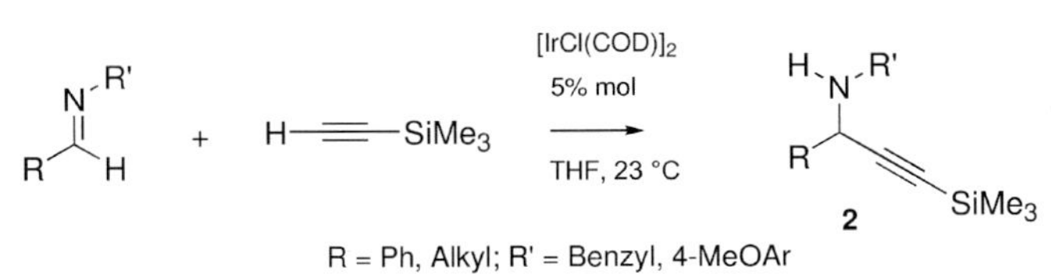

Scheme 6.1 Synthesis of a PLA by use of an Ir(I) catalyst.

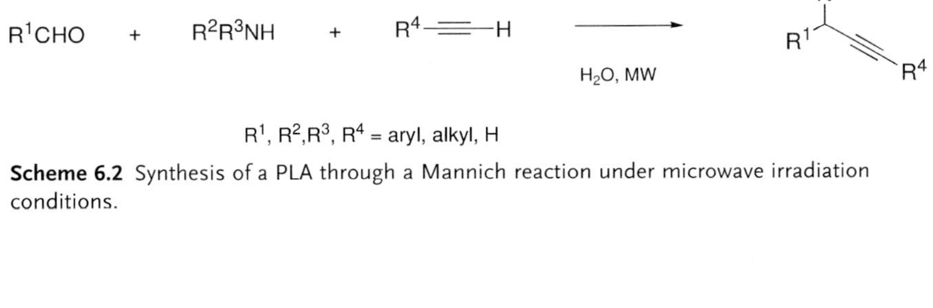

Scheme 6.2 Synthesis of a PLA through a Mannich reaction under microwave irradiation conditions.

speeded up the reactivity by weakening the Cu-alkyne bond, and with chiral ligands the conditions are set for production of a chiral environment around the C–Cu bond, which may give rise to asymmetric induction during the addition to the imine. Several groups have reported excellent enantioselectivity with PYBOX or QUINAP as ligands in the presence of a Cu(I) catalyst (Scheme 6.3). The procedures are simple and applicable to a broad range of aldehydes, amines, and alkynes [10, 11].

If the two appendages on nitrogen are suitably chosen, primary amino alkynes can be obtained under very mild conditions. Knochel has reported (Scheme 6.4) the following procedure, which makes the primary amine ready for further transformations [11].

The enantioselective synthesis of coniine (**3**) from rather cheap starting chemicals represents (Scheme 6.5) an elegant application of chiral synthesis with a PLA [8].

Another attractive approach to generation of a primary amino group in a PLA is based on the use of 4-piperidone as an amino partner in the Mannich reaction, and on its capability to undergo a retro-Michael reaction in the presence of excess of an auxiliary amine (Scheme 6.6). In the present case ammonia does the job perfectly, while the chiral ligand here is PINAP, developed by Carreira [12].

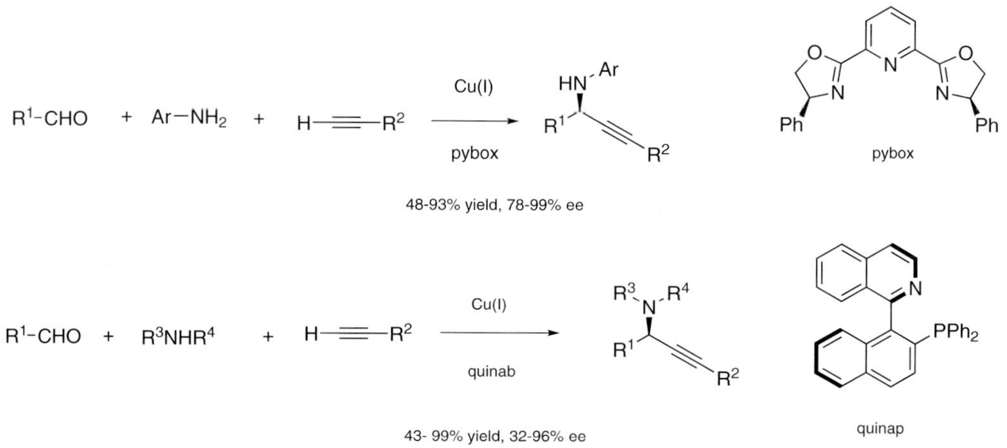

Scheme 6.3 Syntheses of enantioenriched PLAs.

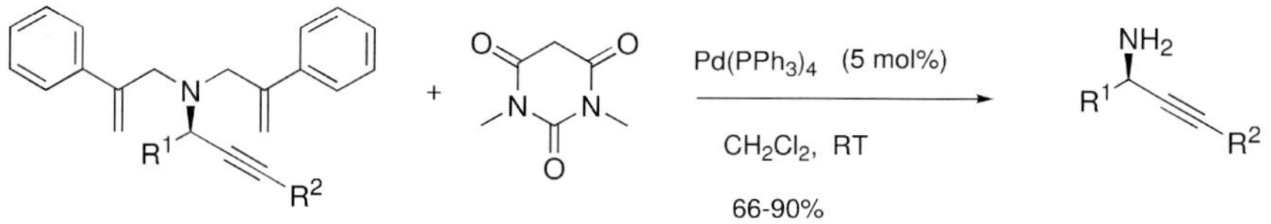

Scheme 6.4 Preparation of PLAs containing unmasked primary amine moieties.

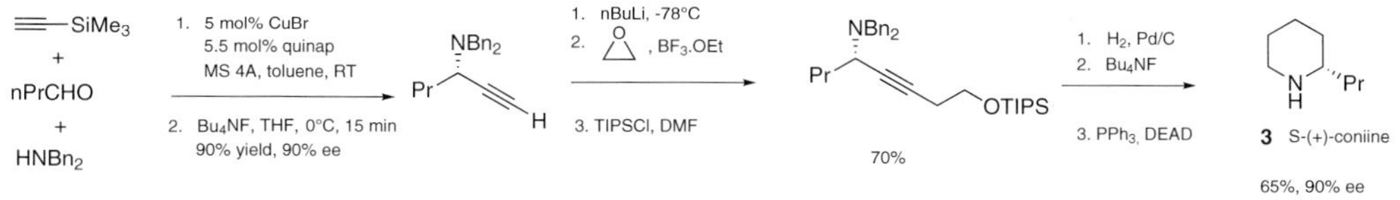

Scheme 6.5 Synthesis of conine (**3**) via a PLA building block.

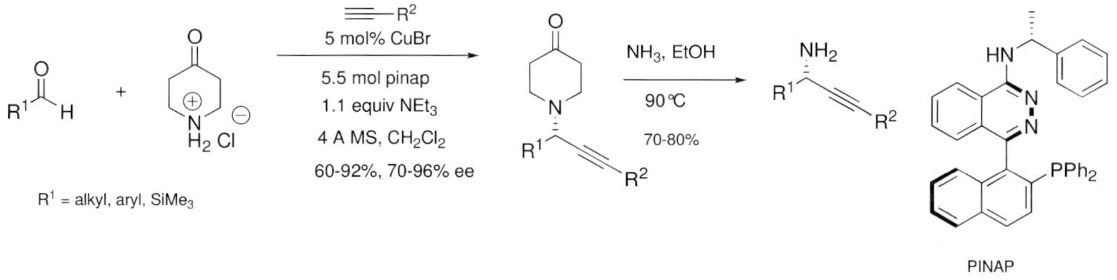

Scheme 6.6 Synthesis of PLAs containing unmasked primary amines.

6.2.2
PLAs in Synthesis

6.2.2.1 PLAs in the Synthesis of Heterocycles

A practical one-pot synthesis of pyridines **4–9**, of varying degrees of complexity, starting from reactions between PLAs and dialkyl acyclic/cyclic ketones, methyl aryl/heteroaryl ketones, or enolizable aldehydes has been reported. Au(III) or Cu(II) are the catalysts of choice for the cyclization of transient *N*-Propargylamine intermediates, which undergo aromatization (Scheme 6.7). When two types of carbonyl function are present, high selectivity in favor of α,β-unsaturated ones has been observed, as exemplified in the case of the Wieland–Miescher ketone [13].

Oxazoles, an important class of heterocycles present in many biomolecules of marine origin, were obtainable through the use of intramolecular cyclization of substituted propargylcarboxamides **10**, assisted by Au(III). The reaction (Scheme 6.8) tolerates various functionalized appendages and is catalytic in Au(III) [14].

6.2.2.2 PLAs in Pd(0)-Catalyzed Processes

Propargylic carbonates are valuable substrate in Pd(0)-catalyzed processes and their reactivity differs from that of simple alkynes since they form allenylpalladium(II) complexes. Thanks to the association of the carbanate and the PLA in the same substrate, it is possible to perform a diastereoselective synthesis, affording azabicyclo[3.1.0] hexanes **11** in moderate to good yields. The final cyclopropane moiety may bear different substituents at all positions, depending on the groups installed in the acyclic ene-yne substrate (Scheme 6.9). The geometry of the double bound

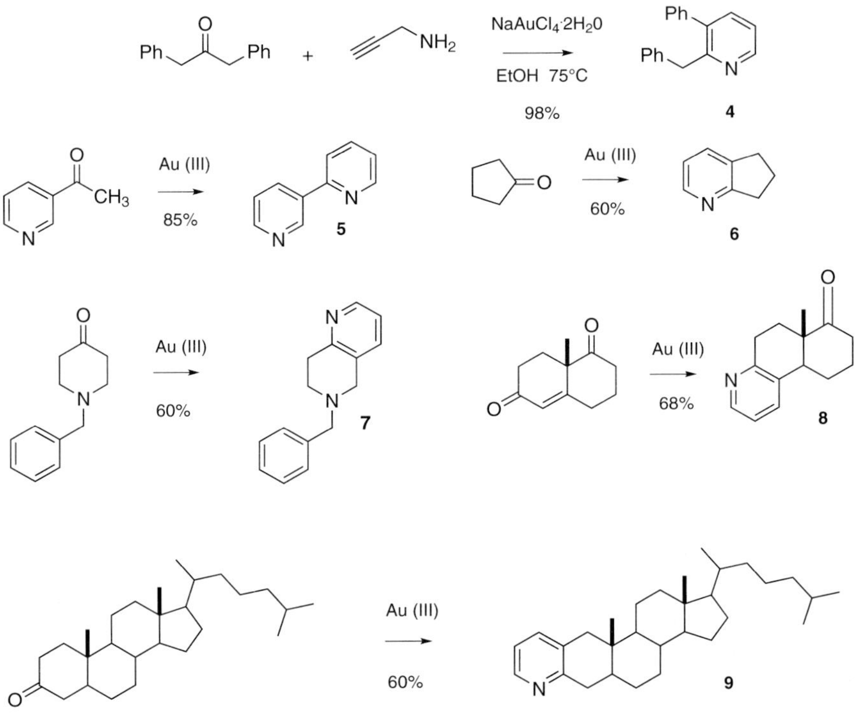

Scheme 6.7 PLAs for the synthesis of the pyridine core with Au(III) as the catalyst.

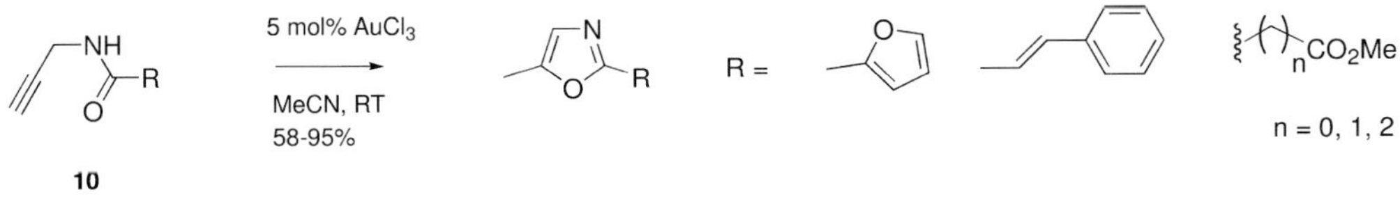

Scheme 6.8 PLAs and Au(III) for the construction of oxazoles.

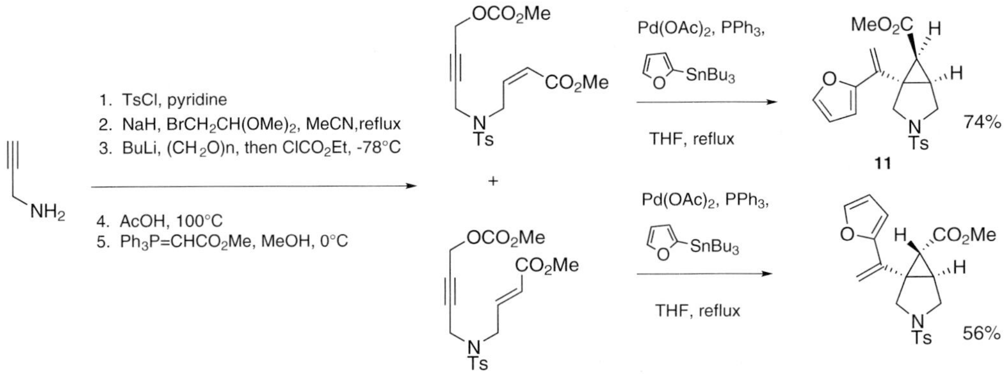

Scheme 6.9 PLA in pallado-catalyzed cascade cyclizations.

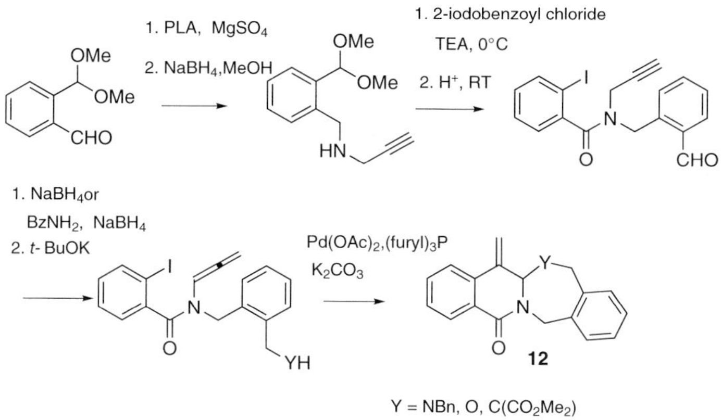

Scheme 6.10 PLAs in intramolecular tandem cylizations.

dictates the stereochemistry on the cyclopropane and 1,3-allylic strain has been invoked in order to explain the observed results [15].

When the PLA nitrogen is part of an amide scaffold, the PLA can be converted into the corresponding allene under basic conditions. A complex intramolecular cascade process, driven by Pd(0), can then operate if complementary functionalities are present in the substrate. The generation of a (π-allyl) palladium intermediate is followed by a regioselective intramolecular nucleophilic addition, producing fused heterocycles **12** in good yields during the reductive elimination step (Scheme 6.10). The nucleophiles may be the oxygen of an alcohol or a basic nitrogen and the adducts are scaffolds that are rather difficult to obtain otherwise [16].

6.2.2.3 **PLAs in Pericyclic Reactions**

PLAs react with 2,5-bis(methoxycarbonyl)-3,4-diphenylcyclopentadienone (a strong Michael acceptor, easily accessible from dimethyl acetonedicarboxylate and benzyl) in pericyclic reactions performed under thermodynamic conditions. The initial adducts were obtained by 1,4 addition, followed by ene cyclization. Depending on the structure of the PLA a further cycloaddition may be achievable, increasing considerably the complexity of the molecular system. After the 1,4 addition, a 1,5-sigmatropic shift of the C–N bond may take place, followed by an intramolecular Diels–Alder reaction and a final dehydrogenation. The examples below (Figure 6.2) illustrate the potential of this powerful sequence for the production of the complex structures **13–17**. Interestingly, the adducts still bear latent functionalities for further utilization [17].

Diversity-oriented synthesis is a popular strategy for the preparation of small libraries. From a common intermediate, the alkynyl-allene **18**, three structurally unique compounds **19–21** are accessible through the use of different sets of reaction conditions. The PLA is now embedded into a cumulene. Depending on the metal (rhodium or molybdenum), the reactivity either of the internal or of the distal double bond in the allene can be activated. The presence of a gas (argon or carbon monoxide) over the reaction mixture can be crucial for the reaction outcome

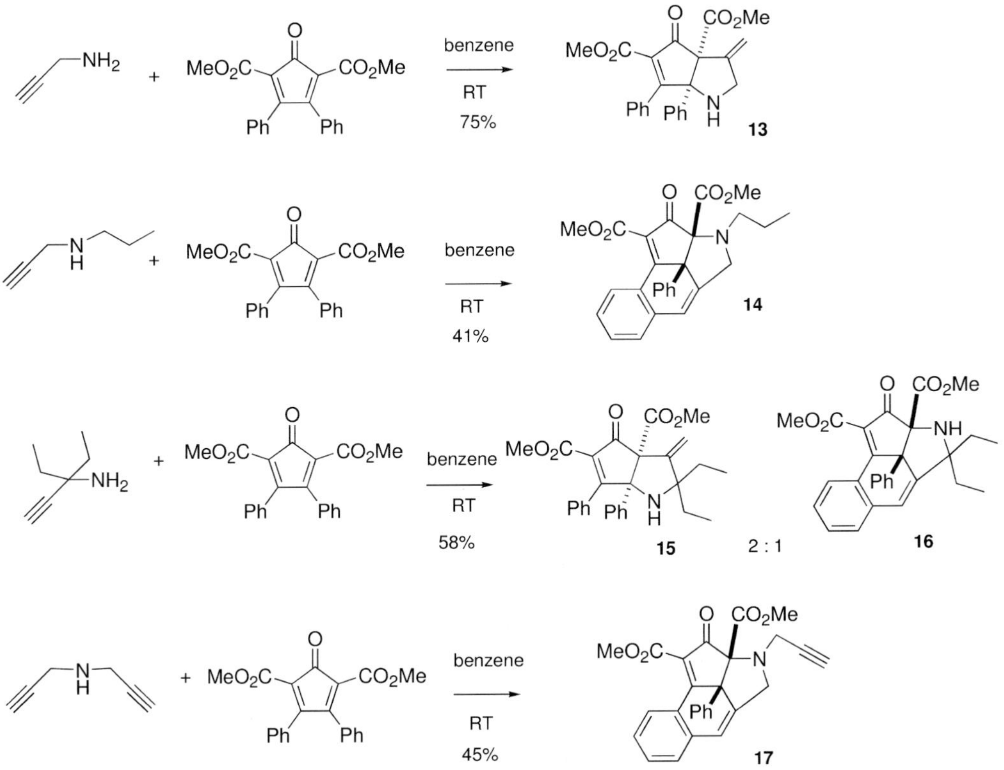

Figure 6.2 Examples of pericyclic reactions under thermal conditions.

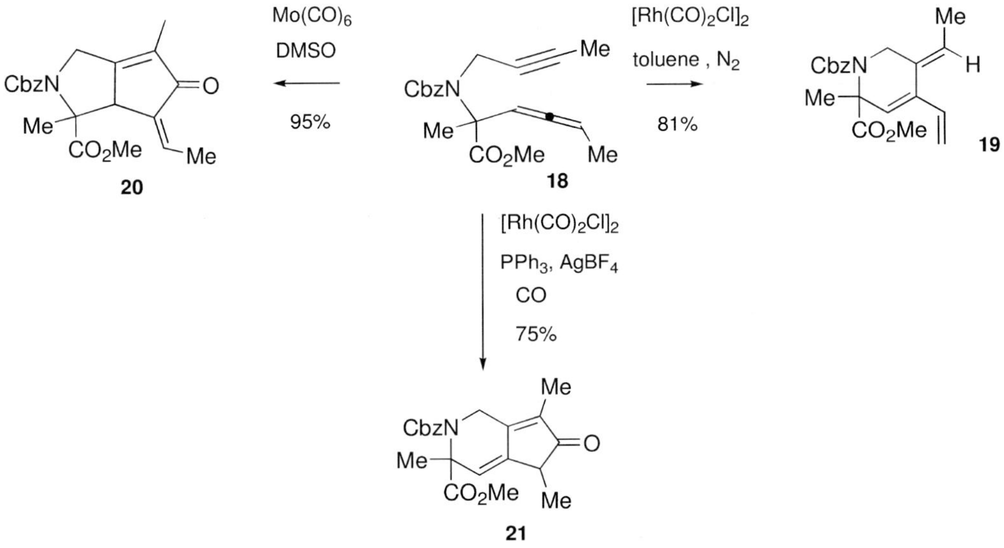

Scheme 6.11 Divergent reactivities of alkynyl-allene **18** with various sources of CO.

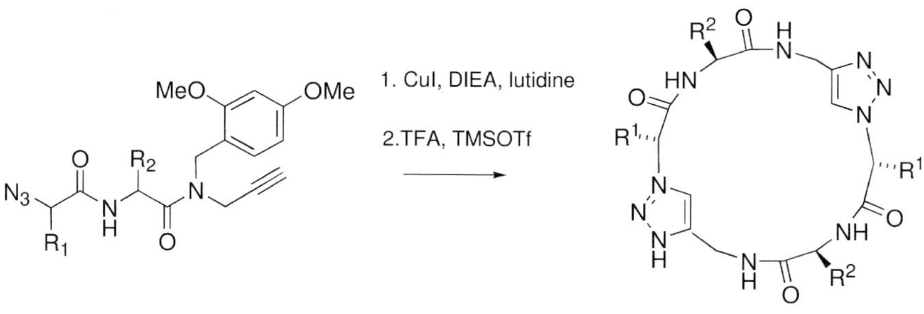

Scheme 6.12 PLAs and click chemistry.

(Scheme 6.11). The starting ene-yne is accessible through a Claisen transposition performed on a propargylic alcohol [18].

A tandem dimerization-macrocyclization approach using the 1,3-dipolar azide-alkyne cycloaddition (Huisgen's reaction) has been employed in facile and convergent solution-phase syntheses of C_2-symmetrical or disymmetrical cyclic peptide scaffolds containing triazole ε-amino acids. The macrocyclic system **22** was designed as a dipeptide surrogate. This chemistry, revisited by Sharpless, gives access (Scheme 6.12) to heterocyclic pseudohexapeptide structures with potential utility in biological and materials settings. By taking advantage of the functional flexibility of the side chains, complex structures can be obtained in a few high-yielding steps [19].

Pyrrolo[3,4]quinolones can be formed through the coupling of anilines with PLAs N-substituted with heterocyclic aldehyde moieties in the presence of mild Lewis acid catalysts. The coupling proceeds through sequential imine formation and an intramolecular aza Diels–Alder reaction (Povarov reaction). This approach has been applied in the formal synthesis of camptothecin (precursor **23**) and in the total synthesis of luotonin A (**24**). This chemistry (Scheme 6.13a, b) underlines the power of inverse-electron demand hetero-Diels–Alder reactions for the preparation of alkaloids, particularly when performed in an intramolecular fashion [20].

6.2.2.4 **PLAs in Multicomponent Reactions (MCRs)**

A multicomponent synthesis based on the sequential use of an Ugi reaction followed by a nitrile oxide cycloaddition has been reported (Scheme 6.14). The final compounds are isoxazolines. With PLA as one of the partners in the Ugi reaction, neither the triple bond nor the nitro function interfere with the reaction sequence. The transformation of the nitro functional group into a nitrile oxide was achieved by treatment with $POCl_3$ in the presence of triethylamine, and several isoxazolines (**25–27**) were obtained in moderate yields by this procedure. A similar sequence has been reported for the synthesis of the triazole analogues; in this case the nitro group was replaced by an azido group that is ready for click chemistry [21].

A simple synthesis of fused imidazoazepine derivatives such as **28** by a van-Leusen/intramolecular enyne metathesis has been reported. The two-step reaction sequence generates compounds of significant molecular complexity from simple

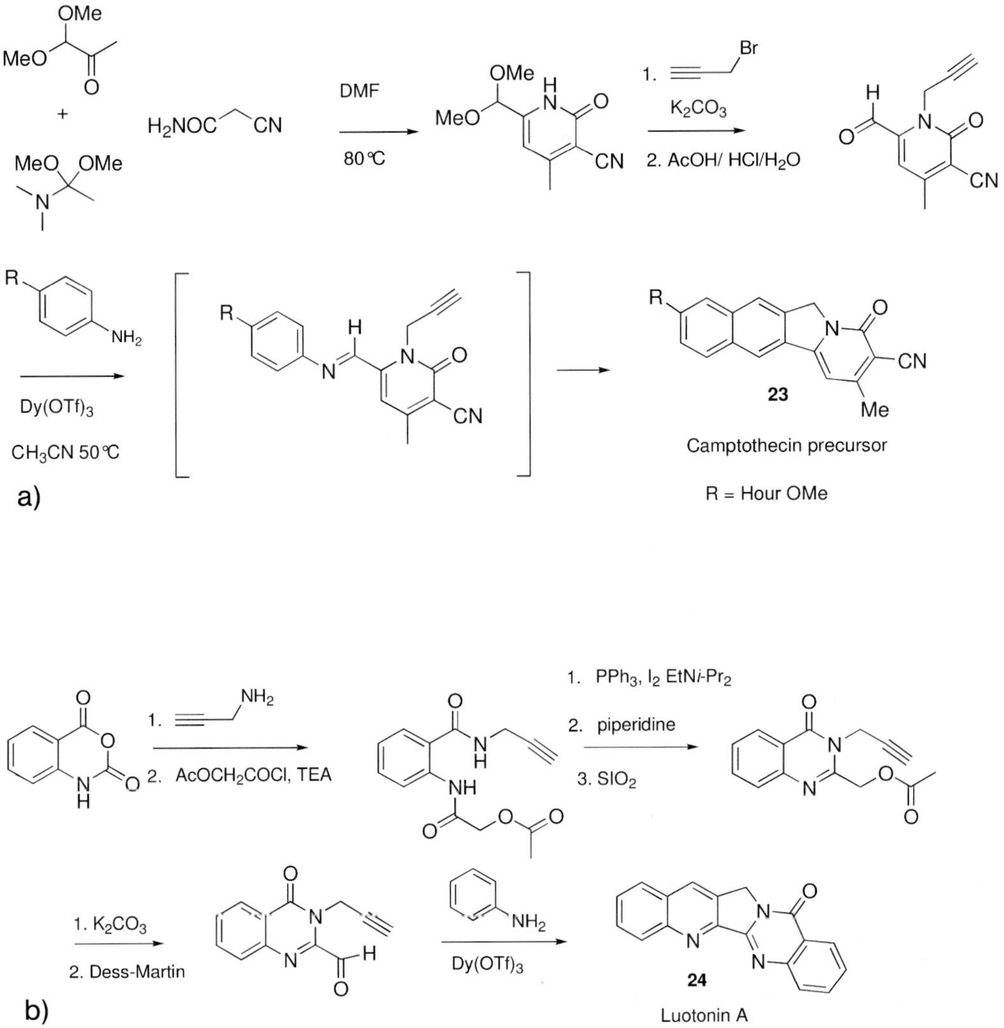

Scheme 6.13 a) Camptothecin analogue from PLA. b) Luotonin A from a salicylamide-containing PLA.

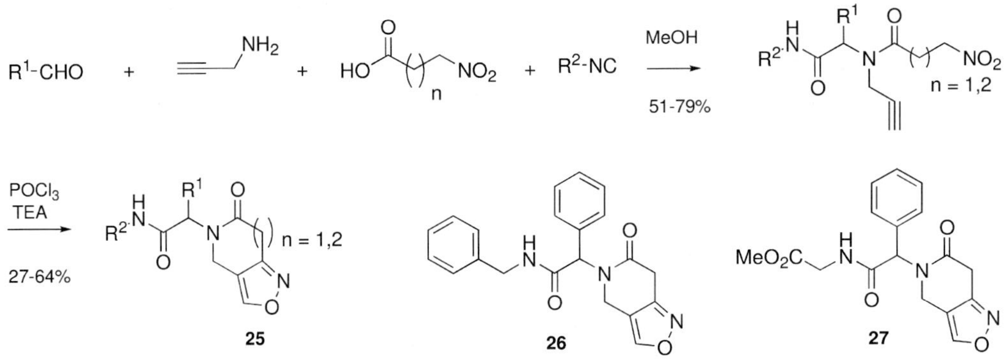

Scheme 6.14 Multicomponent UGI reaction with a PLA.

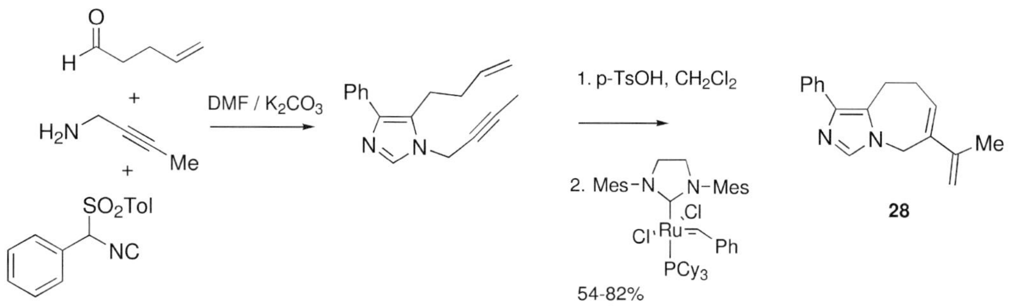

Scheme 6.15 PLAs in metathesis for the construction of aza-heterocycles.

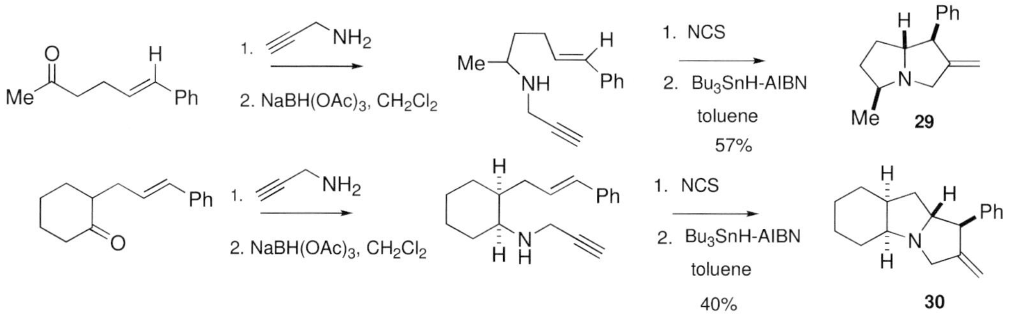

Scheme 6.16 PLA in radical-driven reactions.

starting materials including PLAs, in an expedient fashion with good overall yields. A modification of the van Leusen reaction provided a diene that could be further exploited (Scheme 6.15). It should be noted that the metathesis operated only if an equivalent of *p*-toluensulfonic acid was added [22].

6.2.2.5 **PLA in Radical Reactions**

The generation, with the aid of an *N*-chloroamine, of a radical species in an organic molecule containing a PLA residue and an alkene gives rise to a tandem cyclization with high stereocontrol. The radical reactions are initiated with Bu_3SnH and produce functionalized pyrrolidines. The 5-*exo*,5-*exo*-tandem cyclization of aminyl radicals is a powerful reaction for the formation of complex heterocycles such as **29** or **30** from fairly cheap starting materials (Scheme 6.16). Notably, the reactions produce only single stereoisomers for steric reasons [23].

The use of PLAs in the preparation of biomolecules has therefore attracted new interest since the rediscovery of the chemical potential, especially in intramolecular cascade reactions, of the alkyne functions.

6.3
trans-4-Hydroxy-(S)-proline (HYP)

trans-4-Hydroxy-(*S*)-proline (**31**) is an important component in animal tissues. Indeed, the hydrogen-bonding ability of the 4-hydroxy group is relevant in the structure of collagen, and furthermore, when located in proteins, proline is responsible for steric constraints that stabilize secondary structures. Moreover, with two stereocenters and three versatile functions, HYP is an abundant and attractive chiral building block for a variety of synthetic challenges. Proline and its 4-substituted derivatives have been extensively used in medicinal chemistry, either in the design of peptide ligands, or in incorporation in more elaborated targets. In those cases the hydroxy group at C-4 has been used for direct substitution or has been transformed in order to allow the introduction of substituents at C-3 in five-membered azaheterocycles.

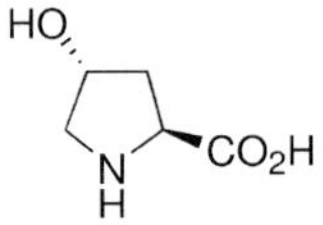

31

(4*R*,2*S*)-4-OH Proline (HYP)

6.3.1
Structural Transformations of HYP

6.3.1.1 C-4 Alkylation of HYP

Fosinopril (**32**), an angiotensin-converting enzyme inhibitor, contains a *trans*-4-cyclohexyl-proline residue **33** obtained from HYP in several steps by Cbz protection of the endocyclic nitrogen, oxidation of the secondary alcohol, and treatment of the resulting ketone with phenylmagnesium bromide. Only the diastereomer resulting from *anti* addition of the Grignard with respect to the bulky carboxylate at C-2 is observed, resulting in a *cis* relationship between the hydroxy group and the carboxylate in the adduct. Dehydration of the benzylic alcohol, followed by chemoselective hydrogenolysis performed either with Pearlman catalyst or with Li/NH_3 produced the *cis*- or the *trans*-4-phenylproline, respectively. Hydrogenation of the remaining aromatic ring with simultaneous removal of the nitrogen protective group then gave the two corresponding cyclohexyl adducts (Scheme 6.17). The final conversion into a fosinopril structure was achieved by classical peptide synthesis [24, 25].

6.3.1.2 C-4 Fluorination and Fluoroalkylation of HYP

When incorporated into collagen mimics, proline derivatives substituted at C-4 have been shown to enhance the thermal stabilities of triple helices; the most striking results were observed when *trans*-4-fluoroproline was embedded into a peptide. However, the introduction of fluorinated appendages at C-4 in proline with defined *cis* or *trans* stereochemistry is not a trivial task, as stereoelectronic participation by the carboxylate at C-1, as a consequence of the puckered ring conformation adopted by the pyrrolidine ring, hampers diastereocontrol. The following three examples illustrate suitable solutions for meeting these challenges. The syntheses of the two C-4 fluoroproline epimers were achieved as follows: *cis*-

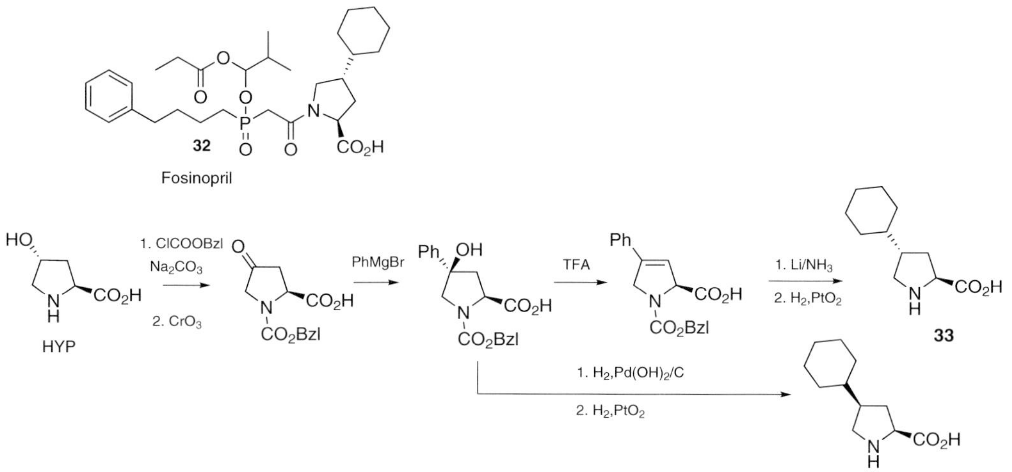

Scheme 6.17 Use of HYP for the synthesis of Fosinopril.

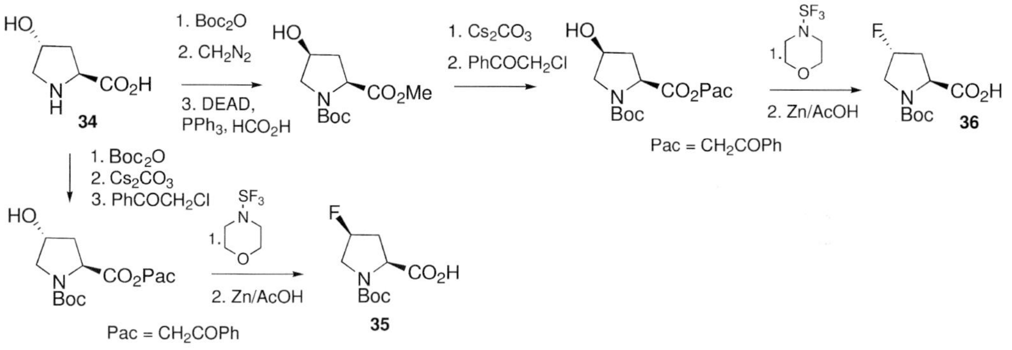

Scheme 6.18 Fluorinated Hyps.

fluoro-HYP (**35**) was obtained by direct fluorination with morpho-DAST on the secondary alcohol **34**, with a clean S_N2 reaction at C-4. Conversely the *trans*-C-4 fluoro analogue **36** was obtained through a Mitsunobu inversion on the protected HYP, followed by fluorination, resulting in a double inversion. It is worth noting that the phenacyl ester on the C-2 carboxylate is important for the success of these transformations (Scheme 6.18) [26].

cis-Difluoro- (**37**) and -trifluoromethyl (**38**) prolines were prepared from *N*-Boc-4-oxo-L-proline, a suitable intermediate for the synthesis of fluorinated prolines [27]. The difluoromethyl group was installed *cis* to the C-1 carboxylate in a two-step sequence in which a modified Wittig reaction, giving the difluoroalkene, was followed by a stereoselective hydrogenation of the *exo* alkene from the less hindered face. By a similar strategy the trifluoromethyl group was established *cis* by use of the Ruppert reagent, followed by a dehydration and then by a facially steered hydrogenation of the *endo* alkene, with simultaneous departure of the Cbz and

Boc protective groups on the carboxylate and the nitrogen (Scheme 6.19). The unusual Boc removal under hydrogenolytic conditions may be accounted for by the acidity of the catalyst.

An interesting approach to both epimers of 4-CF_3-proline analogues has been reported. The strategy to produce the *cis* epimer is based on the steric bias introduced by bulky protection (*tert*-$BuMe_2Si$) on the primary alcohol at C-2 during the hydrogenation step. The introduction of the CF_3 function into 4-oxo-HYP was performed with the Ruppert reagent and the carbinol was then transformed into an *endo* alkene that was subjected to hydrogenation. Two different sets of conditions were used, with or without silyl protection at C-1. In each case a different epimer was obtained: palladium on charcoal yielded the *cis* adduct **38** because of facial selectivity, whereas hydrogenation with the Crabtree catalyst afforded the *trans* epimer **39**, probably as a result of the directing effect of the free hydroxy group during the hydrogenation. In each case the final carboxylate was cleanly restored by oxidation with TEMPO, bleach, and sodium chlorite. The above sequences have been extensively exploited to prepare (Scheme 6.20) both C-4 epimers with various substituents (benzyl, alkyl, methylene carboxylate) on the proline scaffold [28, 29].

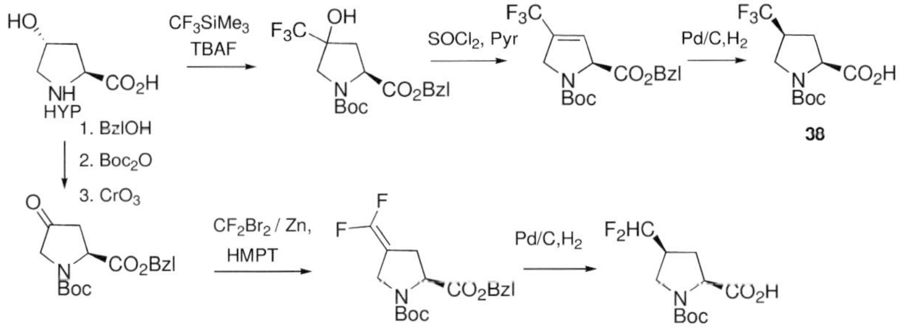

Scheme 6.19 Other fluorinated HYPs.

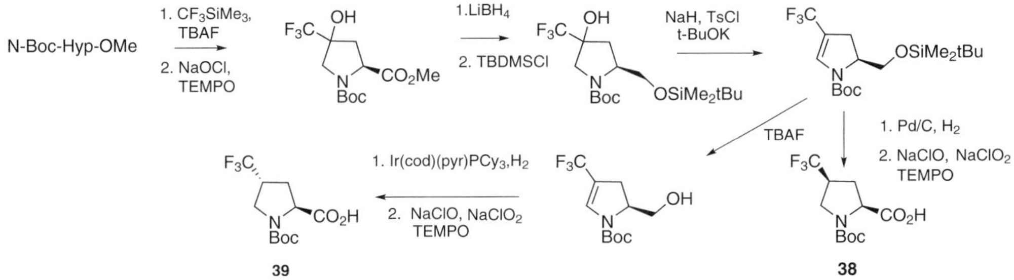

Scheme 6.20 Stereodivergent syntheses of trifluoromethyl-HYP.

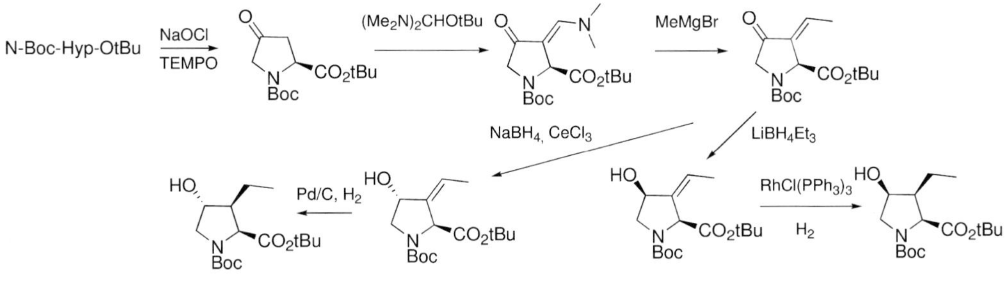

41 40

Scheme 6.21 Alkylated HYPs.

6.3.1.3 C-3 Functionalization of HYP

A practical stereoselective synthesis of 3-alkyl-4-HYP derivatives has recently been reported [30]. The secondary alcohol at C-4 in protected HYP was oxidized to a ketone, the enaminone was subsequently obtained regioselectively by use of Bredereck's reagent, and so the substrate was set up for Michael additions of Grignard reagents. Stereodivergent reduction of the ketone was then performed, either under Luche's conditions or with a superhydride, to provide the two C-4 epimeric alcohols (**40** and **41**, respectively; Scheme 6.21). Reduction of the resulting alkenes was performed under two different sets of conditions depending on the configuration of the secondary alcohol: palladium on charcoal was used for the *trans* hydroxy compound and Wilkinson's catalyst for the *cis* hydroxy compound in order to avoid hydrogenolysis of the C-4 hydroxy group as a side reaction.

6.3.2 HYP in the Synthesis of Biomolecules

6.3.2.1 HYP in the Synthesis of Alkaloids

Weinreb reported an elegant and general strategy, starting from HYP, for the enantiospecific construction of the biologically active *Securinega* alkaloids **42** and **43**, which display an impressive range of biological activities. HYP is transformed into a keto nitrile, which can undergo intramolecular pinacol-type coupling assisted by SmI_2 to yield an azabicyclic α-hydroxyketone. Acylation of the norsecurine A-ring was performed through the intermediacy of an N-acyliminium ion generated regioselectively by a radical-based methodology. Proton abstraction next to the nitrogen was achieved through diazotization of the anthranilic amide, and the resulting aminal was obtained after quenching with methanol. Subsequent Grignard allylation and transformation of the resulting alkene into an alcohol opened the pathway to an intramolecular heterocyclization. Finally, a high-pressure Wittig reaction with the Bestman ylide gave dihydronorsecurinine, the basic core of this class of alkaloids [31]. This sequence was only viable with the use of appropriate protecting groups for the different oxygenated functions (Scheme 6.22). This masterful approach to such an important biomolecule shows that HYP is an useful polyfunctional building block fit for total synthesis.

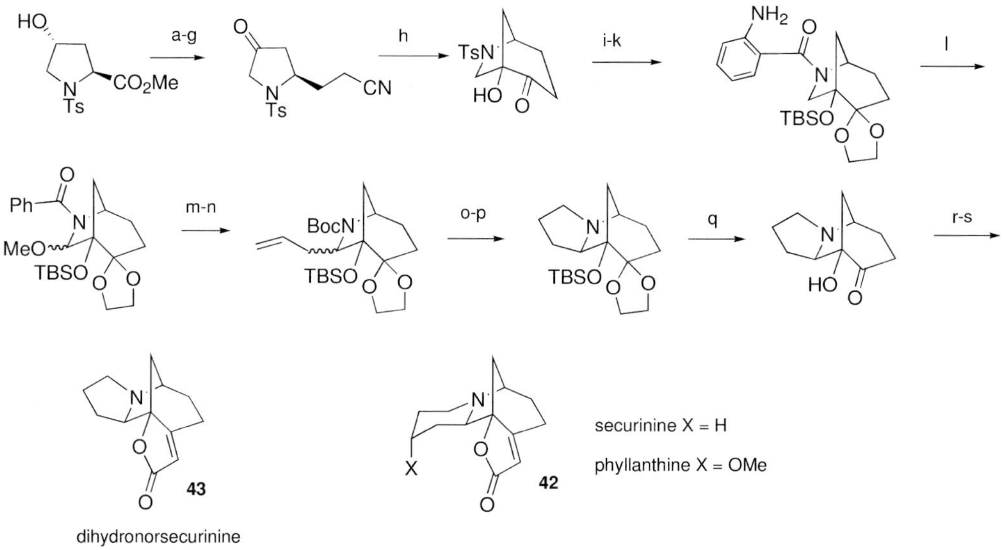

Scheme 6.22 Weinreb's synthesis of securinine sarting from HYP [31]. Reagents: a) TBSCl, imid.; b) DIBALH; c) DMSO, $(COCl)_2$, NEt_3; d) PPh_3=CHCN; e) H_2, Pd/C; f) TBAF; g) Jones reagent; h) SmI_2, MeOH; i) $(CH_2OH)_2$ *p*-TsOH; j) Na, naphthalene, k) isatoic anhydride; l) $NaNO_2$, HCl, CuCl, MeOH; m) allylmagnesium; n) $(Boc)_2O$, NEt_3; o) disiamylborane, H_2O_2, NaOH; p) TsCl, NEt_3; q) HCl 0.8 N, 60 °C MeOH; r) HCl 3 N, 95 °C; s) Ph_3P=C=C=O, 12 kbars, toluene.

6.3.2.2 HYP in the Synthesis of Kainic Acid Derivatives

Kainic acid (**44**) is an important tool for studies in neurochemistry directed towards the mammalian central nervous system. As a constrained analogue of glutamic acid, its activity is assumed to be mediated through the glutamate receptors. There are many reports on the preparation of this relatively simple molecule, but the presence of three contiguous stereocenters on the pyrrolidine ring makes the synthesis challenging. To date the most effective and highly stereoselective synthesis of (–)-kainic acid, starting from inexpensive HYP, requires 14 steps and gives the expected product in 7% overall yield (Scheme 6.23). The key steps are a diastereoselective enolate alkylation at C-3 and the higher-order cyanocuprate substitution of the tosyl-protected alcohol at C-4 with retention of configuration, which ensures the important *cis* relationship between C-3 and C-4. The final step involves the restoration of the correct oxidation state of the side chain at C-3 without affecting the isoprenyl residue at C-4 [32].

6.3.2.3 HYP in the Synthesis of Amino Sugars

Polyhydroxylated N-heterocycles are of great chemical interest because of the wide range of biological activities they display. The iminosugars are strong inhibitors of glycosidases and have been investigated as potential therapeutic agents in the treatment of diabetes, HIV, and cancer. The simplest iminosugars are those based

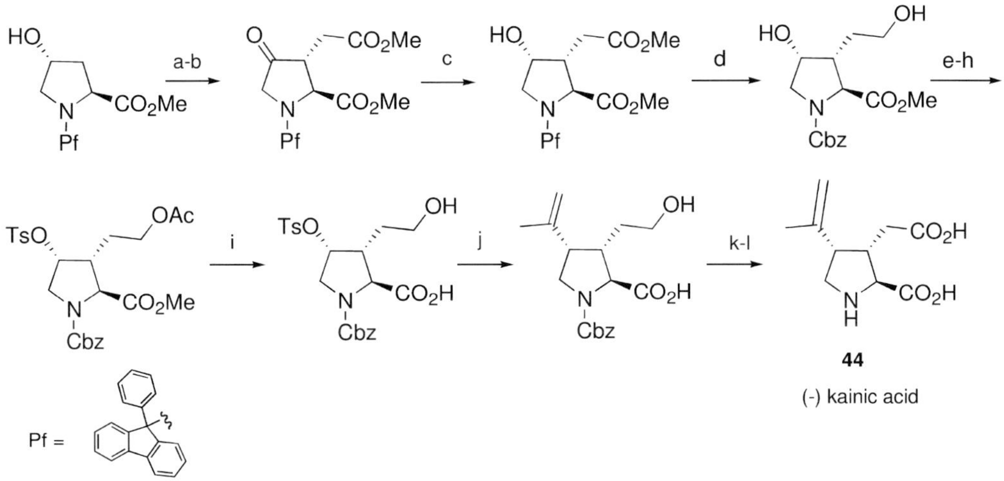

Scheme 6.23 HYP for the preparation of kainic acid.
Reagents: a) IBX, EtOAc; b) BuLi, $BrCH_2CO_2Me$, NaI (cat.); c) L-Selectride; d) $NaBH_4$; e) H_2, Pd/C; f) CbzCOCl; g) AcCl, collidine; h) tosylimidazole, MeOTf; i) NaOH, THF/H_2O; j) $[CH_2{=}C(CH_3)]_2CuCNLi_2$, THF; k) Jones reagent; l) NaOH, reflux.

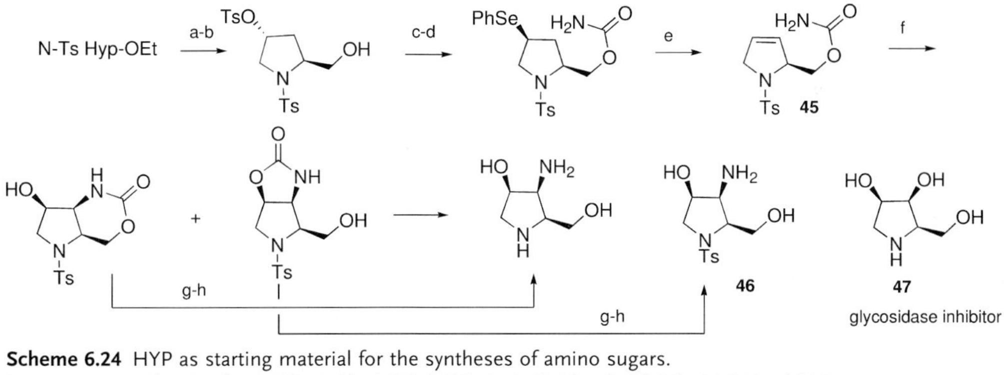

Scheme 6.24 HYP as starting material for the syntheses of amino sugars.
Reagents: a) TsCl, NEt_3; b) $NaBH_4$, LiCl; c) $CCl_3OCN{=}C{=}O$, K_2CO_3; d) $(PhSe)_2$, $NaBH_4$; e) H_2O_2, Pyr.; f) NaOH, *t*BuOCl, *i*Pr_2NEt, Admix; g) LiOH, H_2O/MeOH; h) Na/NH_3, –78 °C.

on the pyrrolidine motif. HYP has been used [33] for the synthesis of the C-3 amino analogue **46** of a known glycosidase inhibitor triol **47** (Scheme 6.24). The chemistry is based on a diastereoselective tethered aminohydroxylation performed on the 3,4-dehydroproline **45**, obtained from HYP by an addition/elimination sequence by the selenide approach.

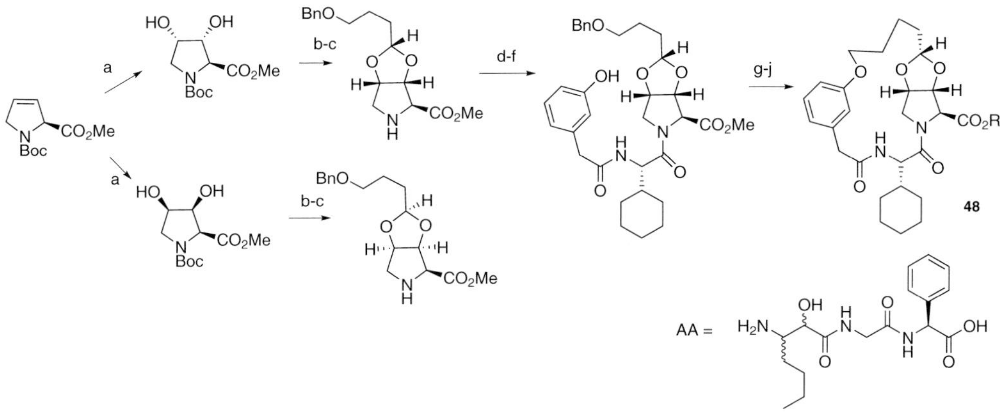

Scheme 6.25 Hydroxy-HYP for the construction of depsipeptides.
Reagents: a) OsO_4, NMO, ^tBuOH; b) $BnO(CH_2)_3CHO$, *p*TsOH, $MgSO_4$; c) HCl 2 N; d) N-Boc-chg-OH, HATU, iPr_2NEt_2; e) HCL (2M), dioxane; f) 3-OH-$PhCO_2H$, HAU, iPr_2NEt, CH_2Cl_2, 0 °C; g) LiOH, MeOH; h) AA, DhBtOH, EDC, NMM; i) Dess–Martin; j) TFA.

6.3.2.4 **Hepatitis C Inhibitors**

A dihydroxyproline fragment has been incorporated into the macrocylic structure **48** for targeting of a serine protease involved in the processing of the structural or nonstructural proteins in the replication of the hepatitis C virus. This enzyme has been crystallized, representing a valuable starting point for the design of new ligands. As shown in its solid-state structure, the catalytic side of the enzyme is shallow and solvent-exposed, which makes the design of small-molecule inhibitors challenging. In order to use most of the structural information extracted from the ligand receptor interaction, a macrocyclic compound based on proline was designed. Indeed the proline fragment is embedded into a macrocycle in a tentative of depeptization of the native substrate of the enzyme. During the synthesis of the target it was observed that the acetalization of the hydroxy-HYP (i.e., dihydroxyproline) selectively gave the *endo* adduct whatever the stereochemistry of the dihydroxyproline. The final compound was constructed by classical peptide chemistry and with use of a Mitsunobu reaction for the macrocyclization (Scheme 6.25). The orientation of the final adduct in the protease has been obtained by X-ray crystal analysis, so better design of these inhibitors can be forecast in the future [34].

6.4
L-Serine (SER)

L-Serine ((*S*)-serine, **49**, SER) the simplest β-hydroxy proteogenic amino acid and the major component of silk fibroin, is also available in the opposite configuration.

The presence of hydroxy functions both in SER and in threonine provides chemists with an additional handle to work with in designing retrosyntheses towards target molecules containing one or more asymmetric centers. The aldehyde derived from SER, the Garner aldehyde, is by far the most popular derivative of L-serine, probably because of its configurational stability. If the Garner aldehyde is used as a starting material for the preparation of amino acid derivatives in their natural configuration, (*R*)-serine has to be used because the initial primary alcohol ultimately becomes the final carboxylate. Some applications of the Garner aldehyde are illustrated below.

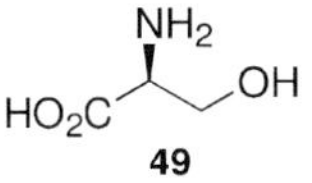

49

(*S*)- Serine (SER)

6.4.1
SER and SER Derivatives in the Synthesis of Biomolecules

6.4.1.1 SER in the Synthesis of Carbolines

Eudistomins **52**, isolated from a Caribbean tunicate, are members of the tetrahydro-β-carboline family of marine alkaloids. Eudistomin C, possessing an oxathiazepine ring, displays potent antiviral activity against DNA and RNA viruses; its total synthesis has been performed by Fukuyama [35]. The key step is the construction of the tetrahydrocarboline framework through a diastereoselective intermolecular Pictet–Spengler reaction between a tryptamine derivative **51** and the Garner aldehyde (**50**). The oxathiazepine ring was obtained by generation of a thiol, which provided (Scheme 6.26) the seven-membered ring through nucleophilic displace-

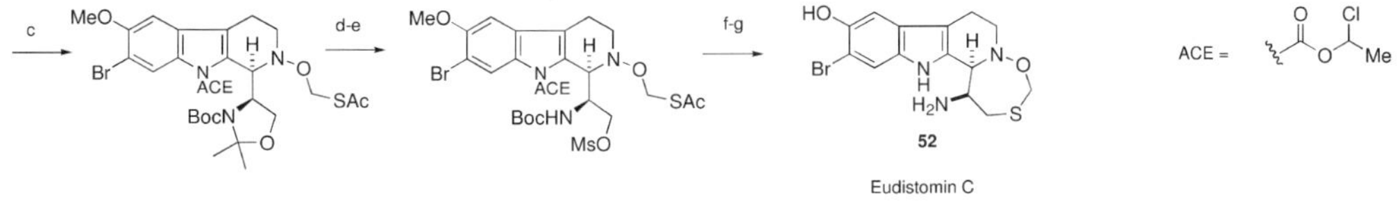

Scheme 6.26 SER as building block for the synthesis of eudistomin C.
Reagents: a) cat Cl_2CHCO_2H, toluene 0 °C (dr 11:1); b) *n*-BuLi, −78 °C, ACE-Cl; c) SO_2Cl_2, −78 °C to RT; AcSH, *i*-Pr_2NEt; d) AcOH aq; e) MsCl, *i*-Pr_2NEt; f) K_2CO_3, MeOH; g) BBr_3, CH_2Cl_2.

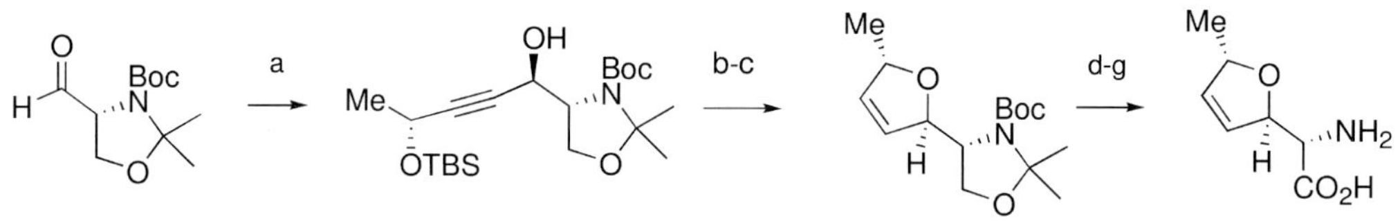

Scheme 6.27 SER as building block for the synthesis of furanomycin.
Reagents: a) (*R*)-MeCH(OTBS)≡CH, BuLi, $ZnBr_2$; b) $LiAlH_4$; c) $AgNO_3$; d) TsOH, MeOH; e) Dess–Martin; f) $NaClO_2$, *t*-BuOH; g) TFA.

ment of a mesylate obtained from the primary alcohol originating from SER. The final steps, as usual, are mainly devoted to the removal of protecting groups.

6.4.1.2 SER in the Synthesis of Furanomycin

Furanomycin (**54**) is a *Streptomyces* metabolite active as an isoleucine substitute in protein translation. Translatable amino acid analogues of this type are of great interest for the preparation of proteins containing unusual amino acids. A concise and modular synthesis starting from the Garner aldehyde **50** and proceeding in seven steps to furanomycin (**54**) has been reported (Scheme 6.27); the key steps include a stereoselective acetylide addition and an Ag^+-mediated cyclization of an α-allenic alcohol **53** to construct the 2,5-dihydrofuran ring [36].

6.4.1.3 SER in the Synthesis of Diketopiperazine Alkaloids

Asperazine (**56**) is a member of a large family of diketopiperazine alkaloids that contain two tryptophan units, but differs from other members of this family by having its tryptophan units linked in an unsymmetrical fashion, generating a diaryl-substituted quaternary stereocenter. Biological evaluation of asperazine revealed no antibacterial or antifungal activity, but significant cytotoxicity towards leukemia. Overman's synthesis was accomplished in 22 steps starting from SER, tryptophan, and phenylalanine as chiral building blocks. In the whole synthetic sequence the only low-yielding step is the hydrogenation of the acylamino function, which produces a mixture of two diastereoisomers in a 4/1 ratio (Scheme 6.28). Building block A was constructed from ethynyloxazolidine **55**, itself obtained from SER (**49**) [37, 38].

6.4.1.4 SER in the Synthesis of Cleomycin

Enantiomerically pure (S)-cleonin (**57**), a key component of the antitumor antibiotic cleomycin, was prepared from SER. Kulinkovich cyclopropanation of the methyl ester of N-Cbz-Ser acetonide afforded [39] the hydrocyclopropyl moiety (Scheme 6.29). An extension of the Kulinkovich cyclopropanation also allowed the preparation of the unnatural substituted cyclopropylglynes **58**.

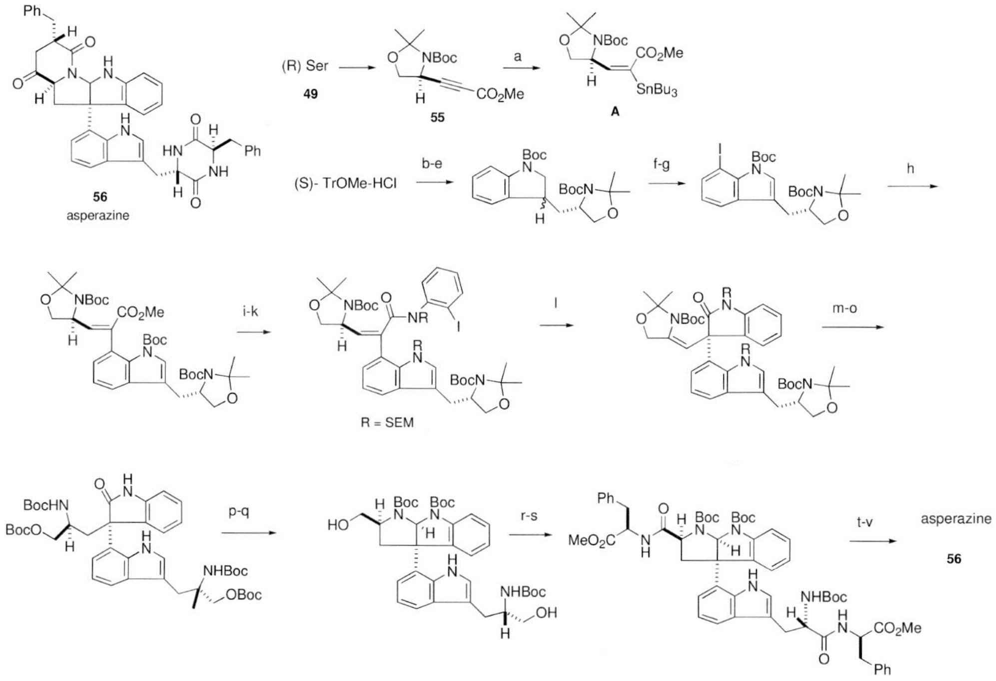

Scheme 6.28 Overman's synthesis of asperazine from SER.
Reagents: a) Bu_3SnH, Pd(0); b) Et_3SiH, TFA; c) Boc_2O, Aq. Na_2CO_3; d) $NaBH_4$, LiCl; e) 2,2 dimethoxypropane, *p*-TsOH; f) *s*-BuLi, ICH_2CH_2I; g) DDQ, K_2CO_3; h) $Pd_2(dba)_3.CHCl_3$, $(2\text{-furyl})_2P$, CuI; i) DMSO, 130 °C; j) 2-iodoaniline, $AlMe_3$, CH_2Cl_2; k) NaH, SEM-Cl; l) $Pd_2(dba)_3.CHCl_3$, $(2\text{-furyl})_2P$, CuI; m) TBAF; n) H_2, Pd/C; o) HCl 1 N, then Boc_2O; 1 N NaOH, Bu_4NHSO_4; p) $LiEt_3BH$; q) KOH, MeOH; r) SO_3, Pyr, DMSO; s) $NaClO_2$, NaH_2PO_4, *t*-BuOH; t) PheOMe-HCl, HATU, Et_3N; u) HCO_2H; v) 0.7 N AcOH, *n*BuOH 120 °C.

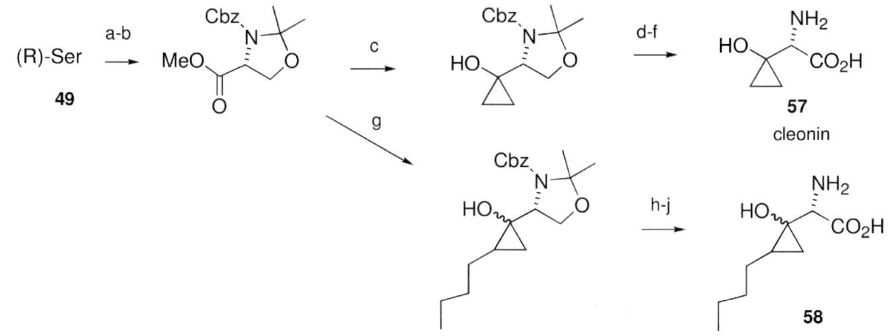

Scheme 6.29 Cleonin from SER through a Kulinkovitch reaction.
Reagents: a) MeOH, $SOCl_2$, then CbzCl, $NaHCO_3$; b) 2,2-dimethoxypropane, $BF_3.Et_2O$; c) EtMgBr, $Ti(O\textit{i}\text{-Pr})_4$, Et_2O; d) PPTS, MeOH, e) PDC, DMF; f) HCO_2NH_4, *i*-PrOH, Pd/C, MW; g) cyclohexyl-MgBr, $ClTi(O\textit{i}\text{-Pr})_3$, hex-1-ene, h) PPTS, MeOH; i) PDC, DMF; j) HCO_2NH_4, Pd/C, is-PrOH.

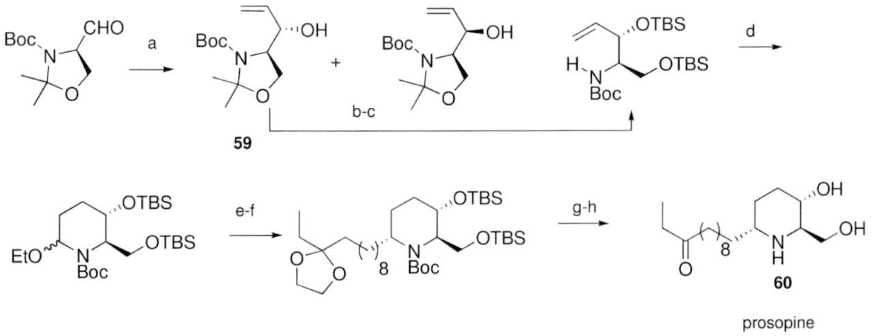

Scheme 6.30 Ojima's synthesis of prosopine starting from SER as building block. Reagents: a) CH_2=CHMgBr; b) *p*TSA, MeOH; c) TBSCl, imide; d) Rh(acac)$(CO)_2$, BIPHEPHOS, H_2/CO (1/1), 4 bars, EtOH; e) EtC(OCH_2CH_2O)$(CH_2)_8CH_2$Br, Li, Et_2O; f) $BF_3.Et_2O$; g) TBAF; h) TFA.

6.4.1.5 SER in the Synthesis of Piperidine Alkaloids

Piperidine alkaloids are commonly found in nature, and many of them exhibit biological activity of interest in medicinal chemistry. (+)-Proposopinine (**60**), a trisubstituted piperidine with antibiotic and anesthetic properties, has attracted considerable interest as a synthetic target. Ojima has reported [40] a very elegant approach by a cyclohydrocarbonylation process. An allylic alcohol **59** derived from the Garner aldehyde was subjected to a regioselective hydroformylation in methanol. The resulting linear aldehyde was trapped by the carbamate nitrogen and the transient immonium species was quenched with methanol. The final side chain was then introduced by using the reactivity of a cuprate versus an acyl-immonium compound, and the synthesis of prosopinine was achieved after deprotection of the hydroxy and amino functional groups (Scheme 6.30). The only bottleneck in the synthesis is the low diastereoselectivity in the vinylation of the Garner aldehyde, but in some cases this stereodivergence turns out to be beneficial, when unnatural analogues are desired.

6.4.1.6 SER in the Synthesis of Nonproteinogenic Amino Acids

The synthesis of nonproteinogenic α-amino acids has been an area of great interest in recent years. Amongst the targets that have attracted substantial interest are amino acids bearing aromatic residues with various physical properties in their side chains; these are employed as probes for biophysical experiments. The use of SER as a starting material for this purpose was an obvious strategy [41, 42]. The availability of the SER-based zinc reagent **61** and use of the Negishi reaction allowed the construction of various phenylalanine analogues **62a–d** bearing either aromatic or heteroaromatic rings. Higher homologues could also be obtained if the zincated serine was first coupled with allyl chloride in order to install a terminal double bond that could be boronylated and then used in Suzuki reactions for the production of many other arylated amino acids (Scheme 6.31). Unlike the use of the glycine strategy for the preparation of unnatural amino acids, the SER strategy features an installed stereocenter from the beginning.

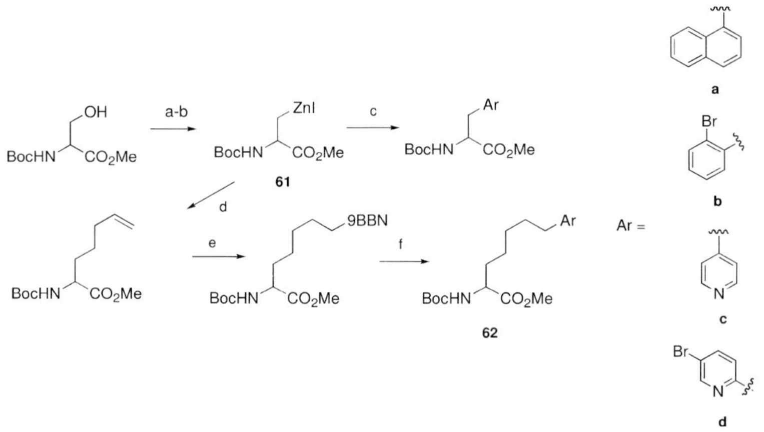

Scheme 6.31 SER: a versatile building block for the preparation of unnatural amino acids. Reagents: a) TsCl, then NaI; b) activated Zn; c) $Pd(PPh_3)_2Cl_2$ then ArI; d) AllylCl, $CuBr\text{-}Me_2S$; e) 9-BBN; f) ArI, $PdCl_2(dppf)$, K_3PO_4.

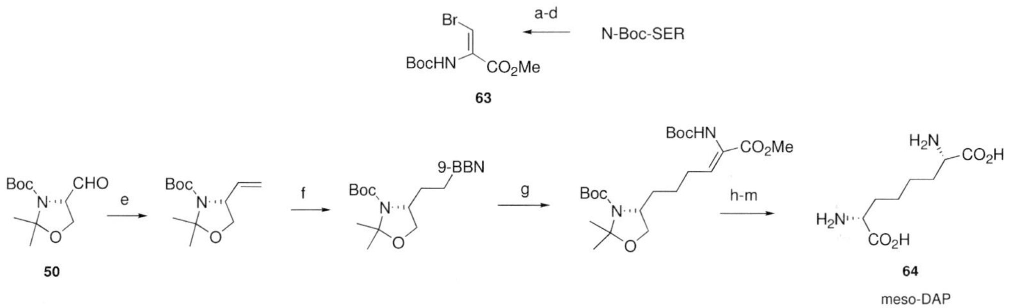

Scheme 6.32 SER as building block for the preparation of *meso*-DAP.
Reagents: a) MeI, K_2CO_3; b) MsCl, Et_3N; c) NBS; d) Et_3N; e) $Ph_3P{=}CH_2$; f) 9-BBN; g) compound **63**, $PdCl_2(dppf).CHCl_3$; h) [(COD)Rh(S,S)-Et-DuPhos]OTf, H_2/250 psi; i) TFA; j) Jones; k) $TMSCHN_2$; l) HCl 5 M; m) propylene oxide.

6.4.1.7 SER in the Synthesis of α,α′-Diaminoacids

meso-DAP (2,6-diaminopimelic acid) is the direct biosynthetic precursor of lysine through the action of a decarboxylase and is essential for the growth and structural rigidity of many bacteria through cross-linking of the polysaccharides of their cell wall peptidoglycan units. These lysine and peptidoglycan activities are not observed in mammalian biochemistry, and this is reflected in interest in utilizing α,α′-diaminoacids as potential antibiotics with low mammalian toxicity. The synthesis reported by Taylor [43] is based both on the Garner aldehyde **50** and on the modified SER **63**. A Suzuki reaction followed by a diastereoselective hydrogenation and a restoration of the amino acid functionality provides a straightforward route (Scheme 6.32) to *meso*-DAP **64**, and also opens the way to many other analogues.

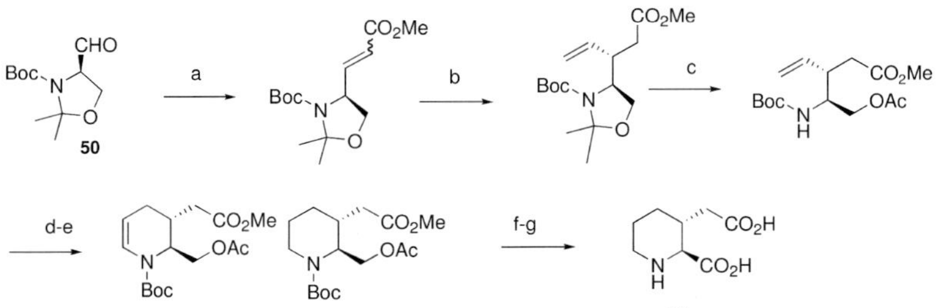

Scheme 6.33 Rigidified analogues of glutamic acid from SER.
Reagents: a) $(MeO)_2POCH_2CO_2Me$, TBAI, K_2CO_3; b) $LiCu(CH{=}CH_2)_2$, TMSCl; c) pTSA, MeOH, then Ac_2O, pyr.; d) $Rh(acac)(CO)_2$, biphephos, CO, H_2 (2 bars); e) Rh-C, H_2; f) DBU, LiBr, TEMPO, NaOCl then CH_2N_2; g) HCl, aq., then propylene oxide.

6.4.1.8 SER in the Synthesis of Rigidified Glutamic Acid

Glutamic acid is an important neurotransmitter that plays various roles in the mammalian brain. The multiple subtypes of glutamatergic transmission have been identified through the synthesis of selective ligands based on conformational rigidification of glutamate acid itself. One attempt was to introduce the glutamate framework in a piperidine ring [44]. The starting material was the Garner aldehyde **50**, which was transformed into an enoate. Subsequent diastereoselective 1,4 addition of a vinyl cuprate offered the opportunity for a cyclohydrocarbonylation. The final compound was a pipecolic derivative **65**, bearing the two acid appendages in a *trans* relationship (Scheme 6.33). This sequence illustrated the relevant potential of hydroformylation for forming azaheterocycles and paves the way for further transformations.

6.5
4-Methoxypyridine (MOP)

Available in bulk from commercial sources, MOP (**66**) is prepared from 4-chloropyridine by nucleophilic substitution with sodium methoxide, but if stored for several days an isomerization to 1-methyl-1*H*-pyridin-4-one occurs. MOP is therefore preferentially stabilized as its *N*-oxide. Many syntheses of azaheterocycles such as substituted piperidines or piperidones have been generated from MOP.

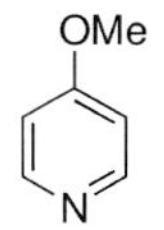

66

4-OMe Pyridine MOP

6.5.1 MOP in the Synthesis of Biomolecules

6.5.1.1 MOP in the Synthesis of Alkaloids

Dipyridopyrazine alkaloids Barrenazine A (**68**) is a cytotoxic alkaloid with a rare dipyridinopyrazine structure, isolated from a tunicate. A strategy frequently used to prepare substituted piperidines and piperidones is based on the addition of a nucleophile to a chiral or achiral *N*-alkyl or *N*-acylpyridinium salt [45, 46]. Starting from readily available pyridines such as MOP (**66**), Charrette has developed an expeditious general methodology for this purpose, making use of a pyridinium salt **67** generated from a secondary amide and triflic anhydride. The presence of a bulky residue attached to the pyridinium nitrogen induced the (*E*) geometry in the imidate, the nitrogen lone pair of which steered the regio- and stereochemistry in the subsequent nucleophilic addition at the pyridine iminium moiety. This pyridinium-imidate approach represents a powerful device for the generation of a stereocenter at C-2 on the corresponding piperidine. The stereoselective addition of Grignard reagents to the chiral pyridinium salts affords piperidones suitable for further transformations (Scheme 6.34). This sequence offers several advantages over the former strategy developed by Comins, which needed the introduction of a sila residue at C-3 for control of the regiochemistry for the construction of similar adducts.

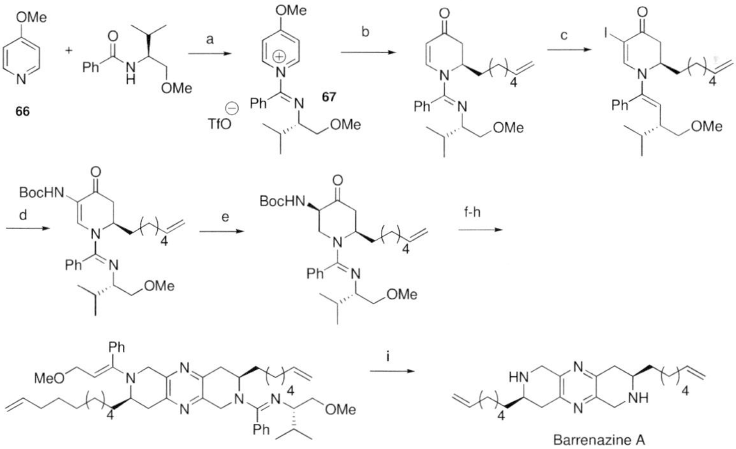

Scheme 6.34 Charette's MOP-based methodology for the construction of barrenazine. Reagents: a) Tf_2O; b) $BrMg(CH_2)_5CH{=}CH_2$; c) I_2, Pyr.; d) NH_2CO_2tBu, Cs_2CO_3, CuI; e) CuBr, LiAl(OtBu)H; f) TFA; g) NaOEt, EtOH, air; h) BBr_3, 2,6-lutidine; i) TFA.

Indole alkaloids An interesting enolate equilibration has been observed during the synthesis of epiuleine, a naturally occurring indole alkaloid. Dihydropyridones **69** were formed from **66**, (Scheme 6.35) by addition of Grignard reagents to an acylpyridinium ion. The kinetic enolate obtained after the 1,4-reduction of the enone underwent equilibration to the corresponding thermodynamic one **70** within 3 hours at 0 °C, and quenching was performed. This equilibration is highly

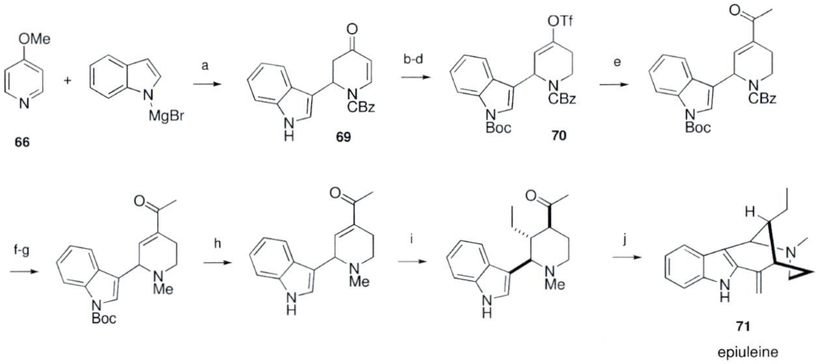

Scheme 6.35 MOP as a building block for the synthesis of epiuleine.

Reagents: a) CbzCl; b) Boc_2O, DMAP; c) L-Selectride; d) 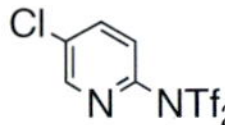

; e) CH_2CHOBu, Pd(II), dppf, then TFA; f) Pd/C, cyclohexadiene; g) CH_2O, $NaCNBH_3$; h) TFA; i) EtMgBr, CuCl; j) TsOH.

dependent on the nature of the substituent introduced with the Grignard reagent. With an indole residue the regioselectivity is 9 to 1 and an elegant preparation of epiuleine **71** after the transformation of the triflate into an acyl residue by use of a Heck vinylation followed by hydrolysis to a methyl ketone (Scheme 6.35) has been disclosed [47].

Pyridine alkaloids Several biologically active naturally occurring alkaloids originating from plants contain pyranopyridine moieties. Pyranoquinoline **73** was shown to inhibit Eg5 proteins and its application in cancer treatment is under investigation. Baldwin has reported [48] (Scheme 6.36) a route to pyrano[3,2-c]pyridines **72a–c** through reverse-electron rearomatizing Diels–Alder cycloaddition with activated alkenes. Interestingly, O-demethylation of pyridine was achieved simply by heating and quenching the dihydropyridine with Boc_2O. The dienophiles were electron-rich enol ethers and the Diels–Alder reaction proceeded under thermal conditions with simultaneous removal of the two Boc protecting groups.

6.5.1.2 MOP in the Synthesis of Plumerinine

Plumerinine (**74**) is the constituent of an Asian plant that has been used in local medicine for its antitubercular properties. Its structure contains a quinolizidine skeleton, four substituents, and five stereogenic centers in a relatively compact molecular framework. A synthesis reported by Comins [49] refers to the reported structure of plumerinine, but the physical data for the synthetic compound did not match the corresponding data for the natural compound. Once again, this discrepancy highlights the decisive value of organic synthesis for the proper

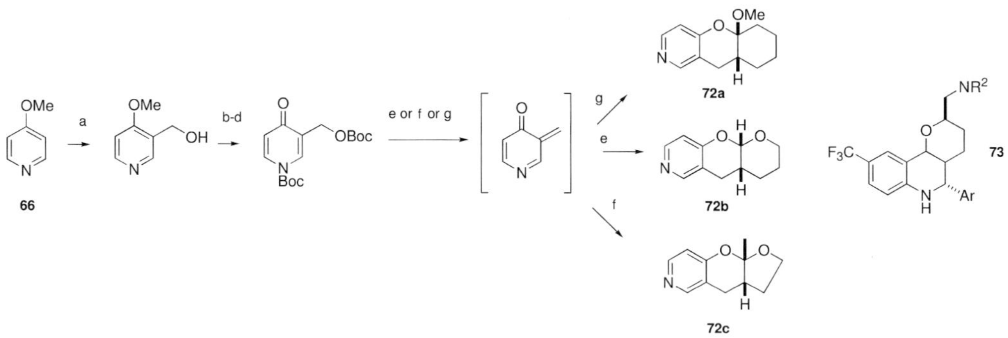

Scheme 6.36 MOP in an unusual Diels–Alder sequence in the synthesis of aza-oxo-heterocycles.
Reagents: (a) *t*-BuLi, then DMF; b) BH_3, DMS; c) $(TMS)_2S$, NaOMe; d) Boc_2O, NEt_3; e) dihydropyran, 130 °C; f) 2-Me dihydrofuran, 130°C; g) 1-MeO-cyclohexene, 130 °C.

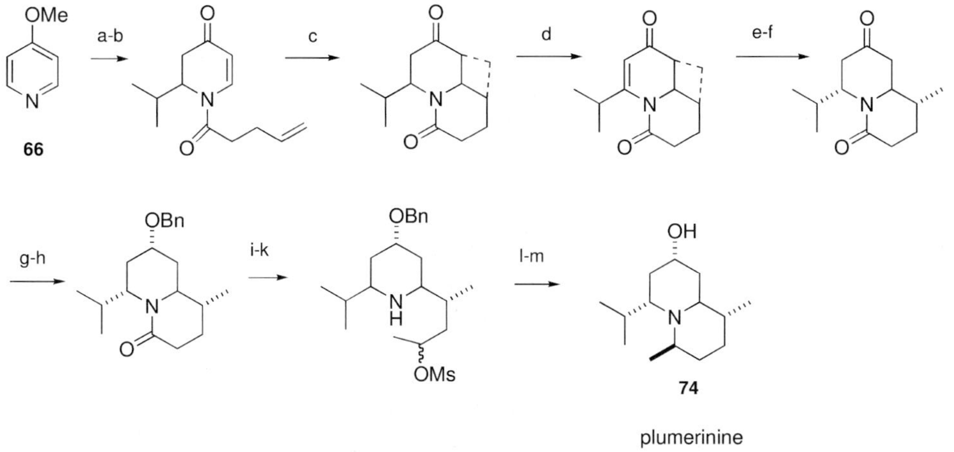

Scheme 6.37 Comins's plumerinine strategy is based on MOP.
Reagents: a) $CH_2{=}CH(CH_2)_2COCl$; b) *i*-PrMgCl, H_3O^+; c) hν, acetone; d) PhSeCl, H_2O_2; e) H_2, Pd/C; f) SmI_2; g) H_2, PtO_2; h) NaH, BnBr; i) MeMgBr; j) $NaBH_4$; k) NaH, MsCl; l) Li_2CO_3; m) H_2, $Pd(OH)_2$, EtOH.

assignment of the chemical structures of natural products. The synthetic plan (Scheme 6.37) involved the use of an N-acylimmonium ion derived from MOP to construct the quinolizidine ring. The key step is a highly face-selective intramolecular [2+2] photocyclization of a 2,3-dihydro-4-pyridone. In the final stage two mesylates were formed but only one cyclized to the desired quinolizidine, making the final purification more straightforward.

6.5.1.3 MOP in the Synthesis of 2,4-Disubstituted Piperidines

The piperidine ring continues to be a recurring scaffold in medicinal chemistry research. Not surprisingly, 1,4-disubstitution in the piperidine ring dominates, due to the easy synthesis and to the absence of difficult stereochemical issues. Conversely, 2,4-disubstituted piperidines reveal a paucity of interest, though to fill the gap a diastereoselective synthesis of this class of compounds has been reported. The chemistry is based on the following observations. If substituents are present both at C-2 and at the nitrogen, pseudoallylic strain ($A^{1,3}$) is observed, but in the case of a free nitrogen this strain is removed and a reversal of the stereocontrol can be expected in a thermodynamically controlled pathway. This interpretation has been successfully exploited [50] (Scheme 6.38) in a stereodivergent synthesis of pairs of 2,4-disubstituted piperidines **75a/b**, together with the subsequent transformations to quinolizinones **76a/b**.

6.5.1.4 MOP in the Synthesis of Toxins

Cylindrospermopsin is a toxin isolated from a fresh water alga and its complex structure has been elucidated by extensive NMR investigations. Weinreb completed the synthesis of epicylindrospermopsin **77** and confirmed the assigned structure, with the exception of one misattributed stereocenter. The approach (Scheme 6.39) is based [51] on a stereospecific intramolecular [4+2] cycloaddition of an N-sulfinylurea and on the use of a new and efficient synthesis of uracil. The starting material was MOP (**66**) and a series of transformations were performed on the piperidine ring. The synthetic scheme presented below refers only to the key Diels–Alder adduct, leaving out the remaining steps.

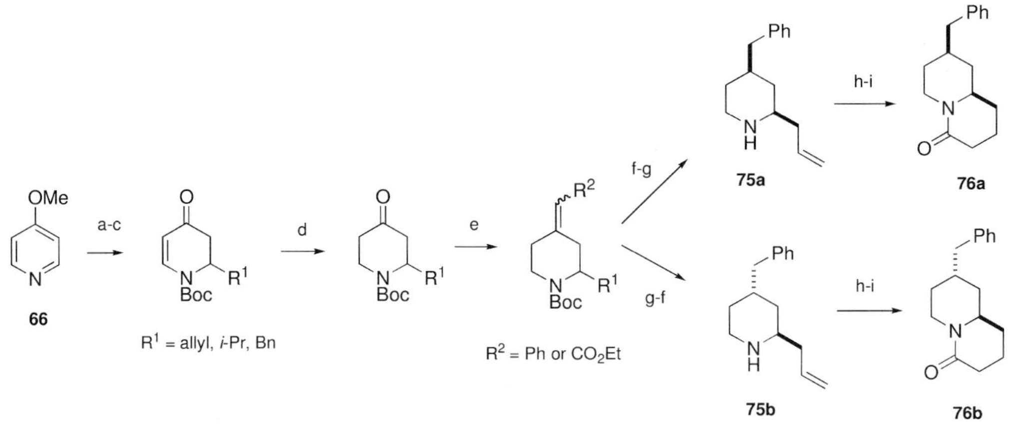

Scheme 6.38 MOP in the preparation of quinolizinones.
Reagents: a) PhOCOCl, RMgCl; b) MeONa; c) Boc_2O; d) Zn, AcOH; e) *tert*-BuOK, $PhCH_2PPh_3Cl$ or $(Ph)_3P{=}CHCO_2Et$; f) Li, NH_3; g) TFA; h) $CH_2{=}CHCOCl$, Na_2CO_3 (aq); i) $[(C_6H_{12})_3]Cl_2Ru(IV){=}CHPh$ (Grubbs I).

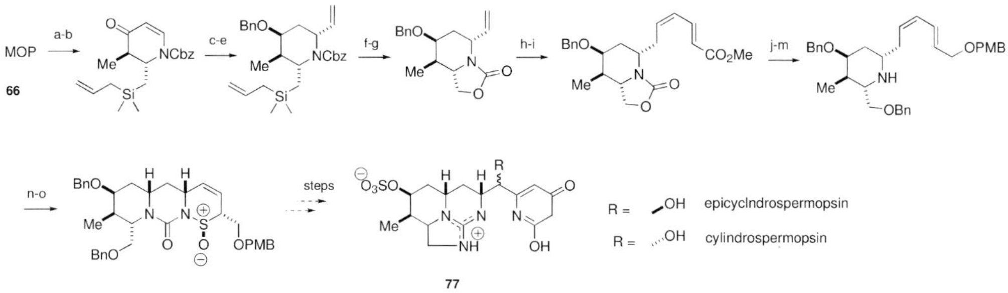

Scheme 6.39 Multistep synthesis of Cylindrospermopsin from MOP.
Reagents: a) BnOCOCl, then CH_2=$CHCH_2Si(Me)_2CH_2MgBr$ and HCl; b) NaHMDS, MeI; c) CH_2=CHMgBr, CuI; d) L-selectride; e) BnBr, NaH; f) KHF_2, TFA; then MeOH, $NaHCO_3$, H_2O_2; g) Sia_2BH, H_2O_2, NaOH; h) $(COCl_2)_2$, DMSO; i) $(EtO)_2POCH_2CH$=CO_2Et, LiOH; j) DibalH, $BF_3.Et_2O$; k) NaH, PMBCl; l) NaOH, H_2O; m) NaH, BnBr; n) KOCN, HOAc; o) $SOCl_2$, imidazole.

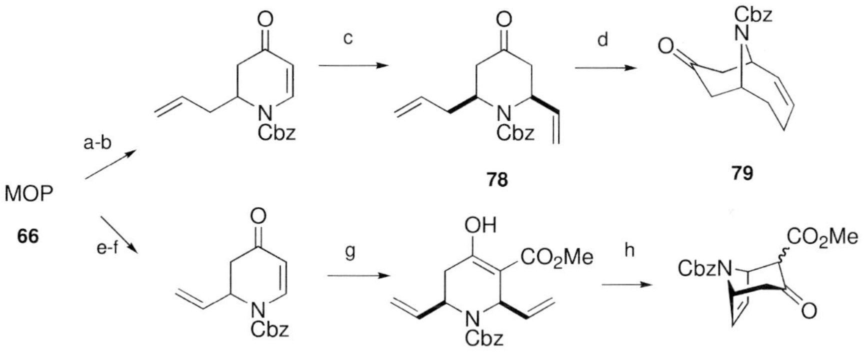

Scheme 6.40 MOP as a building block for the preparation of tropanes.
Reagents: a) CH_2CHCH_2MgBr, $ZnCl_2$; b) H^+; c) $CH_2CHMgBr$, MeLi, CuCN; d) [Ru] Grubbs I; e) $CH_2CHMgBr$, $ZnCl_2$; f) H^+; g) $CH_2CHMgBr$, MeLi, CuCN, then MeO_2CCN; h) [Ru] Grubbs I.

6.5.1.5 MOP in the Synthesis of Tropanes

A strategy for the synthesis of azabicyclo[*m.n.1*]alkanes **79** involving the ring-closing metathesis of *cis*-2,6-dialkyl-*N*-acylpiperidine derivatives **78** has been reported (Scheme 6.40) by Martin [52]. The requisite 2,6-dialkylpiperidines could readily be prepared in three steps from **66**. Practical applications were the production of functionalized tropanes. The preferred chair conformation of the *N*-acylpiperidine with the two alkenyl chains in axial orientations was decisive for a successful outcome of the metathesis.

6.6
Aziridines (AZIs)

An aziridine is a saturated three-membered heterocycle containing one nitrogen atom. Like the other three-membered ring systems (cyclopropanes or epoxides), aziridines are highly strained and their chemistry is dominated by ring-opening reactions in which the driving force is the relief of ring strain. Through suitable choice of substituents on the carbon and nitrogen atoms, excellent stereo- and regiocontrol can be achieved in ring-opening reactions with a wide variety of nucleophiles, including organometallic reagents, making aziridines useful as substrates or building blocks for the synthesis of important biomolecules including alkaloids, amino acids, and β-lactams. If enantiomerically pure or enantioenriched aziridines are used, substrate-controlled enantioselective syntheses are conceivable. In this section several typical reactions of aziridines are presented, some of them using aziridines as building blocks and others installing the aziridine ring during the synthesis and exploiting its reactivity.

Pr = acivating group

Aziridine (AZI)

6.6.1
AZIs in the Synthesis of Biomolecules

6.6.1.1 AZIs in the Synthesis of 1,2-Diamines

Optically active 1,2-diamines are versatile chiral building blocks in organic synthesis, and the best method to approach this class of compounds is to use the desymmetrization of *meso*-AZIs by nitrogen nucleophiles such as TMSCN or $TMSN_3$. Only one catalytic enantioselective method – from Jacobsen's group – has been reported in this field, and it uses a tridentate Schiff base Cr(III) complex as a catalyst [53]. Shibasaki has proposed (Scheme 6.41) an elegant solution to the problems associated with this challenging reaction with the design and development of an efficient catalytic system. The catalyst is a polygadolinium complex derived from glucose and formed with a chiral phosphine oxide as an additive. It is believed to operate in the transition state as a bimetallic complex, which chelates the aziridine and the nucleophile in a chiral environment. The ring-opening then takes place in a highly enantioselective fashion through transfer of an azido moiety. The enantioselective ring-opening of the 1,2-*meso*-AZI of cyclohexadiene was used as a starting point for the preparation of Tamiflu (oseltamivir, **81**), an orally active antiinfluenza drug that inhibits neuramidase. Furthermore, to install the pent-3-enol side chain in one of the final steps of the synthesis a transient AZI

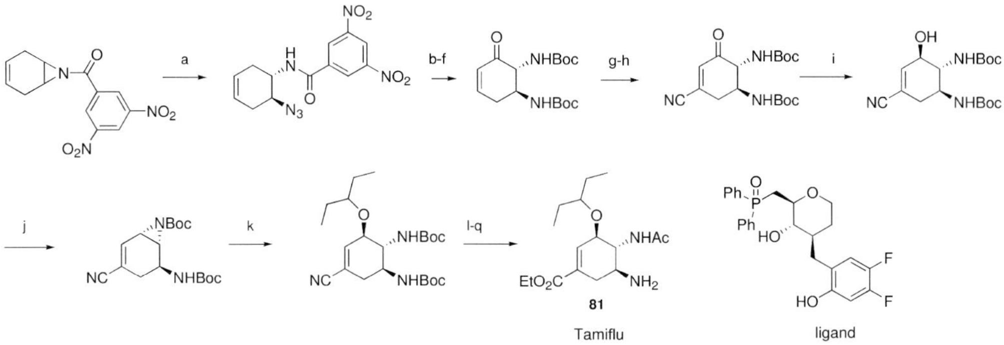

Scheme 6.41 An AZI in the preparation of Tamiflu.
Reagents: a) $Gd(OiPr)_3$, ligand, $TMSN_3$; b) Boc_2O, DMAP; c) NaOH; d) PPh_3, H_2O; e) Boc_2O, Et_3N; f) SeO_2, Dess–Martin; g) $Ni(COD)_2$, TMSCN; h) NBS, Et_3N; i) $LiAlH(OtBu)_3$; j) DEAD, PPh_3; k) pentan-3-ol, $BF_3.Et_2O$; l) TFA; m) Boc_2O, Et_3N; n) Ac_2O, DMAP, Pyr; o) HCl/EtOH; q) H_3PO_4, crystallization.

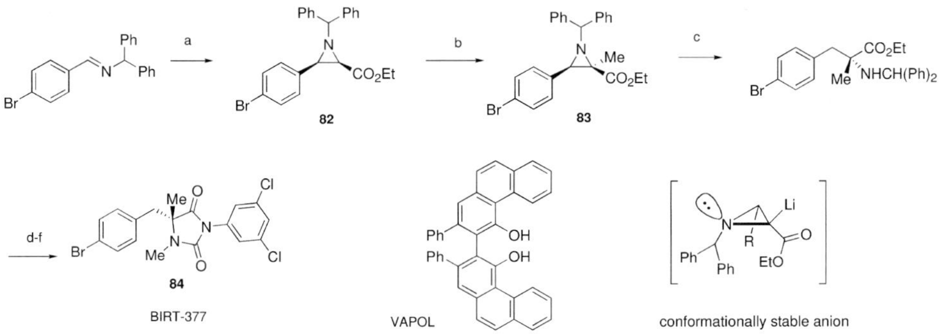

Scheme 6.42 An AZI as building block for the synthesis of BIRT-377.
Reagents: a) N_2CHCO_2Et, VAPOL, $(PhO)_3$; b) LDA, then MeI; c) BH_3, Me_3N, TFA; d) EtSiH, TFA; e) 3,5-dichlorophenyl isocyanate; f) NaHMDS, MeI.

was designed. This synthesis of Tamiflu exemplifies the utility of AZIs when heteroatoms have to be installed on vicinal carbons [54].

6.6.1.2 AZIs in the Synthesis of α-Amino Acids

A large number of methods for the preparation of tetrasubstituted α-amino acids – popular tools used to control conformation, and hence biological properties, in peptides – have been described. AZIs have recently been used (Scheme 6.42) to provide a rapid means of entry to this class of compounds. Indeed, AZI-2-carboxylates can be alkylated next to the ester function (**83**) without self-condensation with the aid of a suitable substituent attached to the AZI nitrogen atom. The benzhydryl residue proved to be well suited to allow easy substitution at the carboxylate; the formation of a stable *C*- or *O*-enolate has been proposed to explain the success of this transformation. Interestingly, the anionic intermediate does not epimerize during the process, and the stereochemistry of the AZI is maintained in the final adducts. The chiral starting AZI **82** was obtained through the addition of a carbene to an imine in the presence of a chiral boronylated ligand derived from BINOL. An application of this methodology was the preparation of BIRT-377 (**84**), a potent inhibitor of the interaction between intercellular adhesion molecule-1 (ICAM-1) and lymphocyte function-associated antigen-1 (LFA-1), a potential treatment for immune diseases [55].

6.6.1.3 AZI in the Synthesis of Ferruginine, an Acetylcholine Receptor

Ferruginine (**88**) is a potent neurotoxin, but it is also used as a ligand for the nicotinic acetylcholine receptor (nAChR). Its synthesis has attracted considerable attention, an elegant synthesis having been reported (Scheme 6.43) by Bäckwall through the BF_3-induced rearrangement of an N-protected cyclohexane AZI-cyclopropane. The key AZI **85** was obtained from cyclohexadiene in a few steps and the ring expansion was then performed in a mixture of dichloromethane and nitromethane, to avoid the formation of fluorinated side-products. The sulfone was transformed into compound **86**, with an acetyl side chain, in a sequence of interesting steps such as the Michael addition of nitroethane, an acyl anion equivalent, while a subsequent benzenesulfinic acid elimination provided **87**. This chemistry provides a novel and efficient new route to the tropane alkaloids and their analogues [56].

6.6.1.4 AZIs in the Synthesis of Tryptophan Derivatives

Reactions between simple AZIs and 3-unsubstituted indoles in the presence of Lewis acid catalysts give tryptophan derivatives or 3-substituted indoles. When 3-substituted indoles are used, the reactions take place similarly. A concise synthesis

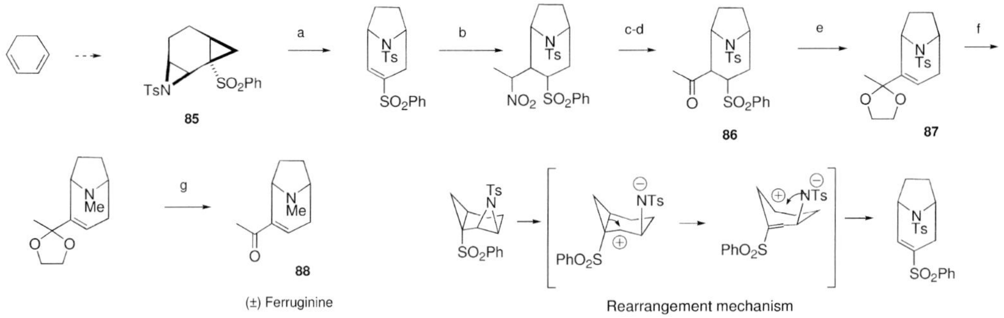

Scheme 6.43 The use of an AZI in the synthesis of ferruginine.
Reagents: a) BF_3, CH_2Cl_2 and CH_3NO_2; b) $EtNO_2$, DBU; c) H_2O_2, K_2CO_3; d) *tert*-BuOK; e) $HOCH_2CH_2OH$, Me_3SiOTf; f) Mg, MeOH; g) CH_2O, $NaCNBH_3$, AcOH.

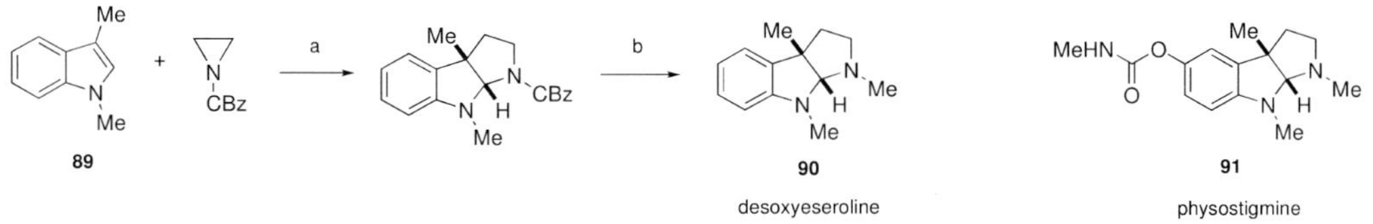

Scheme 6.44 An AZI as building block for the synthesis of physostigmine. Reagents: a) $Sc(OTf)_3$, TMSCl; b) RedAl, toluene.

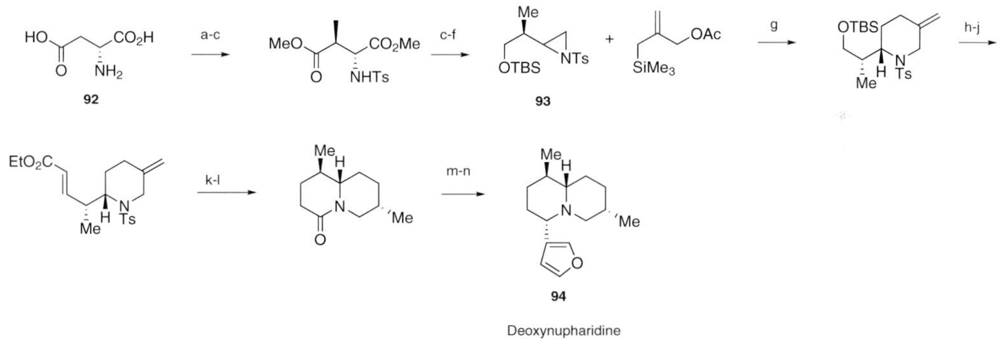

Scheme 6.45 AZIs in [3+3] cycloaddition reaction.
Reagents: a) MeOH, SO_2Cl_2; b) TsCl, Et_3N, c) LiHMDS, then MeI; d) $LiAlH_4$; e) PBu_3, ADDP, Tol; f) TBDMSCl, imidazole, DMF; g) $Pd(OAc)_2$, DPPP; h) TBAF; i) Swern; j) $Ph_3P{=}CHCO_2Et$; k) Mg, MeOH; l) H_2, Pd/C; m) 3-Li-furan; n) BH_3, DMS.

of physostigmine alkaloids (Scheme 6.44) starting from skatole (**89**) and a simple aziridine has been reported [57]. If an excess of $Sc(OTf)_3$ was used the reaction proceeded in acceptable yield and (±) desoxyeseroline (**90**) was obtained in 80% overall yield. This sequence is also applicable to the synthesis of physostigmine (**91**).

6.6.1.5 **AZIs in the Synthesis of Functionalized Piperidines**

The preparation of functionalized piperidines through formal [3+3] cycloaddition reactions between *N*-tosyl-AZIs and Pd-trimethylmethane complexes has been reported [58, 59, 60]. This sequence proved to be an efficient and expeditious method for the synthesis of 2-substituted piperidines in enantiomerically pure form (Scheme 6.45). A remarkable feature of this process is that the product is decorated with a readily functionalizable exocyclic alkene moiety. The starting AZI **93** is available in a few steps from aspartic acid (**92**). Along similar lines, a synthesis of desoxynupharidine (**94**) of the *Nuphar* genus has been described. All Nuphar alkaloids show the presence of one or more furan rings and of a quinolizidine core. The same group, using a similar strategy, has proposed a very elegant synthesis of perhydrohistrionicotoxin.

6.6.1.6 An AZI in the Synthesis of the Alkaloid Pumiliotoxin

An efficient asymmetric synthesis of natural pumiliotoxin, a decahydroquinoline alkaloid, has been reported (Scheme 6.46). The nitrogen atom and the alkyl side chain were introduced through the regioselective ring-opening of the AZI **96** derived from Norleucinol **95**. A remarkable feature is the sequence of three reactions performed in liquid ammonia in one-pot fashion: introduction of a methyl residue on an alkyne, reduction of the alkyne to the alkene, and removal of the tosyl group on the nitrogen. Then, after the construction of a dienyl-amine **97**, a thermal intramolecular Diels–Alder reaction gives rise to the quinoline core. The stereochemistry of the fused rings is *cis*, as is that of the methyl group, resulting from an *exo* transition state during the electrocyclization. Some amount of the isomeric adduct was also obtained, since the isobutyryl amide did not completely prevent the occurrence of a complementary transition state. Finally, after acyl deprotection, pumiliotoxin (**98**) was obtained [61] in nine high-yielding steps.

6.6.1.7 An AZI in the Synthesis of Phenylkainic Acid

Phenylkainic acid (**101**), a kainoid in which the replacement of the isoprenyl group at C(4) by a phenyl residue considerably enhances the biological potency, has been prepared by a regioselective ring-opening of an *N*-activated phenyl-AZI. Cyclopent-2-enylsilane **99** in the presence of a Lewis acid opens the AZI ring at the most sterically hindered benzylic carbon, but the reaction is not enantioselective when a homochiral phenyl-AZI is used (Scheme 6.47). This drawback is explained by the possibility that the phenyl-AZI might behave as a 1,3-dipole in the presence of a Lewis acid, and several arguments seem to support this hypothesis. Intramolecular cyclization assisted by $Pd(OAc)_2$ quantitatively yields the azabicycles **100** as an easily separable diastereomeric mixture. Oxidation of the double bond to the diacid, followed by epimerization with KH and 18-crown-6 and deprotection, finally [62] afforded the target compound **101** in a few conventional steps.

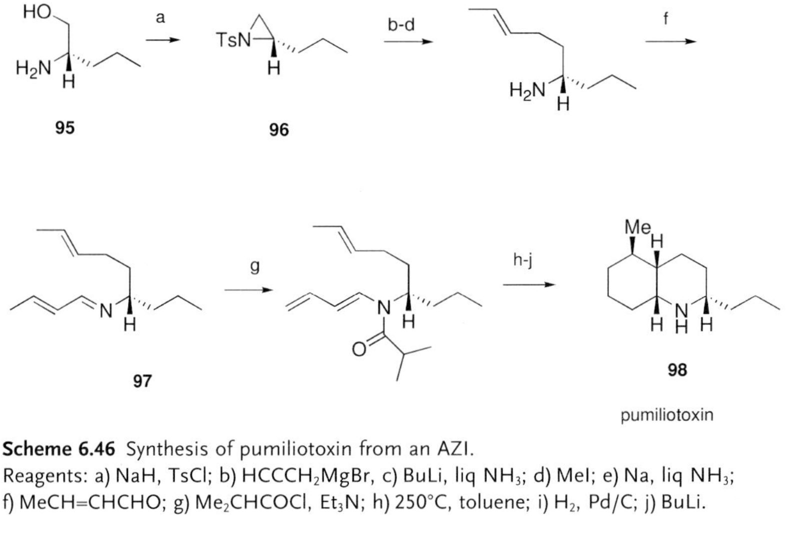

95 96 97 98

Scheme 6.46 Synthesis of pumiliotoxin from an AZI.
Reagents: a) NaH, TsCl; b) $HCCCH_2MgBr$, c) BuLi, liq NH_3; d) MeI; e) Na, liq NH_3; f) MeCH=CHCHO; g) $Me_2CHCOCl$, Et_3N; h) 250°C, toluene; i) H_2, Pd/C; j) BuLi.

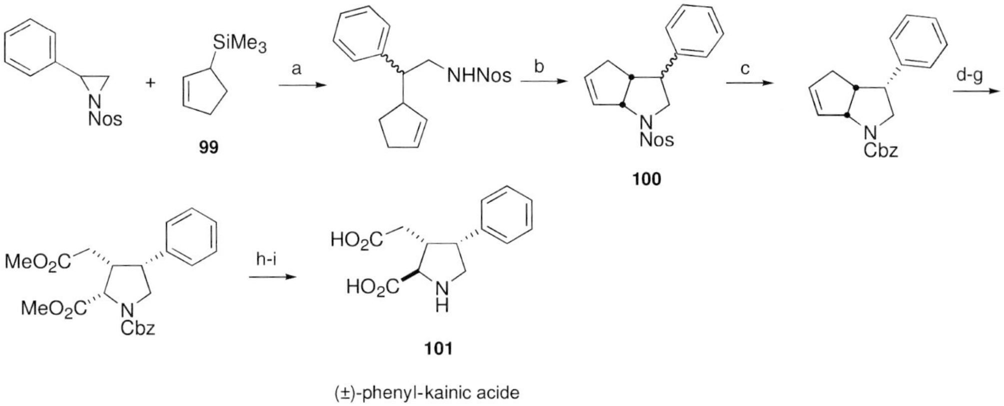

Scheme 6.47 A phenyl-AZI as building block for kainoids.
Reagents: a) $BF_3.Et_2O$; b) $Pd(OAc)_2$, O_2, DMF; c) diastereomer separation by chromatography; d) PhSH; e) Cbz_2O, K_2CO_3; f) Sharpless oxidation; g) CH_2N_2; h) KH, 18-crown-16, benzene; i) HCl, AcOH reflux.

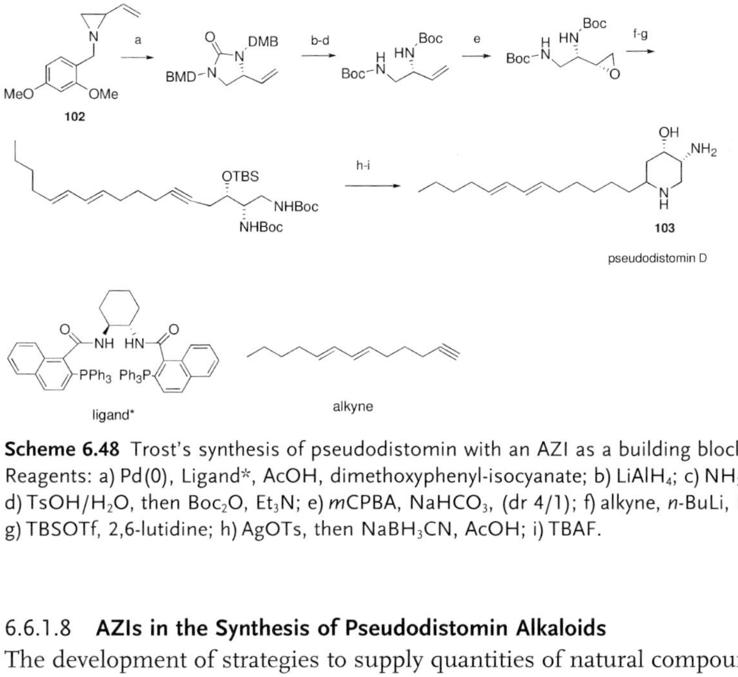

Scheme 6.48 Trost's synthesis of pseudodistomin with an AZI as a building block.
Reagents: a) Pd(0), Ligand*, AcOH, dimethoxyphenyl-isocyanate; b) $LiAlH_4$; c) NH_2OH; d) $TsOH/H_2O$, then Boc_2O, Et_3N; e) *m*CPBA, $NaHCO_3$, (dr 4/1); f) alkyne, *n*-BuLi, Me_2AlCl; g) TBSOTf, 2,6-lutidine; h) AgOTs, then $NaBH_3CN$, AcOH; i) TBAF.

6.6.1.8 **AZIs in the Synthesis of Pseudodistomin Alkaloids**

The development of strategies to supply quantities of natural compounds for biological evaluation remains an important arena in organic synthesis. Pseudodistomins, isolated from tunicates, exhibit calmodulin-antagonistic activities and potent cytotoxicity against both leukemia and human epidermoid carcinoma KB cells. However, for extensive pharmacological studies to be performed these compounds have to be prepared on multigram scales. Trost has designed [63] (Scheme 6.48) an elegant and concise strategy aimed at the synthesis of a particular member of this class: pseudodistomin D (**103**), which contains a unique piperidine core. The synthesis was based on the ring-opening of vinyl-AZI **102** by use of a dynamic kinetic asymmetric cycloaddition of an isocyanates, followed by an intramolecular Ag(I)-catalyzed hydroamination of an alkyne to construct the piperidine core.

6.7 Homoallylamine (HAM)

A large body of work has recently been devoted to the synthesis and use in organic synthesis of homoallylamines (HAMs), probably because they are one of the main substrates for the construction of azaheterocycles. As in the case of propargylamines (PLAs), homoallylamines have chemical potential thanks to the presence of a double bond, allowing a number of chemical manipulations such as ring-closing metathesis and oxidative hydroboration. This section presents a short overview of the most practical accesses to HAM and describes the potential of this versatile building block through several applications.

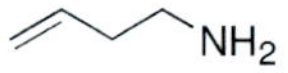

104

Homoallylamine HAM

6.7.1 Synthesis of HAMs

The mostly common procedure for the preparation of HAMs is the allylation of an imine with an organometallic reagent. Imines are poor electrophiles and enolizable imines have tendency to give deprotonation rather than addition. An elegant solution to avoid these drawbacks is provided in the work of Kobayashi [64, 65], who recently reported a short and elegant approach to various HAMs based on reactions between an aldehyde, ammonia, and an allylboronate (Scheme 6.49). This sequence has a wide scope, being suitable for all kind of aldehydes. This preparation represents the simplest way to obtain HAMs, since no protection at nitrogen is required, and the primary amino function comes out directly from the reaction mixture. Unfortunately, though, the stereochemistry at the carbon bearing the nitrogen cannot be controlled properly in this process. When (*E*)- or (*Z*)-crotylboronates are used, *anti*- or *syn*-crotylated products are obtained, respectively, with high diastereoselectivities. The relatively poor stabilities of the allylboronates are the only minor drawback of this protocol.

A fairly similar approach to HAMs using potassium allyltrifluoroborates **105** was disclosed at the same time by Batey [66] (Scheme 6.50). In the presence of Lewis acids, allyltrifluoroborates react smoothly with aromatic or aliphatic aldehydes to give HAMs in excellent yields and, in the case of crotyl derivatives, with very good diastereoselectivities. When the substrate was the chiral benzaldehyde-

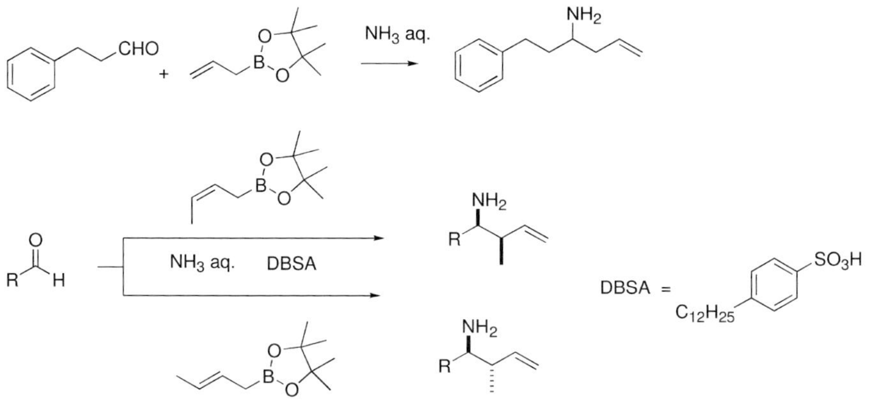

Scheme 6.49 Facile access to HAMs.

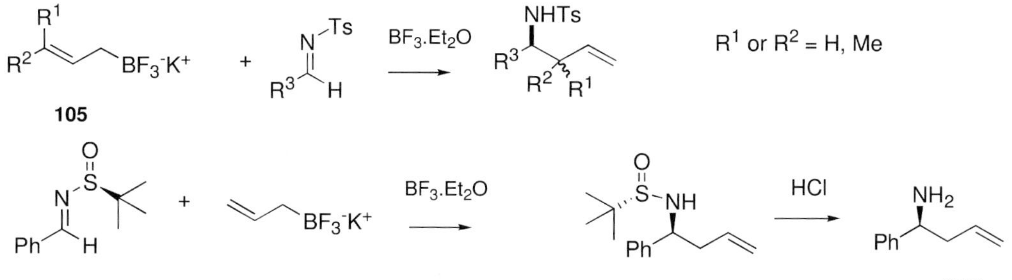

Scheme 6.50 Enantioselectively prepared HAMs.

derived sulfinylimine **106** (Ellman imine), excellent enantioselectivity was achieved with potassium allyltrifluoroborate. Finally the acid-labile chiral auxiliary was readily cleaved to give the desired unprotected amines in excellent yields.

To obtain chiral HAMs, phenylglycidol-derived chiral aldimines **107** were diastereoselectively alkylated [67] with allyl bromide and indium powder in methanol as solvent (Scheme 6.51). Both aromatic and aliphatic aldimines provided good yields and excellent diastereoselectivities. The chiral auxiliary could be removed in a self-immolative procedure to provide the free primary amines ready for further transformations. The Barbier conditions used in this procedure for the preparation of HAMs offer many advantages over similar methods.

6.7.2
HAMs in the Synthesis of Biomolecules

6.7.2.1 HAM in the Synthesis of Imidazoazepines

Fused imidazoazepine derivatives **109** were synthesized [68] by a sequential intramolecular van Leusen/Heck approach (Scheme 6.52). The combination of an aldehyde with a vinylogous bromide, TOSMIC, and HAM furnished a functionalized

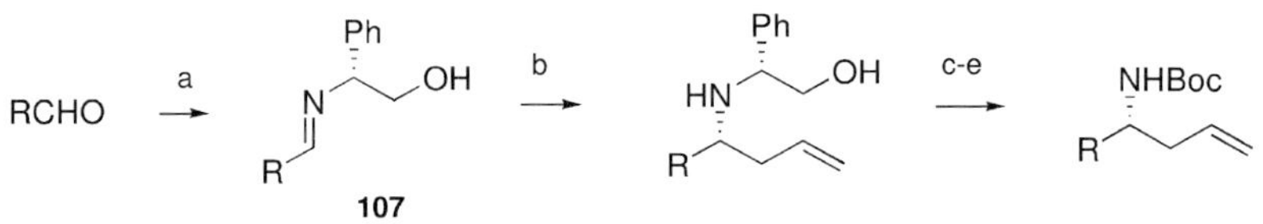

Scheme 6.51 HAMs made via the chiral pool.
Reagents: a) phenylglycidol, MS; b) allyl bromide, In, EtOH; c) $Pb(OAc)_4$; d) NH_2OH,HCl; e) Boc_2O, Et_3N.

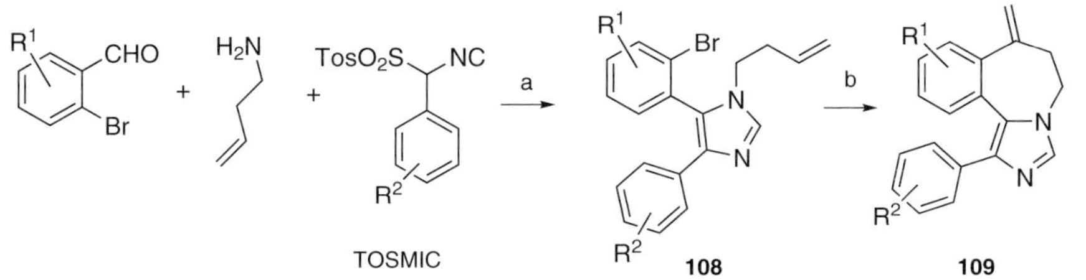

Scheme 6.52 HAMs for diversity-oriented syntheses.
Reagents: a) K_2CO_3, DMF; b) $Pd(OAc)_2$, $P(O\text{-tolyl})_3$, CH_3CN, Et_3N.

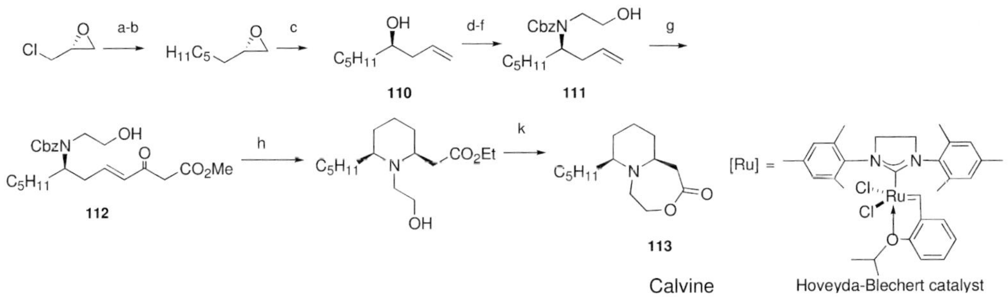

Scheme 6.53 A HAM used in metathesis.
Reagents: a) *n*BuMgCl, CuCN (10 mol %); b) NaOH, THF/H_2O; c) vinyl bromide, CuCN, (10 mol %); d) TsCl, DMAP; e) $NH_2CH_2CH_2OH$; f) CbzCl, K_2CO_3; g) [Ru], $CH_2 = CHCOCH_2CO_2Me$; h) Pd/C, H_2; k) PTSA, benzene.

imidazole **108**, which in turn generated the expected heterocycles **109** on subjection to an intramolecular carbon-carbon bond-forming reaction (Heck coupling). It is noteworthy that the molecular diversity in the final compounds is available in only two steps.

6.7.2.2 HAMs in the Synthesis of Alkaloids

Synthesis of 'Calvine' Alkaloids Calvine (**113**) is the major alkaloid found in several ladybirds, and a convergent synthesis starting from (*R*)-epichlorohydrin, a precursor of a substituted HAM, has been described (Scheme 6.53). The key strategies included an initial copper-catalyzed oxirane ring-opening to produce a homoallyl alcohol **110**, and an S_N2 reaction with an amine derivative then gave rise to HAM **111**, which was subjected to a metathesis to yield the keto ester **112**, ready for hydrogenative cyclization. The final step was an acid-catalyzed ring-closure to generate the seven-membered lactone **113**. The synthesis was completed in nine steps [69]. The construction of the acrylic keto esters through a cross-metathesis reaction is totally unprecedented.

Synthesis of indolizidine alkaloids The indolizidine alkaloids represent challenging targets for organic chemists involved in the application of new methodologies [70, 71] or sequences for the construction of important biomolecules. A practical synthesis of indolizidine 209-D starting from the chiral HAM **114** has been reported (Scheme 6.54). The terminal alkene was boronylated with 9-BBN and directly used in a Suzuki reaction in order to provide compound **115**, a polyfunctional linear adduct with an *exo* methylene function. Oxidation of the alkene afforded a ketone **116**, well suited to produce the piperidine ring **117**, with the two α,α′-substituents in a *cis* relationship, under reductive conditions. The final bicyclic adduct was obtained by substitution of the secondary nitrogen with the latent mesylate, which had survived during all the previous processes. This sequence affording indolizidine **118** is original and has also been applied with some variation to the preparation of monomorine (**119**).

Synthesis of piperidine and pyrrolidine alkaloids An enantiomeric synthesis of piperidine and pyrrolidine alkaloids present in tobacco leaves, such as nicotine (**121**) or anabasine (**122**), using a chiral HAM derived from pyridine-3-carboxaldehyde, has been reported [72, 73] (Scheme 6.55). An intramolecular hydroboration/cycloalkylation of an homoallylic azide intermediate **120** and a ring-closing

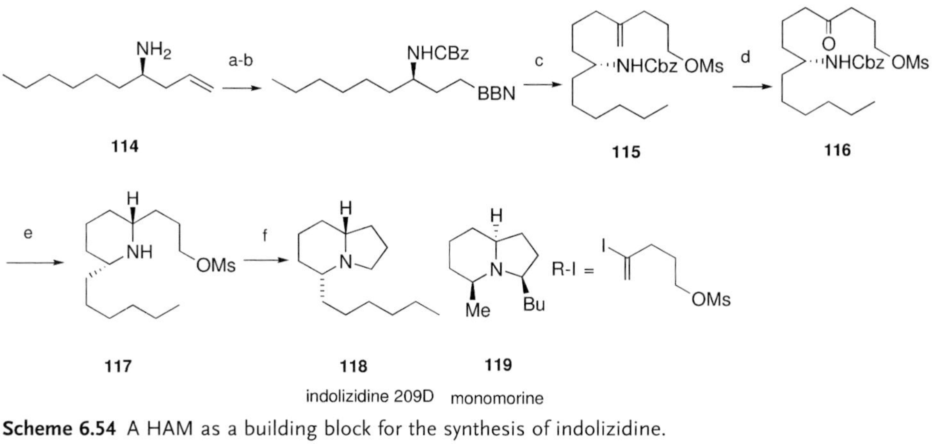

Scheme 6.54 A HAM as a building block for the synthesis of indolizidine.
Reagents: (a) CbzCl, K_2CO_3; b) 9-BBN-H; c) R-I, $Pd(dppf)Cl_2$, $AsPh_3$, $CsCO_3$, DMF; d) O_3, CH_2Cl_2-MeOH then Me_2S; e) H_2, Pd/C; f) Et_3N, MeOH.

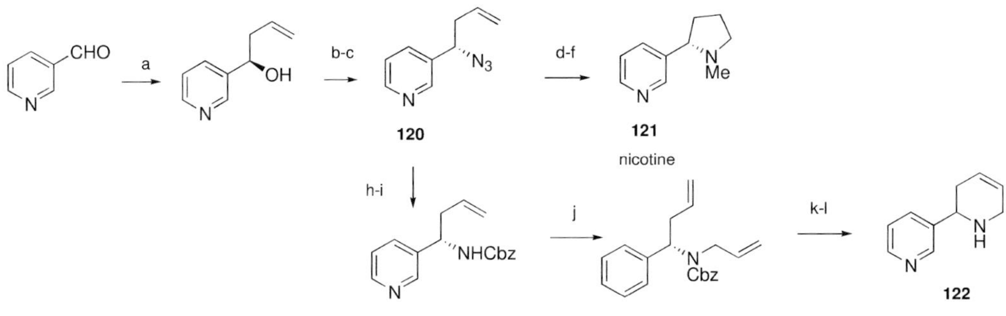

Scheme 6.55 HAMs as building blocks for the synthesis of nicotine analogues.
Reagents: a) Ipc_2BAll, THF; b) MsCl, Et_3N; c) NaN_3, DMF; d) BH_3, Me_2S; e) $ClCO_2Et$, K_2CO_3; f) $LiAlH_4$; h) $SnCl_2$, H_2O; i) CbzCl, K_2CO_3, j) NaH, AllylBr; k) HCl, metathesis [Ru]; l) $BF_3.Et_2O$, Me_2S.

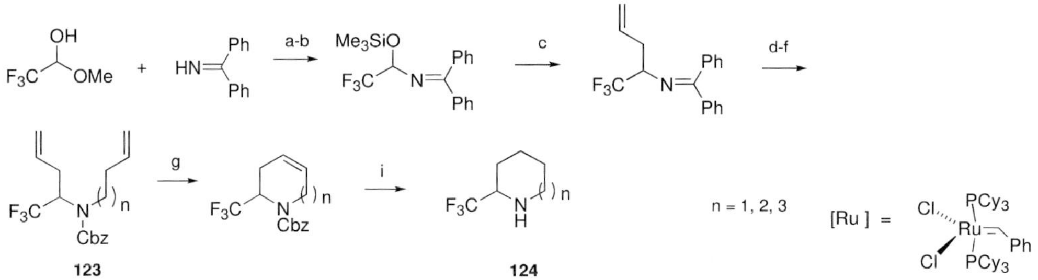

Scheme 6.56 HAMs for the construction of piperidines by metathesis.
Reagents; a) MS, 4 A; b) $ImSiMe_3$; c) AllylTMS, $BF_3.Et_2O$; d) aq. HCl; e) CbzCl, $NaHCO_3$; f) NaH, AllylBr or butenyl-Br; g) [Ru]; i) Pd/C, H_2.

metathesis reaction of a diethylenic amine intermediate are the key steps in the pyrrolidine and in the piperidine ring formation, respectively. Since these two heterocycles represent the most important structural elements present in biologically active compounds, the chemistry outlined below may represent a general route to this class of compounds.

6.7.2.3 **HAMs in the Synthesis of Piperidine Derivatives**

Aza-heterocycles are important structural elements of many natural products and drug candidates. Among them the piperidine ring continues to be extensively used in pharmaceutical research, while at the same time organofluorine compounds are gaining increasing popularity thanks to their unique properties. Trifluoromethyl-substituted molecules constitute a special class of compounds because of their specific properties such as high lipophilicity conferred by the CF_3 moiety. Efficient syntheses of α-CF_3 nitrogen heterocycles **124** have been achieved [74] (Scheme 6.56) from the readily available α-CF_3 HAM **123** and subsequent metathesis.

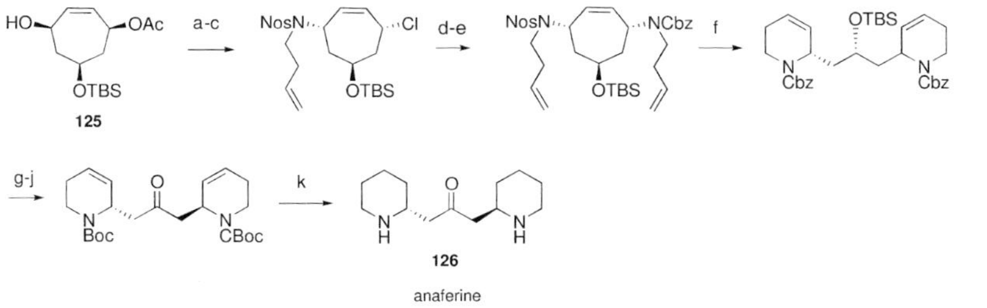

Scheme 6.57 A HAM as a versatile building block for the metathetic construction of piperidine alkaloids.
Reagents: a) *N*-Nosyl-*N*-butenylamine, PPh_3, DEAD; b) KCN, MeOH; c) MsCl, pyr.; d) butenylamine, K_2CO_3; e) PhSH, K_2CO_3, then CbzCl; f) [Ru] Grubbs I; g) H_2, Pd/C; h) HCl, EtOH; i) Boc_2O, Et_3N; j) PCC, CH_2Cl_2; k) HCl, MeOH.

6.7.2.4 HAMs in the Synthesis of Chiral Heterocycles

Synthesis of anaferine The development of methods for the preparation of chiral heterocyclic rings is a challenge in natural product synthesis and the impact of ring-closing and ring-opening metathesis in this field is impressive. The combination of the two pathways was used by Blechert in the preparation of anaferine (**126**); special care has to be taken during the synthesis because anaferine can undergo a retro-Michael reaction under neutral conditions [75]. The starting material **125** is derived from tropone and the synthesis of (–)-anaferine is completed in eleven steps in 23% overall yield (Scheme 6.57). The length of the sequence is only due to the need for manipulations of the protecting groups on the two nitrogen atoms.

Synthesis of martinellines A variety of interesting heterocycles can be formed by Lewis acid-mediated intramolecular cyclizations of imines. The synthesis of the martinellines, natural products that possess interesting biological activities and intriguing chemical structures, takes place [76] (Scheme 6.58) via in situ formation of an imine such as **127**, with an allylsilane side chain. The imine, converted into an acyliminium ion with the assistance of a Lewis acid, cyclized to a pyrrolidine with simultaneous displacement of *tert*-butanol to provide the tricyclic urea **128** as an inseparable mixture of diastereoisomers. The tricyclic adduct was obtained in good yields over the three-step, one-pot sequence. Finally an advanced pyrroloquinoline intermediate **129** for the synthesis of martinellines such as **130** was obtained by a haloamidation reaction.

6.8 Indole (IND)

Indole (**131**) and its derivatives are heterocyclic compounds widely distributed in nature. Tryptophan is an essential amino acid constituent of most proteins and also serves as a biosynthetic precursor for a wide variety of tryptamine-, indole-,

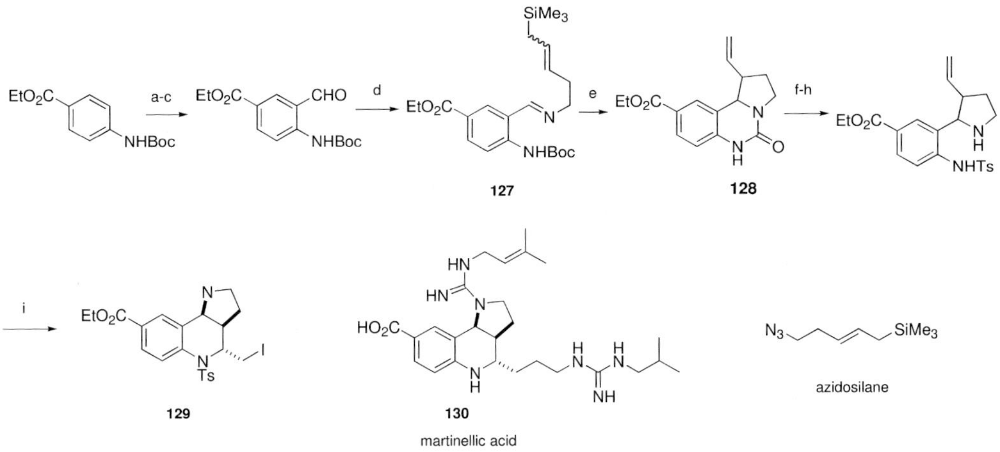

Scheme 6.58 HAMs in the synthesis of martinellic acid and derivatives.
Reagents: a) Br_2,AcOH; b) $nBu_3SnCH{=}CH_2$, $(Ph_3)_4Pd$; c) O_3 then DMS; d) azidosilane, PPh_3; e) $TiCl_4$; f) NaH, TsCl; g) NaOH aq 50%; h) H_2SO_4, MeOH; i) I_2.

and 2,3-dihydroindole-containing secondary metabolites. Serotonin (5-OH-tryptamine) is a very important neurotransmitter, while melatonin is a hormone controlling the circadian rhythm of physiological functions.

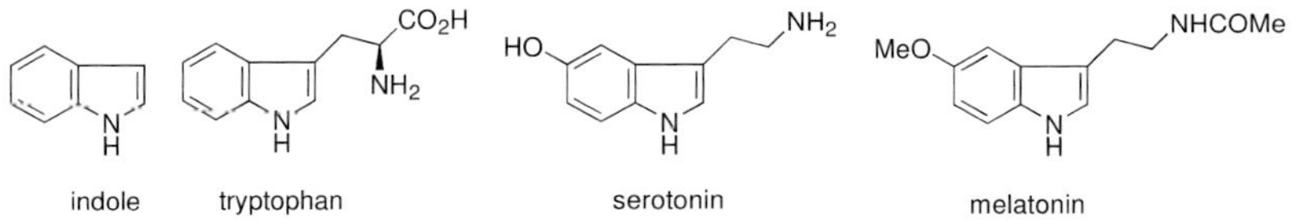

Furthermore, the indole alkaloids from the vincristine family are valuable drugs for fighting cancer, while lysergic acid plays an important role in the identification of human brain functions. Indomethacin, a drug used against rheumatoid arthritis, is also derived from an indole nucleus. The construction of substituted indole rings and their elaborations to natural products remains a hot topic in organic synthesis. One important feature of indole chemistry is the easy electrophilic substitution at the more electron-rich heterocyclic ring on the two carbons at C-2 or C-3 next to the nitrogen, the regioselectivity depending on the reaction conditions and on the substitution pattern. This section reports some recent methods for the preparation of indoles (INDs) and presented some syntheses of IND-containing biomolecules.

6.8.1
Synthesis of Indoles

The very well known Fischer indole synthesis is a two-step sequence consisting of the formation of a hydrazone followed by a [3,3]-sigmatropic rearrangement, and is still a standard method for the synthesis of the IND core. A general method was developed for the synthesis of highly functionalized INDs [77]; starting from simple, commercially available aryl hydrazines, and cyclic enol ethers or enol lactones, the one-pot Fischer IND synthesis afforded substituted IND acetic acids or IND propionic acids. This procedure yielded 2,3-disubstituted indoles as single regioisomers from suitably substituted enol ether lactones. The importance of this new preparation of substituted indoles was highlighted in the syntheses (Scheme 6.59) of two important drugs: the antiinflammatory drug indomethacin (**132**) and the antimigraine drug sumitriptan (**133**).

Eilbracht reported a shortcut synthesis of indoles [78] in which a sequence of hydroformylation and Fischer indolization starting from amino olefins (precursors of aldehydes) and arylhydrazines was elaborated. In a convergent manner, the two units bearing pharmacologically relevant substituents were assembled in the final indolization step. This modular and diversity-oriented approach to tryptamine and homotryptamine can be conducted in water and allowed the preparation of both branched and nonbranched tryptamines, as well as tryptamine-based pharmaceuticals such as the 5-HT1D agonist L775606. This new strategy (Scheme 6.60) is well suited to the rapid preparation of libraries of tryptamines. The number of steps is reduced to a minimum thanks to the cascade sequence and the complexity is introduced from easily available starting materials. The three examples below relate to the synthesis of compounds under clinical trial. It is noteworthy that the

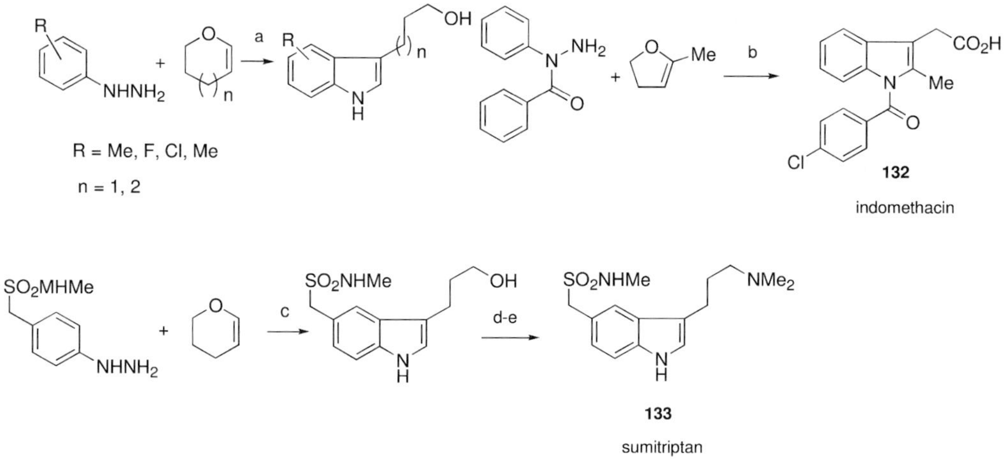

Scheme 6.59 Preparation of INDs by the Fischer synthesis.
Reagents: a) H_2SO_4 (4%), DMA; b) MeCN, 1 eq H_2SO_4; c) H_2SO_4, CH_3CN; d) MsCl, Et_3N; e) NaI, $HNMe_2$.

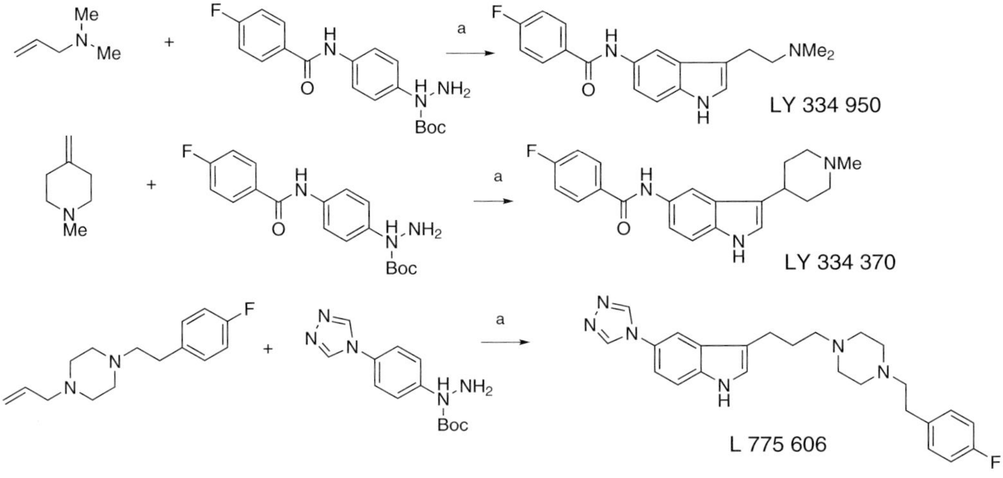

Scheme 6.60 Preparation of IND in one-pot hydroformylation reactions.
Reagents: a) H_2:CO 1/1, 20 bars, Rh(acac)CO_2(1% mol) Xantphos (5 mol %) then H_2SO_4 (4% weight) 100 °C.

hydroformylation step is tolerant of so many functional groups. The Boc protection on the hydrazine components was achieved through their preparation by Gomberg reactions and is cleaved in the reaction mixture.

The synthesis of the indole core from a nitroarene and an excess of a Grignard reagent involves a Bartoli reaction followed by a sigmatropic [3,3]-rearrangement and is probably the shortest route to indoles. At the beginning the reaction was limited to the preparation of indoles bearing a substituent at C-7. To overcome that limitation during the Bartoli reaction, nitroarenes containing an *ortho* bromine substituent were transformed (Scheme 6.61) into 7-bromoindoles **134** and subsequent radical debromination opened avenues to a variety of indoles [79, 80, 81].

6.8.2
INDs in the Synthesis of Biomolecules

An (azacycloalkyl)-bis-indolylmaleimide (LY 317615) **138**, deriving from the anti-angiogenic agent ruboxistaurin (**139**), has been identified as a potent inhibitor of PKC, and evaluated in clinic trials for the treatment of cancer. To prepare this compound, two special IND-based scaffolds **135** and **136** – one bearing a basic chain on the indolyl nitrogen, and the other the amide from indolacetic acid – were constructed (Scheme 6.62). The reactivity of the β-carbon of the indole was exploited in a Friedel–Crafts reaction with oxalyl chloride to provide the corresponding ester **137**, and the final ligation of the indoles to the succinimide was performed in basic media [82, 83]. A scale-up optimization has been published by Eli Lilly chemists.

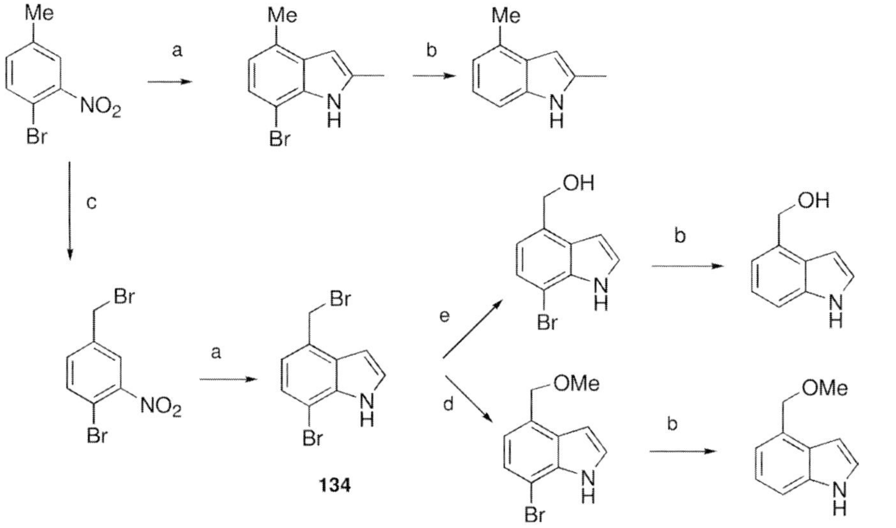

Scheme 6.61 INDs by Bartoli reactions.
Reagents: a) HC=C(Me)MgBr; b) Bu_3SnH, AIBN; c) NBS, benzoyl peroxide; d) NaOMe; e) H_2O.

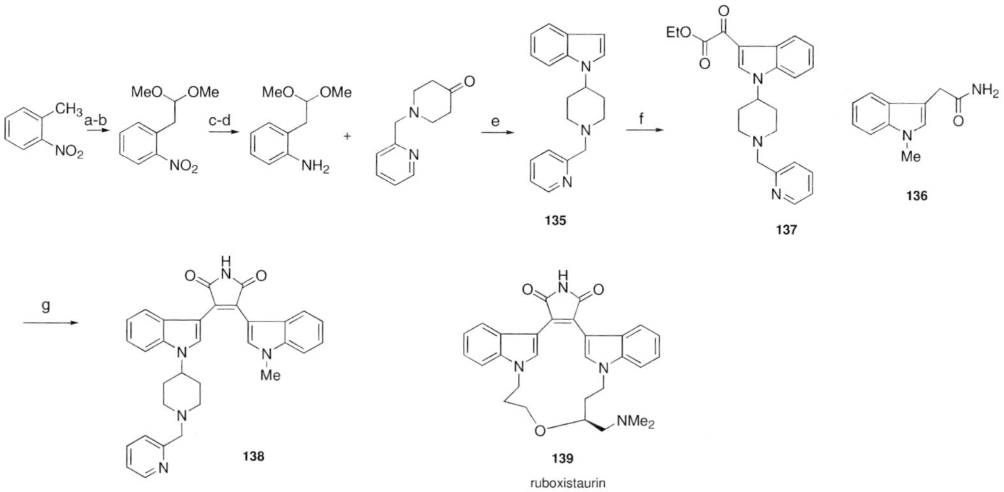

Scheme 6.62 INDs as building blocks in the construction of biomolecules.
Reagents: a) $(MeO)_2NMe_2$; b) TMSCl, MeOH; c) Pd/C, H_2; d) $NaBH(OAc)_3$; e) TFA; f) $(ClCO)_2$, then MeOH; g) **136** and *t*BuOK.

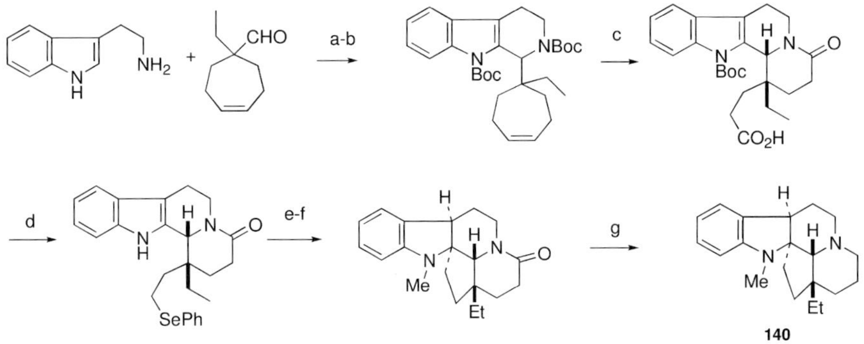

140

vallesamidine

Scheme 6.63 A HAM (tryptamine) for the construction of Eburna alkaloids.
Reagents: a) TFA, CH_2Cl_2; b) Boc_2O, DMAP; c) $KMnO_4$, Bu_4NBr, then MeOH, *p*TSA; d) $ClCO_2Et$, *N*-hydroxypyridine-2-thione, then $(PhSe)_2$, hν; e) nBu_3SnH, AIBN; f) HCHO, $NaBH_3CN$, AcOH; g) $LiAlH_4$, THF.

A recurring feature that appears in the design of many elegant syntheses in the natural products arena is the employment of cyclic precursors associated with the elaboration of chain segments or other types of ring structures, and the consideration of symmetry for the choice of the starting materials. This concept was used in work (Scheme 6.63) directed towards the synthesis of vallesamidine (**140**), an eburna alkaloid. The chemistry is based on an intermolecular Pictet–Spengler reaction performed on a cyclic aldehyde bearing a latent double bond, which upon oxidation gives two carboxylates, one of which collapses into an amide, whereas the other, transformed into a radical precursor, takes part in the final cyclization on the indole skeleton [84, 85]. The only tedious step is the separation of diastereomers, but the whole sequence is illustrative of the design of a new synthesis suited for application to a large family of indole alkaloids.

The transformation of carbolines into pyrroloquinolones by an oxidative process, discovered by Winterfeldt, is an important reaction for the generation of new heterocycles. The pyrroloquinolones **141–143** are substrates for the phosphodiesterases, important enzymes in signal transduction. A modification of the original procedure with use of potassium superoxide (KO_2) as the oxidative reagent has been reported (Scheme 6.64), and gives access to the quinolones from a variety of diverse carbolines even in the presence of sensitive substituents [86, 87]. Some examples of the scope of this transformation are reported below.

The indoloquinolizine ring system is of great interest and significance since this template is found within a plethora of bioactive alkaloids. A general method to generate an amide intermediate that can be converted into the chiral natural compounds in a few steps has been developed (Scheme 6.65). The key compound is generated through an intramolecular Pictet–Spengler reaction, and conditions providing only one of the two possible diastereomers have been identified. Decarboxylation and amide reduction then give the indoloquinolizine. This sequence has been applied [88] to the synthesis of deplancheine (**144**).

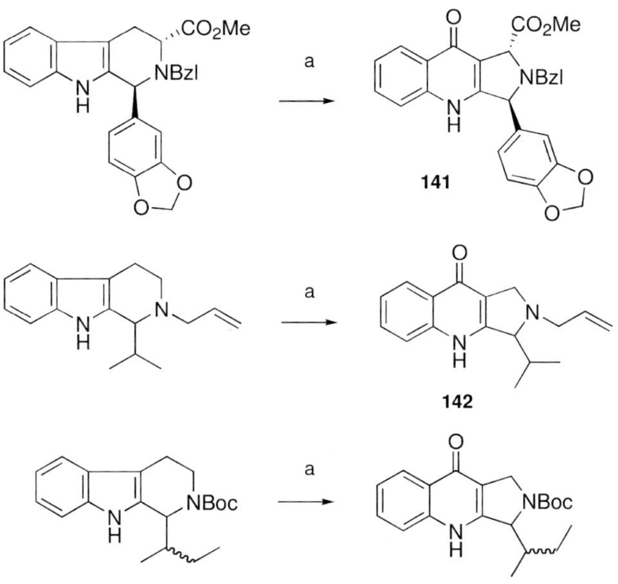

143

Scheme 6.64 Transformations of INDs into quinolones.
Reagents: a) KO_2, 18-crown, DMF, 25 °C.

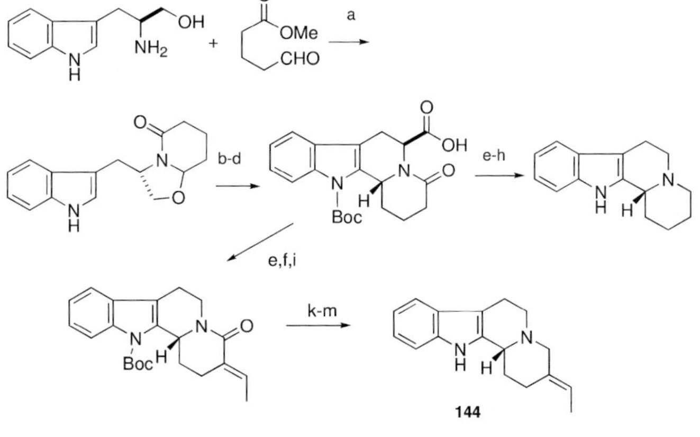

deplancheine

Scheme 6.65 INDs as building blocks for the preparation of indolyl alkaloids.
Reagents: a) toluene, reflux; b) HCl in EtOH; c) IBX; d) $(Boc)_2O$, Et_3N; d) $NaClO_4$, NaH_2PO_4; e) $(PhSe)_2$, PBu_3; f) nBu_3SnH AIBN; g) TBAF; h) $LiAlH_4$; i) LDA, EtCHO; j) Et_3N, MsCl then DBN; k) TBAF; l) Me_3OBF_4, 2,6-*tert*-Bu-pyr; m) $NaBH_4$, MeOH.

6.9 Conclusion

The practice of organic chemistry has today reached an outstanding level, complex target molecules now being accessible in multistep sequences through the use of an arsenal of highly efficient transformations. No doubt the identification, the design, and the obtainment of building blocks has been decisive for achieving important breakthroughs in the field of organic synthesis. As highlighted in the above survey, nitrogen-containing building blocks play a major role in the development of new strategies for the construction of biomolecules. Finally, Barry Sharpless's statement about the use of and need for building blocks for future strategies in medicinal chemistry gives a strong indication of the future need for the identification of novel and even more efficient and flexible core units: "*Nature's giant molecules are constructed from a set of small building blocks. If we are to build small molecules which interact with these large and diverse structures we need more components than nature has developed. While new building blocks will be welcome, we expect that a good set of 500 or so will prove effective for many targets*" [1].

References

1 H. C. Kol, M. G. Finn, K. B. Sharpless *Angew. Chem. Int. Ed.* 2001, *40*, 2004.
2 A review on the Suzuki reaction: A. F. Abdel-Magid, S. J. Mehrman *Org. Proc. & Dev.* 2006, *10*, 971.
3 A review on reductive amination: F. Bellina, A. Carpita, R. Rossi *Synthesis*, 2004, 2419.
4 Y. Herzig, L. Lerman, W. Goldenberg, D. Lerner, H. E. Gottlieb, A. Nudelman *J. Org. Chem.* 2006, *71*, 4130.
5 T. J. Tucker, T. A. Lyle, C.M. Wiscount, S. F. Britcher, S. D. Young, W. M. Sanders, W. C. Lumma, M. F. Goldman, G. A. O'Brien, R. G. Ball, C. F. Homnick, W. A. Schleif, E. A. Emini, J. R. Huff, P. S. Anderson *J. Med. Chem.* 1994, *37*, 2437.
6 M. Konishi, H. Ohkuwa, T. Tsuno, T. Oki *J. Am. Chem. Soc.* 1990, *112*, 3715.
7 C. Fischer, E. M. Carreira *Org. Lett.* 2001, *3*, 4319.
8 L. Shi, Y.-Q. Tu, M. Wang, F.-M. Zhang *Org. Lett.* 2004, *6*, 1001.
9 L. W. Bieber, M. F. Da Silva *Tetrahedron Lett.* 2004, *45*, 8281.
10 C. Wei, Z. Li, C.-J. Li *Synlett.* 2004, 1472.
11 N. Gommermann, P. Knochel *Chem.: Eu. J.* 2006, *12*, 4380.
12 P. Aschwanden, C. R. J. Stephenson, E. M. Carreira *Org. Lett.* 2006, *8*, 2437.
13 G. Abbiati, A. Arcadi, G. Bianchi, S. Di Giuseppe, F. Marinelli, E. Rossi *J. Org. Chem.* 2003, *68*, 6959.
14 A. S. K. Hashmi, J. P. Weyrauch, W. Frey, J. W. Bats *Org. Lett* 2004, *6*, 4391.
15 J. Böhmer, R. Grigg, J. D. Marchbank *Chem. Com.* 2002, 768.
16 R. Grigg, I. Köppen, M. Rasparini, V. Sridharan, *Chem. Comm.* 2001, 964.
17 Y. Yoshitake, K. Yamaguchi, C. Kai, T. Akiyama, C. Handa, T. Jikyo, K. Harano *J. Org. Chem* 2001, *66*, 8902.
18 K. M. Brummond, B. Mitasev *Org. Lett.* 2004, *6*, 2245.
19 J. H. vanMaarseveen, W. S. Hornc, M. R. Ghadiri *Org. Lett.* 2005, *7*, 4503.
20 H. Twin, R. A. Batey *Org. Lett.* 2004, *6*, 4913.
21 I. Akritopoulos-Zanze, V. Gracias, J. D. Moore, S. W. Djuric *Tetrahedron Lett.* 2004, *45*, 3421.
22 V. Gracias, A. F. Gasiecki, S. W. Djuric *Tetrahedron Lett.* 2005, *46*, 9049.
23 H. Hasegawa, H. Senboku, Y. Kajizuka, K. Orito, M. Tokuda *Tetrahedron* 2003, *59*, 827.
24 J. Krapcho, C. Turk, D. W. Cushman, J. R. Powell, J. M. DeForest, E. R. Spitzmiller, D. S. Karanewsky, M. Duggan, G. Rovnyak, J. Schwartz, S. Natarajahn, J. D. Godfrey, D. E. Ryono, R. Neubeck, K. S. Atura, E. W. Petrillo *J. Med. Chem.* 1988, *31*, 1148.
25 M. Tamak, G. Han, V. J. Hruby *J. Org. Chem.* 2001, *66*, 3593.
26 M. Doi, Y. Nishi, N. Kiritoshi, T. Iwata, M. Nago, H. Nakano, S. Uchiyana, T.

Nakazawa, T. Wakamiya, Y. Kobayashi *Tetrahedron* 2002, *58*, 8453.
27 X. I. Qiu, F. I. Qing *J. Org. Chem.* 2002, *67*, 7162.
28 J. R. Del Valle, M. Goodman *Angew. Chem. Int. Ed* 2002, *41*, 1600.
29 J. R. Del Valle, M. Goodman *J. Org. Chem.* 2003, *68*, 3923.
30 P. Chabaud, G. Pépé, J. Courcambeck, M. Camplo *Tetrahedron* 2005, *61*, 3725.
31 G. Han, M. G. Laporte, J. J. Folmer, K. M. Werner, S. M. Weinreb *J. Org. Chem.* 2000, *65*, 6293.
32 J. F. Poisson, A. Orellana, A. E. Greene *J. Org. Chem.* 2005, *70*, 10860.
33 K. L. Curtis, J. Fawcett, S. Handa *Tetrahedron Lett.* 2005, *46*, 5297.
34 K. X. Chen, F. G. Njoroge, B. Vibulbhan, A. Prongay, J. Pichardo, V. Madison, A. Buevich, T.-M. Chan *Angew. Chem. Int. Ed.* 2005, *44*, 7024.
35 T. Yamashita, N. Kawai, H. Tokuyama, T. Fukuyama *J. Am. Chem.* 2005, *127*, 15038
36 M. P. Van Brunt, R. F. Standaert *Org. Lett.* 2000, *2*, 705.
37 G. Reginato, A. Mordini, M. Caracciolo *J. Org. Chem.* 1997, *62*, 6187.
38 S. P. Govek, L. E. Overman *J. Am. Chem. Soc.* 2001, *123*, 9468.
39 A. Esposito, P. P. Piras, D. Ramazzotti, M. Taddei *Org. Lett.* 2001, *3*, 3273.
40 I. Ojima, E. S. Vidal *J. Org. Chem.* 1998, *63*, 7999.
41 A. Rodriguez, D. D. Miller, R. F. W. Jackson *Org. Biomol. Chem.* 2003, *1*, 973;
42 S. Tanabella, I. Valanogne, R. F. Jackson *Org. Biomol. Chem.* 2003, *1*, 4254.
43 P. N. Collier, I. Patel, R. J. K. Taylor *Tetrahedon Lett.* 2001, *42*, 5953.
44 W. H. Chiou, A. Schoenfelder, A. Mann, I. Ojima, *229th ACS National Meeting* 2005, San Diego, CA (USA).
45 T. Focken, A. B. Charrette *Org. Lett.* 2006, *8*, 2985.
46 D. L. Comins, J. J. Sahn *Org. Lett.* 2005, *7*, 5227
47 E. S. Tasber, R. M. Garbaccio *Tetrahedron Lett.* 2003, *44*, 9185.
48 K. Tchabanenko, M. G. O Taylor, R. M. Adlington, J. E. Baldwin *Tetrahedron Lett.* 2006, *47*, 39.
49 D. L. Comins, X. Zheng, R. R. Goehring *Org. Lett.* 2002, *4*, 1611.
50 P. S. Watson, B. Jiang, B. Scott *Org. Lett.* 2000, *2*, 3679
51 G. R. Heintzelmann, W-K. Fang, S. P. Keen, G. A. Wallace, S. M. Weinreb *J. Am. Chem. Soc.* 2001, *123*, 8851.
52 C. E. Neipp, S. F. Martin *J. Org. Chem.* 2003, *68*, 8867.
53 Z. Li, M. Fernandez, E. N. Jacobsen *Org. Lett.* 1999, *1*, 1611.
54 Y. Fukuta, T. Mita, N. Fukuda, M. Kanai, M. Shibasaki *J. Am. Chem. Soc.* 2006, *128*, 6312.
55 A. P. Patwardhan, V. R. Pulgam, Y. Zhang, W. D. Wulff *Angew. Chem. Int. Ed* 2005, *44*, 6169.
56 S. Y. Jonsson, C. M. G. Löfström, J. E. Jan-E. Bäckvall *J. Org. Chem.* 2000, *65*, 8454.
57 M. Nakagawa, M. Kawahara *Org. Lett.* 2002, *2*, 953.
58 W. J. Moran, K. M. Goodenough, P. Raubo, J. P. A. Harrity *Org. Lett.* 2003, *5*, 3427.
59 O. Y. Provoost, S. J. Hadley, A. J. Hazelwood, J. P. A. Harrity *Tetrahedron Lett.* 2006, *47*, 331.
60 Y. C. Hwang, M. Chu, M. Fowle *J. Org. Chem.* 1985, *50*, 3885.
61 W. Oppolzer, E. Flaskamp, L. W. Bieber *Helv. Chim. Acta* 2001, *84*, 141.
62 M. R. Schneider, P. Klotz, I. Ungureanu, A. Mann, C. G. Wermuth *Tetrahedron Lett.* 1999, *40*, 3873.
63 B. M. Trost, D. R. Fandrick *Org. Lett.* 2005, *7*, 823.
64 S. Kobayashi, K. Hgirano, M. Sugiura *Chem. Commun.* 2005, 14
65 M. Sugiura, K. Hirano, S. Kobayashi *J. Am. Chem. Soc.* 2004, *126*, 7182
66 S.-W. Li, R. A. Batey *Chem. Commun.* 2004, 1382.
67 T. Vilaivan, C. Winotapan, V. Bauphavichit, T. Shimada, Y. Ohfune *J. Org. Chem.* 2005, *70*, 3464
68 X. Beebe, Y. Gracias, S. W. Djuric *Tetrahedron Lett.* 2006, *47*, 3225.
69 P. Dewi-Wülfing, J. Gebauer, S. Blechert *Synlett* 2006, 487.
70 G. Kim, S. D. Jung, W. J. Kim *Org. Lett.* 2001, *3*, 2985 and

71 G. Kim, S. D. Jung, E. J. Lee, N. Kim *J. Org. Chem.* 2003, *68*, 5395.

72 F. X. Felpin, S. Girard, G. Vo-Thanh, R. J. Robins, J. Villeras, J. Lebreton *J. Org. Chem.* 2001, *66*, 6305.

73 For a related synthesis see: C. Welter, R. M. Moreno, S. Streiff, G. Helmchen *Org. Biomol. Chem.* 2005, *3*, 3266.

74 S. Gille, A. Ferry, T. Billard, R. B. Langlois *J. Org. Chem.* 2003, *68*, 8932

75 S. Blechert, C. Stapper *Eu. J. Org. Chem.* 2002, 2855.

76 K. Frank, J. Aubé *J. Org. Chem.* 2000, *65*, 655.

77 K. R. Campos, J. C. S. Woo, S. Lee, R. D. Tillyer *Org. Lett.* 2004, *6*, 79.

78 A. M. Schmidt, P. Eilbracht *J. Org. Chem.* 2005, *70*, 5528.

79 A. Dobbs *J. Org. Chem.* 2001, *66*, 638.

80 G. Bartoli, G. Palmieri, M. Bosco, R. Dalpozzo *Tetrahedron Lett.* 1989, *30*, 2129,

81 K. Knepper, S. Bräse *Org. Lett.* 2003, *5*, 2829 for the supported version of the Bartoli reaction

82 S. Boini, K. P. Moder, R. K. Vaid, M. E. Kopach, M. E. Kobierski *Org. Proc. & Dev.* 2006, *10*, 1205.

83 M. M. Faul, J. L. Grutsch, M. E. Kobierski, M. E. Kopach, C. A. Krumrich, M. A. Staszak, U. Udodong, J. T. Vicenzi, K. A. Sullivan *Tetrahedron* 2003, *59*, 7215.

84 T.-L. Ho, C.-K. Chen *Helv. Chim. Acta* 2006, *89*, 249.

85 H. Tanino, K. Fukuichi, M. Ushiyama, K. Okade *Tetrahedron* 2004, *60*, 3273

86 W. Jiang, X. Zhang, Z. Sui *Org. Lett.* 2003, *5*, 43

87 E. Winterfeldt *Liebigs Ann. Chem.* 1971, *745*, 23

88 S. M. Allin, C. I. Thomas, J. E. Allard, K. Doyle, M. R. K. Elsegood *Eu. J. Org. Chem.* 2005, 4179.

7
Aminated Sugars, Synthesis, and Biological Activity

Francesco Nicotra, Barbara La Ferla, and Cristina Airoldi

7.1
Biological Relevance of Aminated Sugars

The amino group is widely present in carbohydrates, often in acetylated form, and it is extensively involved in the recognition phenomena that nature entrusts to this important class of natural compounds [1]. As one example, the difference between the blood group determinants A and B consists of the presence of one hydroxy group in the terminal D-Gal unit in group B in place of the D-GalNAc unit in group A (Figure 7.1). This small difference makes the group B blood incompatible (not self) for an organism with group A, and an immunoresponse immediately occurs [2].

7.1.1
N-Acetylneuraminic Acid

Because of the great number of biological functions that it carries out, *N*-acetylneuraminic acid (Figure 7.2), commonly called sialic acid, is one of the eight essential saccharides needed for optimal health and functioning in humans [3]. Widely distributed throughout the tissues of the body (brain, adrenal glands, and the heart) and in many fluids (saliva, urine, cerebrospinal fluid, amniotic fluid, and breast milk), *N*-acetylneuraminic acid is found mainly in glycoproteins and glycolipids. Sialic acid is important for brain development, learning, memory, and cognitive performance, is also important for cellular communication, and is an immune system modulator. It affects the viscosity of mucus, which in turn repels viruses, bacteria, and other pathogens, and has also been shown to fight flu viruses A and B more effectively than most prescription medications.

These viruses can also cause cold sores, hepatitis, and viral pneumonia, as well as the common cold. As high levels are found in the human brain and kidney, it is speculated that sialic acid may play key roles in brain development and learning, and also in lessening the risk of kidney stone formation. Animal studies indicate that this essential saccharide does appear to improve both memory and cognitive performance. *N*-Acetylneuraminic acid is also found in such other tissues such as

Amino Group Chemistry. From Synthesis to the Life Sciences. Edited by Alfredo Ricci

ISBN: 978-3-527-31741-7

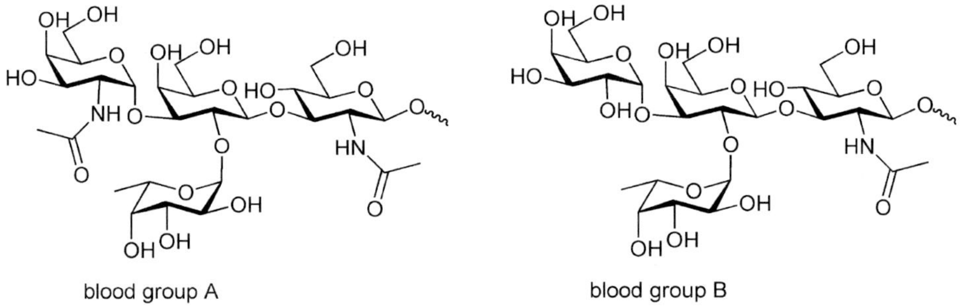

Figure 7.1 Blood group determinants A and B Type I.

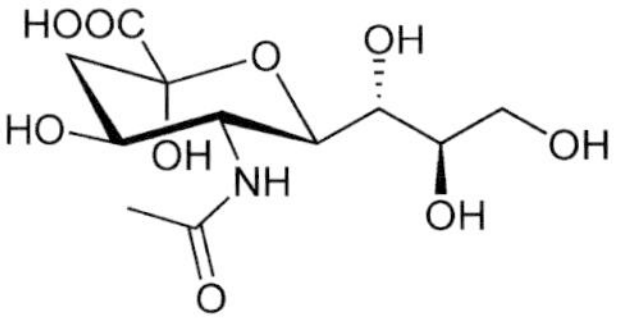

Figure 7.2 Structure of *N*-acetylneuraminic (sialic) acid.

the skin and testes, resulting in speculation that disruptions such as skin diseases and reproductive problems might be reversible through supplementation of this essential sugar. Sialic acid also influences blood coagulation and cholesterol levels, lowering LDL (bad cholesterol). Abnormalities in its metabolism are seen in infants who fail to grow, who regress in development, who have enlarged livers and/or spleens, who show coarsening of facial features, and who display failure to produce pigmentation of the skin and hair. Sialic acid levels are increased during pregnancy and lactation, indicating the developing infant's need for this sugar, both for establishing immunity and for physical and mental development. In severely ill patients, sialic acid levels are markedly decreased in the upper airway cells, where it would otherwise constitute an important barrier for preventing opportunistic respiratory infections seen in many of these patients. Sialic acid levels are markedly reduced in sufferers from rheumatoid arthritis, confirming that this saccharide plays an important part in the immune system. In addition, it has been demonstrated that sialic acid blocks the release of histamine, thus decreasing the severity of allergic reactions and asthmatic bronchial spasms.

7.1.2
Sialyl Lewis X

Like the other sialic acids, *N*-acetylneuraminic acid is typically found located at terminal positions in different glycoconjugates. In particular, it is one of the monosaccharidic units that make up sialyl Lewis X (Figure 7.3). As a consequence of its capability to bind E- and P-selectins, this molecule plays an important role

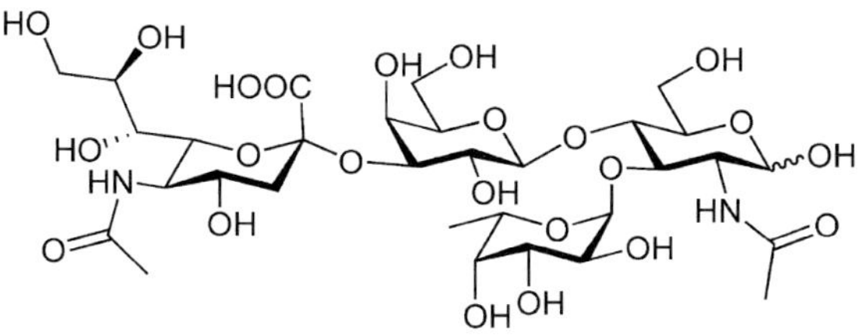

Figure 7.3 Sialyl Lewis X structure.

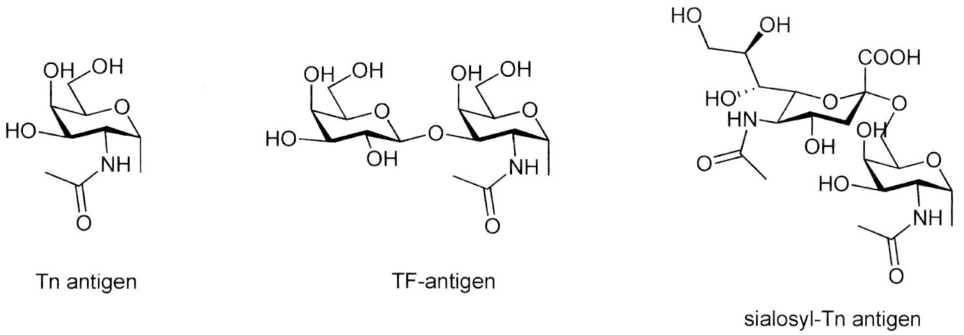

Figure 7.4 Tumor-associated antigens.

in the process of recruiting neutrophils and monocytes from the blood into damaged tissue [4].

Sialyl Lewis X is also a tumor-associated antigen and is in fact present to abnormal extents in certain tumors [5]. In particular, tumor cells containing sialyl Lewis X bind to regions of the vascular endothelium that express E-selectin. As this adhesion is thought to be involved at the origin of tumor metastasis formation, its inhibition could give rise to a new therapeutic approach. Different therapeutic uses of sialyl Lewis X – in particular with regard to its potential antiinflammatory activity – are being explored, but because of the complexity of its chemical synthesis, mimics of the natural compound have proven more easily obtainable [6].

7.1.3 Tumor-Associated Antigens

Important tumor-associated molecules are the Thomsen–Friedenreich antigen (Gal-GalNAc), which is assumed to be one of the few chemically well defined antigens with a proven association with malignancy, the TF-antigen (GalNAc), the sialosyl-Tn antigen (sialyl-GalNAc), and the sialosyl-TF antigen (sialyl-Gal-GalNAc), the cryptic form of the Thomsen–Friedenreich antigen (Figure 7.4) [7]. Many malignant cells (such as those found in breast and stomach cancer) develop the Thomsen–Friedenreich antigen, which serves as a tumor marker. TF antigen is suppressed in normal healthy cells and only becomes "unsuppressed" as a cell moves towards malignancy. It is so rare to find the TF antigen in healthy tissue that we actually have antibodies against it. It is even rarer to find a Tn antigen in a healthy cell [8].

The TF antigen and the Tn antigen show some structural similarity to the A antigen (see Figures 7.1 and 7.4). Not surprisingly, blood type A individuals have the least aggressive antibody immune response against the TF and probably Tn antigens. In fact, the TF and Tn antigens and blood type A antigens are actually immunologically considered to be quite similar because of their shared terminal sugar (*N*-acetylgalactosamine), and so might readily be confused by the immune systems of blood type A individuals [9].

7.1.4
Chitin and Chitosan

In oligosaccharides, the presence of acetamido groups also confers unique chemico-physical properties, such as in chitin, the β-1,4-poly-*N*-acetylglucosamine compound that makes up the insect exoskeleton (Figure 7.5) [10].

Particularly interesting from an applicative point of view is the simplest chitin derivative chitosan, the deacetylated form of chitin, obtainable by heating the natural polymer in alkaline media. Chitosan has interesting properties such as biocompatibility, biodegradability, and low immunogenicity. In addition, the presence of the reactive amino groups in the chitosan backbone allows the covalent conjugation of other molecules showing interesting biological activities. Jeong and co-workers, for example, linked glycol-chitosan to doxorubicin (GC-DOX) and 5β-cholanic acid to obtain micelle-forming polymer–anticancer drug conjugates, with the aim of developing an efficient strategy for passive tumor targeting [11].

7.1.5
Bacterial Polysaccharides

Another example of the structural role played by some aminosugars is represented by bacterial peptidoglycan, formed by repetitive units of *N*-acetylglucosamine and *N*-acetylmuramic acid linked through β-1,4 bonds. *N*-acetylmuramic acid is functionalized with a tetrapeptide consisting of L-alanine, D-glutamic acid, lysine or diaminopimelic acid, and D-alanine (Figure 7.6) that is fundamental for the cross-linkage of different linear polysaccharide chains.

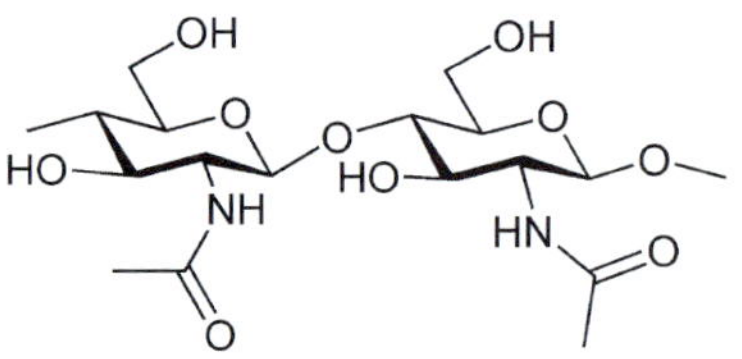

Figure 7.5 Structure of chitin.

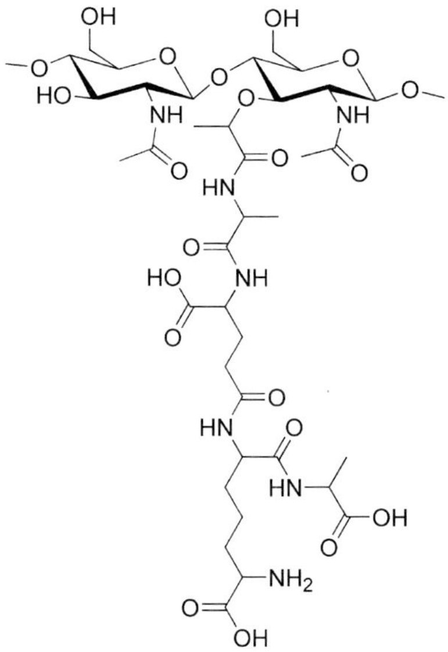

Figure 7.6 Glycan-tetrapeptide, the repetitive unit that constitutes the bacterial cell wall.

N-Acetylglucosamine is also a constituent of the Gram-negative bacterial lipopolysaccharide (LPS) present both in the core polysaccharide and in the lipid A portions. LPSs are the main targets of the antibodies produced by the immune system as a consequence of bacterial infections. A very important biological feature of this structure is its toxicity for mammals: some LPS can cause septic shock in humans.

7.1.6 Glycosaminoglycans

Sugars containing amino groups are also present in the extracellular matrices of eukaryotic cells, made up of heteropolysaccharides (glycosaminoglycans) and proteins [12]. In particular, glycosaminoglycans (GAGs) are primary components of the cell surface and the cell–extracellular matrix interface. Their distribution at the cell–extracellular matrix interface is ubiquitous and they interact with numerous proteins that modulate their activities. For these reasons, they are involved in fundamental biological processes such as development [13], angiogenesis [14], axonal growth [15], cancer progression [16], microbial pathogenesis [17], anticoagulation [18], and immunity [19]. From the structural point of view, glycosaminoglycans are a family of linear polymers made up of a repetitive disaccharidic unit (Figure 7.7).

One of the two monosaccharide is *N*-acetylglucosamine or *N*-acetylgalactosamine; the other is represented by uronic acid or glucuronic acid. In some glycosaminoglycans one or more hydroxy groups can be functionalized with a sulforic group. As a consequence of the combination of sulforic and carboxylic groups,

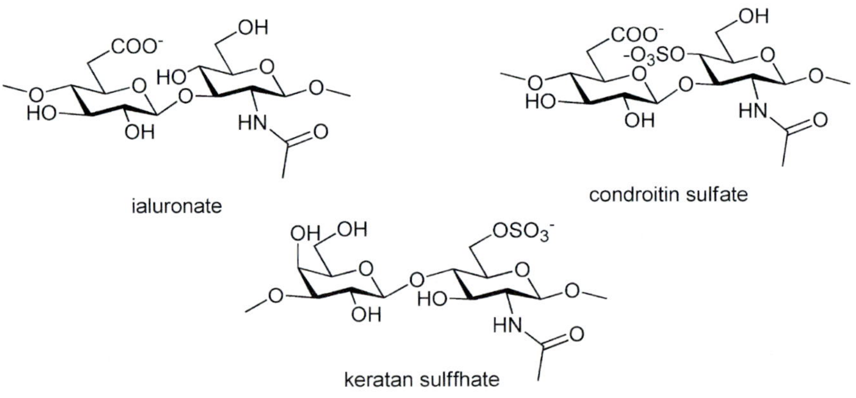

Figure 7.7 Structures of some glycosaminoglycans.

glycosaminoglycans present high densities of negative charges. In order to minimize these repulsive strengths, these polymers assume extensive conformations in solution, so these solutions have high degrees of viscosity. Glycosaminoglycans bind extracellular proteins, generating glycoconjugates known as proteoglycans. They play a critical role in assembling protein–protein complexes such as growth factor receptor or enzyme inhibitor aggregates on cell surfaces and in the extracellular matrices that are directly involved in initiating cell signaling events or inhibiting biochemical pathways. In addition, extracellular glycosaminoglycans can potentially sequester proteins and enzymes and present them to the appropriate site for activation. The specificity of these biomolecular interactions depends on the glycosaminoglycan primary sequence; thus, for a given high-affinity glycosaminoglycan–protein interaction, the positioning of the protein-binding oligosaccharide motifs along the glycosaminoglycan chain determines whether an active signaling complex is assembled at the cell surface or an inactive complex is sequestered in the matrix. The natures of glycosaminoglycan–protein interactions, together with their sequence diversity, makes glycosaminoglycans able to fine-tune protein activity.

The capability to obtain glycosaminoglycan oligosaccharides or oligosaccharide mimics with defined primary sequences has enabled detailed investigations into the biochemical and structural basis of glycosaminoglycan–protein interactions. Specific glycosaminoglycan oligosaccharides are obtained by isolation from biological sources by chemical/enzymatic depolymerization combined with chromatographic purification techniques. Important development in chemical and chemoenzymatic synthesis have also resulted in valuable model compounds that act as glycosaminoglycan mimetics [20].

7.1.7
Iminosugars

Iminosugars are "aminated" carbohydrates in which the nitrogen atom is part of the ring skeleton and substitutes the sugar's endocyclic oxygen. Because of the presence of this nitrogen atom, these molecules are potent glycosidase and

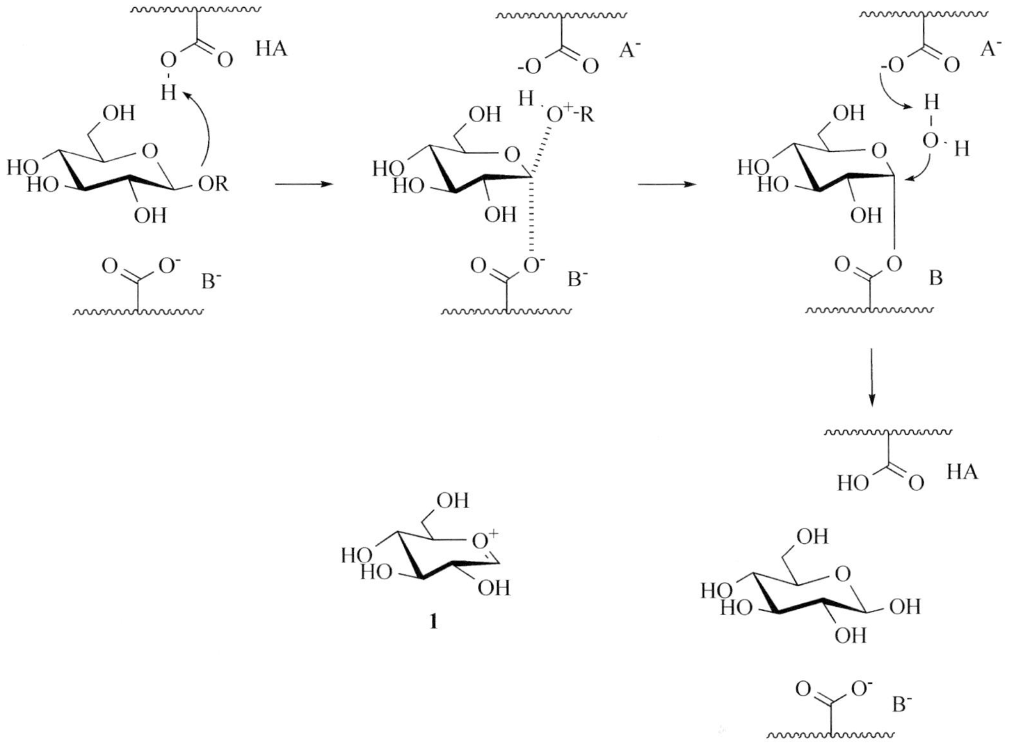

Figure 7.8 The mechanism of glycosidases.

glycosyltransferase inhibitors, as the charged iminosugar mimics the transition states of reactions catalyzed by carbohydrate processing enzymes. Figure 7.8 illustrates the proposed mechanism for retaining glycosidases: the aglycon is activated by coordination of the exocyclic oxygen atom with an enzymatic acidic function (catalytic acid HA), which converts it into a leaving group with formation of a cation/oxocarbenium ion intermediate similar to **1**. This ion reacts with the other catalytic carboxylate (catalytic nucleophile B^-), located on the opposite side of the plane of the pyranose ring, to form an intermediate glycosyl ester. The aglycon then diffuses away from the active site and is replaced by a water molecule. This water molecule, partially deprotonated by the conjugate base of the catalytic acid, attacks the anomeric carbon, and cleaves the newly formed glycosyl ester bond. The double inversion results in overall retention of configuration [21].

The discovery of the first iminosugar – 1-deoxynojirimycin – in 1966 (Figure 7.9) and of its biological activity as a potential inhibitor of α-glucosidase gave rise to the interest in these compounds.

In recent years iminosugars have been investigated as antiviral [22] and antitumor [23] agents, and iminosugar-based drugs are already on the market as antidiabetic [24] agents and in advanced clinical trials for the treatment of the Gaucher's disease [25], a lysosomal storage disorder.

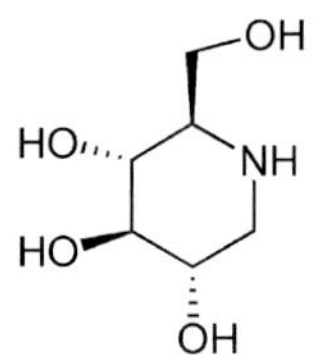

1-deoxynojirimycin (DNJ)

Figure 7.9 Structure of 1-deoxynojirimycin.

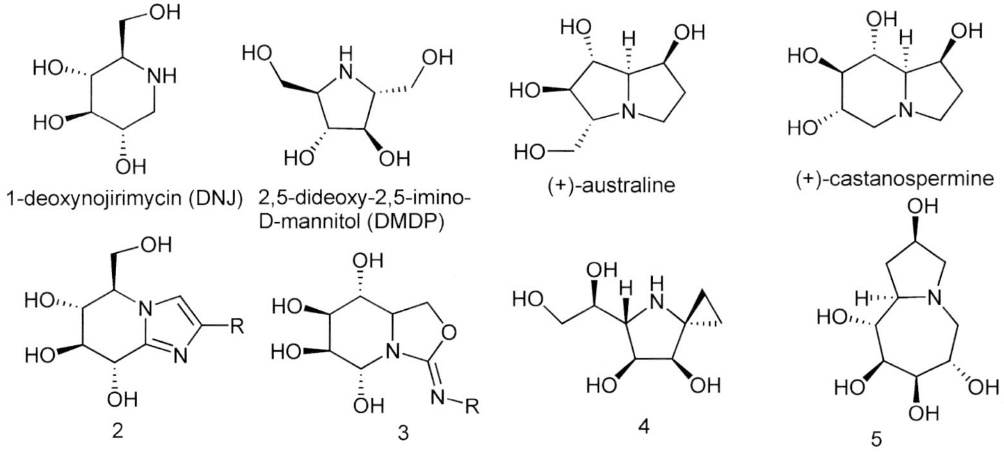

Figure 7.10 Examples of iminosugars.

From a structural point of view, the term "iminosugar" relates to a very wide and heterogeneous class of compounds. The main core structures are the monocyclic piperidine and pyrrolidine, the fused bicyclic pyrrolizidine and indolizidine, and the bridged bicyclic nortropane. Novel structures have recently been synthesized, with the introduction of new functional features such as the bicyclic tetrahydropyridoimidazole (**2**), -triazole, and -tetrazole derivatives [26], isourea-type indolizidines (**3**) [27], spirocyclopropyl (**4**) [28], and azaazulenes (**5**) [29], to give just a few examples (Figure 7.10).

Despite this great heterogeneity there are some common features present in all iminosugars: a) the presence of a ring nitrogen, and b) the presence of a polyhydroxylated structure that provides a resemblance to carbohydrates.

7.1.8
Sugar Amino Acids

Of great interest from a synthetic point of view is a particular class of aminated sugars represented by sugar aminoacids (SAAs). Designed and synthesized as new non-peptide peptidomimetics, by use of carbohydrates as peptide building blocks,

SAAs present both amino and carboxylic acid functional groups that allow their employment as amino acids in peptidic synthesis, but the introduction of these functions onto a rigid scaffold has a great conformational influence on the peptide backbone. In particular, SAAs can be employed to induce β-turns and γ-turns. The introduction of SAAs into peptides by condensation through the amino acid moieties has provided a new class of glyconjugates that can present important structural and biological features, according to the amino acidic sequence selected. The first example of a SAA as a potential peptidomimetic was furnished by Kessler and co-workers [30], who prepared the dipeptide isostere **6** as a mimetic of the Pro-Phe dipeptide sequence in the somatostatin analogue cyclo-(-Pro-Phe-D-Trp-Lys-Thr-Phe-) **7**, obtaining the potent mimetic **8** (Figure 7.11).

Another example of a SAA was proposed in 1997 by Nicolau and coworkers [31], who designed carbohydrate mimics of the cyclic peptide cRGDfV , an antagonist of vitronectins $\alpha_v\beta_3$, natural ligands to integrins, a class of extracellular proteins that facilitate cell–cell recognition (Figure 7.12).

A relatively recent strategy is based on the design and development of foldamers, polymers that present tendencies to adopt specific compact conformations [32].

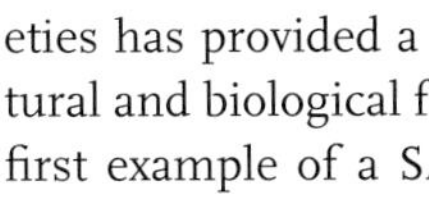

6

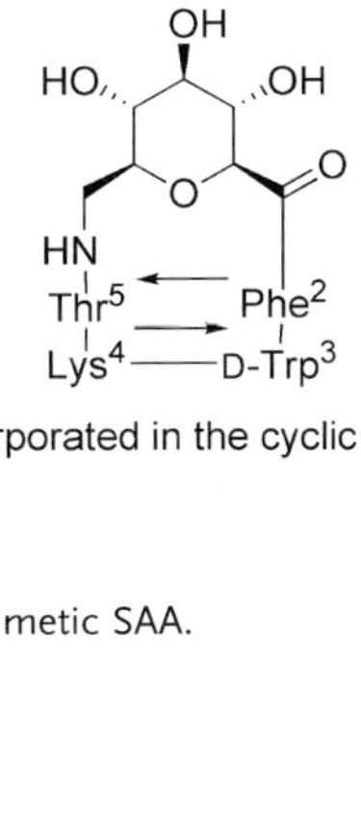

8 incorporated in the cyclic peptide

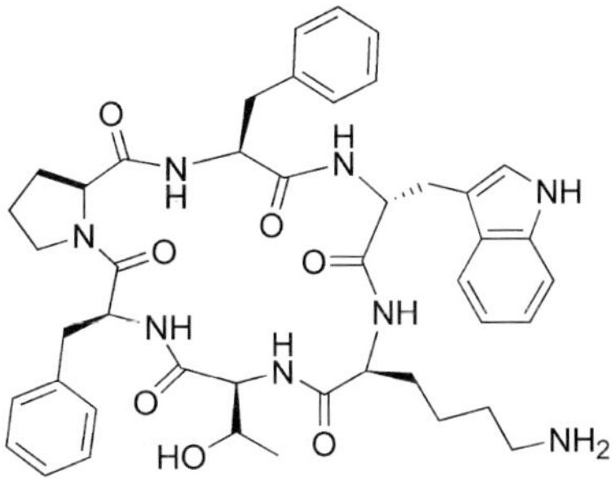

7

Figure 7.11 Kessler's peptidomimetic SAA.

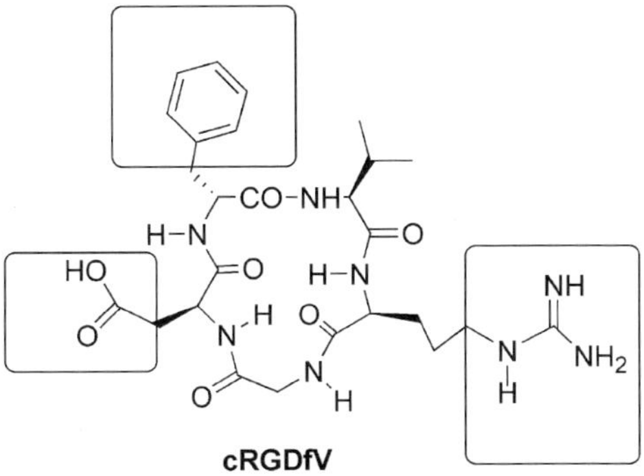

cRGDfV

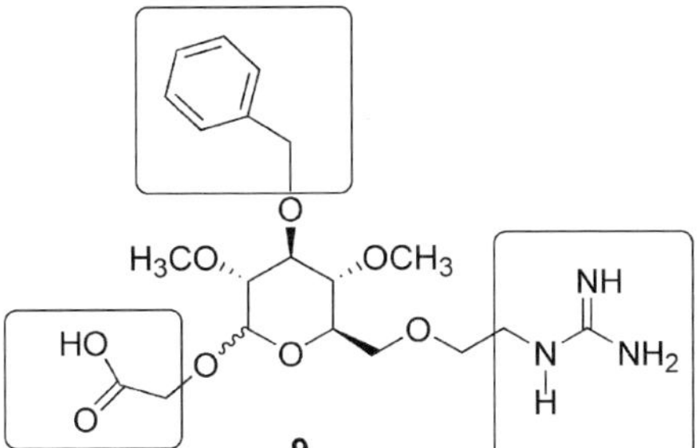

9

carbohydrate-based mimic

Figure 7.12 cRGfV peptide and a carbohydrate-based mimic.

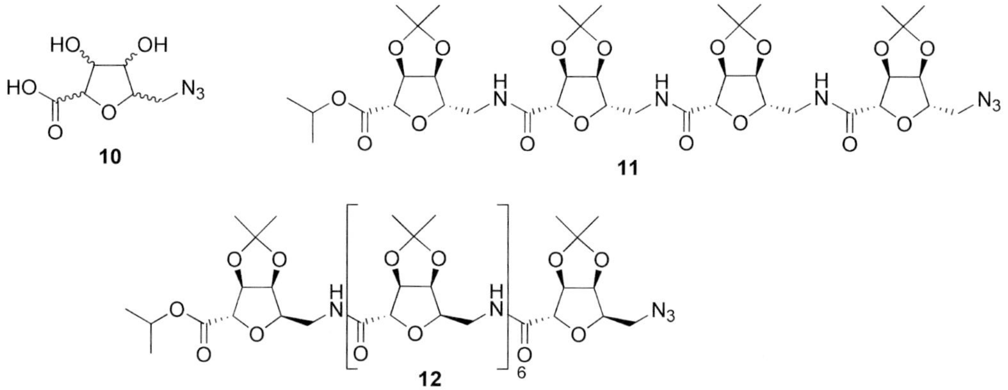

Figure 7.13 Sugar amino acids and their oligomers.

In 1999, Fleet and co-workers reported the synthesis of oligomers capable of producing significant secondary structures, starting from 5-azidomethyl-tetrahydrofuran-2-carboxylates such as **10**, that can be regarded as dipeptide isosteres (Figure 7.13) [33]. In particular, the reported tetramer **11** was demonstrated to induce a repeating β-turn-type conformation. On the other hand, NMR conformational studies revealed that octamer **12** (Figure 7.13) adopts an α-helix structure.

7.2
Synthesis of Aminated Sugars

The relevance of the amino group as a substituent in the carbohydrate structure, together with the fact that aminosugars are not very abundant or easily accessible from natural sources, makes synthetic methods to aminate carbohydrates relevant.

These methods can be divided into methods that insert the amino group at the anomeric center, in which the carbonyl function of that center is involved, and methods that allow the amino group to be inserted in other positions.

7.2.1
Amination at the Anomeric Center

The anomeric center in a carbohydrate has a peculiar reactivity with respect to the other positions of the sugar, and reactions at this center must therefore be treated separately. Amination at the anomeric center can be performed chemoselectively by exploiting the carbonyl function in **16**, which exists in equilibrium with the hemiacetal **13**, or the oxonium ion **15**, easily generated by treatment with Lewis acids (Scheme 7.1).

The insertion of a nitrogen nucleophile at an anomeric center is useful for two main reasons: a) to generate *N*-glycosides of biological relevance, among which

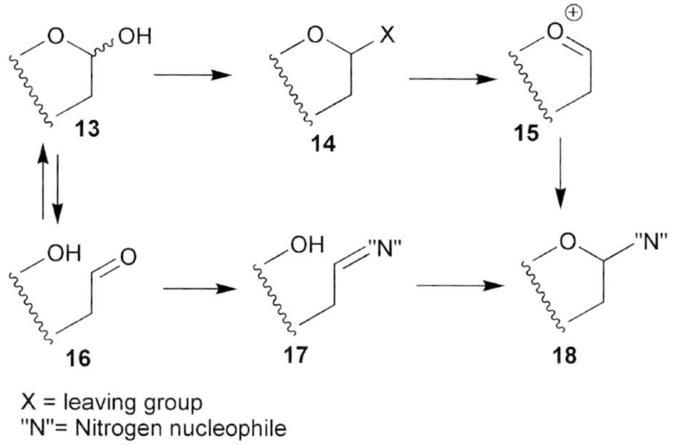

Scheme 7.1

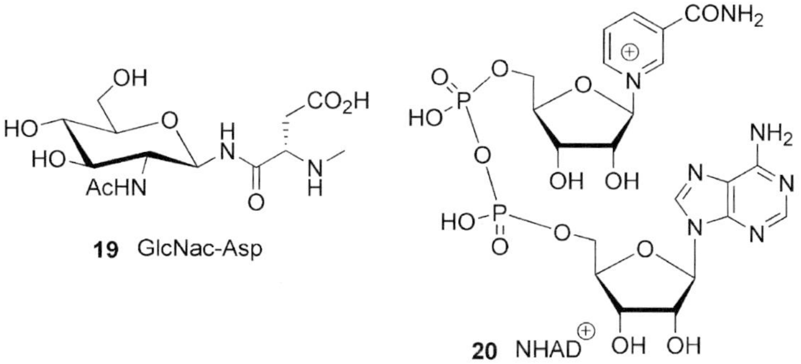

Figure 7.14 Examples of naturally occurring *N*-glycosides

N-linked glycoproteins and nucleosides must be mentioned, and b) as an easy instrument to generate amino sugars or iminosugars of different types after further chemical transformations.

In general, glycosyl amines are not very stable, due to equilibration with the corresponding imine, which undergoes hydrolysis with generation of the corresponding carbonyl compound. *N*-Glycosides do exist in nature, in particular as *N*-glycosyl asparagine in *N*-linked glycoproteins and as *N*-glycosyl heteroaromatic compounds in nucleosides or in NAD^+ (Figure 7.14). In all these cases, however, the delocalization of the electron pair of the nitrogen atom stabilizes the *N*-glycosidic bond.

The lability of glycosyl amines must be seriously considered when an anomeric amination is performed.

7.2.1.1 **Amination Exploiting Carbonyl Reactivity**

The anomeric center of an aldose or ketose can be aminated by treatment with ammonia, a primary amine, or a hydroxylamine. The carbonyl function will react with ammonia or a primary amine to generate an imine that will spontaneously undergo cyclization, but the main problem is that the obtained glycosyl amine will not be very stable and will tend to be hydrolyzed into the starting materials. The process has found some application, however, provided that the glycosyl amine is further converted into a more stable compound.

One important application involves the generation of neoglycopeptides and neoglycoproteins [34] through reactions between oligosaccharides and the amino group of lysine and reduction with $NaBH_3CN$ [35, 36]. The obtained glycopeptide or glycoprotein **22** includes an open-chain sugar linking the oligosaccharide to the peptide chain, but this does not compromise the recognition phenomena, provided that the oligosaccharide chain is long enough (Scheme 7.2).

The same approach has been used to link sugars, or clusters of sugars, to therapeutic agents in order to target them to specific receptors [37]; an interesting example is given in Scheme 7.3. Lactose **23** was reductively aminated with tyramine, providing the *N,N*-dilactitol-tyamine conjugate **24**, which was labeled with ^{125}I. The primary hydroxy groups in the galactose components were then oxidized with galactose oxidase, and the obtained aldehydes in **25** were used to link a protein by reductive amination.

A method that allows *N*-glycoconjugates to be generated by exploitation of carbonyl reactivity with concomitant restoration of the sugar cycle is shown in Scheme 7.4. The method is based on the chemoselective reaction of a substituted hydroxylamine with the carbonyl function of the sugar, affording an intermediate iminium ion **28** that spontaneously cyclizes to afford the cyclic *N*-glycoside **29** (Scheme 7.4) [38].

This approach has been exploited for the chemoselective synthesis of glycoconjugate and oligosaccharide mimics (Scheme 7.5) [39, 40], including a bioactive mimic **34** of lipid A. (Figure 7.15) [41].

An example in which the formation of a glycosyl amine serves as a means, after further manipulation, to insert the amino group in the sugar skeleton is shown

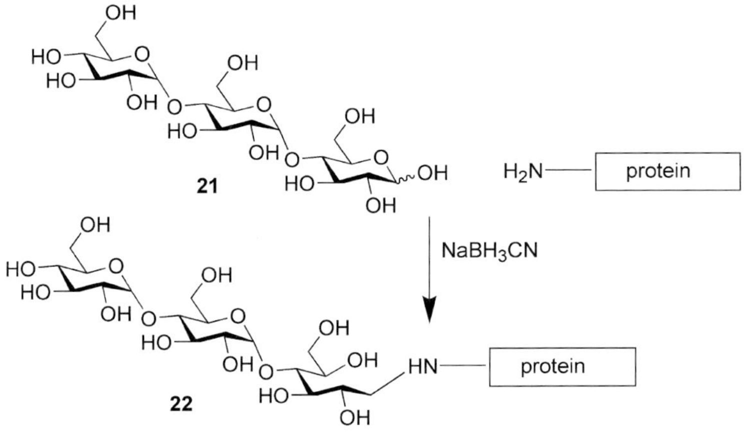

Scheme 7.2

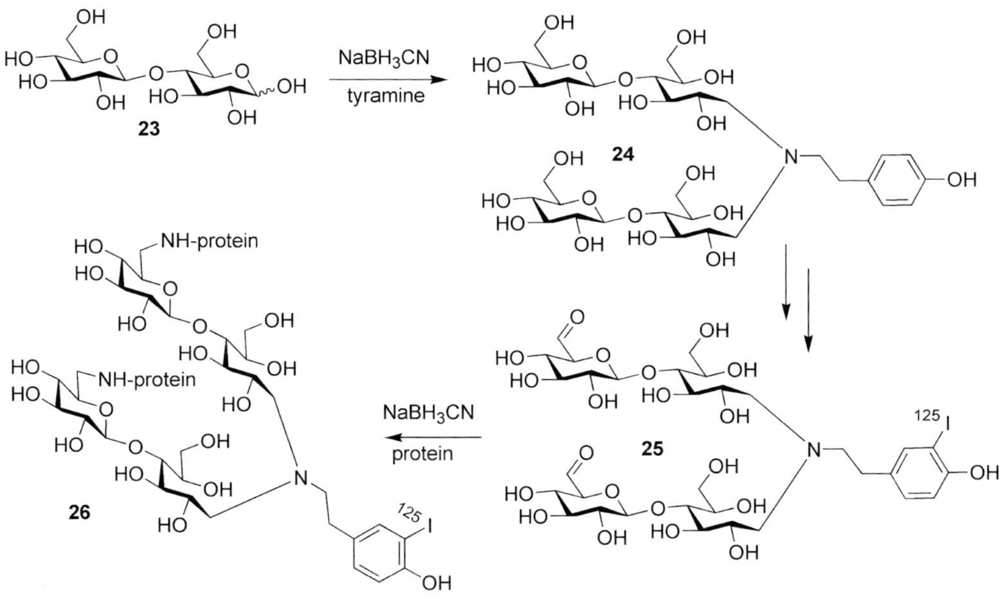

Scheme 7.3

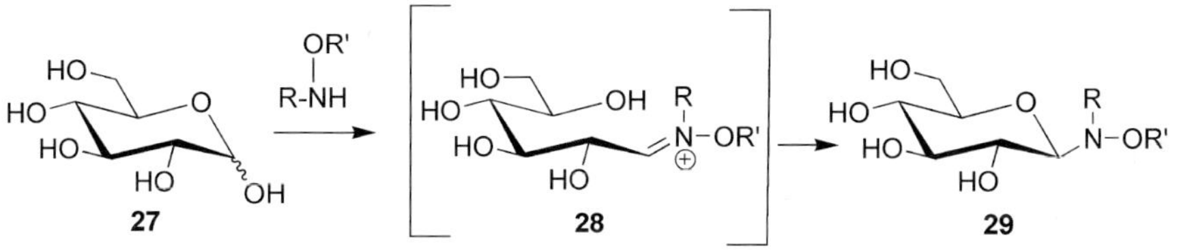

Scheme 7.4

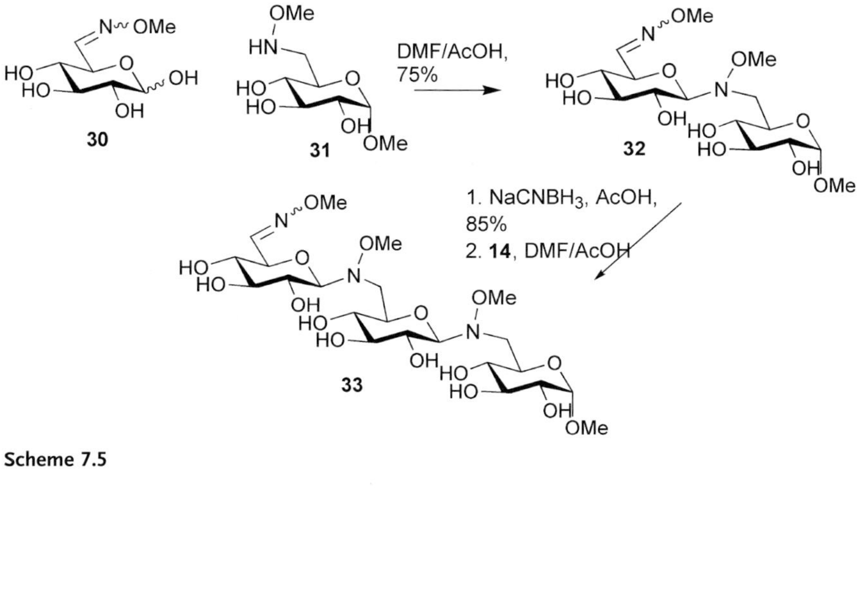

Scheme 7.5

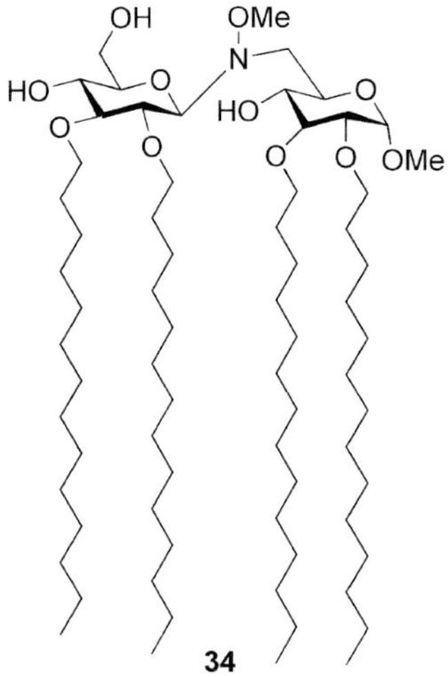

34

Figure 7.15 A lipid A mimic.

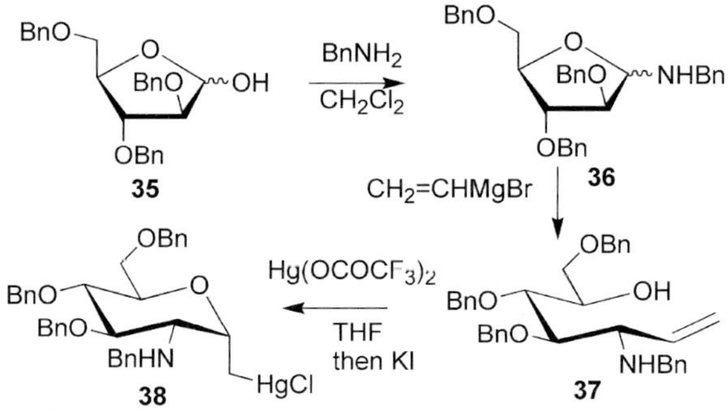

35, 36, 37, 38

Scheme 7.6

in Scheme 7.6. Tri-*O*-benzyl-D-arabinofuranose (**35**) was converted into the corresponding *N*-benzyl glycosyl amine **36**, which was treated with vinylmagnesium bromide to elongate the sugar skeleton. The reaction stereoselectively afforded (88% d.e.) the *gluco* isomer **37**. The obtained *gluco*-aminoheptenitol was mercuriocyclized, stereoselectively affording (60% d.e.) the *C*-glycoside **38** of glucosamine (Scheme 7.6) [42].

7.2.1.2 **Amination Exploiting Oxonium Ion Reactivity**

The insertion of a nitrogen nucleophile at the anomeric center can also be performed by displacement of a leaving group, favored by the stability of the oxonium ion intermediate.

Glycosyl halides have been displaced with azide, with formation of glycosyl azides that can easily be reduced to the corresponding amines. In the example

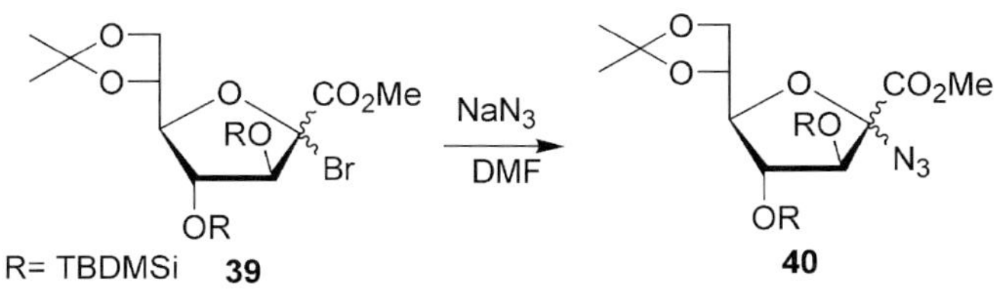

Scheme 7.7

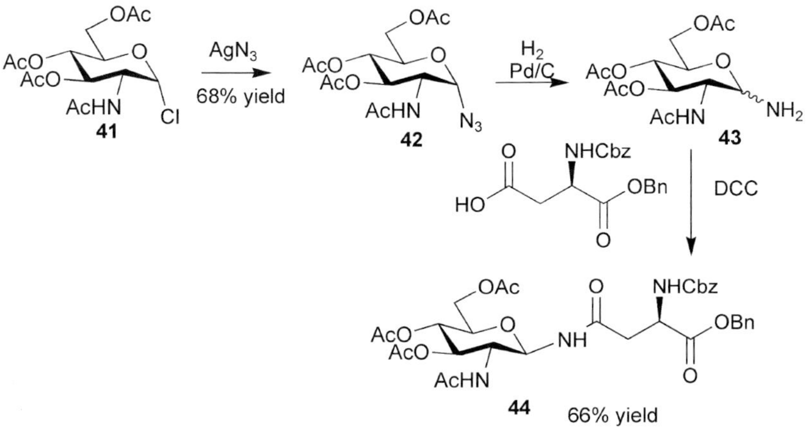

Scheme 7.8

given in Scheme 7.7, a tetrasubstituted glycofuranosyl bromide, treated with NaN_3, gave the corresponding azide by a S_N1 process [43].

2-Acetamido-3,4,6-tri-*O*-acetyl-α-D-glucopyranosyl chloride **41** reacted with silver azide to afford the corresponding β-glucopyranosyl azide **42**, hydrogenation of which over Pd/C gave the glycosyl amine **43** as mixture of anomers. Condensation of the anomeric mixture of glycosyl amines **43**, which exist in equilibrium, with suitably protected aspartate resulted in the stereoselective formation of a single glycoconjugate **44** with the β-anomeric configuration (Scheme 7.8) [44].

In place of glycosyl halides, less reactive glycosides can also generate glycosyl azides, provided that trimethylsilyl azide ($TMSN_3$) is used. This is probably the most efficient method to generate a glycosyl amine. The reaction proceeds in the presence of a Lewis acid as catalyst, through an oxonium ion intermediate, and does not necessarily require a strong leaving group such as a halide. In the case shown in Scheme 7.9, for example, the leaving group is an acetate, and the reaction was performed under solid-phase conditions in a reiterative way, to generate a dimeric sugar-amino acid (SAA) on which the peptide Leu-Phe-Gly-Gly-Tyr was attached in order to generate a carbohydrate-modified encephalin **49** (Scheme 7.9) [45].

The lability of glycosyl amines has also been circumvented by insertion of a carbon atom between the anomeric center and the amino group. Two main

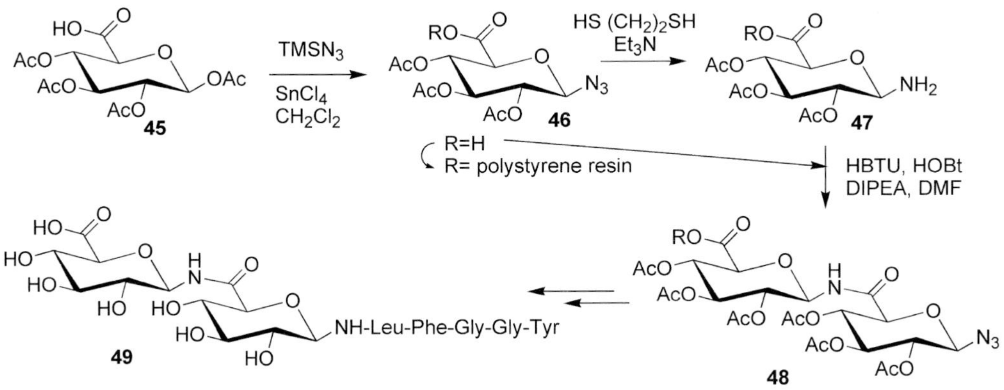

Scheme 7.9

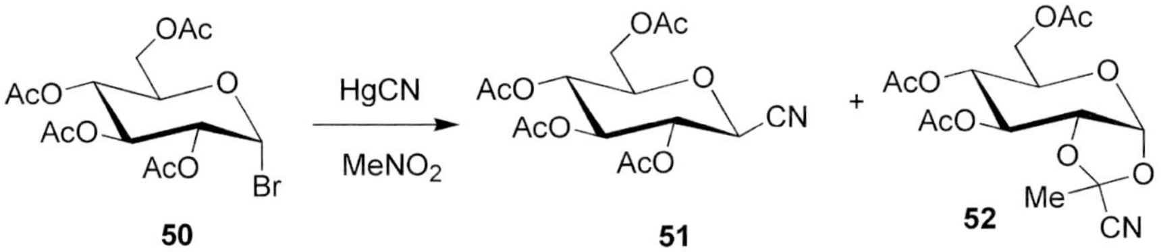

Scheme 7.10

procedures to synthesize this type of aminated carbohydrate mimics, members of the class of *C*-glycosides, have been adopted. The first procedure takes advantage on the nucleophilicity of the cyanide ion, while the second exploits that of the anion of nitromethane.

The observation that glycosyl halides react with HgCN with the formation of a glycosyl cyanide is very old [46, 47]. As a matter of fact, the first observation of this reaction was serendipitous: HgCN was used as co-reagent in the Koenigs–Knorr glycosidation procedure to capture the halide ions liberated in the reaction. In the absence of a glycosyl donor, the cyanide ion can act as nucleophile, and can consequently generate different glycosyl cyanides. The reaction, performed on 2,3,4,6-tetra-*O*-acetyl-α-D-glucopyranosyl bromide (**50**; Scheme 7.10) [48] or on 2,3,4,6-tetra-*O*-acetyl-α-D-galactopyranosyl bromide [49] stereoselectively afforded the β-glycopyranosyl cyanide. The stereochemical outcome of the reaction is determined by the participation of the acetoxy group at C-2, as evidenced by the formation of the cyanoethylidene derivative **52** as a side product.

More recently, HgCN as a cyanogenic reagent has been substituted by trimethylsilyl cyanide (TMSCN), which reacts with glycosides (not necessarily glycosyl halides) in the presence of Lewis acids to afford glycosyl cyanides [50]. Treatment of ethyl 2,3,4-tetra-*O*-benzyl-1-thio-β-D-glucopyranoside **53** with TMSCN and MeOTf in Et_2O gave the α-glucopyranosyl cyanide **54** as the sole stereoisomer in 72% yield (Scheme 7.11) [51].

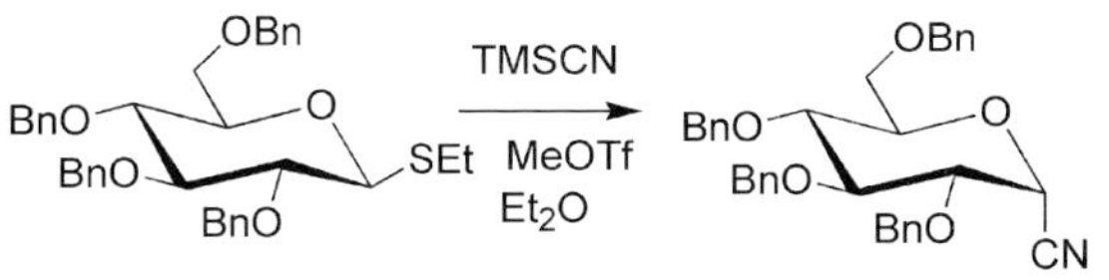

53   54 72% yield

Scheme 7.11

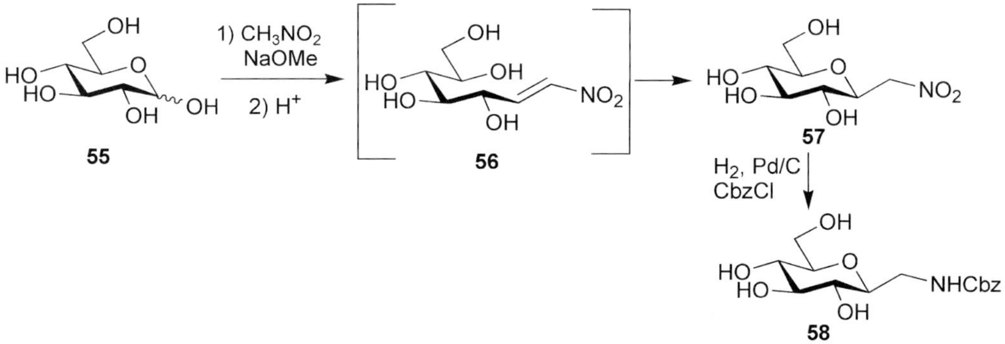

55   56   57   58

Scheme 7.12

The second procedure exploits the nitromethane anion, which reacts with an aldose to generate a *C*-glycoside by an addition/elimination/Michael cyclization process as shown in Scheme 7.12 [52]. The cyclization reaction proceeds under thermodynamic control, therefore affording the more stable anomer, which in the case of *C*-glucopyranosidic structures is the β anomer. From the stereochemical point of view, *C*-glycosidation with TMSCN or with the nitromethane anion are therefore complementary, affording α- and β-*C*-glucopyranosides, respectively.

7.2.2 Amination in the Sugar Chain

To insert an amino group into a sugar chain, the following main strategies have been used: a) displacement of a hydroxy group with a nitrogen-containing nucleophile, usually an azide, b) oxidation of a hydroxy group followed by reductive amination, c) amination of a double bond in a glycal, and d) opening of an epoxide with an azide ion.

7.2.2.1 Amino Sugars by Nucleophilic Displacement

The conversion of a hydroxy group in the sugar into an amino group can be performed through transformation of the hydroxy group into a good leaving group, followed by displacement with a nitrogen-containing nucleophile such as ammonia, hydrazine, phthalimide, or azide ion, which is the most efficient and the most widely used.

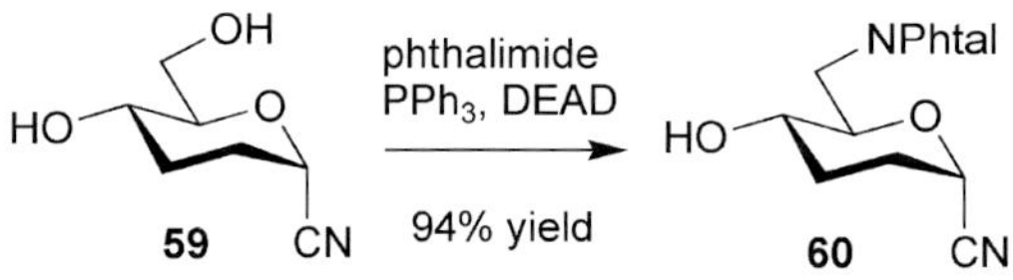

Scheme 7.13

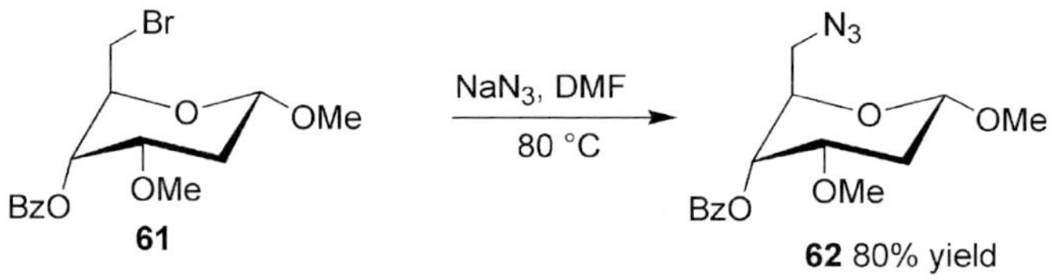

Scheme 7.14

An example of the use of phthalimide is shown in Scheme 7.13; the primary hydroxy group was selectively displaced in the presence of a secondary hydroxy group, through the use of the Mitsunobu approach (PPh_3, DEAD) [53].

The nitrogen nucleophile most widely used in the amination of carbohydrates is the azide anion – which has the advantage of being small – to generate a functional group that can be regarded as a protected primary amino group. NaN_3, LiN_3, and more recently Bu_4N_3 have found application in such azidolyses; the first two reagents require the use of polar aprotic solvents such as DMF, whereas with Bu_4N_3 the reaction can also be performed in toluene [54].

The efficiency of the process depends on the nature of the leaving group, primary ones being of course more reactive towards S_N2 displacement than their secondary counterparts, but is also strongly influenced by the position and orientation (axial or equatorial) of the leaving group in the sugar cycle, and by the orientations of the adjacent substituents. In other words, conformational and steric factors influence the success of the reaction.

Displacement of primary hydroxy groups by conversion into the corresponding chlorides or bromides and subsequent treatment with azide have been reported. An example is shown in Scheme 7.14 with the synthesis of methyl 6-azido-4-*O*-benzoyl-2,6-dideoxy-3-*O*-methyl-α-D-ribopyranoside (**62**) [55].

If a sulfonate is used as leaving group it is possible to discriminate between a primary and a secondary sulfonate, the first being selectively displaced under appropriate experimental conditions [56]. These displacements have also been performed on nucleosides; 5'-azido nucleosides, for example, have been obtained by sulfonylation and displacement of the obtained sulfonates with metal azides [57]. A one-pot procedure in which the nucleoside is treated with triphenylphosphine, carbon tetrahalide (CCl_4, CBr_4 or CI_4), and lithium azide has also been developed [58]. The most effective combination was $PPh_3/CBr_4/LiN_3$, which chemoselectively converted the primary hydroxy group of thymidine (**63**) into an azido group in 90% yield (Scheme 7.15). The same authors also reported the conversion

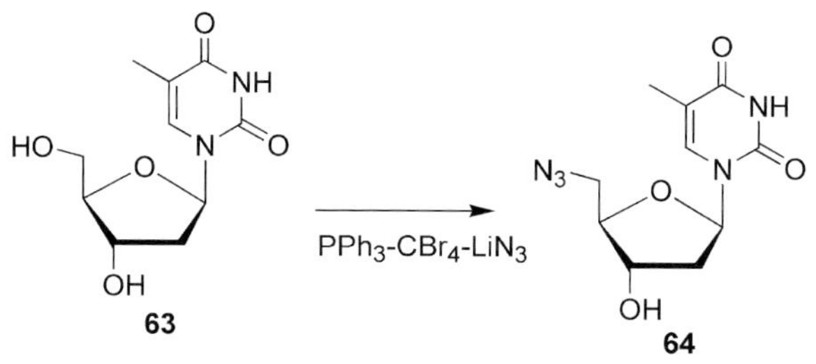

Scheme 7.15

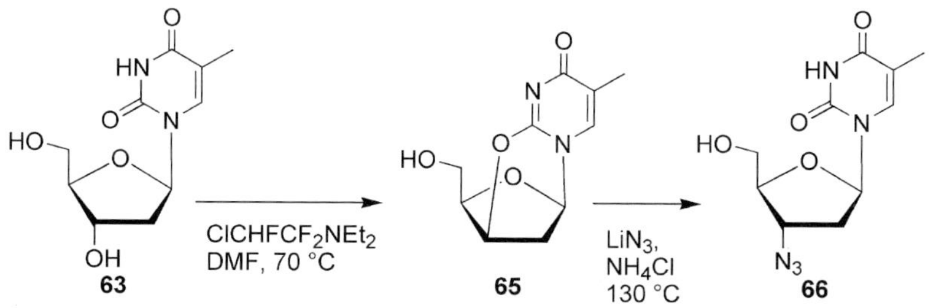

Scheme 7.16

of the secondary hydroxy group in **63** into an amino group under the same experimental conditions, once the primary hydroxy group had been protected as a trityl ether. This reaction was much less efficient, generating the 3′-azido derivative in 45% yield under the best experimental condition. The reaction was also performed on the C-3′ epimers, affording predominantly the 3′-azido-3′-deoxythymidine in 67% yield.

In general, for the conversion of a secondary hydroxy group into an amine, mesylates [59, 60], tosylates [61], triflates [62], or other types of sulfonates such as sulfonylimidazoles [63] are more efficient leaving groups than the halides. Mesylates and tosylates have the advantage of higher stabilities, which allow them to be isolated and stored more easily, whereas triflates are 10^3 times more reactive [64, 65] and generally give superior results [66], but must be handled with care and immediately subjected to the subsequent reaction. Imidazolylsulfonates are relatively stable, and offer reactivity towards nucleophiles comparable to that of triflates.

An interesting example of a nucleophilic substitution in which an azido group is introduced is reported in the synthesis of AZT (**66**) shown in Scheme 7.16 [67]. In this case, interestingly, thymine participates in the reaction, first as a nucleophile with generation of compound **65**, and then as leaving group.

Particular attention has been devoted to the introduction of amino groups at C-2, 2-amino-2-deoxy sugars being the most relevant amino sugars. In the case of β-D-glucopyranosides, for example, once a hydroxy group at C-2 has been converted into a good leaving group such as a triflate, nucleophilic substitution with an azido ion occurs in good yields, through axial attack of the nucleophile, affording the 2-azido-2-deoxy *manno* derivative **68** (Scheme 7.17) [68].

The difficulty in generating β-mannosidic linkages in glycosidation reactions makes the synthesis of oligosaccharides containing β-mannosamine units particularly difficult. A solution to this problem was reported in the synthesis of the repeating unit of the capsular polysaccharide of *Streptococcus pneumoniae* type F19, β-D-ManNAc-(1–4)-α-D-Glc-(1–2)-α-L-Rha. The adopted strategy consisted of a stereoselective formation of a β-glycosidic linkage between two D-gluco units formed by exploiting the acetate protecting group at C-2 of **69**. This was then in turn stereoselectively removed, and the unmasked hydroxyl group was converted into a sulfonylimidazole and displaced by the azido ion with inversion of the configuration (Scheme 7.18) [69].

In the case of glucopyranosides it is interesting to note that, if the anomeric substituent is α-oriented, the displacement of the equatorial hydroxy group at C-2 is less favorable. A C-glycoside of α-D-mannosamine **73**, for example, was obtained [70] by a low-yielding (31%) nucleophilic displacement of the free hydroxy group, regioselectively generated from the tetrabenzyl derivative **73** by iodocyclization-reductive elimination (Scheme 7.19).

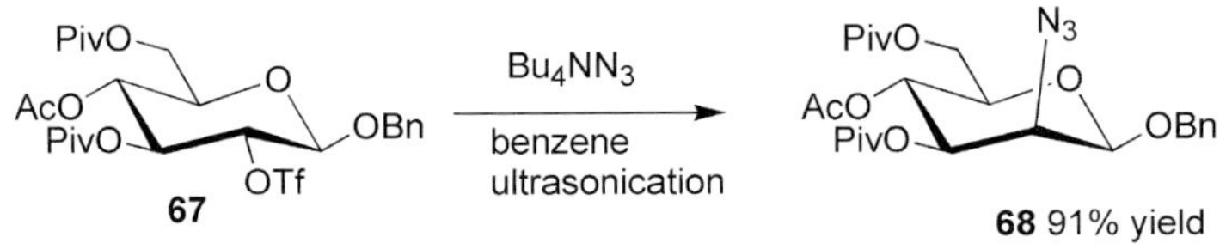

Scheme 7.17

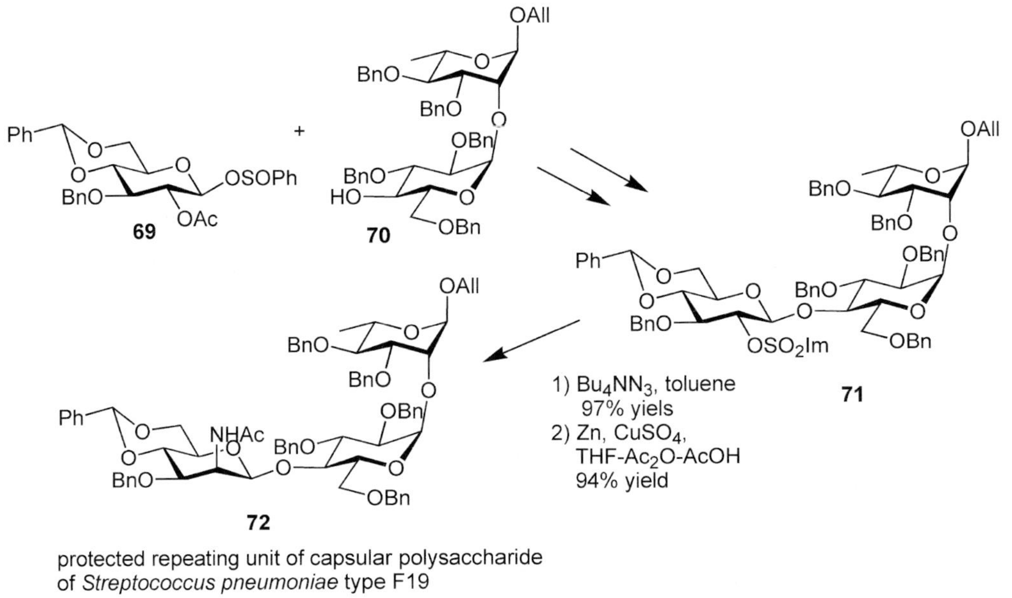

Scheme 7.18

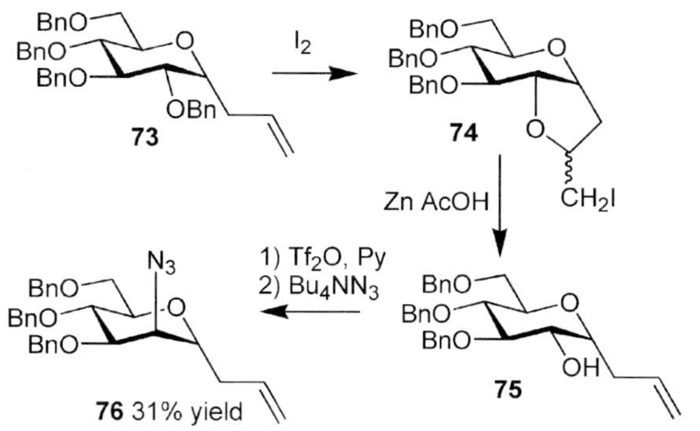

Scheme 7.19

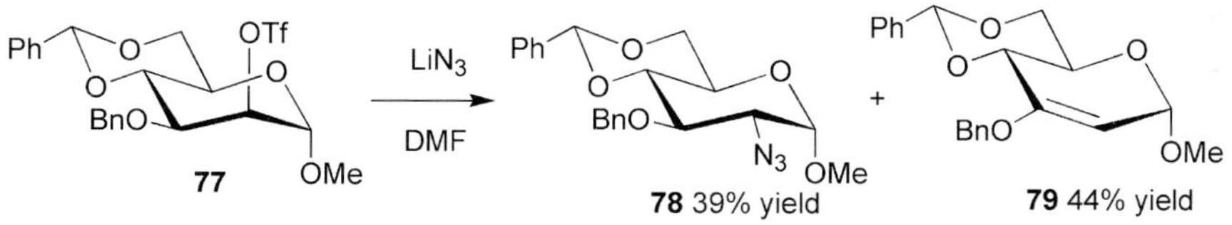

Scheme 7.20

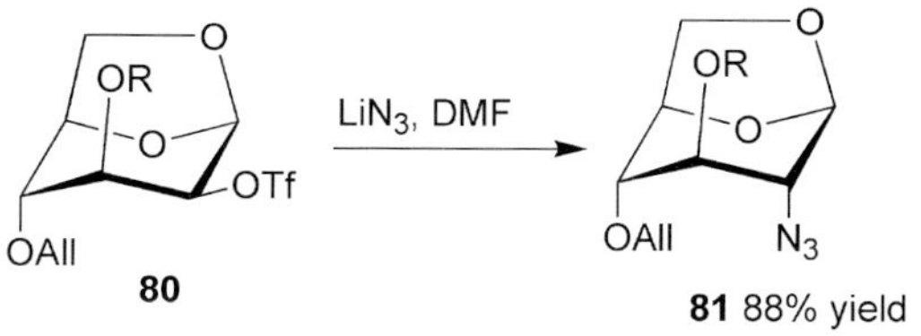

Scheme 7.21

Equatorial attack by the azido ion is in general less easy; the reaction between the triflate **77** (Scheme 7.20) and LiN_3, for example, resulted in a mixture of substitution and elimination products **78** and **79**, due to a transition state disfavored by dipolar and steric interactions.

It is interesting to note that, if the conformation is blocked with a 1,6-anhydro bridge, as in the mannoderivative **80** (Scheme 7.21), the nucleophilic displacement of the triflate, which is fixed equatorial, occurs efficiently, affording the azide **81** in high yields (88%).

The anomeric configuration also has a great influence in this case; in β-mannosides – stereoelectronically disfavored compounds – nucleophilic displacement at C-2, which affords much more stable β-glucopyranosides, occurs in higher yields. One example is shown in Scheme 7.22, in which the axial triflate in the β-mannopyranosidic structure **82** is displaced by LiN_3 to afford the azide **83** in 80% yield [71].

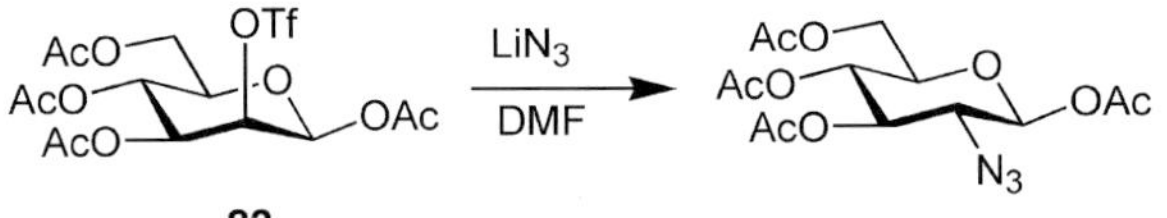

Scheme 7.22

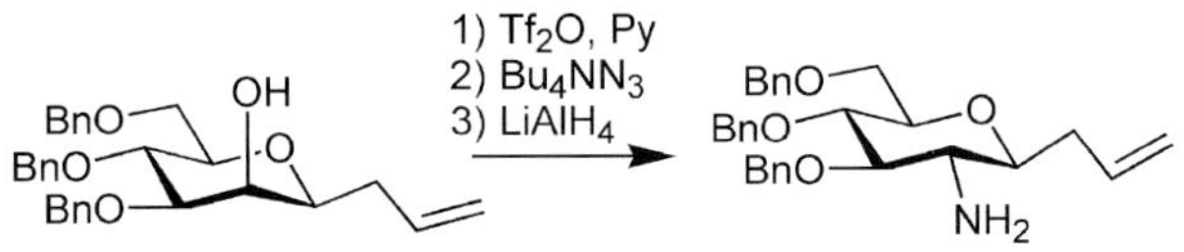

Scheme 7.23

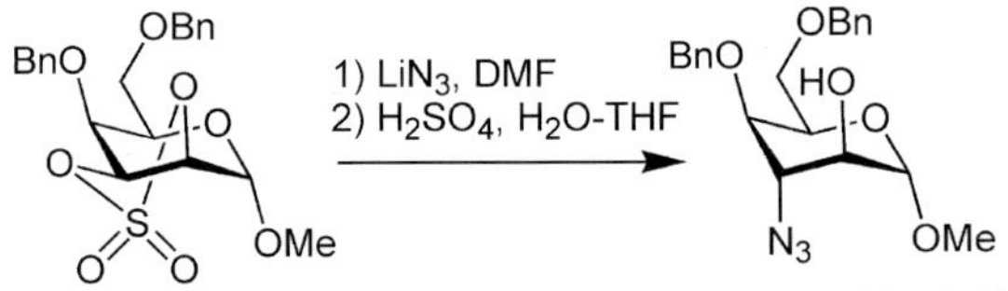

Scheme 7.24

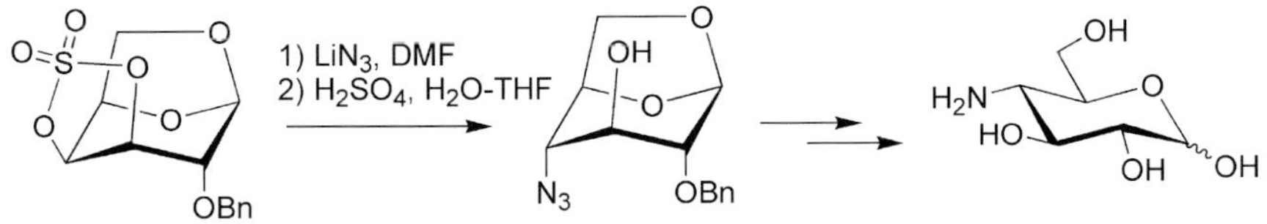

Scheme 7.25

In the case of the synthesis of the *C*-allyl β-D-glucosaminoside **85** shown in Scheme 7.23, the conversion of the axial hydroxy group of the mannoderivative **84** into the corresponding triflate, displacement of the triflate with Bu_4NN_3, and the final reduction of the azide with $LiAlH_4$ proceeds in 57 % yield over three steps [70].

Cyclic sulfites [72], sulfates [73], and sulfamidates [74] undergo nucleophilic displacement with azide. The process occurs regioselectively, mainly affording the *trans*-diaxial products, such as **87** in Scheme 7.24.

If a pyranosidic *trans*-diequatorial product is required, the process can be performed on a 1,6-anhydropyranoside such as compound **88** in Scheme 7.25, in which the 1,6-anhydro bridge constrains the molecule in the otherwise disfavored $^1C_4$ conformation. Once the displacement of the cyclic sulfate has generated the diaxial

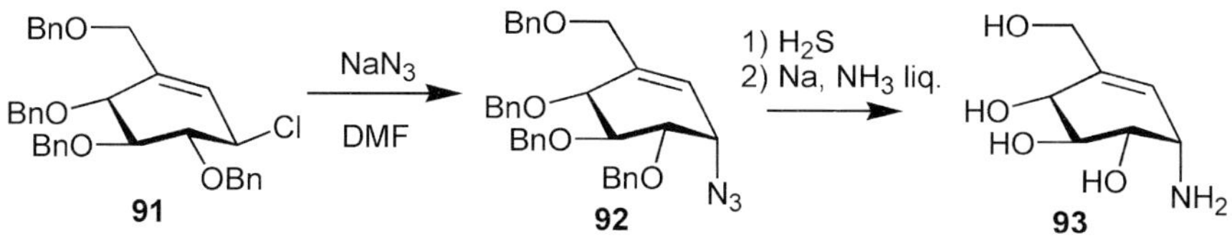

91 92 93

Scheme 7.26

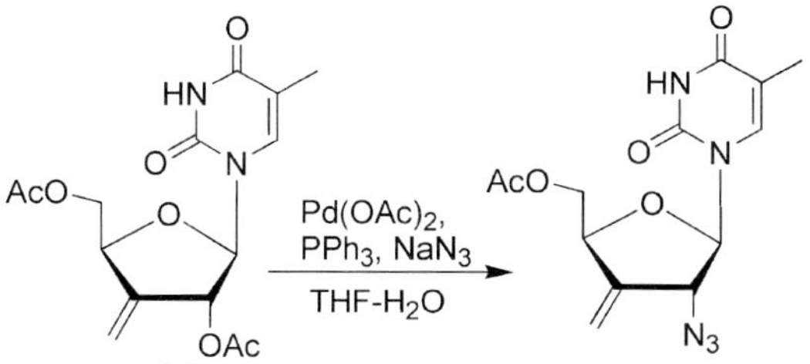

94 95 65% yield

Scheme 7.27

azido alcohol, hydrolysis of the acetalic 1,6-anhydro bridge, generates a $^4C_1$ pyranoside in which both the hydroxy and the azido groups are axial (Scheme 7.25)

The nucleophilic displacement of an allylic halide has been used in the synthesis of valienamine (**93**), an aminocyclitol present in the antibiotic validamycin (Scheme 7.26) [75].

The palladium-catalyzed displacement of an allylic acetate has been reported in, for example, the synthesis of the 2-azido thymidine derivative **95** shown in Scheme 7.27, in which the reaction occurred stereoselectively with retention of the configuration [76].

7.2.2.2 Amino Sugars through Intramolecular Displacements

A carbohydrate molecule containing a leaving group and a nitrogen nucleophile in suitable positions and orientations can undergo an intramolecular displacement to insert the nitrogen in the carbohydrate skeleton. To achieve this process, the nitrogen can be inserted at the anomeric center and the nucleophilic displacement can occur either at the adjacent position (C-2), or at C-4, provided that the two groups are *trans*-related (Scheme 7.28) [77, 78].

7.2.2.3 Amino Sugars by Reductive Amination

One of the most efficient methods to transform a hydroxy group into an amino group is by oxidation followed by reductive amination.

The oxidation can be performed with different reagents, among which are the well known pyridinium chlorochromate (PCC) and the Swern oxidation reagent, which involves the use of dimethylsulfoxide (DMSO) in the presence of a strong electrophile such as acetic anhydride, dicyclohexylcarbodiimide/H_3PO_4, or P_2O_5, Once the carbonyl compound has been generated, the reductive amination can be

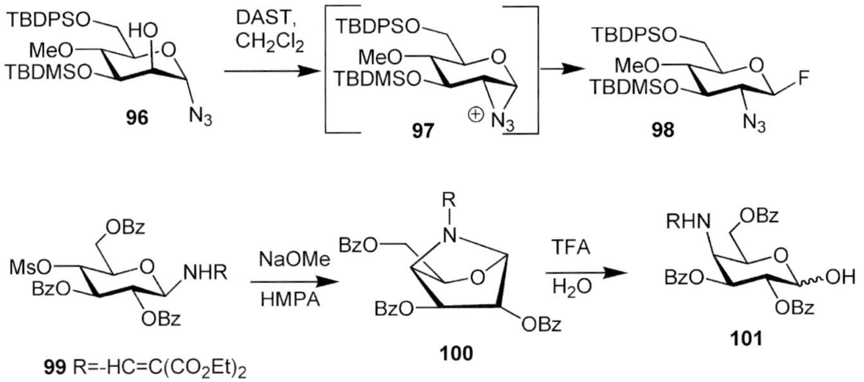

Scheme 7.28

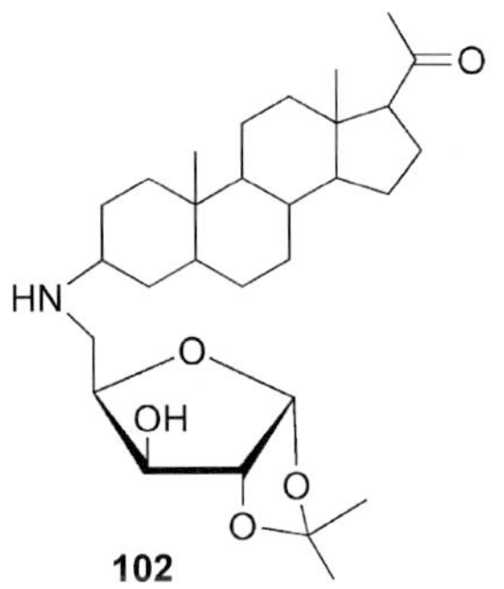

Figure 7.16 A synthetic glyco-steroid.

effected by addition of a primary or secondary amine with selective reduction of the iminium ion obtained at equilibration. This approach, however, is useful only if a glycol derivative (or glycoconjugate) containing a specific substituent at the nitrogen is required. An example is that of the reported glyco-steroid [79] (Figure 7.16).

In general, aminated carbohydrates contain the –NHAc or the $-NH_2$ groups, so reductive amination via an oxime is much more efficient than the use of ammonia or ammonium salts. $HONH_2$, $AcONH_2$, and CH_3ONH_2 have been widely used to generate the corresponding oximes, and the reduction has been performed in different ways, such as by catalytic hydrogenation (Pd/C, H_2) or by reduction with boranes or with $LiAlH_4$. The literature on the topic is very wide; one of the first examples was reported by Collins and Overend in 1965 [80]. Particular attention must be devoted to the stereochemistry of the reaction, as the reducing agent can attach the sugar from either the α- or the β-face, generating two possible epimers. The stereochemical outcome of the reaction can be influenced by many factors, among them the orientations and natures of the neighboring groups and the nature of the reducing agent. An interesting example showing the effect of the nature of the neighboring groups was reported by Lemieux [81], who observed that, depending on the nature of the alkyl group present at the anomeric position of the tri-*O*-acetyl-2-oximino-α-D-*arabino*-hexopyranosiduloses **103** shown in Scheme 7.29, the *gluco/manno* ratio of the products changed from 62% *gluco* in the case of a propyl group to 80% *gluco* if R was an isopropyl. The nature of the reducing agent is also relevant in the stereochemistry of the reduction.

Maintaining the α-oriented isopropyl group at the anomeric center, it has been observed that the reduction of the 2-oximino-α-D-*arabino*-hexopyranosidulose **106** with borane or with H_2 over Pd/C stereoselectively afforded the *gluco* isomer **107** (*gluco/manno* ratios 92:8 and 70:30, respectively), whereas with $LiAlH_4$ the stereoselection was reduced and the *manno* isomer **108** was formed predominantly (*gluco/manno* ratio 40:60) (Scheme 7.30) [82].

An interesting application of this procedure has been reported in the conversion of easily available ketoses into 2-amino-2-deoxy-aldoses, including disaccharides, taking advantage of the already existing keto function in the starting material [83]. By this approach, which consists of a revisited Heyns rearrangement [84], the lactosamine **110** was synthesized from lactulose **109** in a simple two-step procedure shown in Scheme 7.31.

On changing of the orientation of a single substituent, such as the hydroxy group at C-4, the stereochemical behavior described above is strongly perturbed. The reduction of isopropyl 2-oximino-α-D-*arabino*-hexopyranosidulose (**111**) either with borane or with $LiAlH_4$ proceeded with negligible stereoselection, affording a *galacto/talo* ratio around 6:4 (Scheme 7.32).

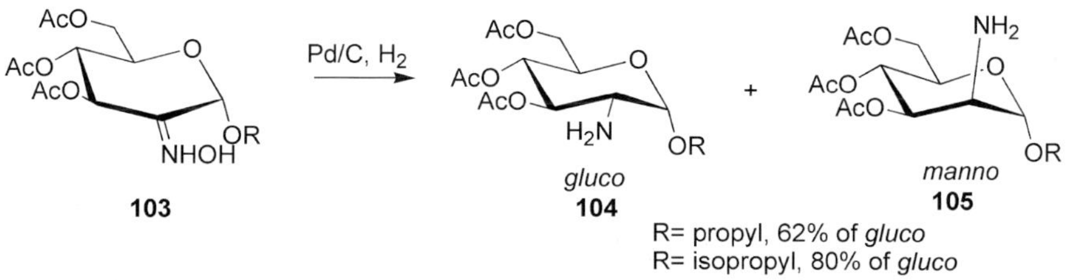

Scheme 7.29

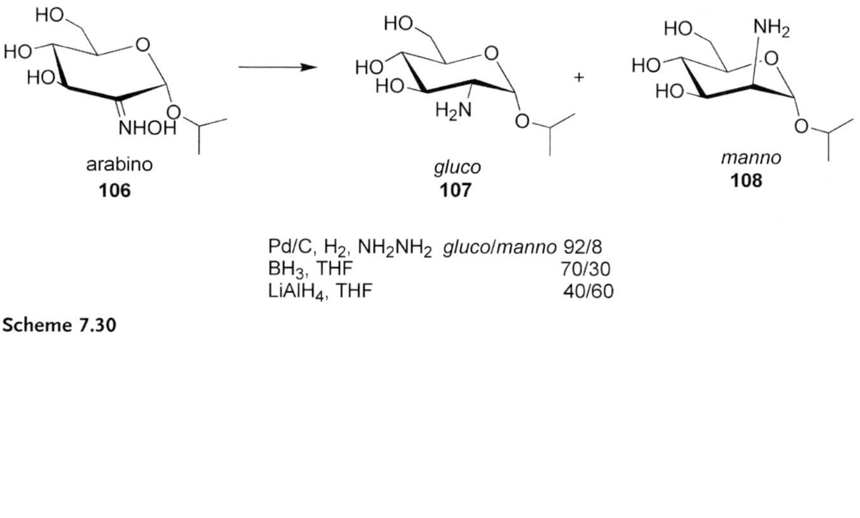

Scheme 7.30

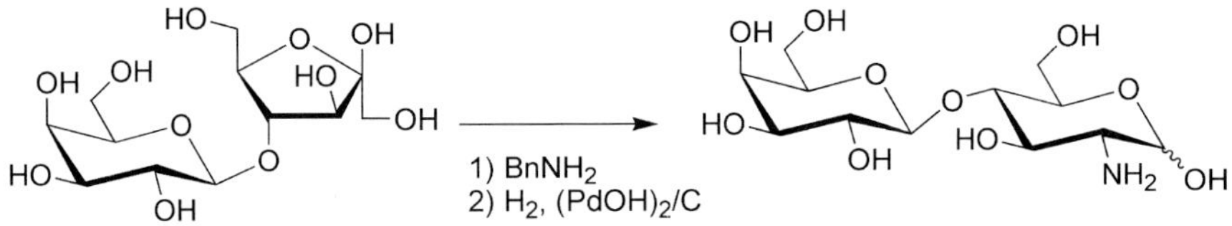

Scheme 7.31

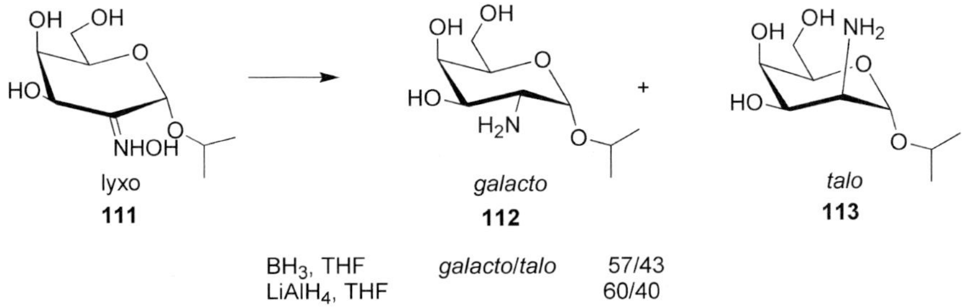

Scheme 7.32

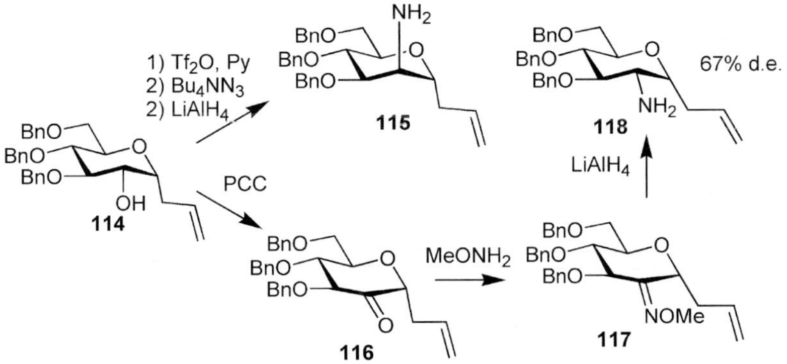

Scheme 7.33

The reductive amination procedure has also been applied in the synthesis of 2-amino-2-deoxy-*C*-glycosides, and in some cases it is complementary and synergic with the nucleophilic displacement method. From the *C*-allyl α-D-glucopyranoside **114**, selectively deprotected at C-2, it is possible to generate either the mannosamino derivative **115**, as shown in Scheme 7.33, through an S_N2 displacement of the triflate, or alternatively the glycosamino derivative **118** by exploiting the reductive amination procedure. The $LiAlH_4$ reducing agent attacks the methoxyimine **117** from the less hindered β-face, generating an equatorial amine **118** (D-*gluco*) in 67% diastereomeric excess.

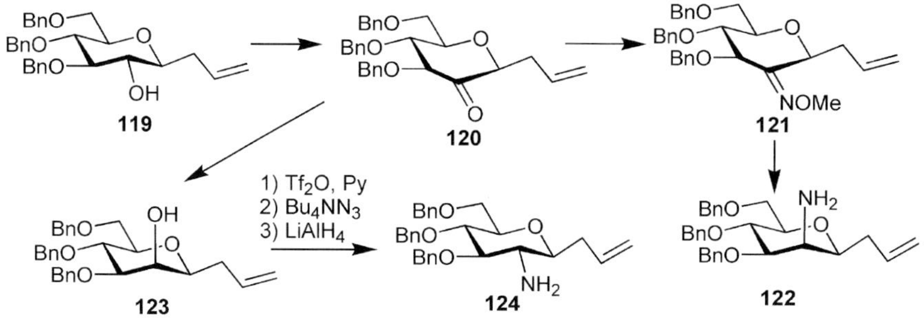

Scheme 7.34

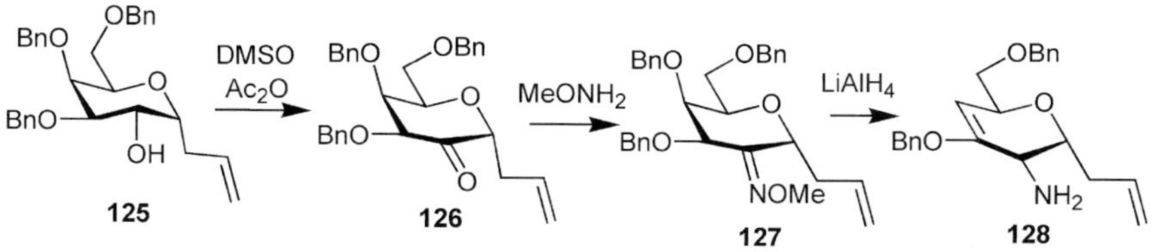

Scheme 7.35

Also interesting is the synergy in the use of the two procedures for the synthesis of the aminated *C*-allyl β-D-glycopyranosides shown in Scheme 7.34. By starting from the ketone **120**, and by taking into account that the reducing agent preferentially attacks from the α-face, due to the β-orientation of the allyl group, reductive amination via the methyloxime **121** afforded the mannosamino derivative **122**, whereas reduction of the keto function of **120** followed by nucleophilic displacement of the obtained axial hydroxy group of **123** stereoselectively afforded the glucosamino derivative **124**.

The synthesis of the corresponding *C*-glycoside of *N*-acetyl-α-D-galactosamine, which is particularly important for the generation of stable mimics of the Tn tumor-associated antigen (GalNAc-α-Ser/Thr), was particularly problematic. In this case the reduction of the methyloxime from the β-face was prevented with the aid of a benzyloxy group present at C-4, and under drastic conditions with $LiAlH_4$ the elimination product was obtained (Scheme 7.35) [85].

In order to obtain the desired product, it was necessary to synthesize the glucosamino derivative and then to invert the stereochemistry at C-4 through displacement of a triflate by intermolecular attack of the pivaloyl group at C-3 (Scheme 7.36).

7.2.3 Amination of Glycals

Azidonitration of glycals is an old reaction that allows a nitrogen to be inserted at C-2 in a sugar ring. The reaction was first performed on galactal with cerium

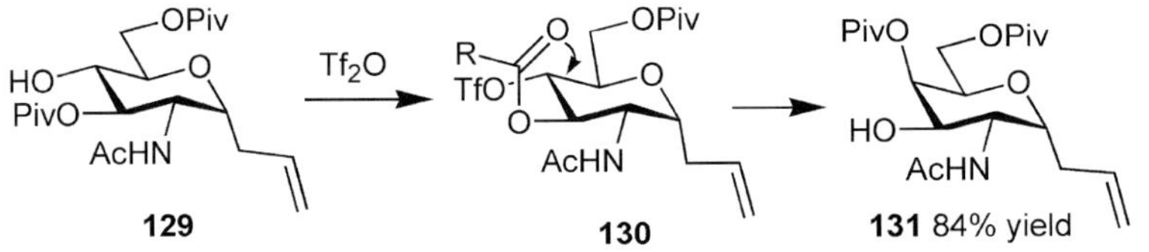

Scheme 7.36

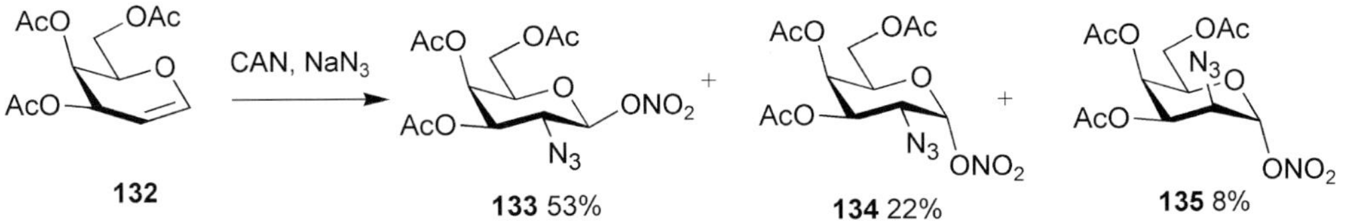

Scheme 7.37

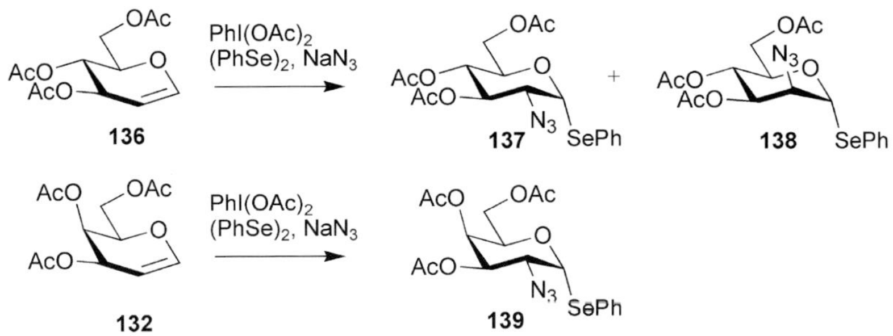

Scheme 7.38

ammonium nitrate and sodium azide, and resulted in the insertion of a nitrate at the anomeric center and the azide at the adjacent position (Scheme 7.37). Galactal **132** afforded mainly the two azides of the *galacto* series (53% β and 22% α), together with traces of the α-*talo* isomer [86].

Amination of glycals can also be performed with (diacetoxyiodo)benzene, phenylselenyl chloride, and sodium azide [87]. Under these conditions tri-*O*-acetyl-D-glucal **136** (Scheme 7.38) afforded a 60:40 mixture of the α-*gluco* and α-*manno* azides, whereas the tri-*O*-acetyl-D-galactal afforded the α-*galacto* compound as the sole product.

The insertion of nitrogen at the double bonds in glycals has also been performed by transition metal-promoted formation of aziridines [88]. The reaction was performed on different glycals, affording the corresponding 2-aminosugars, as trifluoroacetates, in acceptable yields and with high stereoselection (Scheme 7.39).

Another interesting methodology for the conversion of glycals into amino sugars is based on the [4 + 2] cycloaddition of azodicarboxylate reported by Leblanc et al. [89]. The glycal in Scheme 7.40, for example, was irradiated (350nm) for 18 h with

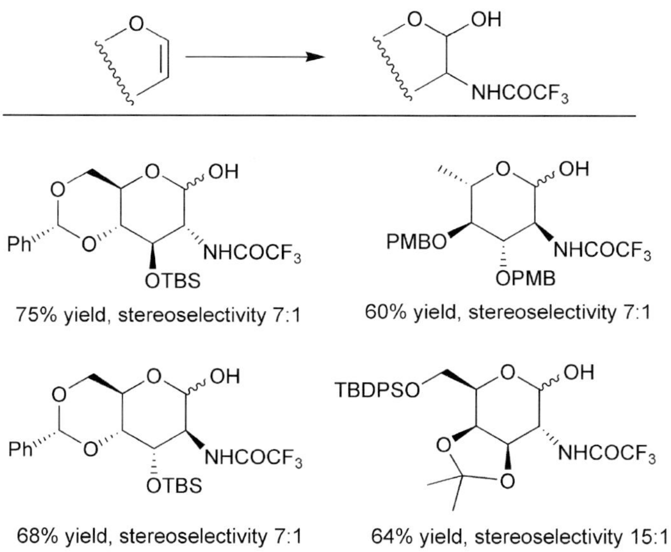

Scheme 7.39

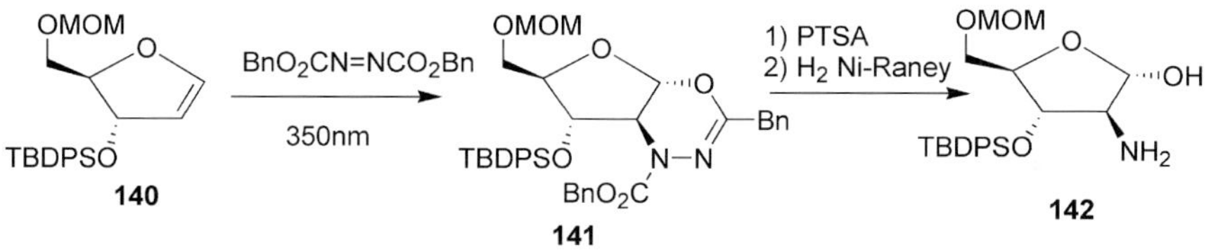

Scheme 7.40

dibenzyl azodicarboxylate in cyclohexane/dichloromethane, affording the single cycloaddition product in 73% yield. The diastereofacial selectivity of the cycloaddition is apparently controlled by the stereochemistry at C-3. Hydrolysis of the cycloadduct with *p*-toluenesulfonic acid, followed by hydrogenation with Raney nickel, generates the amino sugar.

Some procedures for amination at C-2 in glycals are based on insertion of a nitrogen-containing substituent at the adjacent position, followed by an intramolecular rearrangement.

One example of this strategy is shown in Scheme 7.41 [90, 91]. Treatment of tri-*O*-benzyl-glucal with benzenesulfonamide and iodonium di-*syn*-collidine perchlorate (IDCP) generated the α-iodo-sulfonamide **144**. This compound easily underwent displacement of the iodide by intramolecular attack of the nitrogen and formation of an aziridine intermediate that was capable of affording either the 2-sulfonamido-2-deoxy-glucopyranose **146** by nucleophilic attack of water, or eventually a disaccharide through reaction with a glycosyl donor in the presence of lithium tetramethylpiperidine and silver triflate.

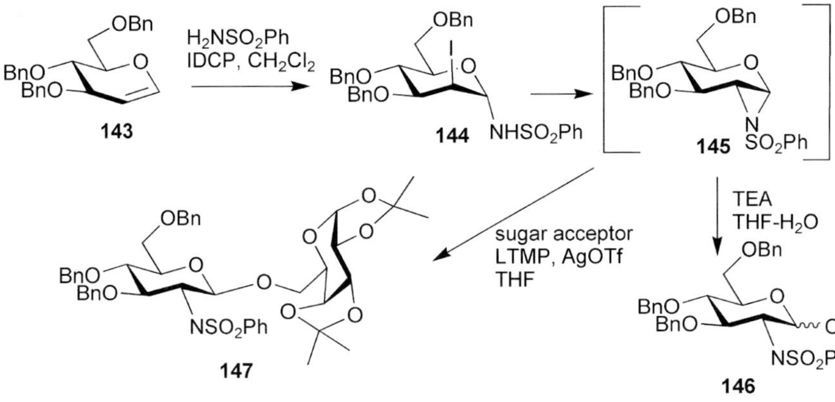

Scheme 7.41

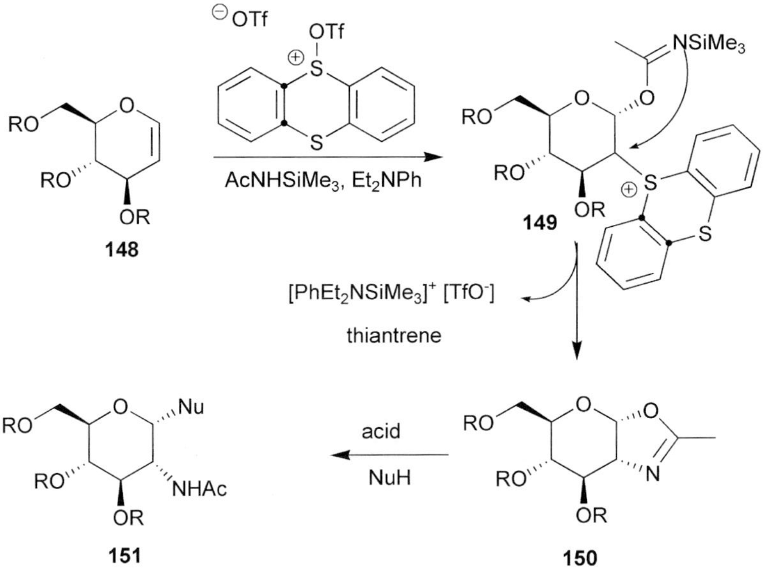

Scheme 7.42

A second example is based on treatment of glycals with thianthrene 5-oxide, Tf_2O, $AcNHSiMe_3$, and Et_2NPh, which results in the insertion of an acetamido group at C-2 by the mechanism shown in Scheme 7.42 [92]. Interestingly, the acid-mediated ring-opening of the intermediate oxazoline allows the insertion of different alcohols, including glycosyl acceptors, at the anomeric center. The whole process therefore represents an innovative method for the generation of glycosides of 2-acetamido-2-deoxy-sugars **151**, including disaccharides.

A third process that allows glycals to be aminated at C-2 in a stereoselective fashion through intramolecular substitution is shown in Scheme 7.43 [93]. The

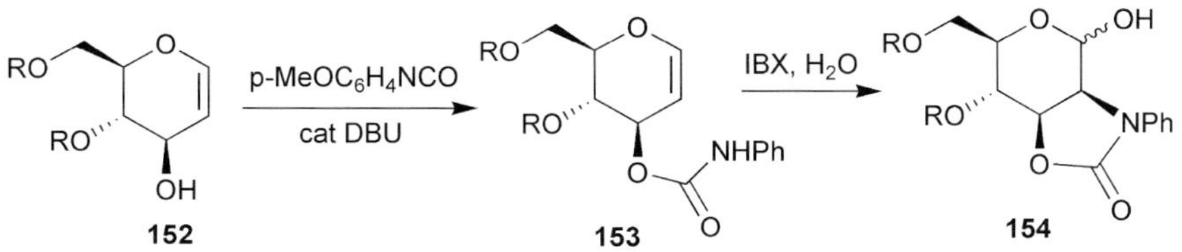

Scheme 7.43

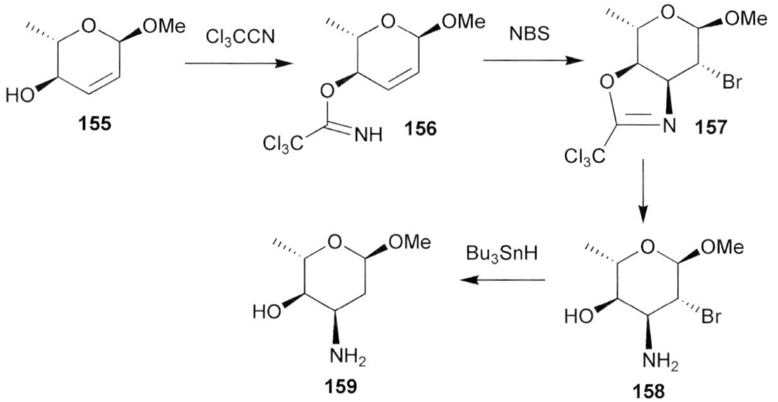

Scheme 7.44

free allylic hydroxy group is treated with an isocyanate in order to obtain a glycal carbamate, which undergoes a single electron transfer process on treatment with IBX, with formation of the cyclic oxazolidinone **154**.

Intramolecular amination has also been applied to unsaturated sugars other than glycals. Cardillo et al. [94] reported an intramolecular amination of a 2,3-unsaturated sugar involving the insertion of an amino group at C-3 in 1983. Treatment of the unsaturated sugar **155** (Scheme 7.44) with trichloroacetonitrile afforded the trichloroacetimidate **156**, the nitrogen of which, once the double bond has been activated with *N*-bromosuccinimide (NBS), entered into an intramolecular cyclization to afford the oxazoline **157**, which spontaneously generated the amino sugar, which was in turn dehalogenated with Bu_3SnH to afford methyl L-ritrosaminide.

7.2.4
Amination through Ring-Opening of Epoxides

Ring-opening of sugar epoxides with azide provides another route to generate amino sugars. The stereochemistry of the reaction occurs with inversion of the configuration, consistently with the S_N2 mechanism of nucleophilic epoxide ring-opening, whereas the regiochemistry in general favors the *trans*-diaxial product. The two isomeric glycopyranoside epoxides shown in Scheme 7.45, for example,

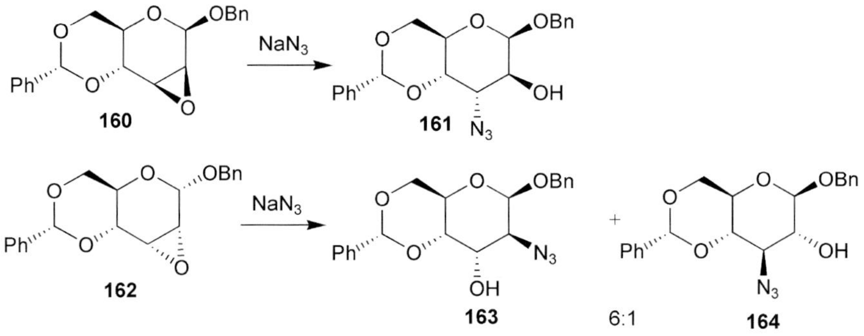

Scheme 7.45

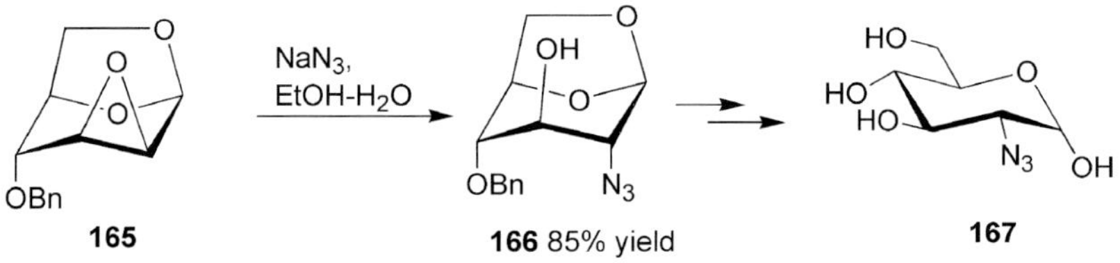

Scheme 7.46

underwent ring-opening with opposite regiochemistries to afford in the first case the *trans*-diaxial azido alcohol as the sole product and in the second case its regioisomer as the predominant (6:1) product [95].

In order to generate amino sugars of interest with diequatorial substituents, such as glucosamine, the trick of inverting the conformation by use of a 1,6-bridge (1,6-anhydrosugars) was exploited (Scheme 7.46) [96, 97].

7.3 Synthesis of Iminosugars

Thanks to the structural analogy between iminosugars and commercially available carbohydrates, these have represented the starting materials of choice for a great number of synthesis. As a matter of fact, sugars are not only rich in hydroxy groups, but these are present in many combination of stereochemical disposition, according to the sugar, so that they represent a pool of highly chiral compounds. The methodologies for iminosugar synthesis have been exhaustively reviewed both by us [98], and by others [99]; this section focuses on the introduction of the amino group into a carbohydrate skeleton and its manipulation in order to generate iminosugars.

The synthetic steps necessary for the conversion of sugars into iminosugars are essentially the introduction of the amino function (as already described above) and

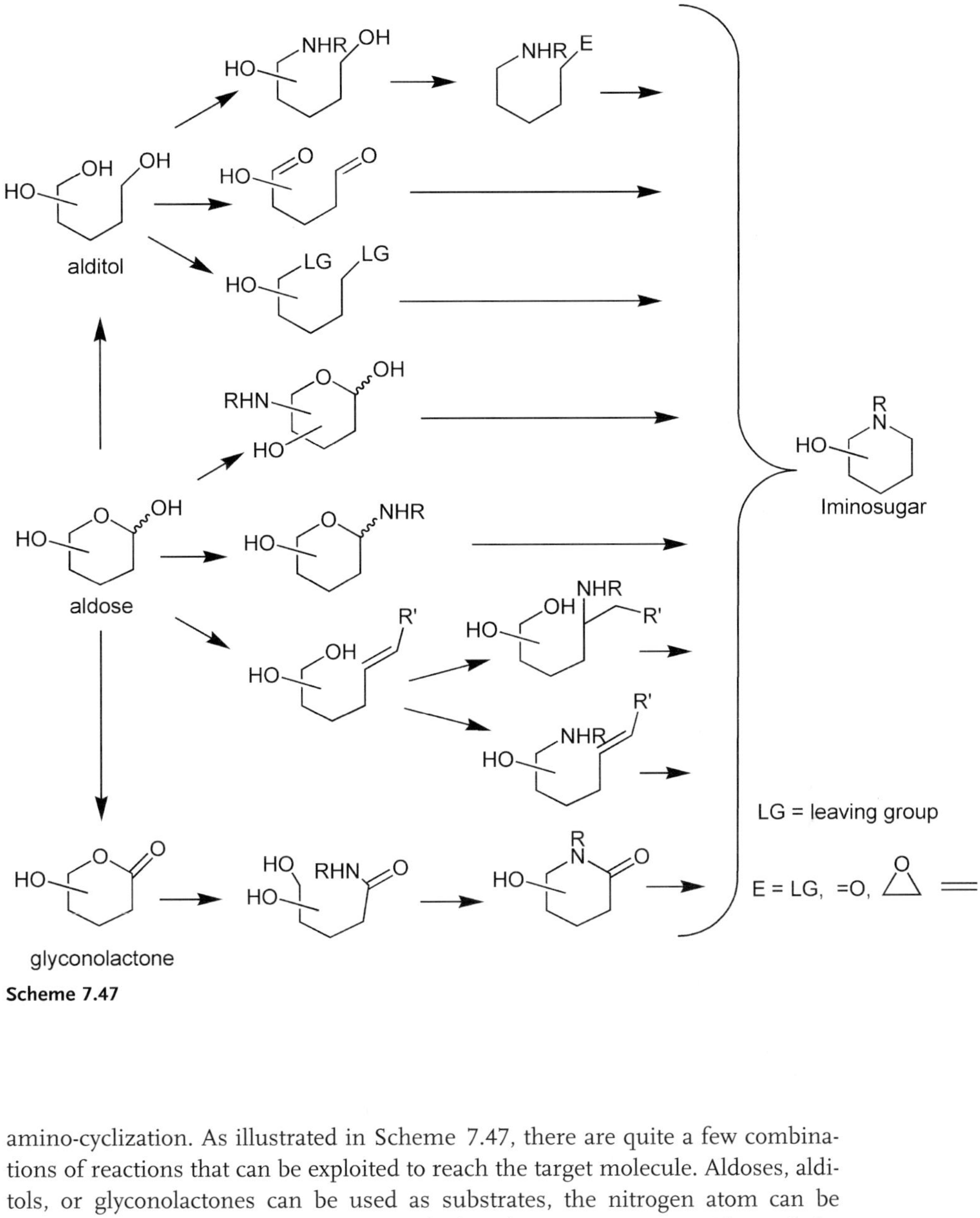

Scheme 7.47

amino-cyclization. As illustrated in Scheme 7.47, there are quite a few combinations of reactions that can be exploited to reach the target molecule. Aldoses, alditols, or glyconolactones can be used as substrates, the nitrogen atom can be introduced variously as an amine, as a hydroxylamine, or as an azide, and the amino-cyclization can be performed by exploiting an intramolecular nucleophilic displacement, a reductive amination, an epoxide ring-opening, or a nucleophilic attack on an activated double bond. Although the scheme is very general, it does not summarize all the strategies reported in the literature.

7.3.1
Amination at the Anomeric center with Subsequent Cyclization

7.3.1.1 Exploitation of the Reactivity of the Carbonyl Function

The hemiacetal functionality of aldoses suggests the easy introduction of the amino group as discussed in Section 7.2.1. Commercial 2,3,4,6-tetra-*O*-benzyl-D-glucopyranose (**168**; Scheme 7.48), for example, has been easily converted to the emiaminal **169** [100], which exists in equilibrium with the corresponding open chain imine **170**, by treatment with a primary amine. Treatment with Grignard reagents afforded an intermediate amino alcohol **171** with a free hydroxy group that could be exploited for the cyclization. In this example the cyclization was achieved through an intramolecular reductive amination carried out on the oxidized compound **172**. This method afforded as main product compound **173**, with a retained configuration at the CH_2OBn group with respect to the parent sugar.

A similar approach was reported for the synthesis of the pyrrolidine iminosugar **177** (Scheme 7.49) from 2,3,5-tri-*O*-benzyl-D-arabinose (**174**) [101]. In this case the cyclization was carried out by treatment with trifluoroacetic anhydride, thus transforming the free hydroxy group into a good leaving group, which was directly displaced by the secondary amine. In this second example the cyclization afforded

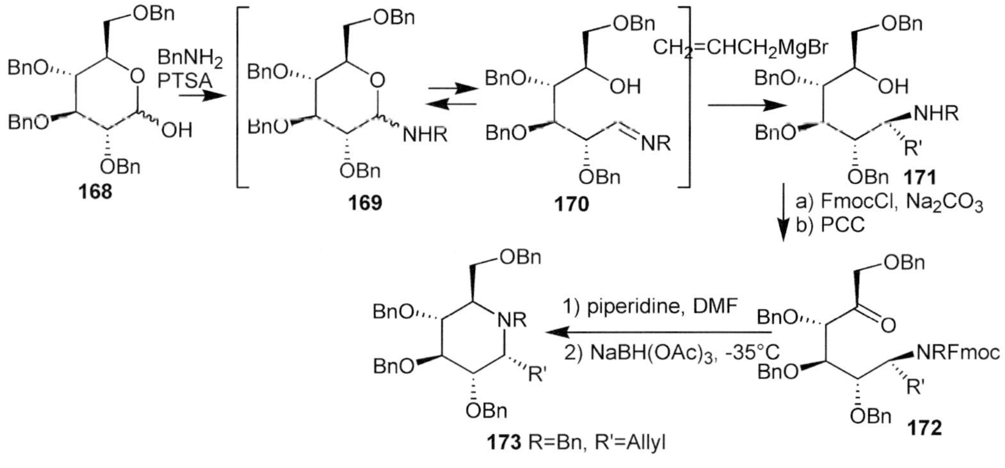

Scheme 7.48

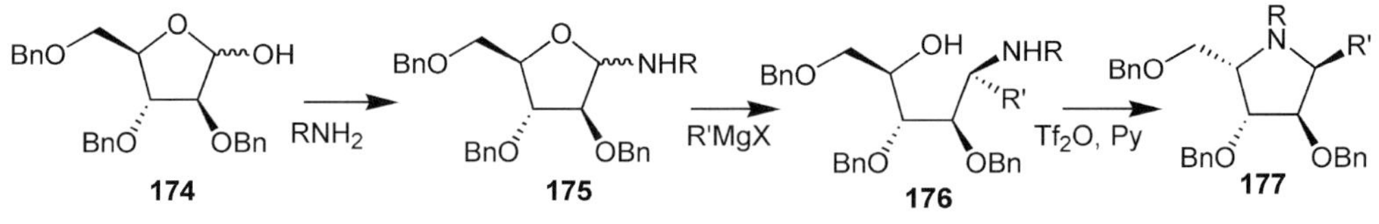

Scheme 7.49

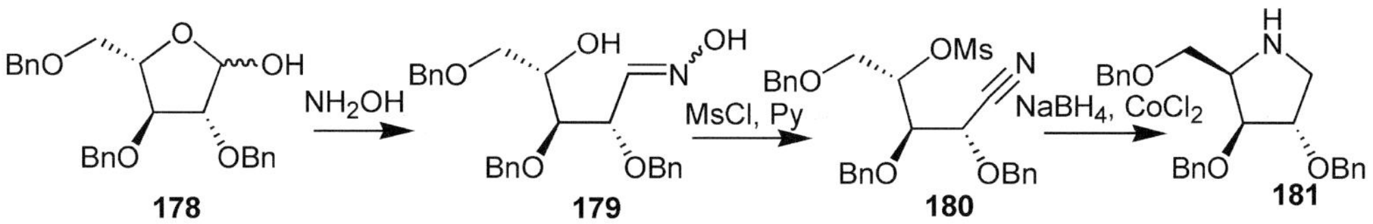

Scheme 7.50

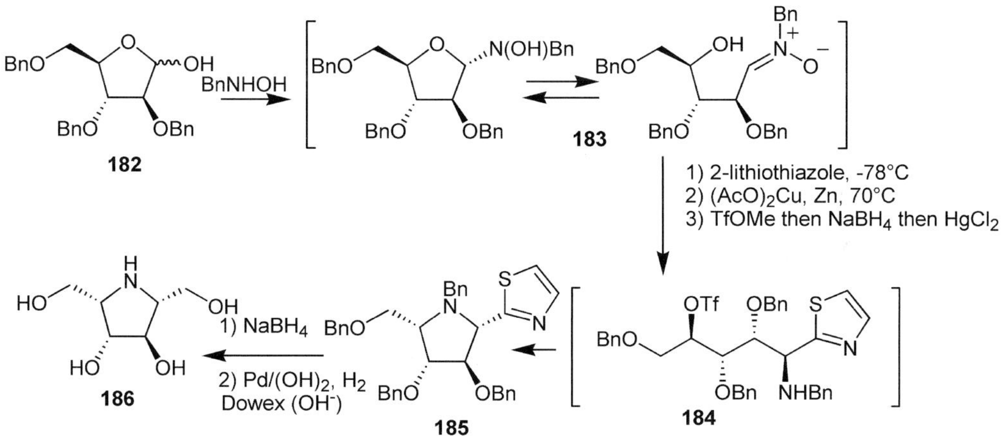

Scheme 7.51

the final iminosugar with inversion of configuration at the cyclization center, with respect to the parent sugar.

The anomeric center has also been exploited for the introduction of the nitrogen atom by treatment with hydroxylamines; two examples are described in Schemes 7.50 and 7.51. Treatment of 2,3,5-tri-*O*-benzyl-L-arabinose (**178**) with hydroxylamine [102] (Scheme 7.50) afforded oxime **179**. This compound was treated with mesyl chloride, thus converting the free hydroxy group into a leaving group and, at the same time, transforming the oxime into a nitrile. Selective reduction of the nitrile then afforded the amine, which directly cyclized with inversion of configuration at C4.

Treatment of benzylated D-arabinose **182** with an alkylated hydroxylamine such as *N*-benzylhydroxylamine afforded a nitrone intermediate **183** (Scheme 7.51) [103]. Treatment of **183** with 2-lithiothiazole afforded the open chain thiazole-amine, which was converted into the corresponding triflate **184**. The displacement of the triflate by the amine afforded the cyclized compound **185**, which was then converted to the final iminosugar **186**.

7.3.1.2 Exploitation of the Reactivity of Lactones

As illustrated in the summarizing Scheme 7.47, glyconolactones have been used for the synthesis of iminosugars through their conversion into the corresponding

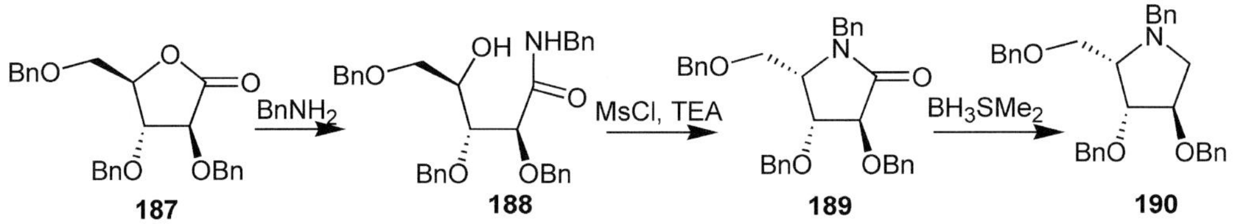

Scheme 7.52

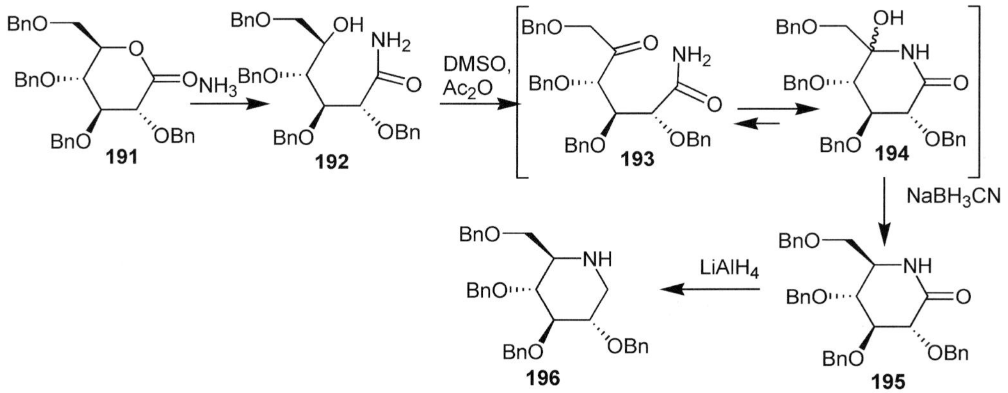

Scheme 7.53

open-chain amides, followed by cyclization and lactam reduction. The following Schemes 7.52 and 7.53 show two examples of such strategies. In the first case, 2,3,5-tri-*O*-benzyl-D-arabinolactone (**187**) [104] was converted into the amide **188** by treatment with benzylamine. The amide was then transformed into lactam **189** by mesylation of the free hydroxy group followed by intramolecular displacement. Final lactam reduction afforded the desired iminosugar **190**.

In the second example [105] a different strategy was used in the lactam formation, to retain the stereochemical configuration on C5. In this case the free hydroxy group in amide **192** was oxidized to the corresponding ketone, which was attacked by the amidic nitrogen to afford compound **194**. Reduction with $LiAlH_4$ gave protected DNJ **196**.

7.3.1.3 Insertion of a New Electrophile

The carbonyl group at the anomeric center can be exploited to introduce a different functional group, which can undergo nucleophilic attack by an amine. An example is shown in Scheme 7.54 [106], in which treatment of 2,3,4,6-tetra-*O*-benzyl-D-glucopyranose (**197**) with $Ph_3P{=}CHCO_2R$ afforded the α,β-unsaturated ester **198**, which underwent a Michael addition on treatment with a primary amine, affording compound **199** as a mixture of diastereoisomers. Each separated diastereoisomer was converted into ketone **200**, which was exploited in the cyclization achieved through an intramolecular reductive amination, again affording a mixture of

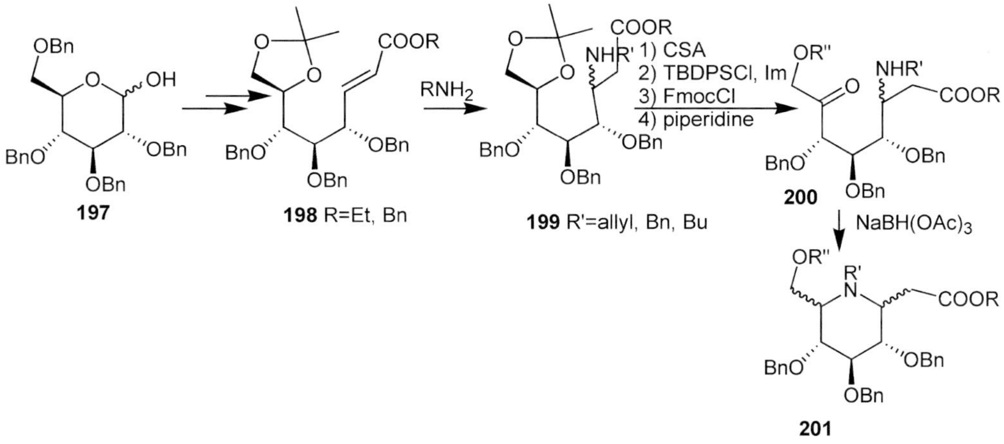

Scheme 7.54

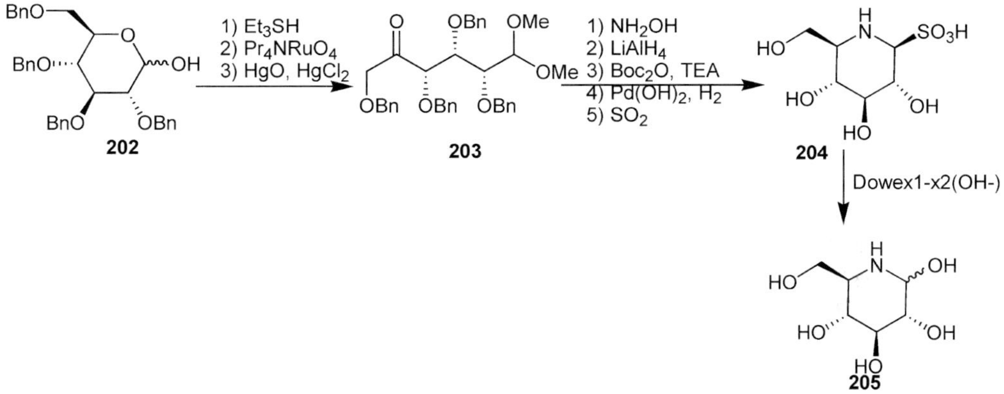

Scheme 7.55

diastereoisomers. Since this strategy was designed for the preparation of a library of compounds, the lack of stereoselectivity in both the Michael addition and the reductive amination allowed the introduction of sources of diversity on the two carbon atoms next to the ring nitrogen.

7.3.2
Amination at the Carbohydrate Chain and Subsequent Cyclization

Another approach for the synthesis of an iminosugar from a carbohydrate starting material consists of the introduction of an amino function by chain amination and subsequent cyclization on the anomeric carbon. This approach requires a preliminary manipulation of the starting sugar for temporary protection or masking of the anomeric function. Schemes 7.55 and 7.56 show two applications of such an

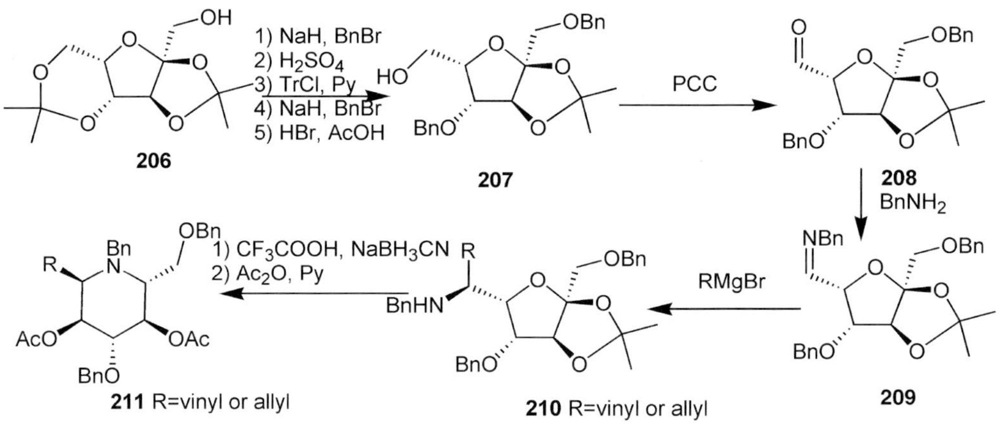

Scheme 7.56

approach. 2,3,4,6-Tetra-*O*-benzyl-D-glucopyranose (**202**) [107] was converted into ketone **203** by conversion of the aldehyde of the parent sugar into a dimethyl acetal and oxidation of the C5 hydroxy group. The carbonyl group was used for the introduction of the nitrogen atom, which was achieved through a reductive amination with hydroxylamine. Reduction of the hydroxylamine and hydrolysis of the acetal afforded the cyclized product, which was completely deprotected and treated with sulfur dioxide to yield the sulfonic acid **205**, a direct precursor of the final (+)-nojirimycin.

Scheme 7.56 shows a similar approach for the synthesis of C glycosides of DNJ [108]. Starting from protected sorbose **206**, a few protecting group manipulations afforded alcohol **207**, which was used for amination via intermediate oxidation. In this case the intermediate imine was not reduced but was treated with different Grignard reagents in order to generate the amine **210**. Finally, regeneration of the anomeric carbonyl function, followed by intramolecular reductive amination, afforded the final products **211**.

An example of a synthesis of an iminosugar by chain amination, in which the anomeric carbonyl group of the starting sugar is not involved in the cyclization step, but is exploited for a chain elongation, is shown in Scheme 7.57. L-Arabinose [109] was converted into the selectively protected intermediate **213**, in which the sugar carbonyl is in the thioketal form. The amino function was introduced as a phthalimido group under Mitzunobu conditions. Regeneration of the aldehyde was followed by chain elongation by treatment with a chiral allylborane. Alcohol **215** was then mesylated in order to form a first iminosugar ring **216** after phthalimide deprotection, and final hydroboration-oxidation of the double bond of **216** and further mesylation and amine displacement gave rise to the second ring of the target molecule with an indolizidine structure.

In the example reported in Scheme 7.58 [110], developed for the synthesis of α- and β-homogalactonojirimycin **222** and **225**, the olefin introduced on the anomeric center is activated to react intramolecularly with an amino group present in

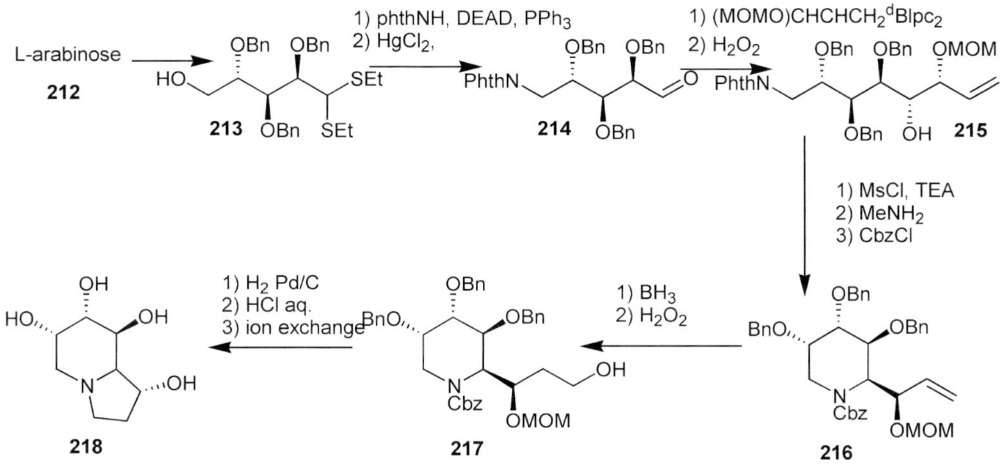

Scheme 7.57

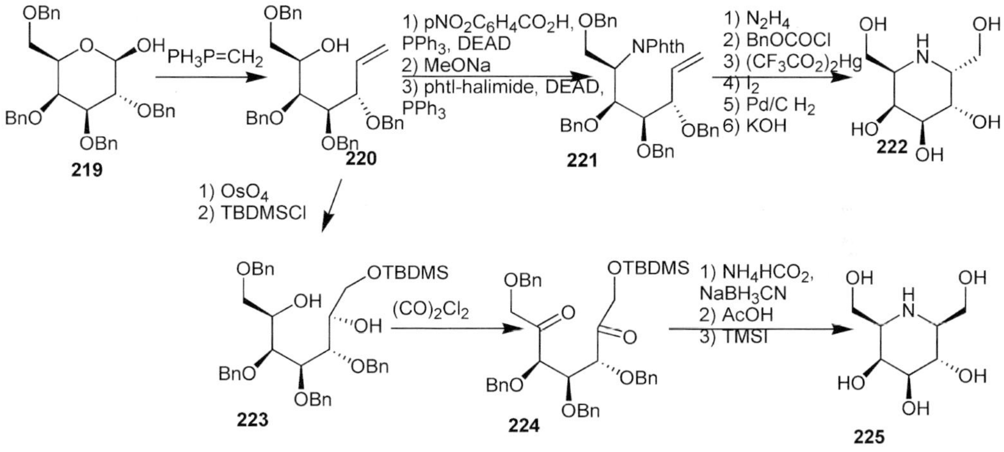

Scheme 7.58

the chain. 2,3,4,6-Tetra-*O*-benzyl-β-D-galactopyranose (**219**) was treated with methylenetriphenylphosphorane to generate olefin **220**, and the amino function was then introduced as a phthalimide with double inversion of the free hydroxy group. In this example, activation of the double bond for the cyclization step was carried out with mercuric salts, and the intermediate mercurial compound was converted into the alcohol.

An alternative activation of the double bond is its conversion into an epoxide. This method allows cyclization by intramolecular epoxide opening by an amino group and concomitant hydroxylation and was adopted for the synthesis of the pyrrolidine derivative compound **226** [111] (Scheme 7.59).

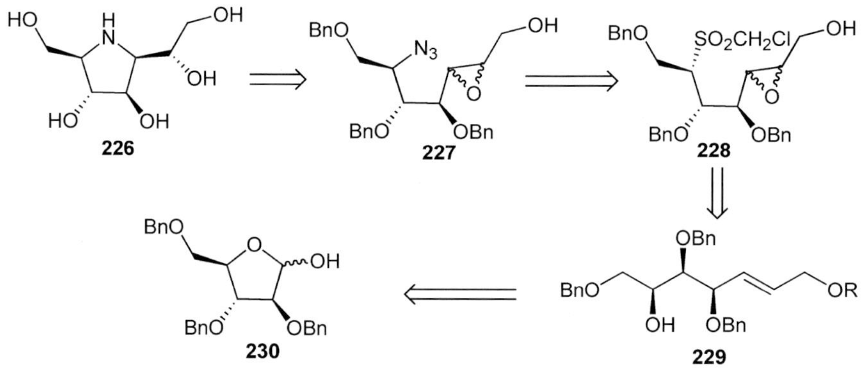

Scheme 7.59 Retrosynthesis of a pyrrolidine by epoxide ring-opening

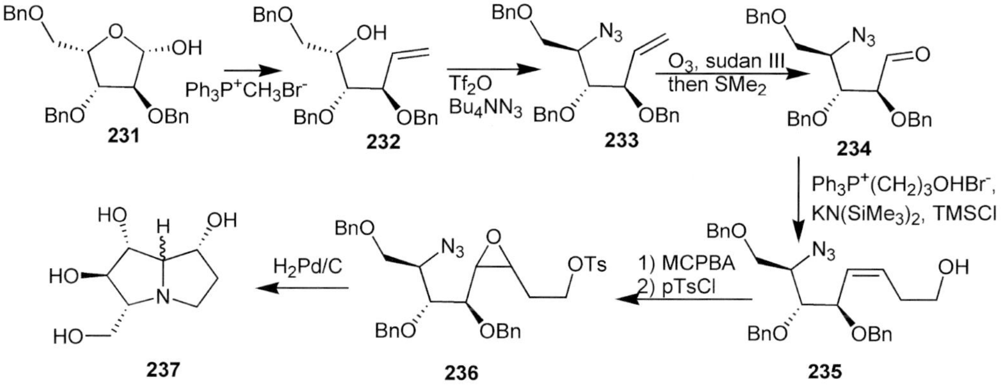

Scheme 7.60

As illustrated in the retrosynthetic Scheme 7.59, a Wittig reaction at the anomeric center generates the olefin derivative with a free hydroxy group. The nitrogen atom is introduced in the form of an azido group with a double inversion of configuration, and a Sharpless epoxidation then generates the epoxide, which is opened by an intramolecular nucleophilic attack of the amino group obtained by reduction of the azide.

Epoxide opening was used for the synthesis of a bicyclic pyrrolizidine iminosugar **237** [112], shown in Scheme 7.60. In this case a first Wittig reaction afforded olefin **232** with a free hydroxy group, which was once more exploited for the introduction of an azido group with inversion of configuration. The double bond was converted into an aldehyde, which was subjected to a second Wittig reaction to afford olefin **235**. Epoxidation of **235** gave the azido-epoxy-tosylate **236**, which upon reduction of the azido group afforded the bicyclic structure through a simultaneous epoxide ring-opening and tosylate displacement by the formed amino function.

7.3.3
Concomitant Insertion of Nitrogen at Both Carbon Atoms

A great number of synthesis of iminosugars start from alditols, sometimes generated by reduction of aldoses. The conversion of alditols into iminosugars requires the reaction of two hydroxy groups with a single nitrogen nucleophile with generation of a cycle. This can be performed step by step, by subsequent amination of the two positions, or at the same time, in a concerted reaction. A simultaneous amination approach requires the formation of intermediate diketone, dimesylate (ditosylate), disulfate, or dihalogen compounds. A diketone intermediate was generated for a straightforward synthesis of DNJ (**240**) and of its *N*-butyl derivative **241** [113] (Scheme 7.61).

Reduction of 2,3,4,6-tetra-*O*-benzyl-D-glucopyranose (**238**) followed by oxidation afforded intermediate diketone **239**, which was subjected to a double reductive amination to generate the target iminosugars. Formation of a diketone intermediate was also adopted for the synthesis of β-homogalactonojirimicyn **225** (illustrated in Scheme 7.58 above). Another example involves the synthesis of iminosugar **245** [114] (Scheme 7.62), obtained from mannitol, which was converted into 1:2,3:4-di-O-benzylidene-D-mannitol (**242**) and then oxidized to the diketone, which exists in the hydrated cyclic diketal form **243**. Reductive amination and deprotection converted **243** into the final iminosugar **245**.

An example of a dimesylate activation is shown in Scheme 7.63. As in the previous case, the starting alditol **246** was first converted into the protected dibenzylidene derivative **247**, which was activated as a dimesylate and subjected to a double displacement by an amine to yield the cyclic iminosugar **249** [115].

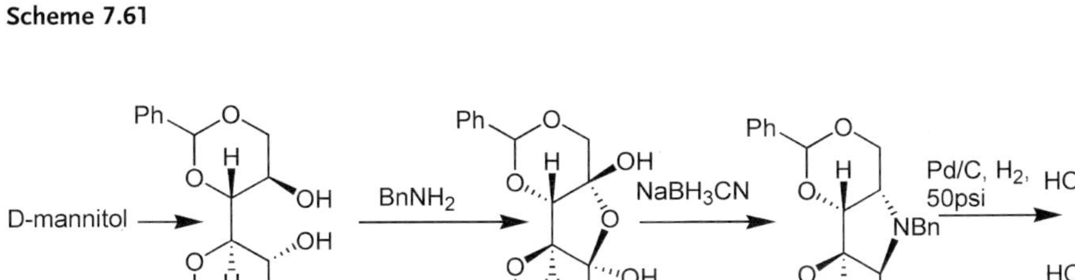

Scheme 7.61

Scheme 7.62

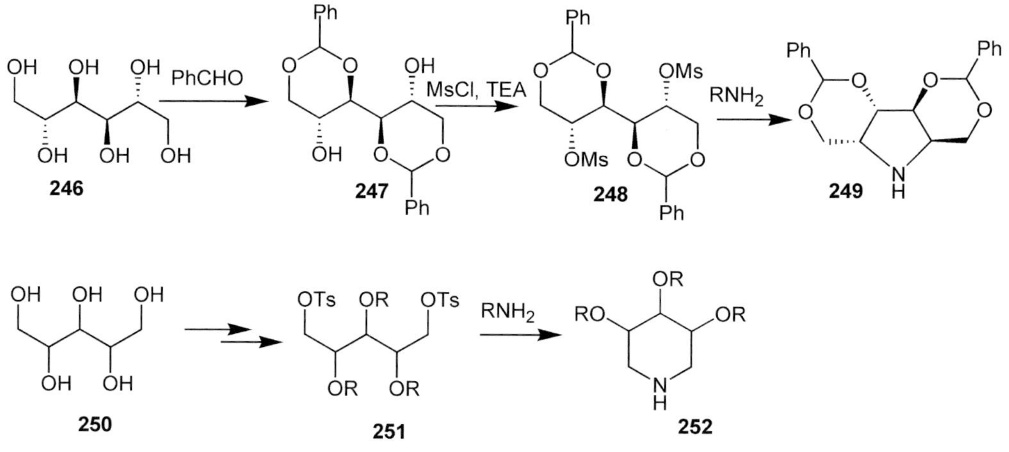

Scheme 7.63

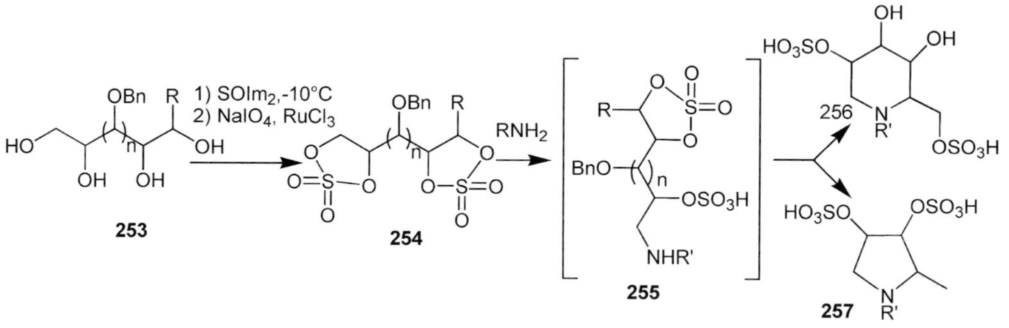

Scheme 7.64

In a similar way, D-xylitol, L-arabinitol, and D-arabinitol (general structure **250**; Scheme 7.63) were also converted into iminosugars. In these cases the alditols were selectively protected and then converted into the corresponding ditosylates **251**. Treatment with primary amines afforded the corresponding iminosugars **252** by displacement of the tosyl groups [116].

Alditol activation can also be achieved through cyclic sulfates, as shown for compound **254** (Scheme 7.64).

In this case a primary amine first reacted with one of the sulfates to afford intermediate **255**, and intramolecular attack on the second sulfate then afforded the two products **256** and **257**, with a piperidine and a pyrrolidine structure, respectively [117]. The cyclic sulfate **259** [118] (Scheme 7.65) was also used in a two-step strategy for the preparation of iminosugar **261**.

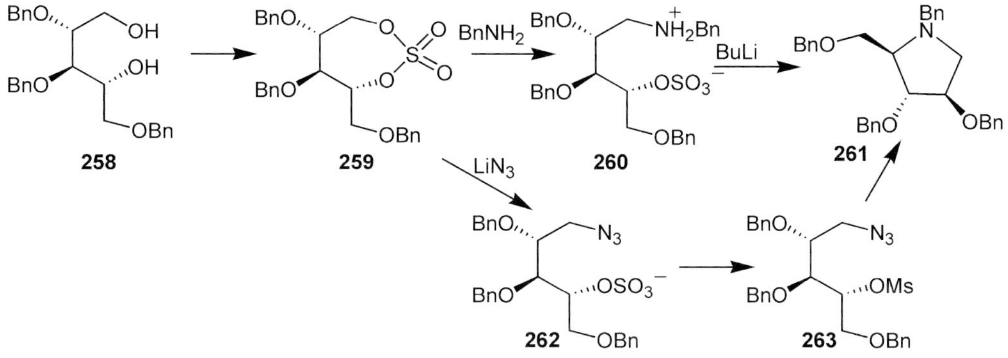

Scheme 7.65

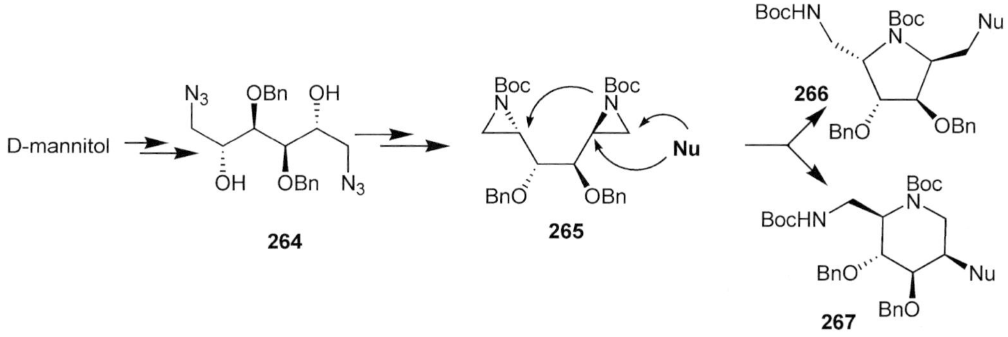

Scheme 7.66 Synthesis of iminosugars via a bisaziridine intermediate.

The cyclic sulfate was treated either with a primary amine or with an azide, affording the sulfate intermediates **260** and **262**, respectively. Treatment of **260** with BuLi directly afforded iminosugar **261** by amine displacement of the sulfate, while the sulfate group of **262** was first converted into the mesylate and then displaced by the amine formed from azide reduction.

Another proposed strategy that starts from alditols is based on the formation of bisaziridine intermediates. D-Mannitol was first transformed into the bis-azido compound **264** [119] (Scheme 7.66), which was in turn converted into the bisaziridine derivative **265**. Nucleophilic ring-opening of one aziridine ring afforded an intermediate amine, which reacted with the second aziridine to afford the five-membered or six-membered ring iminosugars.

Alditols can be activated as dihalogenated compounds, treatment of which with amines affords iminosugars. This approach is particularly interesting since it does not require protecting steps; an example is illustrated in Scheme 7.67 [120]. The dihalogenated compound **269**, easily obtained from the corresponding glyconolactone **268**, was directly converted into the iminosugar **270** by treatment with aqueous ammonia.

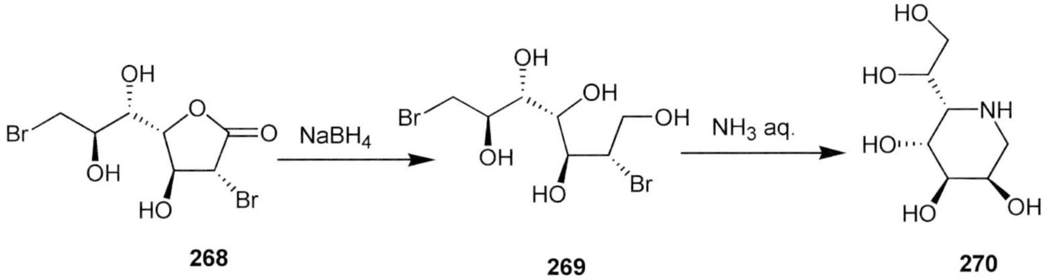

Scheme 7.67

7.4 Conclusions

From the examples reported here, we can conclude that the amino group has a key importance in glyco-structures, tuning the chemico-physical properties and actively participating in molecular recognition phenomena and, in the case of iminosugars, in enzymatic inhibition. Besides the relevance of aminated sugars, the poor accessibility of these compounds requires the development of efficient sugar amination procedures. We have tried to schematize the different synthetic approaches, and our report is certainly not exhaustive of all the examples reported in the literature. In particular, for each synthetic approach, we have arbitrarily chosen only one or a few of the so many examples that have been published. Our intent has been to give a general overview on the topic of why and how to generate aminated sugars.

References

1 V. Ginsburg, P. Robbins, *Biology of Carbohydrates, Vols. 1 and 2*, 1984, Wiley Intersciences, New York.

2 a) R. U. Lemieux, *Chem. Soc. Rev.* 1978, *7*, 423; b) M. H. Yazer, M. L. Olsson, M. M. Palcic, *Transfus Med Rev.* 2006, *20*, 3, 207–217.

3 a) S. Roseman, D. G. Comb, *J. Am. Chem. Soc.* 1958, *80*, 497; b) D. G. Comb, S. Roseman, *J. Am. Chem. Soc.* 1958, *80*, 3166.

4 F. Lehmann, E. Tiralongo, J. Tiralongo, *Cell. Mol. Life Sci.* 2006, *63*, 1331–1354.

5 F.-G. Hanisch, C. Hanski, A. Hasegawa, *Cancer Res.* 1992, *52*, 3138–3144.

6 R. Dwek, *Chem. Soc. Rev.* 1996, *96*, 683–720.

7 W. Dippold, A. Steinborn, K.-H. Meyer zum Buschenfelde, *Environmental Health Perspectives* 1990, *88*, 255–257.

8 G. F. Springer, *Crit. Rev. Oncog.* 1995, *6*, 57–85.

9 O. Kurtenkov, K. Klaamas, L. Miljukhina, *Int. J. Cancer* 1995, *60*, 781–785.

10 G. O. Aspinall, *The Polysaccharides, Vols. 1–3*, 1985, Academic Press, Inc., New York.

11 a) Y. J. Son, J. S. Jang, J. W. Cho, H. Chung, R. W. Park, I. C. Kwon, I. S. Kim, J. Y. Park, S. B. Seo, C. R. Park, S. Y. Jeong, *J. Control. Release* 2003, *91*, 135–155; b) J. H. Park, S. Kwon, M. H. Chung, J. H. Kim, Y. S. Kim, R. W. Park, I. S. Kim, S. B. Seo, I. C. Kwon, S. Y. Jeong, *Biomaterials* 2006, *27*, 119–126.

12 R. Sasisekharan, R. Raman, V. Prabhakar, *Ann. Rev. Biomed. Eng.* 2006, *8*, 181–231.

13 a) X. Lin, *Development* 2004, *131*, 6009–6021; b) N. Perrimon, M. Bernfield, *Nature* 2000, *404*, 725–728.

14 a) B. Casu, M. Guerrini, A. Naggi, M. Perez, G. Torri, D. Ribatti, P. Carminati, G. Giannini, S. Penco, C. Pisano, M. Belleri, M. Rusnati, M. Presta, *Biochemistry*, 2002, *41*, 10519–10528; b) I. Vlodavsky, O. Goldshmidt, E. Zcharia, R. Atzmon, Z. Rangini-Guatta, M. Elkin, T. Peretz, Y. Friedmann, *Semin. Cancer Biol.* 2002, *12*, 121–129.

15 a) D. Carulli, T. Laabs, H. M. Geller, J. W. Fawcett, *Curr. Opin. Neurobiol.* 2005, *15*, 116–120; b) L. A. Cavalcante, J. Garcia-Abre, V. Moura Neto, L. C. Silva, G. Weissmuller, *An. Acad. Bras. Cienc*, 2002, *74*, 691–716; c) M. Inatani, F. Irie, A. S. Plump, M. Tessier-Lavigne, Y. Yamaguchi, *Science* 2003, *302*, 1044–1046.

16 a) R. Sasisekharan, Z. Shriver, G. Venkataraman, U. Narayanasami, *Nat. Rev. Cancer 2* 2002, 521–528; b) G. W. Yip, M. Smollich, M. Gotte, *Mol. Cancer Ther.* 2006, *5*, 2139–2148.

17 a) E. E. Fry, S. M. Lea, T. Jackson, J. W. Newman, F. M. Ellard, W. E. Blakemore, R. Abu-Ghazaleh, A. Samuel, A. M. Q. King, D. I. Stuart, *EMBO J.* 1999, *18*, 543–554; b) V. K. Ganesh, S. A. Smith, G. J. Kotwal, K. H. Murthy, *Acad. Sci. USA* 2004, *101*, 8924–8929.

18 a) B. Casu, M. Guerrini, G. Torri, *Curr. Pharm. Des.* 2004, *10*, 939–949; b) M. Sundaram, Y. Qi, Z. Shriver, D. Liu, G. Zhao, G. Venkataraman, R. Langer, R. Sasisekharan, *Acad. Sci. USA* 2003, *100*, 651–656.

19 K. R. Taylor, R. L. Gallo, *FASEB J.* 2006, *20*, 9–22.

20 a) A. Naggi, B. De Cristofano, A. Bisio, G. Torri, B. Casu, *Carbohydr. Res.* 2001, *336*, 283–290; b) A. Naggi, G. Torri, B. Casu, O. P, Zoppetti, J. P. Li, U. Lindahl, *Semin. Thromb. Hemost.* 2001, *27*, 437–443.

21 T. D. Heightman, A.T. Vasella, *Chem. Int. Ed.* 1999, *38*, 750–770.

22 a) L. Ratner, N. V. Heyden, D. Dedera, *Virology* 1991, *181*, 180–192 and references cited therein; b) F. Chery, L. Cronin, J. L. O'Brien, P. V. Murphy, *Tetrahedron* 2004, *60*, 6597–6608; c) P. Greimel, J. Spreitz, A. E. Stütz, T. M. Wrodnigg, *Curr. Topics Med. Chem* 2003, *3*, 513–523; d) J. Alper, *Science* 2001, *291*, 2338–2343; e) D. Pavlovic, D. C. A. Neville, O. Argaud, B. Blumberg, R. A. Dwek, W. B. Fischer, N. Zitzmann, *Acad. Sci. USA* 2003, *100*, 6104–6108; f) D. Durantel, S. Carrouée-Durantel, N. Branza-Nichita, R. A. Dwek, N. Zitzmann, *Antimicrob. Agents Chemother.* 2004, *48*, 497–504; g) S.-F. Wu, C.-J. Lee, C.-L. Liao, R. A. Dwek, N. Zitzmann, Y-L. Lin, *J. Virol.* 2002, *76*, 3596–3604.

23 a) H. Paulsen, I. Brockhausen, *Glycoconj. J.* 2001, *18*, 867–870; b) P. E. Gross, M. A. Baker, J. P. Carver, J. W. Dennis, *Clin. Cancer Res.* 1995, *1*, 935–944.

24 a) P. B. Anzeveno, L. J. Creemer, J. K. Daniel, C.-H. R. King, P. S. Liu, *J. Org. Chem.* 1989, *54*, 2539–2542; b) J. A. Balfour, D. McTavish, *Drug* 1993, *46*, 1025–1054.

25 a) T. D. Butters, R. A. Dwek, F. M. Platt, *Glycobiol.* 2005, *15*, 43R–56R; b) T. M. Cox, J. M. F. G. Aerts, G. Andria, M. Beck, N. Belmatoug, B. Bembi, R. Chertkoff, S. Vom Dahl, D. Elstein, A. Erickson, M. Giralt, R. Heitner, C. Hollak, M. Hrebicek, S. Lewis, A. Mehta, G. M. Pastores, A. Rolfs, M. C. S. Miranda, A. Zimran, *J. Inherit. Metab. Dis.* 2003, *26*, 513–526; c) T. D. Butters, L. A. G. M van den Broek, G. W. J. Fleet, T. M. Krulle, M. R. Wormald, R. A. Dwek, F. M. Platt, *Tetrahedron: Asymm.* 2000, *11*, 113–124; d) T. D. Butters, R. A. Dwek, F. M. Platt, *Chem. Rev.* 2000, *100*, 4683–4696.

26 K. Tatsuta, S. Miura, S. Ohta, H. Gunji, *J. Antibiot.* 1995, *48*, 286–288; b) T. Granier, N. Panday, A. Vasella, *Helv. Chim. Acta* 1997, *80*, 979–987; c) T. Aoyagi, H. Suda, K. Uotani, F. Kojima, T. Aoyama, K. Horiguchi, M. Hamada, T. Takeuchi, *J. Antibiot.* 1992, *45*, 1404–1408.

27 a) M. I. García-Moreno, P. Díaz Pérez, C. Ortiz Mellet, J. M. García-Fernández, *Chem. Commun.* 2002, *8,* 848–849; b) M. I. García-Moreno, P. Díaz Pérez, C. Ortiz Mellet, J. M. García-Fernández, *J. Org. Chem.* 2003, *68,* 8890–8901.

28 C. Laroche, J.-B. Behr, J. Szymoniak, P. Bertus, C. Schütz, P. Vogel, R. Plantier-Royon, *Biorg. Med. Chem.* 2006, *14,* 4047–4054.

29 M. I. Torres-Sánchez, P. Borrachero, F. Cabrera-Escribano, M. Gómez-Guillén, M. Angulo-Alvarez, J. M Diánez, D. Ma Estrada, A. López-Castro, S. Pérez-Garrido, *Tetrahedron: Asymm.* 2005, *16,* 3897–3907.

30 a) E. Graf von Roedern, E. Lohof, G. Hessler, M. Hoffmann, H. Kessler, *J. Am. Chem. Soc.* 1996, *118,* 10156–10167; b) E. Graf von Roedern, H. Kessler, *Angew. Chem. Int. Ed. Engl.* 1994, *33,* 687–688.

31 K. C. Nicolau, J. I. Trujillo, K. Chibale, *Tetrahedron* 1997, *53,* 8751–8778.

32 S. H. Gellman, *Acc. Chem. Res.* 1998, *31,* 173–180.

33 a) D. D. Long, N. L. Hungerford, M. D. Smith, D. E. A. Brittain, D. G. Marquess, T. D. W. Claridge, G. W. J. Fleet, *Tetrahedron Lett.* 1999, *40,* 2195–2198; b) T. D. W. Claridge, D. D. Long, N. L. Hungerford, R. T. Aplin, M. D. Smith, D. G. Marquess, G. W. J. Fleet, *Tetrahedron Lett.* 1999, *40,* 2199–2202.

34 Neoglycoconjugates, Preparation and Applications, Eds. J. C. Lee, T. R. Lee, Academic Press, San Diego, 1994.

35 P. Wang, T. G. Hill, C. A. Wartchow, M. E. Huston, L. M. Oehler, M. Bradley Smith, M. D. Bednarski, M. R. Callstrom, *J. Am. Chem. Soc.* 1992, *60,* 2216–2226.

36 C. A. Wartchow, P. Wang, M. D. Bednarski, M. R. Callstrom, *J.Org. Chem.* 1995, *60,* 2216–2226.

37 G. Molema, D. K. F. Meijer, *Adv. Drug. Delivery Rev.* 1994, *14,* 25–50.

38 F. Peri, P. Dumy, M. Mutter, *Tetrahedron* 1998, *54,* 12269–12278.

39 F. Peri, A. Deutman, B. La Ferla, F. Nicotra, *Chem. Commun.* 2002, 1504–1505.

40 F. Peri, F. Nicotra, *Chem. Commun.* 2004, 623–627.

41 F. Peri, C. Marinzi, M. Barath, F. Granucci, M. Urbano, F. Nicotra, *Bioorg Med Chem.* 2006, *14,* 190–199.

42 M. Carcano, F. Nicotra, L. Panza, G. Russo, *Chem. Commun.* 1989, 297–298.

43 T. W. Brandstetter, C. de la Fuente, Y. Kim, R. L. Cooper, D. J. Watkin, N. G. Oikonomakos, L. N. Johnson, G. W. J. Fleet, *Tetrahedron* 1996, *52,* 10711–10720.

44 C. H. Bolton, R. W. Jeanloz, *J. Org. Chem.* 1963, *28,* 3228–3230.

45 B. Drouillat, B. Kellam, G. Dekany, M. S. Starr, I. Toth, *Bioorg. Med. Chem. Lett.* 1997, *7,* 2247–2250.

46 N. Constantzas, J. Kocourek, *Collection Czech. Chem. Commun.* 1959, *24,* 1099–1104.

47 B. Helferich, K. L. Bettin, *Chem. Ber.* 1961, *94,* 1159–1160.

48 H. G. Fletcher, Jr,B. Coxon, *J. Am. Chem. Soc.* 1963, *85,* 2637–2642.

49 B. Coxon, H. G. Fletcher, Jr, *J. Am. Chem. Soc.* 1964, *86,* 922–926.

50 F. G. De la Heras, P. Fernandez-Resa, *J. Chem. Soc. Perkin I,* 1982, 903–907.

51 Y. Igarashi, T. Shiozawa, Y. Ichikawa, *Bioorg. Med. Chem. Lett.* 1997, *7,* 613–616.

52 E. G. van Roedern, H. Kessler, *Chem. Int. Ed.* 1994, *33,* 687–689.

53 H. S. Overkleeft, S. H. Verhelst, E. Pieterman, N. J. Meeuwenoord, M. Overhand, L. H. Cohen, G. A. van der Merel, J. H. Van Boom, *Tetrahedron Lett.* 1999, *40,* 4103–4106.

54 S. Hanessian, J.-M. Vatèle, *Tetrahedron Lett.* 1981, *22,* 3579–3582.

55 C. Monneret, R. Garnet, J.-C. Florent, *Carbohydr. Res.* 1993, *240,* 313–322.

56 D. H. Ball, F. W. Parrish, *Adv. Carb. Chem. Biochem.* 1969, *24,* 139–197.

57 J. P. Horwitz, A. J. Tomson, J. A. Urbanski, J. Chua, *J. Org. Chem.* 1962, *27,* 3045–3048.

58 I. Yamamoto, M. Semine, T. Hata, *J. Chem. Soc. Perkin I* 1980, 306–310.

59 N. A. L. Al-Masuadi, N. J. Tooma, *Carbohydr. Res.* 1993, *239,* 273–278.

60 M. P. Watterson, A. A. Edwards, J. A. Leach, M. A. Smith, O. Ichiara, G. W. J. Fleet, *Tetrahedron Lett.* 2003, *44,* 5853–5857.

61 A. Malik, N. Afza, W. Voelter, *J. Chem. Soc. Perkin I* 1983, 2103–2109.
62 W. Karpeisiuk, A. Banaszek, A. Zamojski, *Carbohydr. Res.* 1989, *186*, 156–161.
63 E. Bousquet, M. Khitri, L. Lay, F. Nicotra, L. Panza, G. Russo, *Carbohydr. Res.* 1998, *311*, 171–181.
64 L. D. Hall, D. C. Miller, *Carbohydr. Res.* 1976, *47*, 299–305.
65 R. D. Howells, J. D. McGown, *Chem Rev.* 1977, *77*, 69–85.
66 H. H. Baer, Y. Gan, *Carbohydr. Res.* 1991, *210*, 233–245.
67 G. Kowollik, K. Gaetner, P. Langen, *P. Tetrahedron Lett.* 1969, 3863–3865.
68 K.-I. Sato, A. Yoshimoto, *Chem. Lett.* 1985, 39–42.
69 E. Busquet, M. Khitri, L. Lay, F. Nicotra, L. Panza, G. Russo, *Carbohydr. Res.* 1998, *311*, 171–181.
70 L. Cipolla, L. Lay, F. Nicotra, *J. Org. Chem.* 1997, *62*, 6678–6681.
71 V. Pavliak, P. Kovak, *Carbohydr. Res.* 1991, *210*, 333–337.
72 A. Guiller, C. H. Gagnieu, H. Pacheco, *Tetrahedron Lett.* 1985, *26*, 6343–6344.
73 K. Vanhessche, E. Van der Eycken, M. Vandewalle, *Tetrahedron Lett.* 1990, *31*, 2337–2341.
74 B. Aguilera, A. Fernandez-Mayoralas, *Chem. Commun.* 1996, 127–128.
75 F. Nicotra, L. Panza, F. Ronchetti, G. Russo, *Gazzetta Chim. Ital.* 1989, *119*, 577–579.
76 S. Czernecki, A. Ezzitouni, P. Crausz, *Synthesis* 1990, 651–653.
77 K. C. Nicolau, T. Ladduwahetty, J. L. Randall, A. Chucholowski, *J. Am. Chem. Soc.* 1986, *108*, 2466–2467.
78 M. A. Prodera, D. Olano, F. Fuentes, *Tetrahedron. Lett.* 1995, *36*, 8653–8656.
79 J. M. Tronchet, B. Bachler, J.-B. Zumwald, *Helv. Chim. Acta.* 1977, *60*, 1932–1934.
80 P. M. Collins, W. G. Overend, *J. Chem. Soc.* 1965, 3448–3456.
81 R. U. Lemieux, S. W. Gunner, *Can. J. Chem.* 1968, *46*, 397–400.
82 R. U. Lemieux, K. James, T. L. Nagabhushan, Y. Ito, *Can. J. Chem.* 1973, *51*, 33–41.
83 T. Wrodnigg, A. E. Stutz, *Chem. Int. Ed.* 1999, *38*, 827–828.
84 K. Heyns, W. Koch, *Naturforsch.* 1952, *7*, 486–488.
85 L. Cipolla, B. La Ferla, L. Lay, F. Peri, F. Nicotra, *Tetrahedron Asymm.*, 2000, *11*, 295–303.
86 R. U. Lemieux, R. M. Ratcliffe, *Can. J. Chem.* 1979, *57*, 1244–1251.
87 S. Czernecki, E. Ayadi, D. Randiamandimby, *Chem. Commun.* 1994, 35–36.
88 J. Du Bois, C. S. Tomooka, J. Hong, E. M. Carreira, *J. Am. Chem. Soc.* 1997, *119*, 3179–3181.
89 B. J. Fitzsimmons, Y. Leblanc, J. Rokach, *J. Am. Chem. Soc.* 1987, *109*, 285–286.
90 D. A. Griffith, S. A. Danishefsky, *J. Am. Chem. Soc.* 1990, *112*, 5811–5819.
91 D. A. Griffith, S. A. Danishefsky, *J. Am. Chem. Soc.* 1991, *113*, 5863–5864
92 V. Di Bussolo, J. Liu, L. G. Huffman, Jr., D. Y. Gin, *Chem. Int. Ed.* 2000, *39*, 204–207.
93 K. C. Nicolau, P. S. Baran, Y.-L. Zhomg, J. A. Vega, *Chem. Int. Ed.* 2000, *39*, 2525–2529.
94 A. Bongini, G. Cardillo, M. Orena, S. Sandri, C. Tomasini, *Tetrahedron*, 1983, *39*, 3801–3806.
95 H. Gnichtel, D. Rebentisch, *J. Org. Chem.* 1983, 2691–2697.
96 H. Paulsen, H. Koebernick, W. Stenzel, P. Koll, *Tetrahedron Lett.* 1975, *18*, 1493–1494.
97 H. P. Wessel, L. Labler, T. B. Tschopp, *Helv. Chim Acta* 1989, *72*, 1268–1277.
98 B. La Ferla, F. Nicotra, "Synthetic Methods for the Preparation of Iminosugars", *Iminosugars as Glycosidase Inhibitors;* A. Stütz Ed. WILEY-VCH *Verlagsgesellschaft mbH*, 68–92, 1998; b) L. Cipolla, B. La Ferla, F. Nicotra, *Current Topics in Medicinal Chemistry*, 2003, *3*, 485–511; c) L. Cipolla, B. La Ferla, M. Gregori, *Combinatorial Chemistry & High-throughput screening*, 2006, *9*, 571–582.
99 a) Y. Nishimura, *Heterocycles* 2006, *67*, 461–488; b) T. Ayad, Y. Genisson, M. Baltas, *Curr. Org. Chem.* 2004, *8*, 1211–1233; c) P. Compain, O. R. Martin, *Current Topics in Medicinal Chemistry* 2003, *3*, 541–560.

100 L. Cipolla, B. La Ferla, F. Peri, F. Nicotra, *Chem. Commun.* 2000, 1289–1290.

101 L. Cipolla, L. Lay, F. Nicotra, C. Pangrazio, L. Panza, *Tetrahedron* 1995, *51*, 4679–4690.

102 J. G. Buchanan, K. W. Lumbard, R. J. Sturgeon, D. K. Thompson, R. H. Wightman, *J. Chem. Soc. Perkin Trans I* 1990, 699–706.

103 A. Dondoni, D. Perrone, *Tetrahedron Lett.* 1999, *40*, 9375–9378.

104 Q. Meng, M. Hesse, *Helv. Chim. Acta* 1991, *74*, 445–450.

105 H. S. Overkleeft, J. van Wiltenburg, U. K. Pandit,. *Tetrahedron Lett.* 1993, *34*, 2527–2528.

106 B. La Ferla, P. Bugada, L. Cipolla, P. Peri, F. Nicotra, *Eur. J. Org. Chem.* 2004, 2451–2470.

107 S. Moutel, M. Shipman, *J. Chem. Soc. Perkin Trans I* 1999, 1403–1406.

108 G. Masson, P. Compain, O. R. Martin, *Org. Lett.* 2000, *19*, 2971–2974.

109 K. Burgess, D. A. Chaplin, I. Henderson, Y. T. Pan, A. D. Elbein, *J. Org. Chem.* 1992, *57*, 1103–1109.

110 O. R. Martin, O. M. Saavedra, F. Xie, L. Liu, S. Picasso, P. Vogel, H. Kizu, N. Asano, *Bioorg. Med. Chem.* 2001, *9*, 1269–1278.

111 M. Takebayashi, S. Hiranuma, Y. Kanie, T. Kajimoto, O. Kanie, C.-H. Wong, *J. Org. Chem.* 1999, *64*, 5280–5291.

112 W. H. Pearson, J. V. Hines, *Tetrahedron Lett.* 1991, *32*, 5513–5516.

113 C. R. R. Matos, R. S. C. Lopes, C.C. Lopes, *Synthesis* 1999, 571–573.

114 W. Zou, W. A. Szarek, *Carbohydr. Res.* 1993, *242*, 311–314.

115 A. Esposito, M. Falorni, M. Taddei, *Tetrahedron Lett.* 1998, *39*, 6543–6546.

116 A. E. Mc Caig, B. Chomier, R. H. Wightman, *J. Carbohydr. Chem.* 1994, *13*, 397–407.

117 V. Glacon, M. Benazza, D. Beaupère, G. Demailly, *Tetrahedron Lett.* 2000, *41*, 5053–5056.

118 P. A. M. Van der Klein, W. Filemon, H. J. G. Broxterman, A. T. G. Van der Marel, J. H. Van Boom, *Synt. Commun.* 1992, *22*, 1763–1771.

119 I. McCort, S. Fort, A. Duréault, J.-C. Dépezay, *Bioorg. Med. Chem.* 2000, *8*, 135–143.

120 I. Lundt, R. Madsen, *Synthesis* 1995, 787–793.

8
Selective *N*-Derivatization of Aminoglycosides *en Route* to New Antibiotics and Antivirals

Floris Louis van Delft

8.1
Aminoglycoside Antibiotics

Aminoglycoside antibiotics make up a large class of clinically relevant drugs with broad antibacterial spectra and proven efficacy, particularly against Gram-negative bacteria [1]. Most of the aminoglycosides are naturally occurring substances readily obtained from actinomycetes either of the genus *Streptomyces* (labeled "-mycin") or of *Micromonospora* (labeled "-micin") [2]. From a clinical perspective, the most prominent members of the family are streptomycin, tobramycin, and, in particular, gentamicin, due to its low cost and reliable activity. The clinical efficacy of aminoglycosides is enhanced upon coadministration with penicillin, a particularly useful combination for treatment of patients with infections of unknown origin, thanks to a strong synergistic effect. Despite the apparent advantages, however, extensive clinical use of the aminoglycosides is limited because of their associated nephrotoxicity and ototoxicity, and to a lesser extent neuromuscular blockade [2]. Another, and arguably more alarming, drawback of the aminoglycosides (and, for that matter, antibiotics in general) is the global development of microbial resistance. The most common mechanism is structural modification by bacterial enzymes, as shown in Figure 8.1 for kanamycin B, the aminoglycoside most susceptible to resistance [3–5]. It is not surprising, therefore, that the phenomenon of bacterial resistance has fostered much research into the chemoselective modification of the natural aminoglycosides to afford semisynthetic analogues with superior antibacterial activities.

An overview of natural and semisynthetic (*) aminoglycoside antibiotics is given in Tables 8.1–8.3. The key structural feature of all the aminoglycosides is a tri- or tetrahydroxylated 1,3-diaminocyclohexane unit, termed 2-deoxystreptamine or streptamine, respectively, glycosylated with a (di)aminosugar on 4-OH and on either 5-OH or 6-OH (kanamycin numbering). Semisynthetic derivatives can be obtained from these natural aminoglycosides by two general strategies: a) by deoxygenation (as in dibekacin), or b) by selective *N*-1 modification, such as ethylation (netilmicin) and/or acylation (amikacin, arbekacin, isepamicin). Given the plethora of hydroxy and amino functions in aminoglycosides, the chemistry

Amino Group Chemistry. From Synthesis to the Life Sciences. Edited by Alfredo Ricci

ISBN: 978-3-527-31741-7

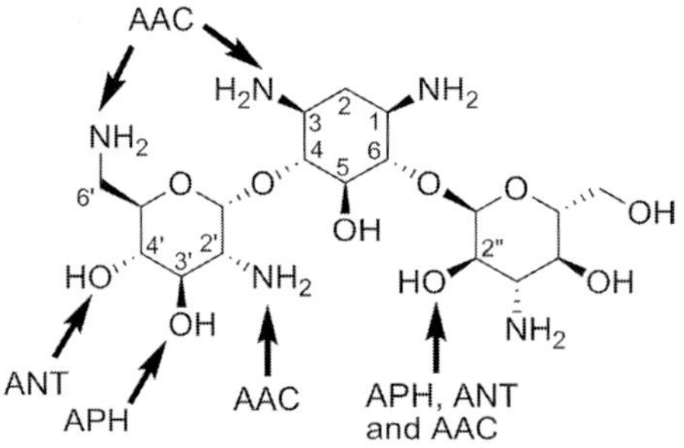

Figure 8.1 Enzymatic modification of kanamycin B by bacterial resistance enzymes aminoglycoside phosphotransferases (APHs), adenylyltransferases (AADs or ANTs), and acetyltransferases (AACs).

Table 8.1 Common 4,6-linked aminoglycoside antibiotics.

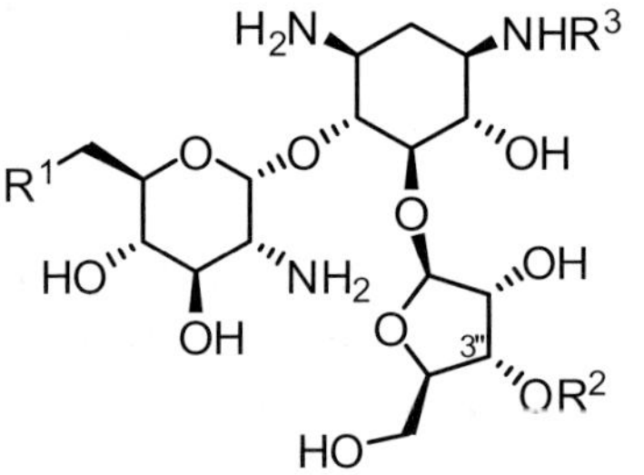

Aminoglycoside	*Year*	*R¹*	*R²*	*R³*
neomycin A	1949	NH_2	H	H
neomycin B	1949	NH_2	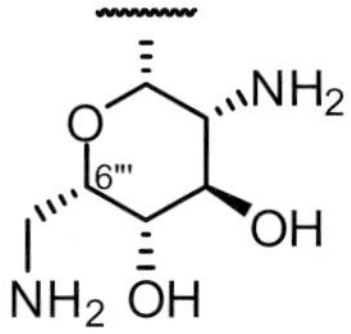	H
paromomycin	1959	OH		H
neomycin C	1949	6‴-epimer of neomycin B		
ribostamycin	1970	NH_2	H	H
butirosin B	1970	NH_2	H	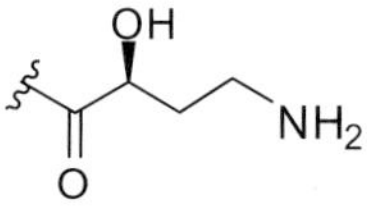
butirosin A	1970	3″-epimer of butirosin B		

Table 8.2 Gentamicin-type 4,5-linked aminoglycoside antibiotics.

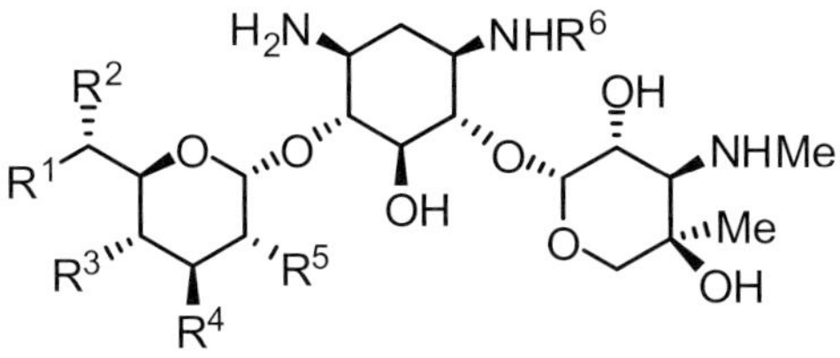

Aminoglycoside	*Year*	R^1	R^2	R^3	R^4	R^5	R^6
gentamicin A	1967	NH_2	H	OH	OH	NH_2	
gentamicin B	1972	NH_2	H	OH	OH	OH	
gentamicin C_1		MeHN	Me	H	H	NH_2	H
gentamicin C_{1a}		NH_2	H	H	H	NH_2	H
gentamicin C_2	1963	NH_2	Me	H	H	NH_2	H
gentamicin C_{2a}		Me	NH_2	H	H	NH_2	H
gentamicin C_{2b}		MeHN	H	H	H	NH_2	H
geneticin	1974	Me	OH	OH	OH	NH_2	H
isepamicin*	1978	NH_2	H	OH	OH	OH	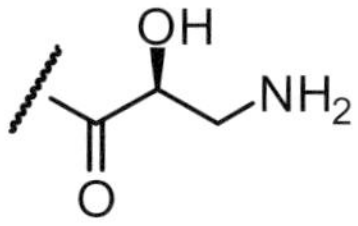
sisomycin	1970	4′,5′-dehydrogentamicin C_{1a}					
netilmicin*	1976	1-*N*-ethylsisomycin					

* Semisynthetic

employed to achieve such regio- and chemoselective modifications is the subject of Section 8.4.1.

Before the chemistry pertaining to the regioselective modification of aminoglycosides is highlighted, however, a more general perspective on the potential of aminoglycosides for drug discovery will be discussed. The mode of action of aminoglycosides, in contrast to those of the vast majority of current drugs, is not through targeting of a bacterial protein, but is rather the effect of selective binding to prokaryotic ribosomal RNA [6]. Apart from that, in the past decade it has become increasingly clear that aminoglycosides show rather promiscuous binding behavior to a variety of nonribosomal RNA structures, such as the HIV Rev response element (RRE) and the hepatitis C IRES, an overview of which is provided in Section 8.2. Specific molecular interactions between aminoglycosides and RNA are discussed in Section 8.3, with a particular focus on the role of the amino functions. The presence of multiple amino functions in the aminoglycosides is essential in RNA binding. At physiological pH values, the amines will be predominantly or fully protonated [7], to give polycationic structures that display favorable electrostatic interactions with the negatively charged phosphate diester backbone of

Table 8.3 Kanamycin-type 4,5-linked aminoglycoside antibiotics.

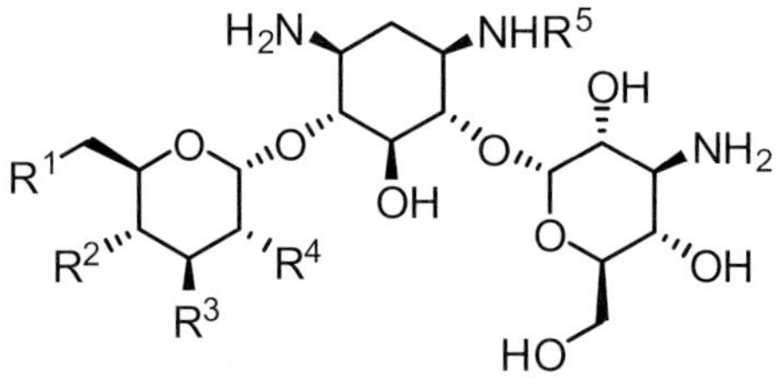

Aminoglycoside	*Year*	*R¹*	*R²*	*R³*	*R⁴*	*R⁵*
kanamycin A		NH_2	OH	OH	OH	H
kanamycin B	1957	NH_2	OH	OH	NH_2	H
kanamycin C		OH	OH	OH	NH_2	H
tobramycin	1967	NH_2	OH	H	NH_2	H
dibekacin*	1971	NH_2	H	H	NH_2	H
amikacin*	1972	NH_2	OH	OH	OH	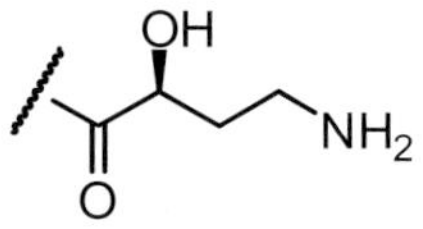
arbekacin*	1973	NH_2	H	H	NH_2	

* Semisynthetic

the ribonucleotide. In designing a novel aminoglycoside-based analogue, it is therefore essential to consider the potential influence of the introduction of *N*-substituents on the starting structure.

The general strategy to apply aminoglycosides as scaffolds for the preparation of new RNA-targeting drugs forms the main body of this chapter and is described in Section 8.4. Some general considerations in the field are discussed first, followed by a more or less chronological overview of the synthetic strategies that have been applied over the years to obtain new aminoglycoside derivatives, subdivided into the regioselective *N*-modification of natural aminoglycosides (Section 8.4.1) and the generation of aminoglycoside-type libraries based on neamine (Section 8.4.2). Particular attention is focused on chemistry developed for discrimination and selective manipulation of a single amino group present in a given aminoglycoside, either directly or through a number of protective group manipulations. Research papers dealing solely with global *N*-protection followed by selective modification of hydroxy functions, such as alkylation or glycosylation, lie outside the scope of this chapter and are therefore omitted.

8.2
RNA Targeting by Aminoglycosides

Historically, pharmaceutical industries have focused on the discovery of compounds that target only the final products of genes; that is, proteins [8]. The intermediary product between gene and protein – RNA – has remained largely unexplored, however, despite the fact that the advantages of targeting of RNA over targeting of proteins are readily recognized. First of all, the major current drug target classes mostly involve functional proteins, such as receptors, enzymes, ion channels, and transporters. However, these functional proteins cover only 17% of the human genome, whereas the remaining part encodes for proteins not readily amenable to high-throughput screening (HTS), such as structural proteins or genes with unknown functions or expressed sequence tags [9]. Unlike that of proteins, however, targeting of RNA has hardly been explored, despite the pivotal role of RNA in cell division, gene expression, and catalysis. At least 15% of all human genetic disorders involve aberrant mRNA processing [10]. In addition to its functional importance, RNA offers further advantages as an attractive drug target for a number of reasons: a) RNA functional domains are highly conserved and perhaps more accessible than the shapes of enzyme active sites, b) the absence of cellular RNA repair mechanisms, and c) the potential for slower development of drug resistance. Accordingly, there is considerable interest in the identification of small molecules that bind tightly to RNA [9, 11]. However, compounds that specifically recognize RNA secondary structures are scarce. One of the notable exceptions is represented by the aminoglycoside antibiotics.

Neomycins bind to functional sites in the 16S ribosomal RNA (Figure 8.2a) and cause miscoding and translocation arrest of the ribosome [6]. The effect is an increase in production of aberrant proteins that accumulate in the bacterial membrane, thereby compromising the barrier function of the membrane and finally inducing cell lysis [12]. Surprisingly, aminoglycoside binding differs only moderately between prokaryotic 16S and human 18S rRNA [13]. Nevertheless, aminoglycoside antibiotics only kill bacterial cells, a chemotherapeutic index that can be explained both by sequence differences [14] and by the occurrence in prokaryotes

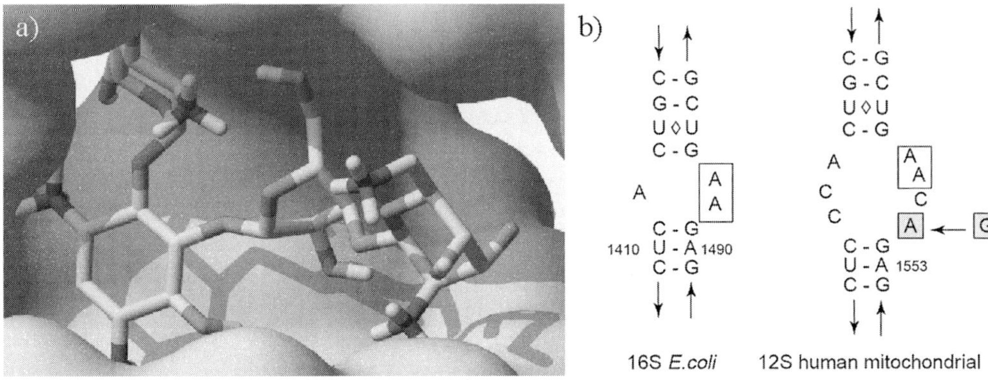

Figure 8.2 a) Graphic representation of neomycin in the 16S bacterial A-site of *E. coli*, based on crystal structure 2ET4.pdb [17]. b) RNA sequences of bacterial A-site and human mitochondrial 12S RNA. A point mutation of adenine 1555 to guanine causes aminoglycoside hypersensitivity (reprinted from reference 9, with permission).

of transporter proteins that actively take up and concentrate aminoglycosides in the cytoplasm [12]. The binding of aminoglycosides appears to depend on the size of an asymmetric interior loop [15], consisting of three nucleotides in *E. coli* 16S rRNA, an assumption corroborated by the fact that aminoglycosides normally do not bind 12S human mitochondrial ribosomes. However, a single point mutation A1555G (Figure 8.2b) causes aminoglycoside hypersensitivity in certain individuals [16], a finding suggestive of a high RNA sequence and structure specificity of aminoglycosides.

Surprisingly, the structure specificities of aminoglycosides are not high at all. Instead, aminoglycosides are known to interact with a large variety of different RNA targets [18], a promiscuity attributable to the large structural flexibility of aminoglycosides. A recent study suggests that the specific prokaryotic cytotoxicity may be determined by aminoglycoside-induced reduction in mobility of A1492 rather than by binding affinity for the A-site [19]. Irrespective of the underlying structural explanation, aminoglycosides – most notably neomycin – have been found to bind in the micromolar range to a wide variety of RNA structures other than the ribosome. For example, aminoglycosides bind to catalytic RNA molecules or ribozymes, including the self-splicing group I introns [20], the hammerhead ribozyme [18, 21], the hepatitis delta virus HDV ribozyme [22], and bacterial RNase P [23]. A variety of RNA targets from HIV were also found to be effectively complexed, including the Rev responsive element (RRE) [24], the trans-activation responsive element (TAR) [25], and the packaging region Ψ-RNA [26]. Apart from that, aminoglycosides also have affinity for tRNA [27] and hairpin loops [28], as well as kissing-loop complexes [29]. The final class of aminoglycoside targets is formed by sequences (aptamers) that have been selected *in vitro* for specific and high-affinity drug binding [30]. As well as ribonucleotides, aminoglycosides have also been found to bind triple helix DNA [31].

8.3
The Role of Amino Functions in RNA Binding

The natural aminoglycosides share the 2-deoxystreptamine (2-DOS) or streptamine core, an all *trans*-configured cyclohexanetriol (Tables 8.1 and 8.2). Moreover, the streptamine core has a 1,3-diamino function, and is further substituted by at least one, but mostly two, aminosugars, which possess one or two additional amino functions. Consequently, aminoglycosides are polycationic oligosaccharides and interactions between polar residues of the aminoglycosides and the RNA backbone and/or heterocyclic bases are likely to occur [18, 32, 33a]. The reported structure/activity relationships for the natural aminoglycosides suggest that electrostatic interactions are important for RNA binding [20, 21, 24]. The most active derivatives contain at least five or six amino groups that at pH 7 are predominantly charged. Indeed, molecular recognition of the aminoglycosides through polar hydrogen bonds is finely tuned by modulation of the basicity of amino groups [33], as a function of monomethylation (as in spectinomycin, and the gentamicins), deoxygenation patterns (as in the gentamicins and tobramycin), or the presence of additional guanidinium substituents (as in streptomycin). Several of these features from natural aminoglycosides were recognized and have been exploited in the synthesis of semisynthetic aminoglycosides, by 1-*N*-ethylation of sisomicin (netilmicin) [34], for example, or by 3′,4′-dideoxygenation of kanamycin B (dibekacin, amikacin, arbekacin) [35–37]. More recently, similar strategies have also been applied in the development of inhibitors for hammerhead ribozyme by deoxygenation of tobramycin [38] or in the synthesis of guanidinoglycosides as novel HIV RRE ligands [39]. However, it must be noted that the increase of compound basicity may result in higher binding affinity, but specificity is lost simultaneously [39b].

It is clear that electrostatic forces play a major role in enhancing the binding affinities of aminoglycosides with (polyanionic) RNA because of favorable enthalpic contributions. However, it has been shown that electrostatic interaction is also responsible, to a remarkable extent, for specific binding [40]. First of all, binding of an ammonium function to a phosphate group is not directed purely by charge attraction, but the polar H-bonds between anionic oxygens as acceptors and positively charged H-donor groups also contribute to the binding energy. Obviously, this force is determined by the orientation of the H-bond, and hence by the orientation of the aminoglycoside with respect to RNA, as revealed by several RNA–aminoglycoside structures in solution [41]. Secondly, the spatial distribution of negatively charged pockets in an RNA fold provides a three-dimensional pattern that is most favorably targeted by ligands exhibiting structural electrostatic complementarity. For hammerhead ribozyme, for example, it has been shown that solution conformers of aminoglycosides provide conformationally constrained scaffolds carrying cationic ammonium groups in the correct spatial orientation for binding in the electronegative pockets of the hammerhead [18], and the binding specificity decreases upon replacement of one of the aminosugars by a more flexible aliphatic chain [40]. For the ribozyme, aminoglycosides in fact compete with catalytic metal ions (Mg^{2+}) for the negative binding pockets. However, the picture of metal ion displacement by positively charged ligands may also hold for noncatalytic RNAs, since cations probably play roles in stabilization of the three-dimensional structures of all RNAs. A recent study based on pH-dependent ^{15}N NMR spectroscopy found that at least five out of six amino functions in the RNA–neomycin complex are protonated, regardless of pH, with similar pictures for paromomycin and lividomycin [7b]. A modified RNA–neomycin structure was proposed on the basis of these findings, since the known aminoglycoside–RNA structures did not permit discrimination between the different protonation states of the amino groups.

Apart from electrostatic interactions, several other factors also contribute to RNA recognition by an aminoglycoside. Through a combination of spectroscopic and calorimetric techniques it was shown that aminoglycoside–rRNA binding is also controlled by a broad range of factors including hydrogen bonding, drug protonation, water displacement, and binding-induced alterations in the structure of host RNA [42]. In general, the RNA binding free energy reflects a balance of two unfavorable and two favorable contributions. The two unfavorable contributions stem from the entropic cost of bimolecular complex formation and conformational changes in the RNA required for formation of the drug binding site, while the favorable contributions stem from the polyelectrolyte effect and, most importantly, noncovalent drug–RNA interactions. Rather surprisingly, binding of dimers of 2-deoxystreptamine to RNA octaloops is largely governed by entropic factors rather than enthalpy, with only a slight contribution from electrostatic interactions [43]. It is clear that these studies contribute to the broad effort to define the molecular recognition patterns that control the affinities and specificities of RNA-binding ligands and will usefully serve in the future design of novel drugs.

8.4
Development of RNA-Targeting Drugs

The development of synthetically accessible analogues of RNA-binding drugs is severely limited by the lack of medicinal chemistry efforts aimed at RNA [8] and, consequently, by the poor understanding of RNA recognition principles. Much effort in the development of novel drugs targeting RNA therefore exploits large molecules, with particular focus on antisense [44], RNA interference [45], and siRNA [46]. However, small molecules can also be RNA-targeting drugs, as is clear from the bactericidal effects both of aminoglycosides and of the recently developed oxazolidinones [47], which also bind to ribosomal RNA. A disadvantage of the aminoglycosides as RNA-targeting drugs lies in their structural complexity, high polarity, and large molecular weights, factors that violate the popular Lipinski rules used to guide medicinal chemistry towards orally bioavailable "drug-like" molecules [11c]. Apart from that, most aminoglycosides are difficult to modify in a regioselective manner because of the plethora of hydroxy and amino substituents. These drawbacks notwithstanding, the vast majority of studies aimed at RNA targeting with novel ligands are still based upon aminoglycoside scaffolds, and a large variety of aminoglycoside derivatives have been prepared in the past decade. One logical reason for such a strategy is to be found in the small number of scaffolds suitable for RNA targeting [48]. Apart from that, aminoglycosides are cheap and readily available, and a great deal of knowledge on the selective functionalization of aminoglycosides exists, mainly stemming from the synthetic efforts aimed at the development of aminoglycosides with improved pharmaceutical profiles in the early 1970s. In a keynote paper published by Wong and coworkers in 1996, it was shown that neomycin may also serve as a versatile scaffold for the combinatorial assembly of a library of ligands directed towards the HIV Rev Responsive Element [49]. After these pioneering studies, synthetic efforts directed towards the chemoselective modification of aminoglycosides, in targeting a wide variety of RNA structures, rapidly increased. In general, three different categories of aminoglycoside derivatives can be recognized: a) those resulting from the chemoselective (combinatorial) modification of an amino function, b) those resulting from the chemoselective (combinatorial) modification of an alcohol [11b, 32b, 40a, 48a, 49, 50], and c) bifunctional structures of aminoglycosides covalently linked to another

RNA ligand, including homodimers [51] and heterodimers [51f, 52]. The first category, the selective modification of one (or more) of amino functions in a natural aminoglycoside, is exhaustively discussed here. The chemistry aimed at the other two classes of aminoglycosides, although of substantial interest, lies outside the scope of this discussion.

8.4.1 Regioselective *N*-Modification of Naturally Occurring Aminoglycosides

The first antibiotic belonging to the class of aminoglycosides, streptomycin, was discovered by Waksman in 1944 [53], and has been widely used for the treatment of bacterial infections, in particular tuberculosis. Only a few years later, Waksman [54] and Umezawa [55] simultaneously discovered neomycin, to be followed by a large numbers of others (Figure 8.3), such as kanamycin [56], gentamicin [57], and butirosin [58], in what may be called the "golden era" of antibiotics. The contemporary finding of other antibacterial compounds, including chloramphenicol and the tetracyclins, together with the availability of penicillin, provided the physician with a large arsenal of antibiotics and, consequently, an enormous reduction in diseases and deaths after bacterial infections.

However, the widespread clinical use of antibiotics also gave rise to the rapid emergence of drug-resistant bacteria, and serious problems were caused by staphylococci and Gram-negative bacteria resistant to all antibiotic agents. Bacterial strains resistant to kanamycin, an antibiotic for treatment of streptomycin-resistant tuberculosis, for example, began to appear as early as 1965, only eight years after the discovery of kanamycin. The biochemical mechanisms of resistance to aminoglycosides, elucidated for the most part by Umezawa et al. [59], involved aminoglycoside modification by three types of bacterial enzymes: aminoglycoside

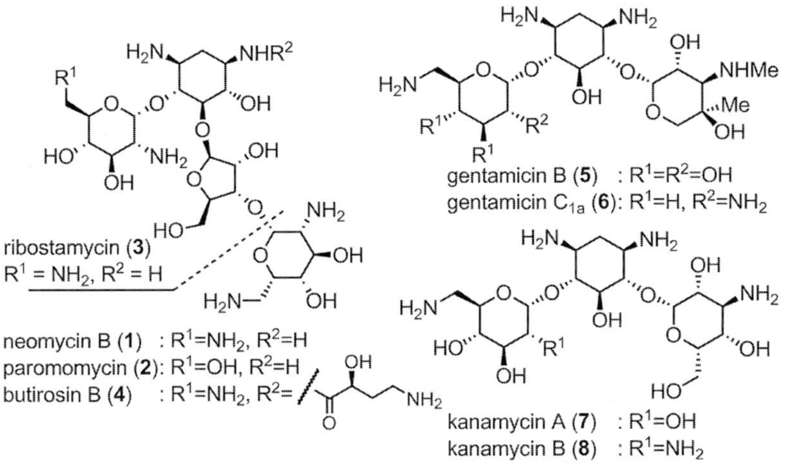

Figure 8.3 Structures of representative aminoglycoside antibiotics.

acetyltransferases (AACs), phosphotransferases (APHs), and adenylyltransferases (AADs) (see Figure 8.1). On the basis of these findings, structures refractory to these enzymes were rationally designed and synthesized, in particular by deoxygenation and/or *N*-modification, resulting in several semisynthetic aminoglycosides for clinical use [34–37].

One of the first reports describing the selective liberation and functionalization of a single amine out of four in a natural aminoglycoside involved the conversion of ribostamycin [60] into butirosin B [61], as depicted in Scheme 8.1. Thus, ribostamycin (**3**) was first converted into alcohol **9**, involving global *N*-protection with benzyloxycarbonyl (Cbz) functions and treatment with 1,1-dimethoxycyclohexane in the presence of *p*-toluenesulfonic acid [62]. Upon subjection of **9** to sodium hydride, the singly unprotected hydroxy group was deprotonated, resulting in the formation of the 1,6-carbamate derivative. The cyclic carbamate could be selectively hydrolyzed in the presence of the other three carbamates by treatment with 1.1 equivalent of barium hydroxide in aqueous dioxane at 80 °C, to give amino alcohol **10** in 70% yield. The monoamino derivative **10** was then condensed with (*S*)-2-hydroxy-4-phthalimidobutyric acid in the presence of DCC and *N*-hydroxysuccinimide, to afford the coupled amide in a yield of 67%. Deprotection (hydrazine, hydrogenation, and acid hydrolysis) gave butirosin B (**4**) in 63% yield over the three steps.

An analogous approach to obtain butirosin analogues by the modification of selectively liberated *N*-1 was based on the use of butirosin itself as a starting material [63]. Compound **11** (Scheme 8.2) was thus prepared by treatment of butirosin B with dimedone (5,5-dimethylcyclohexane-1,3-dione) followed by alkaline hydrolysis with hot barium hydroxide. The vinylogous amides of **11** were found to be conveniently stable to nucleophilic attack, and furthermore showed excellent solubility and chromatographic properties. The resulting tris(dimedone) was then utilized for synthesis of analogues by procedures used for peptides, in particular by condensation with *p*-nitrophenyl and pentafluorophenyl activated esters of Cbz-protected amino acids. After the removal of the Cbz group on the amino acid side-chain, the dimedone groups were removed with chlorine gas at 5 °C, to give the butirosin analogues **12**.

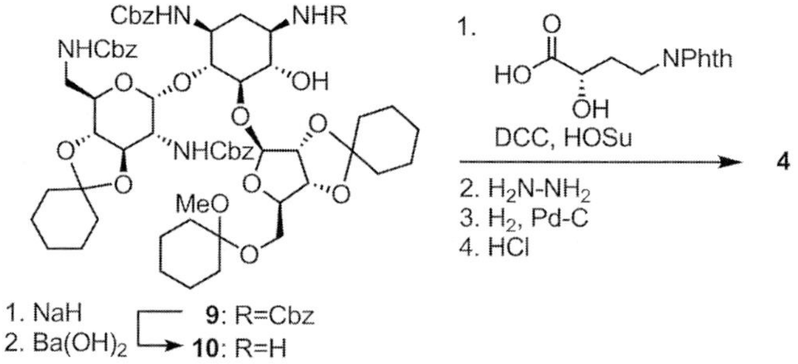

Scheme 8.1 Selective liberation of *N*-1 of ribostamycin via cyclic carbamate formation in the synthesis of butirosin B (**4**).

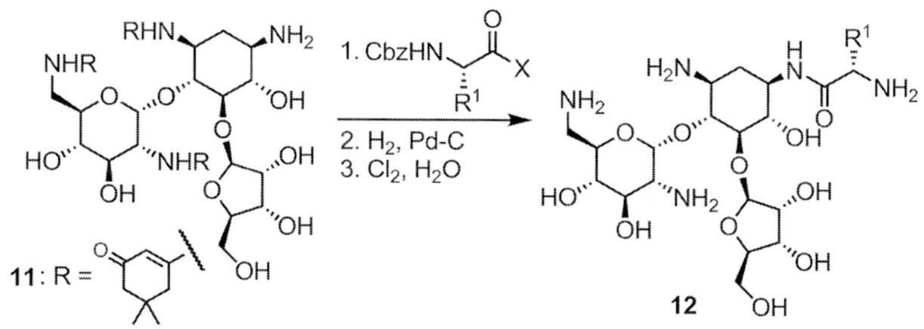

Scheme 8.2 Synthesis of butirosin B analogues from tris(dimedone)-protected ribostamycin.

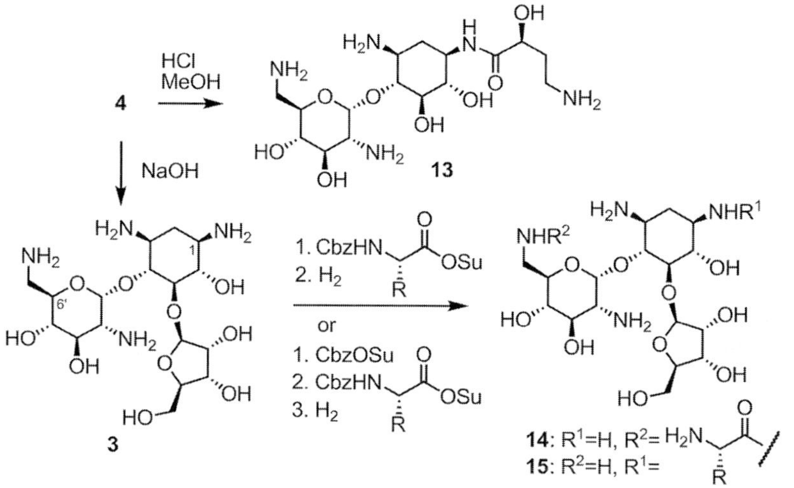

Scheme 8.3 Conversion of butirosin B (4) into a variety of amino acylated analogues either by partial methanolysis or by side-chain removal followed by selective acylation procedures.

Another study based on the strategy of applying natural butirosin as starting material for the assembly of novel analogues [64] involved either the selective hydrolysis of a single glycosidic bond or the regioselective acylation of the most reactive amino function of ribostamycin (**3**). In a procedure based on chemistry developed for neomycin [65], the furanosyl glycosidic bond was chemoselectively cleaved upon subjection of butirosin B to methanolic hydrogen chloride to give 1-*N*-[(*S*)-γ-amino-α-hydroxybutyryl]neamine (**13**), a new antimicrobial compound showing activity similar to but weaker than that of butirosin B (Scheme 8.3).

Alternatively, the butyryl side-chain of butirosin B was hydrolyzed under basic conditions to give ribostamycin (**3**). It was found that regioselective acylation occurred selectively at the most nucleophilic 6′-amino function, similarly to the situation with 4,6-linked aminoglycosides (*vide infra*), upon treatment with *N*-hydroxysuccinimide esters of amino acids (**3**→**14**), but these analogues possessed only weak antibacterial activities. A range of butirosin analogues was prepared in a three-step procedure involving selective blocking of *N*-6′ with Cbz-OSu, prior to acylation of the most nucleophilic amino function at *N*-1, thus allowing the preparation of a range of amino acyl modified ribostamycins **15** after Cbz removal by hydrogenolysis. The antibacterial activities of these analogues, although superior to those of compounds **14**, were less than that of butirosin B. Finally, double *N*-Cbz protection at positions 1 and 6′ (**14** with R^1=R^2=Cbz) followed by amino acylation resulted in monoacylation (not depicted), but the regioselectivity could not be determined.

Whereas these synthetic efforts on the butirosin-type aminoglycosides did not afford clinically relevant compounds, a greater deal of success was met in the rational design and synthesis of analogues of 4,6-linked aminoglycosides. It was hypothesized that bacterial modification with resistance enzymes could be minimized by selective deoxygenation of enzymatically modified hydroxy group(s). Furthermore, it was recognized that one of the clinically most useful antibiotics was gentamicin C, which is in fact a mixture predominantly consisting of three components (C_1: <45%, C_{1a}: <35% and C_2: <30%), but all of these components feature a 3′,4′-dideoxygenation pattern. Furthermore, it was recognized that introduction of the butirosin amino acyl side-chain onto 4,6-linked aminoglycosides might further reduce resistance susceptibility. In view of these considerations, several research groups in Japan embarked on the chemoselective modification of gentamicin and kanamycin to give a variety of highly clinically useful semisynthetic aminoglycosides.

The first successful semisynthetic aminoglycoside derivative was 3′,4′-dideoxykanamycin B, prepared by Umezawa et al. [35]. A monodeoxygenated derivative of kanamycin had been prepared earlier [66], but involved a tedious synthesis with a cumbersome glycosylation step. A much simpler approach involved the direct transformation of kanamycin B by double deoxygenation [35]. To this end, kanamycin B was first converted into the tetraethoxycarbonyl (Cbe), bis(isopropylidene)-protected derivative **16** (Scheme 8.4). In four straightforward steps, including the chemoselective formation of a six-membered isopropylidene acetal, **16** was converted into the dimesylate **17**. The key deoxygenation was now performed by subjection of **17** to sodium iodide in the presence of zinc, giving the 3′,4′-unsaturated intermediate, which was hydrogenated and fully deprotected to give 3′,4′-dideoxykanamycin B (**18**). Compound **18** was later termed dibekacin and constituted the first of a class of semisynthetic aminoglycosides of great clinical relevance. A 3′,4′-dideoxy derivative of neamine was also prepared [67], but has not found clinical application.

Right about the same time, it was observed by Kawaguchi et al. that the amino acyl side-chain of butirosin B (**4**) had a marked positive influence on antibacterial activity by reducing susceptibility to resistance mechanisms. This observation stimulated the investigation of new types of modified aminoglycosides, in particular from kanamycin A (**7**), featuring the (*S*)-γ-amino-α-hydroxybutyryl (AHB) side-chain [36]. Since kanamycin A contains four amino functions, introduction of the AHB substituent solely at *N*-1 required careful design of reaction conditions. It was found that the *N*-6′ position was the most reactive amine, and so could be

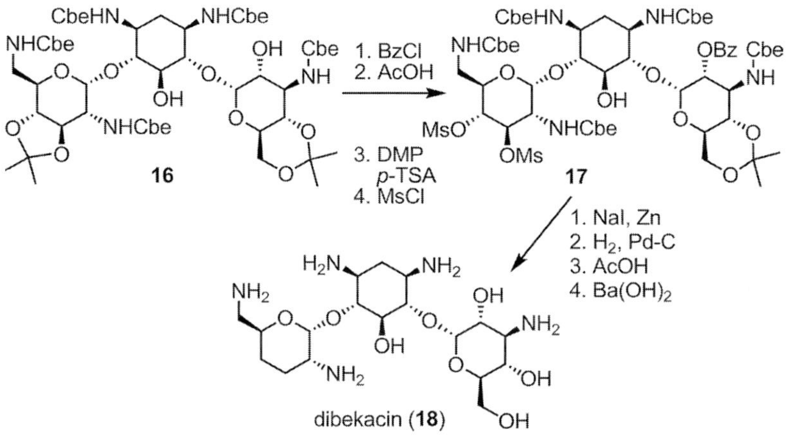

Scheme 8.4 Preparation of dibekacin (**18**) by selective dimesylation/elimination of 3′,4′-dideoxykanamycin B, followed by hydrogenation.

selectively protected (45% yield) with Cbz-OSu. As in the case of ribostamycin, the *N*-1 amino was next in line in terms of nucleophilicity, thereby allowing selective amino acylation with Cbz-protected AHB-OSu ester **19b**, to give **20** (Scheme 8.5). After hydrogenation, the AHB-derivative of kanamycin A (**21**), later known as amikacin, was isolated in 22% yield over the two steps.

Immediately, Umezawa responded [37]. With access to kanamycin B (**8**) and 3′,4′-dideoxykanamycin B (**18**), the same strategy could be applied to these aminoglycosides, although the synthesis was slightly more challenging, due to the presence of an additional amino function at *C*-2′ with respect to kanamycin A. It was discovered (Scheme 8.5) that **8** and **18** behaved similarly in terms of amino group reactivity, giving selective *N*-6′ protection (Cbz or Boc), followed by selective *N*-2′ protection with a Boc or Cbz group upon condensation with $BocN_3$ or CbzCl (not depicted), respectively, along with a trace of positional isomers. Subsequent introduction of the (*S*)-AHB side-chain proceeded basically identically to that reported earlier to give **22**. Full deprotection involving hydrogenation in the presence of methanol and trifluoroacetic acid gave 1-(*S*)-AHB-3′,4′-dideoxykanamycin B (**23**), later known as arbekacin. Arbekacin was found to be refractory to most aminoglycoside-modifying enzymes including APH(3′), AAD(4′,4″), and AAC(6′)/APH(2″) in multiple drug-resistant *Staphylococcus aureus* and *Staphylococcus epidermis* and inhibited not only Gram-negative bacteria, including *Pseudomonas*, but also staphylococci [59b]. Furthermore, in contrast to expectations, arbekacin also retains around 50% activity against resistant strains possessing AAC(2′) or AAC(3″), despite the fact that *N*-2′ or *N*-3″ of arbekacin is effectively acetylated by these enzymes. Such antibacterial activity of enzymatically modified aminoglycosides is not paralleled by any other aminoglycoside antibiotic.

After the success of amikacin and arbekacin, researchers at Schering Corporation explored the possibility that similar derivatives of gentamicin may also

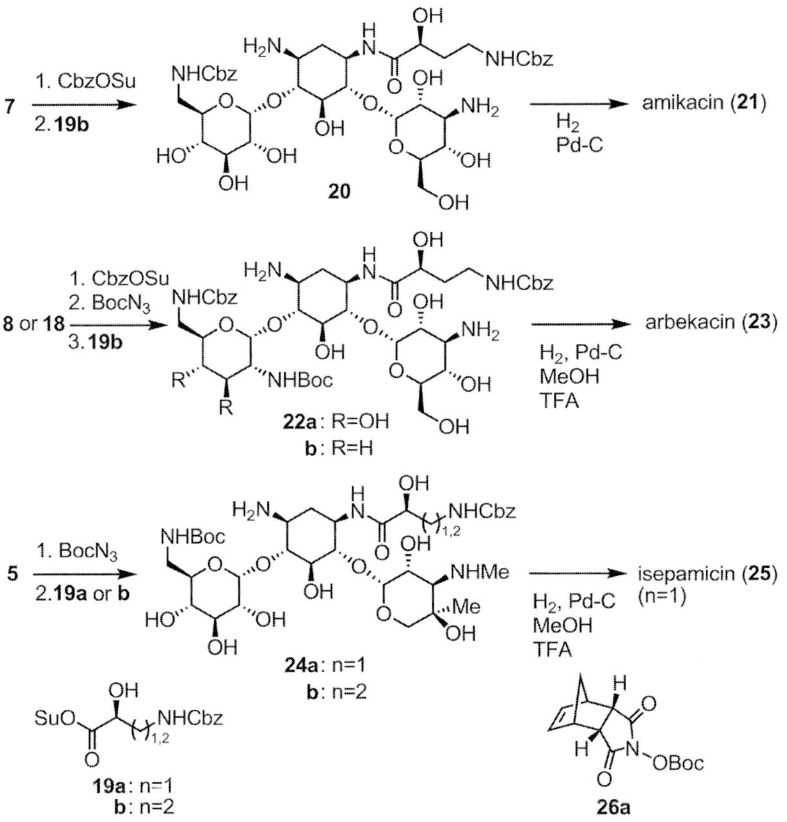

Scheme 8.5 Synthetic routes to semisynthetic aminoglycosides amikacin (**21**), arbekacin (**23**), and isepamicin (**25**).

contribute to the arsenal of semisynthetic aminoglycoside antibiotics. The (*S*)-γ-amino-α-hydroxybutyryl derivative was thus prepared from gentamicin B (**5**) by virtually the same approach, with *N*-6′ Boc protection prior to *N*-1 acylation [68]. At the same time, a truncated variant of the AHB side-chain was also prepared by condensation with **19a** (*n*=1), giving the more potent final product isepamicin (**25**) after global deprotection. Apparently, a direct procedure for the selective *N*-1 acylation of gentamicin C_1 based on the relative acidities of the amines (**6**→**27**; Scheme 8.6), published by the same Schering group two years earlier [69], was inferior for the synthesis of isepamicin with respect to temporary protection of *N*-6′.

The high utility of selective protection of *N*-6′ in aminoglycosides stimulated several other research groups to look into improved procedures. The sterically hindered norbornene-derived reagent **26a** (Scheme 8.5), for example, was applied for the selective introduction of a Boc group, allowing the selective protection of both primary amino functions in isepamicin, in 75% yield [70a].

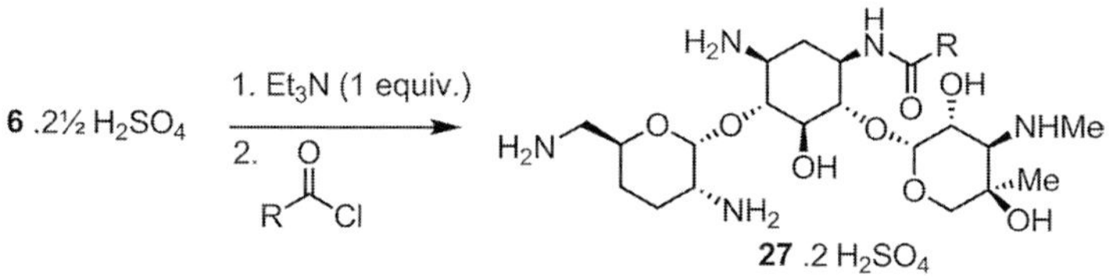

Scheme 8.6 Selective acylation of the most acidic amino function of gentamicin C_{1a}.

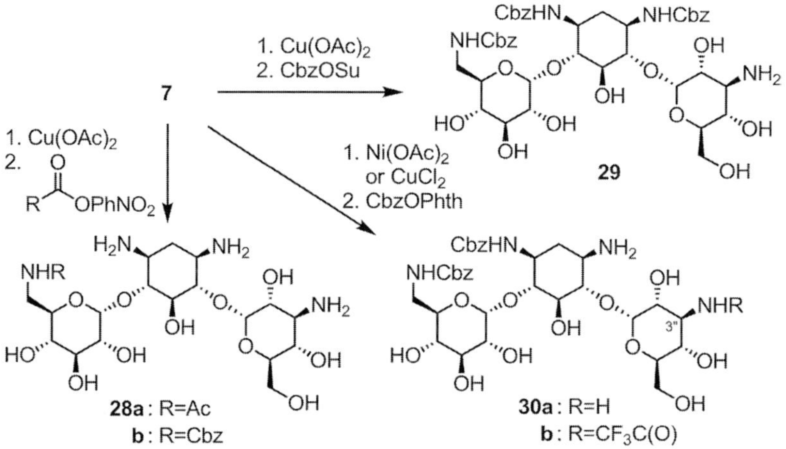

Scheme 8.7 Transition metal-catalyzed carbamoylation of kanamycin A.

A more sophisticated approach was based on the temporary protection of suitably disposed vicinal amino alcohol functions as metal chelates. Hanessian et al. [71], for example, pioneered the selective *N*-Cbz protection of kanamycin A (**7**) in the presence of copper(II) acetate (Scheme 8.7). Treatment of a mixture of kanamycin A (**7**) and $Cu(OAc)_2.H_2O$ in aqueous THF with *p*-nitrophenyl acetate or benzyl *p*-nitrophenyl carbonate gave, after decomposition of the chelate with ammonia, 6′-*N*-acetylkanamycin A (**28a**) or 6′-*N*-Cbz-kanamycin A (**28b**) in yields of 82% and 73%, respectively. Formation of the 1,3,6′-tri-*N*-benzyloxycarbamate **29** was also feasible upon treatment of **7** with 10 equivalents of $Cu(OAc)_2$, followed by addition of 5 equivalents of Cbz-OSu (86% yield). In the same year, it was reported that nonvicinal amino–hydroxy group pair-complexing between the 1 and 2″ positions was also possible in dipolar aprotic solvents [72]. Treatment of kanamycin A with $Ni(OAc)_2$ tetrahydrate in DMSO prior to addition of *N*-CbzPhth, for example, gave the 3,6′-diprotected derivative **30a** in excellent yield (88%). Surprisingly, complete monoprotection at *N*-6′ could also be achieved upon treatment with cupric chloride in DMSO, followed by *N*-CbzPhth. Use of a range of other combinations of aminoglycosides with the transition metal salts Co^{2+}, Ni^{2+}, and Cu^{2+} gave similarly high yields through 1,2″ and 3″,4″ complexing. Of an even wider range of metal ions, including Ca^{2+}, Cr^{3+}, Mn^{2+}, Fe^{3+} and others, Zn^{2+} was identified as the ideal metal for selective 3,6′-di-Cbz protection of kanamycin A to give **30a** (82%), whereas arbekacin gave mostly 3,2′,6′-triprotection in 85% yield [73]. Interestingly, treatment of **30a** with ethyl trifluoroacetate (1.3 equiv.) in DMSO gave the *N*-3″-trifluoroacetylated product **30b** with excellent regioselectivity (95% yield).

A final contribution to the field was provided by selective protection of apramycin with Cbz-OSu in the presence of $Ni(OAc)_2$, $Cu(OAc)_2$, or $Zn(OAc)_2$ [74]. It was shown that the nature of the metal salt had a decisive influence on the regioselectivity of the acylation, selectively giving the *N*-2′- (**32**) or *N*-3-protected (**33**) apramycin in the presence of Ni^{2+} or Cu^{2+}, respectively (Scheme 8.8). Application of the cupric ion procedure to a range of aminoglycosides, in the presence of Boc_2O or the sterically encumbered reagent **26**, as well as an in-depth investigation of the structure of the copper(II)-complexes, was recently reported [70b]. A general review on the chemistry of metal-chelated reactions of polyamino compounds has recently appeared [75].

The fifth and last clinically relevant semisynthetic aminoglycoside derivative, a compound not obtained by deoxygenation, acylation, or a combination thereof, is netilmicin (**35**; Scheme 8.9). Netilmicin is the 1-*N*-ethyl derivative of naturally occurring sisomicin [76] and can be obtained in a single transformation by exploiting the relative difference in reactivity between the amino functions of sisomicin (**34**). As found earlier for gentamicin (Scheme 8.6), the amino function at *C*-1 is the most acidic, thus allowing selective reductive alkylation under conditions of low pH.

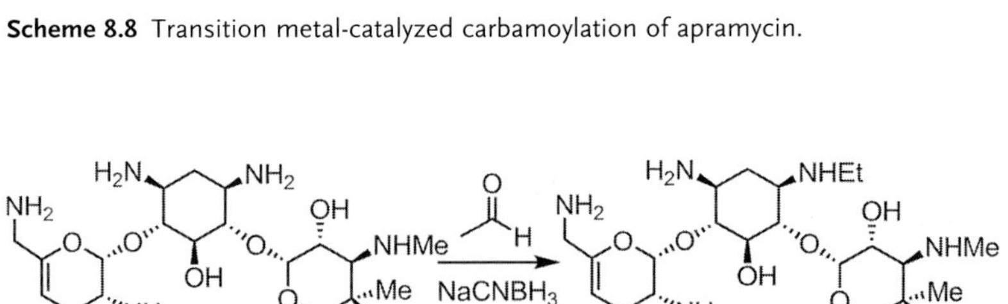

Scheme 8.8 Transition metal-catalyzed carbamoylation of apramycin.

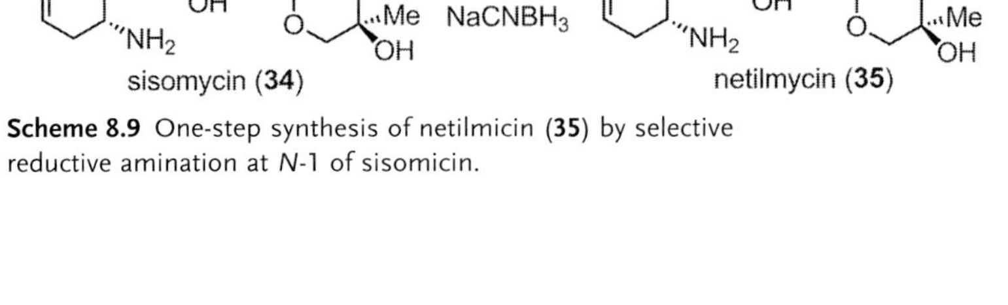

Scheme 8.9 One-step synthesis of netilmicin (**35**) by selective reductive amination at *N*-1 of sisomicin.

8.4.2
Neamine-Based RNA ligands

After the successful efforts in the development of semisynthetic aminoglycosides in the 1970s, interest in the synthesis of novel aminoglycoside analogues gradually increased (Figure 8.4). After some time, however, interest in aminoglycoside chemistry lost momentum when further improvements remained elusive, and because of the general lack of interest in the development of antibiotics, with the idea prevailing that "the battle with microorganisms was fought and won". Consequently, efforts aimed at the development of novel aminoglycoside analogues with improved pharmaceutical profile and reduced resistance susceptibility were confined to a few scattered papers (Figure 8.4).

One of the last of papers of the early era [77] – that is, aiming specifically at aminoglycosides with improved antibacterial profiles – focused on the preparation of a range of *N*-6′ amino acylated derivatives of neamine (**36**), a compound readily obtained by acid methanolysis of neomycin B. It was shown that neamine can be selectively protected at *N*-6′ with *p*-methoxybenzyloxycarbonyl-OSu, to be liberated by hydrogenolysis after a few additional protective group manipulations, giving **37** (Scheme 8.10). Several amino acylated derivatives were obtained in investigations involving coupling and deprotection (not depicted), but antibacterial activities were not reported.

Although conceptually not new, the recognition that neamine (**36**), and not one of the aminoglycosides in clinical use, may serve as a scaffold for the synthesis of aminoglycoside analogues is an idea that has since been adopted by a large

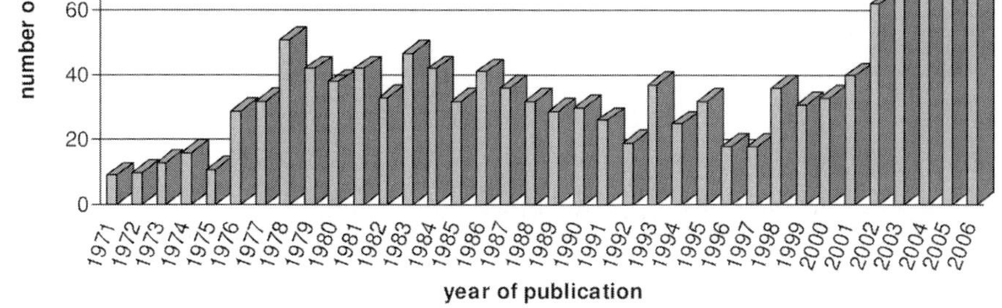

Figure 8.4 Numbers of publications with keywords "synthesis" and "aminoglycosides" in Chemical Abstracts between 1971 and 2006.

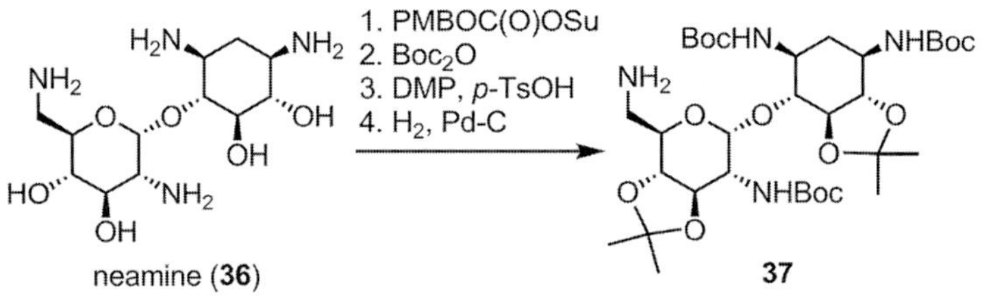

Scheme 8.10 Liberation of *N*-6′ of neamine (**36**) by a four-step protective group protocol.

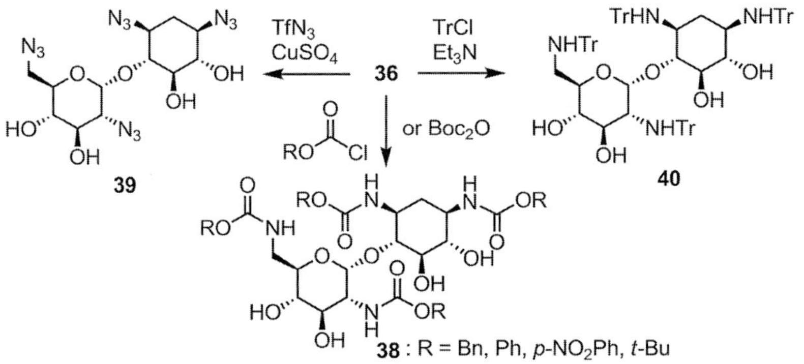

Scheme 8.11 Global transformation of amino functions of neamine into carbamates (**38**) or a tetraazide (**39**) or protection with trityl groups (**40**).

number of other research groups: neamine is readily prepared from (cheap) neomycin, has an acceptable number of four amino and four hydroxy functions, and may be further derivatized to provide either 4,5- or 4,6-type aminoglycosides. A paradigm shift in the potential of aminoglycosides, from antibiotics to general scaffolds for libraries of ligands targeting RNA, was sparked by a seminal paper on the discovery of inhibitors of the HIV Rev Responsive Element [49]. The following section describes synthetic efforts on neamine modification, once again focusing on chemistry pertaining to the selective modification of one (or more) of the amino functions.

Logically, the chemistry of neamine strongly resembles that of natural aminoglycosides and amino alcohols in general. In particular, a large number of alkoxycarbonylation procedures have been applied to neamine, resulting in the carbamates **38** depicted in Scheme 8.11 [52b, 67, 78]. Of more recent date and of more practical value, in particular in terms of solubility and ease of spectral interpretation, is the transition involving metal-catalyzed amine→azido interconversion of neamine to **39** with triflic azide [79]. The global tritylation of neamine to give the tetra-*N*-trityl-protected derivative **40** has also been reported [80].

Although the above procedures are useful practically, a larger synthetic challenge lies in the selective protection or liberation of a single amino function in neamine. Original methods developed to obtain such structures were aimed at application of neamine for the total synthesis of butirosin [78b] or novel *N*-1 aminoacylated aminoglycoside analogues [67] and are depicted in Scheme 8.12. It was shown, for example, that compound **38** (R = phenyl, *p*-nitrophenyl, or benzyl) can be conveniently converted in 86% yield into the cyclic urea **41**, with both *N*-2′ and *N*-6′ free, upon treatment with sodium hydroxide in methanol [78b]. Alternatively, treatment of **38** (R=Ph, *p*-NO_2Ph) with the strongly basic resin Dowex 1X2 (OH^- form) gave the selectively *N*-3-liberated tris(cyclic carbamate) **42** in moderate yield (66%). Thirdly, treatment of the tetracarbamate **38** (R=Bn) with sodium hydride in DMF selectively affords the bis(cyclic carbamate) **43**. Now *N*-1 can be selectively targeted through protection of the 3′- and 5-hydroxy groups by treatment with NaH and benzyl bromide (simultaneously resulting in the *N*-benzylation of linear carbamates), followed by selective hydrolysis of the five-membered carbamate with barium hydroxide, to give amino alcohol **44** [81].

Transition metal protection of vicinal and nonvicinal amino alcohol functions of aminoglycosides (*vide supra*) has also been investigated in detail for neamine [82]. Treatment of neamine (**36**) with Boc_2O in the presence of copper(II) ion thus gave **45**, with *N*-3 and *N*-6′ selectively protected (60% yield), while subsequent treatment of **45** with *N*-benzyloxycarbonyloxy-5-norbornene-*endo*-2,3-dicarboximide (**26b**) in the presence of zinc acetate selectively furnished **46** (Scheme 8.13). Alternatively, neamine can be subjected to **26b** in the absence of metal salts, predominantly affording the monocarbamoylated product **47b** in a yield of 61%. In a slight modification of the procedure, it was recently shown that stirring of the free base neamine and **26b** for a few minutes either in acetone/water or in acetonitrile/water (1 : 1) was sufficient to furnish **46b** in 75% yield after reversed-phase HPLC [83]. A range of other NBD-esters were also evaluated, and afforded the *N*-6′-acylated products in excellent yields (R=Ph: 80%, Naph: 85%, 2-ClPh: 90%), including 2-bromoacetyl-modified neamine **47c**, *en route* to a bisubstrate inhibitor of aminoglycoside

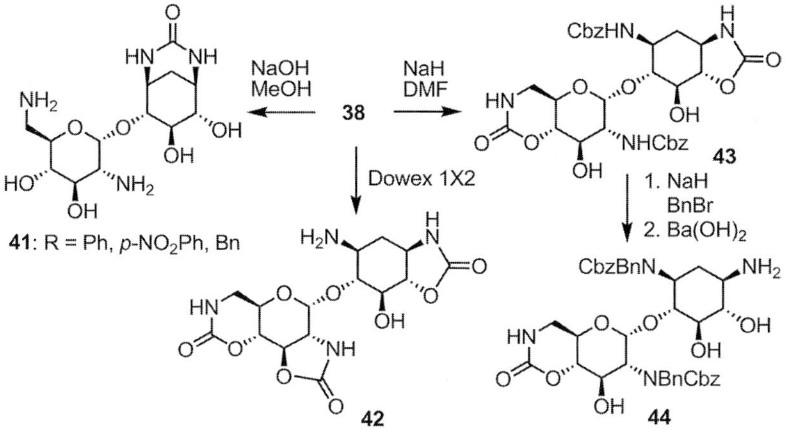

Scheme 8.12 Selective liberation of amino functions of tetracarbamoylated neamine (**38**) under a range of basic conditions.

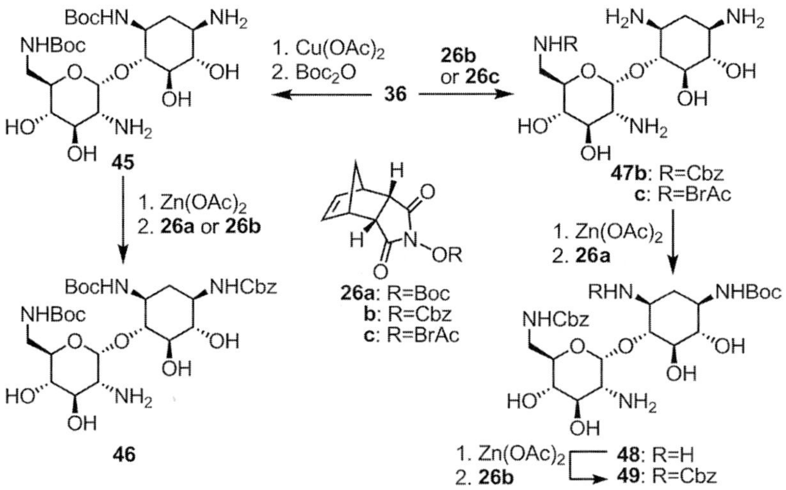

Scheme 8.13 A combination of transient protection of amino alcohols with transition metals and with a sterically hindered carbamoylating reagent (**26**) gives access to a range of selectively protected neamine derivatives.

6′-*N*-acetyltransferases [83]. Compound **47c** had earlier been prepared in lower yield as an irreversible inhibitor of 3′-phosphotransferase type IIa [84]. The *N*-6′ Cbz protected product **47b** can be further converted into the 1,6′-diprotected derivative **48** by an additional condensation with **26a** in the presence of zinc acetate (50% overall). Further selectivity is again achieved by subjection to norbornene-derived reagent **26a** in the presence of zinc, to provide **49** with *N*-2′ selectively liberated in 70% yield.

A fundamentally different approach aims not at the selective protection of an amino function but at the chemoselective reduction of an azido function. There are some examples of such reactions in the literature [85] and the order of reactivity for triphenylphosphine (primary > secondary > tertiary) suggested that steric hindrance played a decisive role in determining the chemoselectivity. Later, however, it was found that the tetraazidoneamine derivative **50a**, obtained in four steps from neomycin (**1**), predominantly gave the 2′-amine **51** by 2′-azido instead of 6′-azido reduction upon subjection of **50a** to one equivalent of trimethylphosphine (Scheme 8.14) [79b]. It was proposed that the main determinant is not steric aspects but electronic factors, resulting in preferential reduction of more electron-deficient azides. This phenomenon was recently elegantly applied in the chemoselective reduction of the *C*-1 azido function of the neamine derivate **50b** [86]. It was found that the deshielding effects of the *p*-chlorobenzoyl groups at *O*-5 and *O*-6 were strongest for the azido function at *C*-1, which could be selectively reduced with PMe_3 in preference to the *C*-3 azide in a ratio of 5 : 1 and 45% combined yield (not depicted). Rather unexpectedly, an even greater chemoselectivity was later observed for the analogous 3′,4′-dehydroneamine [87], which led the authors to conclude

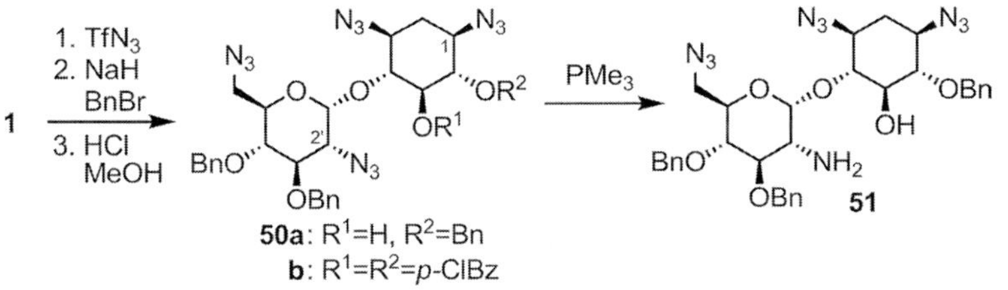

Scheme 8.14 Selective Staudinger reduction of the most electron-deficient azide of neamine with trimethylphosphine.

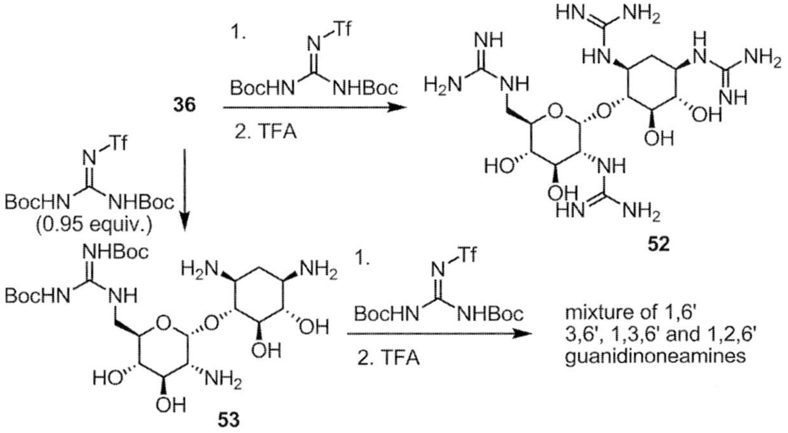

Scheme 8.15 Selective guanidinylation of neamine.

that both steric and stereoelectronic effects govern the selectivity of the Staudinger reaction. In fact, the products prepared by Chang et al. represent one of the few examples of selectively *N*-1-liberated neomycin-type aminoglycosides [62, 63, 88].

The selective guanidinylation of neamine has also proven feasible in a study directed towards the preparation of novel inhibitors for anthrax lethal factor protease [89]. Initially, tetraguanidinoneamine **52** was prepared by condensation of neamine with excess *N*,*N*′-di-(*tert*-butoxycarbonyl)-*N*″-triflylguanidine, followed by TFA deprotection (Scheme 8.15). More challenging was the selective *N*-6′ guanidinylation, which was successfully achieved with use of one molar equivalent of the same reagent, although no yield was provided. Further attempted synthesis of the *N*-1,6′ doubly guanidinylated structure gave a mixture of products.

An elegant study aiming at the determination of electrostatic interactions of individual amino groups on aminoglycosides for binding to RNA was performed by the selective deamination of each of the four amino functions in neamine [82]. To this end, mono *N*-6′ Cbz-protected neamine **47b** (Scheme 8.13) was Boc-protected at the remaining amino functions, followed by liberation of *N*-6′ upon hydrogenation to give **54** (Scheme 8.16). The free amine of **54** was converted into the isocyanide upon formylation/dehydration, furnishing **55**. Radical deamination

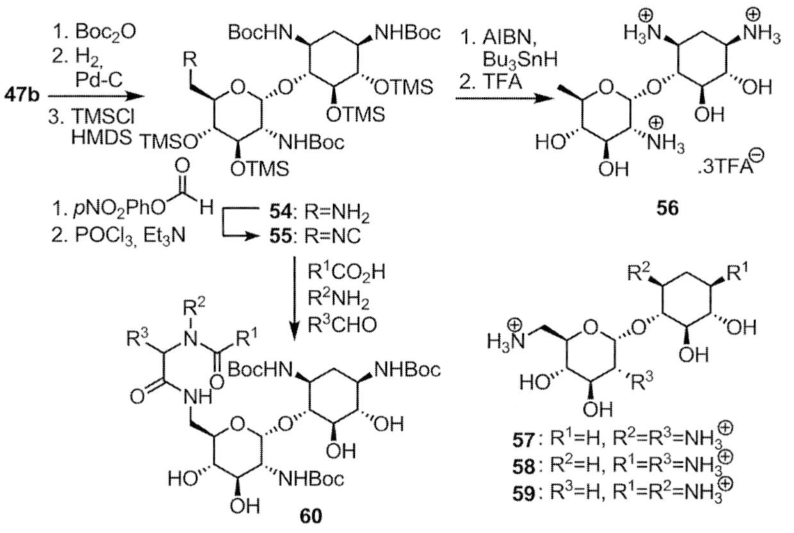

Scheme 8.16 Preparation of a range of deaminated neamine analogues (**56–59**) and Ugi condensation from isonitriles via the selectively liberated amine.

could now be achieved upon treatment of **55** with tributyltin hydride in the presence of AIBN, giving the monodeaminated structure **56** after global deprotection. The positional isomers **57–59** were prepared under identical conditions from mono-Cbz-protected neamines obtained earlier (Scheme 8.13). Herrisford et al. later recognized the potential of isocyanide **55** for the combinatorial assembly of a library of neamine-peptidyl conjugates **60** by an Ugi four-component condensation with a range of aldehydes, amines, and carboxylic acids [90].

In the same paper, neamine was also selectively *N*-6′ alkylated by reductive amination of the 1,3,2′-tri-*N*-Boc protected derivative **61**, obtained by a conventional sequence involving transient trityl protection of the primary 6′-amino function (Scheme 8.17).

Finally, in a sequence of papers from the group of Hamasaki et al. on the synthesis of neamine conjugates containing aromatic DNA intercalators at *N*-6′ for targeting of HIV RNA sequences, no temporary protective groups were employed (Scheme 8.18). Direct condensation of neamine (**36**) with a range of aryl-substituted hydroxysuccinimide esters, for example, gave, after reversed-phase HPLC, the aromatic conjugates **63** [91]. By the same strategy, the pyrene moiety was introduced in a two-step procedure involving peptide coupling between **36** and N_α-Fmoc-protected arginine under the influence of carbodiimide, followed by Fmoc removal and another selective amidation with pyrenecarbonyl succinimidyl ester, affording compound **64a** (R^2 = 1-pyrene). Unfortunately, the level of the remarkable selectivity of the above amidations remains unclear, because no yields are reported [92a]. Most recently, the strategy was extended to targeting of HIV Tar-Tat with derivatives **64a** and **64b**, modified at *N*-6′ with acetic acid functional-

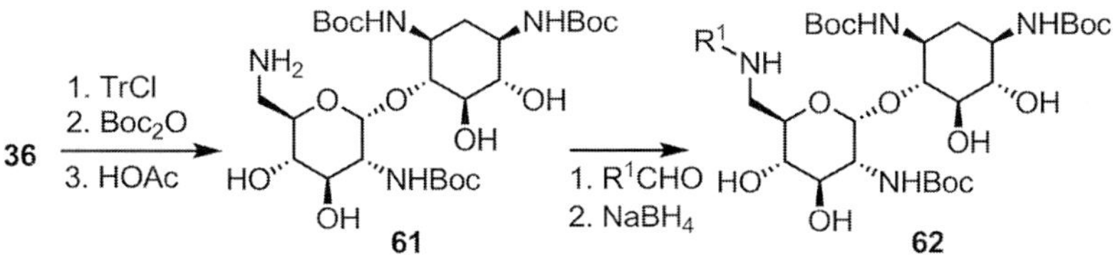

Scheme 8.17 Reductive *N*-6′ amination of neamine through trityl protection.

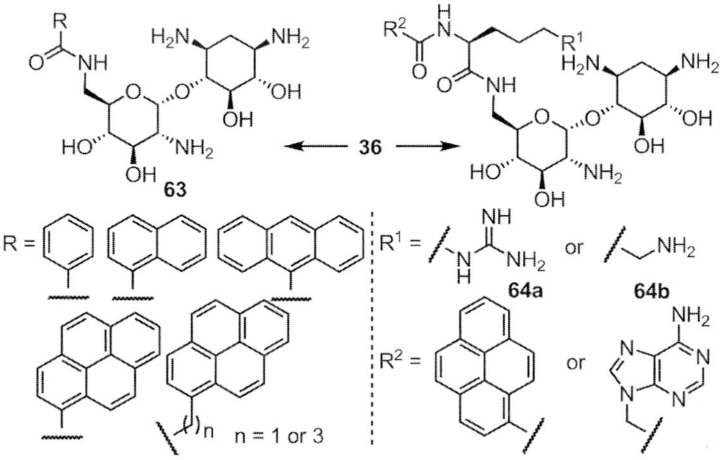

Scheme 8.18 Preparation of a range of *N*-6′ acylated analogues of neamine by selective acylation followed by reversed-phase HPLC.

ized with nucleobases at the α-position [92b]. Condensation of neamine with N_α-adenylyl-derivatized lysine (R^2 = 7-adenylylmethyl), for example, gave a yield of 23% for (protected) **64b** after HPLC.

8.5
Concluding Remarks

The overview of technologies described above for the selective manipulation of aminoglycoside amines can be roughly divided into three categories: a) based on inherent difference in amino nucleophilicity and/or basicity, b) based on protective group strategies employing neighboring hydroxy groups, and c) chemoselective reduction of azido functions.

The surge of novel semisynthetic derivatives of aminoglycosides in the previous century has provided substantial insight into the order of reactivity of amines in 4,6-linked aminoglycosides and neamine, providing the general model for nucleophilicity N-6′ > N-1 > N-3 > N-2′. Inspired by the structure of butirosin, a large number of analogues based on *N*-1 modification of natural aminoglycosides has been prepared through temporary protection of *N*-6′ followed by selective *N*-1 acylation.

The fact that many of the amino functions in aminoglycosides are flanked by a hydroxy group has been employed for the base-induced formation of cyclic carbamates from 1,2-hydroxy carbamates followed by selective hydrolysis over acyclic alkyl carbamates. In this fashion, the liberation of a single specific amine is feasible, in particular *N*-1 in 4,5-linked aminoglycosides and neamine. The presence of hydroxy functions is also taken advantage of in the transient protection of vicinal (and nonvicinal) amino alcohols with transition metal salts. Significant progress in the selective manipulation of aminoglycosides was made, in particular in combination with norbornene-derived carbamoylating reagents. It should be noted, however, that the technology appears to be limited to 4,6-linked aminoglycosides (and neamine), since no examples of 4,5-linked structures are known.

Finally, the exhaustive transformation of amino functions followed by chemoselective reduction presents a novel technology with significant practical advantages, in particular in terms of solubility, ease of purification, and interpretation of spectral data. The selective azido reduction to afford *N*-2′- or *N*-1-liberated neamine, directed by specific hydroxy protective group patterns, has opened up hitherto unprecedented chemical avenues for the preparation of selectively modified neamines [93].

References

1 Beaucaire, G. *J. Chemother.* 1995, *7* (suppl. *2*), 111–123.

2 Zembower, T. R.; Noskin, G. A.; Postelnick, M. J.; Nguyen, C.; Peterson, L. R. *Int. J. Antimicrob. Ag.* 1998, *10*, 95–105.

3 Mingeot-Leclercq, M. P.; Glupczynski, Y.; Tulkens, P. M. *Antimicrob. Agents Chemother.* 1999, *43*, 727–737.

4 Haddad, J.; Vakulenko, S.; Mobashery, S. *J. Am. Chem. Soc.* 1999, *121*, 11922–11923.

5 Neonakis, I.; Gikas, A.; Scoulica, E.; Manios, A.; Georgiladakis, A.; Tselentis, Y. *Int. J. Antimicrob. Ag.* 2003, *22*, 526–531.

6 Moazed, D.; Noller, H. F. *Nature* 1987, *327*, 389–394.

7 a) Botto, R. E.; Coxon, B. *J. Am. Chem. Soc.* 1983, *105*, 1021–1028. b) Kayul, M.; Barbieri, C. M.; Kerrigan, J. E.; Pilch, D. S. *J. Mol. Biol.* 2003, *326*, 1373–1387. c) Barbieri, C. M.; Pilch, D. S. *Biophys. J.* 2006, *90*, 1338–1349. d) Freire, F.; Cuesta, I.; Corzana, F.; Revuelta, J.; González, C.; Hricovini, M.; Bastida, A.; Jiménez-Barbero, J.; Asencio, J. L. *Chem. Commun.* 2007, 174–176.

8 Drews, J. *Science* 2000, *287*, 1960–1964.

9 Zaman, G. J. R.; Michiels, P. J. A.; van Boeckel, C. A. A. *Drug. Discov. Today* 2003, *8*, 297–306.

10 Smith, C. W. J.; Valcárel, J. *Trends Biochem. Sci.* 2000, *97*, 14035–14037.

11 a) Ecker, D. J.; Griffey, R. H. *Drug Discov. Today* 1999, *4*, 420–429. b) Gallego, J.; Varani, G. *Acc. Chem. Res.* 2001, *34*, 836–843. c) Foloppe, N.; Matassova, N.; Aboul-ela, F. *Drug Discov. Today* 2006, *11*, 1019–1027.

12 Davis, B. D.; Chen, L.; Tai, P. C. *Proc. Natl. Acad. Sci. U. S. A.* 1986, *83*, 6164–6168.

13 Ryu, D. H.; Rando, R. R. *Bioorg. Med. Chem.* 2001, *9*, 2601–2608.

14 Recht, M. I.; Douthwaite, S.; Puglisi, J. D. *EMBO J.* 1999, *18*, 3133–3138.

15 Ryu, D. H.; Rando, R. R. *Bioorg. Med. Chem. Lett.* 2002, *12*, 2241–2244.

16 Hutchin, T.; Cortopassi, G. *Antimicrob. Agents Chemother.* 1994, *38*, 2517–2520.

17 François, B.; Russell, R. J. M.; Murray, J. K.; Aboula-ela, F.; Masquida, B.; Vicens, Q.; Westhof, E. *Nucl. Acids Res.* 2005, *33*, 5677–5690.

18 Hermann, T.; Westhof, E. *J. Mol. Biol.* 1998, *276*, 903–912.

19 Kaul, M.; Barbieri, C. M.; Pilch, D. S. *J. Am. Chem. Soc.* 2006, *128*, 1261–1271.

20 a) von Ahsen, U.; Davies, J.; Schroeder, R. *Nature* 1991, *353*, 368–370. b) Liu, Y.; Tidwell, R. R.; Leibowitz, M. J. *J. Euk. Microbiol.* 1994, *41*, 31–38.

21 a) Stage, T. K.; Hertel, K. J.; Uhlenbeck, O. C. *RNA* 1995, *1*, 95–101. b) Clouet-d'Orval, B.; Stage, T. K.; Uhlenbeck, O. C. *Biochemistry* 1995, *34*, 11186–11190.

22 Rogers, J.; Chang, A. H.; von Ahsen, U.; Schroeder, R.; Davies, J. *J. Mol. Biol.* 1996, *259*, 916–925.

23 Mikkelsen, N. E.; Branvall, M.; Virtanen, A.; Kirsebom, L. A. *Proc. Natl. Acad. Sci. USA* 1999, *96*, 6155–6160.

24 a) Zapp, M. L.; Stern, S.; Green, M. R. *Cell* 1993, *74*, 969–978. b) Werstuck, G.; Zapp, M. L.; Green, M. R. *Chem. Biol.* 1996, *3*, 129–137.

25 a) Hamy, F.; Brondani, V.; Flörsheimer, A.; Stark, W.; Blommers, M. J. J.; Klimkait, T. *Biochemistry* 1998, *37*, 5086–5095. b) Mei, H.-Y.; Cui, M.; Heldsinger, A.; Lemrow, S. M.; Loo, J. A.; Sannes-Lowry, K. A.; Sharmeen, L.; Czarnik, A. W. *Biochemistry* 1998, *37*, 14204–14212.

26 McPike, M. P.; Sullivan, J. M.; Goodisman, J.; Dabrowiak, J. C. *Nucl. Acids Res.* 2002, *30*, 2825–2831.

27 a) Kirk, S. R.; Tor, Y. *Bioorg. Med. Chem.* 1999, *7*, 1979–1991. b) Mikkelsen, N. E.; Johansson, K.; Virtanen, A.; Kirsebom, L. A. *Nat. Struct. Biol.* 2001, *8*, 510–514. c) Walter, F.; Pütz, J.; Giegé, R.; Westhof, E. *EMBO J.* 2002, *21*, 760–768. d) Evans, J. M.; Turner, B. A.; Bowen, S.; Ho, A. M.; Sarver, R. W.; Benson, E.; Parker, C. N. *Bioorg. Med. Chem.* 2003, *13*, 993–996.

28 a) Yan, Z.; Baranger, A. M. *Bioorg. Med. Chem. Lett.* 2004, *14*, 5889–5893. b) Verhelst, S. H. L.; Michiels, P. J. A.; van der Marel, G. A.; van Boeckel, C. A. A.; van Boom, J. H. *ChemBioChem* 2004, *5*, 937–942. c) DeNap, J. C. B.; Thomas, J. R.; Musk, D. J.; Hergenrother, P. J. *J. Am. Chem. Soc.* 2004, *126*, 15402–15404. d) Thomas, J. R.; DeNap, J. C. B.; Wong, M. L.; Hergenrother, P. J. *Biochemistry* 2005, *44*, 6800–6808.

29 a) Ennifar, E.; Paillart, J. C.; Marquet, R.; Ehresmann, B.; Ehresmann, C.; Dumas, P.; Walter, P. *J. Biol. Chem.* 2003, *278*, 2723–2730. b) Ennifar, E.; Paillart, J. C.; Bodlenner, A.; Walter, P.; Weibel, J.-M.; Aubertin, A.-M.; Pale, P.; Dumas, P.; Dumas, P. *Nucl. Acids Res.* 2006, *34*, 2328–2339.

30 a) Wang, Y.; Rando, R. R. *Chem. Biol.* 1995, *2*, 281–290. b) Wang, Y.; Killian, J.; Hamasaki, K.; Rando, R. R. *Biochemistry* 1996, *114*, 234–244.

31 a) Arya, D. P.; Lane Coffee, R., Jr. *Bioorg. Med. Chem. Lett.* 2000, *10*, 1897–1899. b) Arya, D. P.; Lane Coffee, R., Jr.; Willis, B.; Abramovitch, A. I. *J. Am. Chem. Soc.* 2001, *123*, 5385–5395.

32 a) Hendrix, M.; Alper, P. B.; Priestley, E. S.; Wong, C.-H. *Angew. Chem. Int. Ed.* 1997, *36*, 95–98. b) Wong, C.-H.; Hendrix, M.; Manning, D. D.; Rosenbohm, C.; Greenberg, W. A. *J. Am. Chem. Soc.* 1998, *120*, 8319–8327.

33 a) Tor, Y.; Hermann, T.; Westhof, E. *Chem. Biol.* 1998, *5*, R277–283. b) Wang, H.; Tor, Y. *Angew. Chem. Int. Ed.* 1998, *37*, 101–111.

34 Wright, J. J. *J. Chem. Soc., Chem. Commun.* 1976, 206–208.

35 a) Umezawa, H.; Umezawa, S.; Tsuchiya, T.; Okazaki, Y. *J. Antibiot.* 1971, *24*, 485–487. b) Umezawa, S.; Umezawa, H.; Okazaki, Y.; Tsuchiya, T. *Bull. Chem. Soc. Jpn.* 1972, *45*, 3624–3628.

36 Kawaguchi, U.; Naito, T.; Nakagawa, S.; Fujisawa, K. *J. Antibiot.* 1972, *25*, 695–708.

37 Kondo, S.; Iinuma, K.; Yamamoto, H.; Maeda, K.; Umezawa, H. *J. Antibiot.* 1973, *26*, 412–415.

38 Wang, H.; Tor, Y. *J. Am. Chem. Soc.* 1997, *119*, 8734–8735.

39 a) Baker, T. J.; Luedtke, N. W.; Tor, Y.; Goodman, M. *J. Org. Chem.* 2000, *65*, 9054–9058. b) Luedtke, N. W.; Baker, T. J.;

Goodman, M.; Tor, Y. *J. Am. Chem. Soc.* 2000, *122*, 12035–12036.

40 a) Alper, P. B.; Hendrix, M.; Sears, P.; Wong, C.-H. *J. Am. Chem. Soc.* 1998, *120*, 1965–1978. b) Greenberg, W. A.; Priestley, E. S.; Sears, P. S.; Alper, P. B.; Rosenbohm, C.; Hendrix, M.; Hung, S.-C.; Wong, C.-H. *J. Am. Chem. Soc.* 1999, *121*, 6527.

41 a) Fourmy, D.; Recht, M. I.; Blanchard, S. C.; Puglisi, J. D. *Science* 1996, *274*, 1367. b) Jiang, L.; Patel, D. J. *Nat. Struct. Biol.* 1998, *5*, 769–774.

42 a) Jin, E.; Katritch, V.; Olson, W. K.; Kharatisvili, M.; Abagyan, R.; Pilch, D. S. *J. Mol. Biol.* 2000, *298*, 95–110. b) Pilch, D. S.; Kaul, M.; Barbieri, C. M.; Kerrigan, J. E. *Biopolymers* 2003, *70*, 58–79.

43 Thomas, J. R.; Liu, X.; Hergenrother, P. J. *Biochemistry* 2006, *45*, 10928–10938.

44 Branch, A. D. *Trends Biochem. Sci.* 1998, *23*, 45–50.

45 Hammonds, S. M.; Candy, A. A.; Hanon, G. J. *Nature Rev. Genet.* 2001, *2*, 111–119.

46 Elbashir, S. M.; Harborth, J.; Lendeckel, W.; Yalcin, A.; Weber, K.; Tuschl, T. *Nature* 2001, *411*, 494–498.

47 Diekema, D. J.; Jones, R. N. *Drugs* 2000, *59*, 7–16.

48 a) Hermann, T. *Angew. Chem. Int. Ed.* 2000, *39*, 1890. b) Sucheck, S. J.; Wong, C.-H. *Curr. Opin. Chem. Biol.* 2000, *4*, 678–686. c) Hermann, T. *Biopolymers* 2003, *70*, 4–18.

49 Park, W. K. C.; Auer, M.; Jaksche, H.; Wong, C.-H. *J. Am. Chem. Soc.* 1996, *118*, 10150–10155.

50 a) Sucheck, S. J.; Greenberg, W. A.; Tolbert, T. J.; Wong, C.-H. *Angew. Chem. Int. Ed.* 2000, *39*, 1080–1084. b) Hanessian, S.; Tremblay, M.; Kornienko, A.; Moitessier, N. *Tetrahedron* 2001, *57*, 3255–3265. c) Haddad, J.; Kotra, L. P.; Llano-Sotelo, B.; Kim, C.; Azucena, E. F., Jr.; Liu, M.; Vakulenko, S. B.; Chow, C. S.; Mobashery, S. *J. Am. Chem. Soc.* 2002, *124*, 3229–3237. d) Wang, J.; Li, J.; Tuttle, D.; Takemoto, J. Y.; Chang, C.-W. T. *Org. Lett.* 2002, *4*, 3997–4000. e) Hanessian, S.; Tremblay, M.; Swayze, E. E. *Tetrahedron* 2003, *59*, 983–993. f) François, B.; Szychowski, J.; Adhikari, S. S.; Pachamuthu, K.; Swayze, E. E.; Griffey, R. H.; Migawa, M. T.; Westhof, E.; Hanessian, S. *Angew. Chem. Int. Ed.* 2004, *43*, 6735–6738. g) Liang, F.-S.; Wang, S.-K.; Nakatani, T.; Wong, C.-H. *Angew. Chem. Int. Ed.* 2004, *43*, 6495–6500. h) Rege, K.; Moore, J. A.; Dordick, J. S.; Cramer, S. M. *J. Am. Chem. Soc.* 2004, *126*, 12306–12315. i) Li, J.; Wang, J.; Czyryca, P. G.; Chang, H.; Orsak, T. W.; Evanson, R.; Chang, C.-W. T. *Org. Lett.* 2004, *6*, 1381–1384. j) Fridman, M.; Belakhov, V.; Lee, L. V.; Liang, F.-S.; Wong, C.-H.; Baasov, T. *Angew. Chem. Int. Ed.* 2005, *44*, 447–452. k) Wang, J.; Li, J.; Chen, H.-N.; Chang, H.; Tanifum, C. T.; Liu, H.-H.; Czyryca, P. G.; Chang, C.-W. T. *J. Med. Chem.* 2005, *48*, 6271–6285.

51 a) Wang, H.; Tor, Y. *Bioorg. Med. Chem. Lett.* 1997, *7*, 1951–1956. b) Tok, J. B.-H.; Huffman, G. R. *Bioorg. Med. Chem. Lett.* 2000, *10*, 1593–1595. c) Sucheck, S. J.; Wong, A. L.; Koeller, K. M.; Boehr, D. D.; Draker, K.; Sears, P.; Wright, G. D.; Wong, C.-H. *J. Am. Chem. Soc.* 2000, *122*, 5230–5231. d) Tok, J. B.-H.; Dunn, L. J.; Des Jean, R. C. *Bioorg. Med. Chem. Lett.* 2001, *11*, 1127–1131. e) Tok, J. B.-H.; Fenker, J. *Bioorg. Med. Chem. Lett.* 2001, *11*, 2987–2991. f) Luedtke, N. W.; Liu, Q.; Tor, Y. *Biochemistry* 2003, *42*, 11391–11403. g) Riguet, E.; Désiré, J.; Boden, O.; Ludwig, V.; Göbel, M.; Bailly, C.; Décout, J.-L. *Bioorg. Med. Chem. Lett.* 2005, *15*, 4651–4655. h) Liang, C.-H.; Romero, A.; Rabuka, D.; Sgarbi, P. W. M.; Marby, K. A.; Duffield, J.; Yao, S.; Cheng, M. L.; Ichikawa, Y.; Sears, P.; Hu, C.; Hwang, S.-B.; Shue, Y.-K.; Sucheck, S. J. *Bioorg. Med. Chem. Lett.* 2005, *15*, 2123–2128. i) Nudelman, I.; Rebibo-Sabbah, A.; Shallom-Shezifi, D.; Hainrichson, M.; Stahl, I.; Ben-Yosef, T.; Baasov, T. *Bioorg. Med. Chem. Lett.* 2006, *16*, 6310–6315.

52 a) Kirk, S. R.; Luedtke, N. W.; Tor, Y. *J. Am. Chem. Soc.* 2000, *122*, 980–981. b) Liu, M.; Haddad, J.; Azucena, E.; Kotra, L. P.; Kirzhner, M.; Mobashery, S. *J. Org. Chem.* 2000, *65*, 7422–7431. c) Charles, I.; Xue, L.; Arya, D. P. *Bioorg. Med. Chem. Lett.* 2002, *12*, 1259–1262. d) Lee, J.; Kwon, M.; Lee, K. H.; Jeong, S.; Hyun, S.; Shin, K. J.; Yu, J. *J. Am. Chem. Soc.* 2004, *126*, 1956–1957. e) Ahn, D.-R.; Yu, J. *Bioorg. Med.*

Chem. 2005, *13*, 1177–1183. f) Hyun, S.; Lee, K. H.; Yu, J. *Bioorg. Med. Chem. Lett.* 2006, *16*, 4757–4759. g) Kaiser, M.; Sainlos, M.; Lehn, J.-M.; Bombard, S.; Teulade-Fichou, M.-P. *ChemBioChem* 2006, *7*, 321–329. h) Blount, K. F.; Tor, Y. *ChemBioChem* 2006, *7*, 1612–1621. i) Charles, I.; Xi, H.; Arya, D. P. *Bioconj. Chem.* 2007, *18*, 160–169.

53 Schatz, A.; Bagie, E.; Waksman, S. A. *Proc. Soc. Exp. Biol. Med.* 1944, *55*, 66–69.

54 Waksman, S. A.; Lechevalier, H. A. *Science* 1949, *109*, 305–307.

55 Umezawa, H.; Tazaki, T.; Okami, Y.; Fukuyama, S. *J. Antibiot.* 1949, *3*, 232–235.

56 Umezawa H.; Ueda M.; Maeda, K.; Yagishita, K.; Kondo S.; Okami Y.; Utahara R.; Osato, Y.; Nitta K.; Takeuchi, T. *J. Antibiot. A* 1957, *10*, 181–188.

57 Weinstein, M. J.; Luedemann, G. M.; Oden, E. M.; Wagman, G. H.; Rosselet, J. P.; Marquez, J. A.; Coniglio, C. T.; Charney, W.; Herzog, H. L.; Black, J. *J. Med. Chem.* 1963, *6*, 463–464.

58 a) Woo, P. W. K.; Dion, H. W.; Coffey, G. L.; Fusari, S. A.; Senos, G., US Patent 3,541,078, Nov. 17th, 1970. b) Woo, P. W. K.; Dion, H. W.; Bartz, Q. R. *Tetrahedron Lett.* 1971, *28*, 2617–2620.

59 a) Umezawa, H.; Kondo, S. in *Handbook of Experimental Pharmacology, Vol. 62, Aminoglycoside Antibiotics*, Umezawa, H., Hooper, I. R. (editors), Berlin Heidelberg New York, Springer-Verlag, 1982, 267–292. b) Kondo, S.; Hotta, K. *J. Infect. Chemother.* 1999, *5*, 1–9.

60 Shomura, T.; Ezaki, N.; Tsuruoka, T.; Niwa, T.; Akita, E.; Niida, T. *J. Antibiotics* 1970, *23*, 155–172.

61 Ikeda, D.; Tsuchiya, T.; Umezawa, S.; Umezawa, H. *J. Antibiot.* 1972, *25*, 741–742.

62 Umezawa, S.; Tsuchiya, T.; Ikeda, D.; Umezawa, H. *J. Antibiot.* 1972, *25*, 613–616.

63 Haskell, T. H.; Rodebaugh, R.; Plessas, N.; Watson, D.; Westland, R. D. *Carbohydr. Res.* 1973, *28*, 263–280.

64 Tsukiura, H.; Fujisawa, K.; Konishi, M.; Saito, K.; Numata, K.; Ishikawa, H.; Miyaki, T.; Tomita, K.; Kawaguchi, H. *J. Antibiot.* 1973, *26*, 351–357.

65 Rhinehart, K. L., Jr. in *The Neomycins and Related Antibiotics*, New York, Wiley and Sons, Inc., 1964, 93.

66 Umezawa, S.; Tsuchiya, T.; Muto, R.; Nishimura, Y.; Umezawa, T. *J. Antibiot.* 1971, *24*, 274–275.

67 Umezawa, S.; Tsuchiya, T.; Jikihara, T. *J. Antibiot.* 1971, *24*, 711–712.

68 Nagabhushan, T. L.; Cooper, A. B.; Tsai, H.; Daniels, P. J. L.; Miller, G. H. *J. Antibiot.* 1978, *31*, 681–687.

69 Cooper, A. B.; Daniels, P. J.; Nagabhushan, T. L.; Rane, D.; Turner, W. N.; Weinstein, J. *J. Antibiot.* 1976, *29*, 714–719.

70 a) Grapsas, I.; Cho, Y. J.; Mobashery, S. *J. Org. Chem.* 1994, *59*, 1918–1922. b) Grapsas, I.; Massova, I.; Mobashery, S. *Tetrahedron* 1998, *54*, 7705–7720.

71 Hanessian, S.; Patil, G. *Tetrahedron Lett.* 1978, *12*, 1035–1038.

72 Nagabhushan, T. L.; Cooper, A. B.; Turner, W. N.; Tsai, H.; McCombie, S.; Mallams, A. K.; Rane, D.; Wright, J. J.; Reichert, P.; Boxler, D. L.; Weinstein, J. *J. Am. Chem. Soc.* 1978, *100*, 5253–5254.

73 Tsuchiya, T.; Takagi, Y.; Umezawa, S. *Tetrahedron Lett.* 1978, *51*, 4951–4954.

74 Kirst, H. A.; Truedell, B. A.; Toth, J. E. *Tetrahedron Lett.* 1981, *22*, 295–298.

75 Lee, S. H.; Cheong, C. H. *Tetrahedron* 2001, *57*, 4801–4815.

76 Weinstein, M. J.; Marquez, J. A.; Testa, R. T.; Wagman, G. J.; Oden, E. M.; Waitz, J. A. *J. Antibiot.* 1970, *23*, 551–554.

77 Georgiadis, M. P.; Constantinou-Kokotou, V. *J. Carbohydr. Chem.* 1991, *10*, 739–748.

78 a) Umezawa, S.; Koto, S.; Tatsuta, K.; Hineno, H.; Nishimura, Y.; Tsumura, T. *Bull. Chem. Soc. Jpn.* 1969, *42*, 537–541. b) Kumar, V.; Remers, W. A. *J. Org. Chem.* 1978, *43*, 3327–3331. c) Tohma, S.; Yoneta, T.; Fukatsu, S. *J. Antibiot.* 1980, *33*, 671–674.

79 a) Alper, P. B.; Hung, S.-C.; Wong, C.-H. *Tetrahedron Lett.* 1996, *37*, 6029–6032. b) Nyffeler, P. T.; Liang, C.-H.; Koeller, K. M.; Wong, C.-H. *J. Am. Chem. Soc.* 2002, *124*, 10773–10778.

80 Riguet, E.; Désiré, J.; Bailly, C.; Décout, J.-L. *Tetrahedron* 2004, *60*, 8053–8064.

81 Sharma, M. N.; Kumar, V.; Remers, W. A. *J. Antibiot.* 1982, *35*, 905–910.

82 Roestamadji, J.; Grapsas, I.; Mobashery, S. *J. Am. Chem. Soc.* 1995, *117*, 11060–11069.

83 Gao, F.; Yan, X.; Baettig, O. M.; Berghuis, A. M.; Auclair, K. *Angew. Chem. Int. Ed.* 2005, *44*, 6859–6862.

84 Roestamadji, J.; Mobashery, S. *Bioorg. Med. Chem. Lett.* 1998, *8*, 3483–3488.

85 a) Knouzi, N.; Vaultier, M.; Carrié, R. *Bull. Chem. Soc. Chim. Fr.* 1985, *5*, 815–819. b) Ariza, X.; Urpí, F.; Viladomat, C.; Vilarrasa, J. *Tetrahedron Lett.* 1998, *39*, 9101–9102.

86 Li, J.; Chen, H.-N.; Chang, H.; Wang, J.; Chang, C.-W. T. *Org. Lett.* 2005, *7*, 3061–3064.

87 Rai, R.; Chen, H.-N.; Czyryca, P. G.; Li, J.; Chang, C.-W. T. *Org. Lett.* 2006, *8*, 887–889.

88 a) Horii, S.; Fukase, H.; Kameda, Y.; Mizokami, N. *Carbohydr. Res.* 1978, *60*, 275–288. b) Torii, T.; Tsuchiya, T.; Umezawa, S. *J. Antibiot.* 1982, *35*, 58–61.

89 Jiao, G.-S.; Simo, O.; Nagat, M.; O'Malley, S.; Hemscheidt, T.; Cregar, L.; Millis, S. Z.; Goldman, M. E.; Tang, C. *Bioorg. Med. Chem. Lett.* 2006, *16*, 5183–5189.

90 Nunns, C. L.; Spence, L. A.; Slater, M. J.; Berrisford, D. J. *Tetrahedron Lett.* 1999, *40*, 9341–9345.

91 Hamasaki, K.; Woo, M.-C.; Ueno, A. *Tetrahedron Lett.* 2000, *41*, 8327–8332.

92 a) Hamasaki, K.; Ueno, A. *Bioorg. Med. Chem. Lett.* 2001, *11*, 591–594. b) Yajima, S.; Shionoya, H.; Akagi, T.; Hamasaki, K. *Bioorg. Med. Chem. Lett.* 2006, *14*, 2799–2809.

93 Li, J.; Chiang, F.-I.; Chen, H.-N.; Chang, C.-W. T. *J. Org. Chem.* 2007, *72*, 4055–4066.

9
Evolution of Transition Metal-Catalyzed Amination Reactions: the Industrial Approach

Ulrich Scholz

9.1
Introduction: First Steps in the Field of Catalytic Aromatic Amination

In 1901, F. Ullmann reported that when 2-bromonitrobenzene is heated in the presence of finely ground copper powder, the metal loses its gloss and turns gray [1]. His group isolated a bromine-free organic compound, quickly identified as 2,2′-dinitrobiphenyl, together with copper bromide (Scheme 9.1).

While it was at that time not a catalytic approach, it was recognized that, with the help of copper, halogen atoms on aromatic system can be replaced more easily than without the metal. Some halogenated aromatic compounds with so-called "movable" halogens, such as dinitrochlorobenzene, picric chloride, or chloronitrobenzoic ester, had been investigated before this, but Ullmann extended the methodology to a much broader scope (i.e., to aromatic compounds that did not bear several electron-withdrawing groups), describing, in addition to bromo arenes, the coupling of iodo and chloro derivatives. Ullmann also found that the nature of the copper powder influenced the ease of the reaction. When his group started out, they preferred molecular copper (i.e., copper sulfate that was reduced with zinc just before using it). They later found out that commercial copper bronze does not need prior treatment with a reducing agent, nor was a washing procedure necessary for preactivation.

The order of reactivity of halogenated aromatics was also already recognized in their publication; to produce 2,2′-dinitrobiphenyl on small scales, 2-bromonitrobenzene was a good starting material, whereas for performing the same reaction on larger scales, the group suggested the use of 2-chloronitrobenzene as starting material, since with the bromo compounds the strongly exothermic nature of the reaction could no longer be controlled. When turning to less powerfully activated starting materials, such as 3-nitroarenes, Ullmann suggested the use of iodo derivatives as starting materials, iodine being more easily replaceable than bromine.

In subsequent years the group extended their work to the synthesis of diaryl ethers [2], in which they recognized the important role of a base for complete conversion of the starting materials [3]. When bromobenzene and phenol were

Amino Group Chemistry. From Synthesis to the Life Sciences. Edited by Alfredo Ricci

ISBN: 978-3-527-31741-7

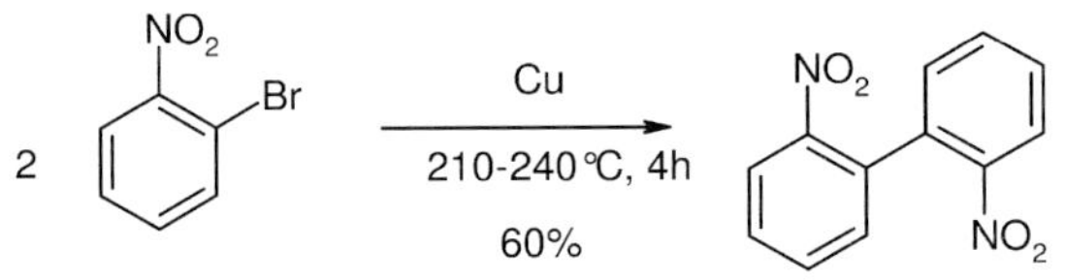

Scheme 9.1 Ullmann biphenyl synthesis of 1901.

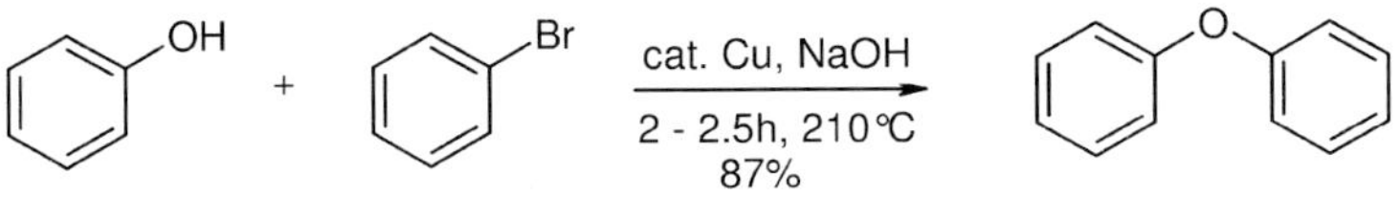

Scheme 9.2 Ullmann diphenylether synthesis of 1905.

mixed in the presence of one equivalent of sodium hydroxide alone, only traces of product were isolated, even after elongated times and high temperatures. When a catalytic amount of copper was added, however, the product was generated within hours and could be isolated by distillation in high yield (Scheme 9.2).

They also recognized the influence of the nature of the base, since on replacement of sodium hydroxide with potassium metal the isolated yield went from 87% to 90%. As the use of solvents was not usually the first option in those days, all these reactions had been run by mixing the neat starting materials, so these old procedures represent not only a very practical but also an economical approach to biphenyl and aryl ether compounds and have been used in many applications since then.

The first important milestone for the synthesis of arylamines with the use of copper as a mediator, though still with stoichiometric amounts of the metal, was again presented by Ullmann [2].

In close cooperation with Ullmann, Irma Goldberg published the first application of a copper-catalyzed amidation of an aryl bromide with the synthesis of 2-hydroxy-*N*-phenyl-benzamide [4]. While an explanation for the ease of this reaction (i.e., the roles of salicylic amide not only as the starting material but also as a chelating and activating ligand for the copper catalyst) was provided much later, her synthesis of 2-hydroxy-*N*-phenylbenzamide has to be regarded as one of the first examples of a modern, copper-catalyzed arylamide synthesis (Scheme 9.3).

The first procedure for a copper-catalyzed aryl amination was again presented by Goldberg and Ullmann, in 1905. The described conditions almost resemble today's methods for aryl amination for the synthesis of arylanthranilic acid [5] (Scheme 9.4).

It is therefore important to note that the scope for improvement in this reaction might have already seemed limited even in those days. Highly specific starting materials undergo halogen–amine exchange in high yields. Still, for this reaction to be developed into a modern, reliable, broadly applicable, and simple to run process, almost one hundred years and countless laboratory hours still had to pass.

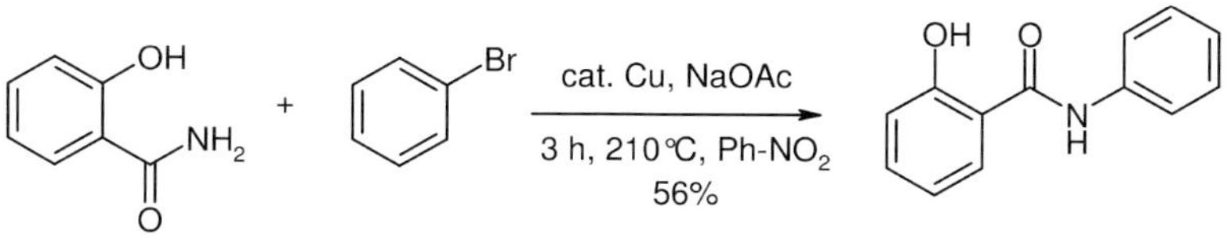

Scheme 9.3 Goldberg's 2-hydroxyphenylbenzamide synthesis of 1906.

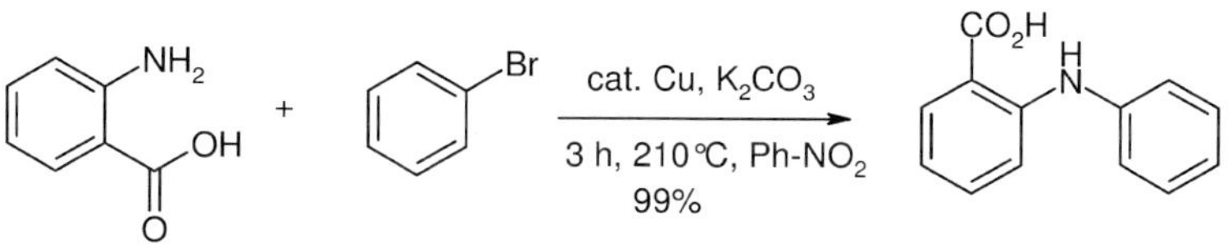

Scheme 9.4 Goldberg's arylanthranlic acid synthesis of 1905.

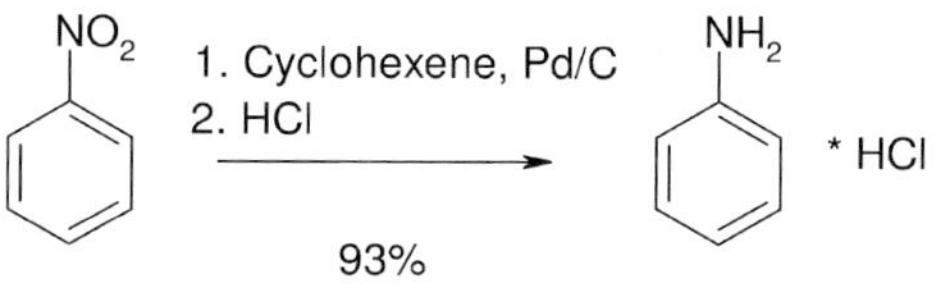

Scheme 9.5 Synthesis of aniline hydrochloride by transfer hydrogenation.

9.2
Alternatives to Transition Metal-Catalyzed Arylamination

9.2.1
Reduction of Nitroarenes

Arylamine synthesis is still today often focused on the catalytic reduction of nitroarenes and subsequent alkylation with alkyl halides. Since nitration of aromatic compounds is a well established, economical, and reliably controllable process, this approach has much more importance for the synthesis of arylamines than for the synthesis of alkylamines, since alkyl-nitro compounds are far less easily prepared.

Hundreds of techniques for the reduction of nitroarenes are available; a few prominent ones are listed below.

9.2.1.1 Transfer Hydrogenation

Nitroarenes are easily reduced to anilines by transfer hydrogenation. With the help of a transition metal catalyst, hydrogen is effectively transferred from a hydrogen source, such as cyclohexene, to, for example, nitrobenzene [6] (Scheme 9.5).

Alternative catalysts are, for example, platinum catalysts in combination with ammonium formate or formic acid as hydrogen source [7]. In most cases transfer hydrogenation is preferred if the use of hydrogen is unfavorable due to safety issues, supply constraints, or the absence of high-pressure equipment.

9.2.1.2 Direct Hydrogenation

Even more common – and from an industrial viewpoint one of the reactions in which isolated yields below 90% are already considered to be rather poor – the reduction of nitroarenes through the use of a transition metal catalyst in combination with molecular hydrogen is extremely common. Hydrogen is still the cheapest reducing agent available per kilogram, and the catalysts can often be reused or, if not, recycled. Disadvantageous, of course, is the need for a safe infrastructure for this highly combustible gas and the need for high-pressure equipment. Since the reduction of one nitro group consumes three equivalents of hydrogen and reactions are usually run at concentrations above 20%, a constant supply of hydrogen, rapid transport of the reactive gas into the reaction medium, and the removal of the enormous reaction heat from the reactor have to be allowed for to guarantee a process with high catalytic activity and few side products. The stepwise nature of the reaction – from nitro to nitroso to hydroxylamine to amine – is mentioned even in 19th century textbooks [8–10] and by clever design of the reaction, each of these intermediates could, if desired, be isolated as the product.

Common metals are platinum, nickel, chromium, or palladium, while rhodium, iridium, and ruthenium are also used quite often. The transition metal catalyst is applied either as a soluble complex, a metal sponge, or a skeletal or supported catalyst. Supports are numerous, and include charcoal, silica, or modern materials such as mesoporous structures. Each support and the way the catalyst is manufactured change the reactivity and selectivity of the catalyst. Modern catalyst manufacturers offer compatibility lists and allow the precise planning of a synthesis that will reduce the nitro group and leave other functionalities, even those sensitive to reduction, completely intact. More information about these catalysts can be found on the homepages of typical catalyst vendors, such as Engelhard (now BASF), Heraeus, Johnson Matthey, or Degussa.

Since these catalysts often lose only part of their activity during a reduction process or can be reactivated by a simple washing procedure, larger scale industrial processes usually work by keeping 90% of the catalyst inside the reactor after a complete reaction run; 10% is taken out and 10% of fresh catalyst is added. The 10% of the "spent" catalyst are then recycled. Therefore only the rework of these 10% (plus the loss of precious metal in the rework) has to be included in the economic consideration of this reaction. For this reason even the use of costly metals such as platinum or rhodium is still regarded as economical, even for bulk chemicals.

High selectivities are possible; just one selective method is the reduction of a nitro group in the presence of a cinnamic acid as described by Silverman in 1944, by use of a Raney nickel skeletal catalyst [11] (Scheme 9.6).

9.2.1.3 Other Methods for Nitro Reductions

The use of iron in acidic media as a reducing agent for aromatic nitro compounds, either as metal or as salt, is often referred to as the Bechamp reduction [12].

Alternatives in alkaline media often show higher selectivities [13] and methods are available in which iron(III) chloride can be applied in catalytic

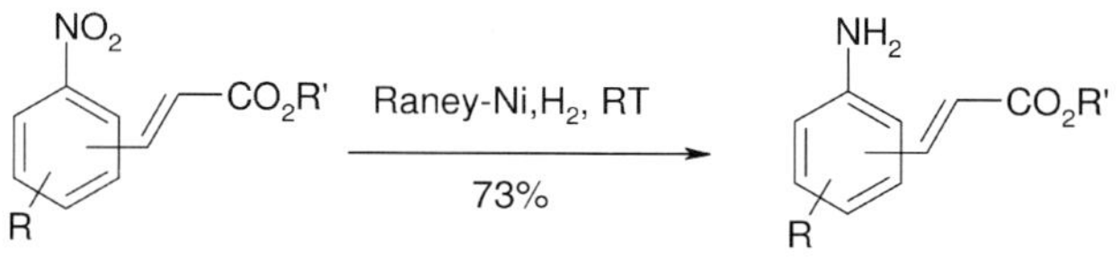

Scheme 9.6 Selective reduction of nitro groups.

amounts, while hydrazine is added in equimolar amounts as the stoichiometric reducing agent.

Other publications use sodium sulfide for the reduction of nitro groups, though disulfide dimers are often observed [14]; they can sometimes be circumvented by the use of additives [15]. When polynitro compounds have to be partly reduced, sulfur-based reducing agents also show interesting properties [16], such as in the reduction of 1,3-dinitrobenzene to 3-nitroaniline. Other prominent reducing agents include tin or tin chlorides [17], hydrazine either alone [18] or with catalytic amounts of zinc [19], or complex metal hydrides such as the DIBAH/DMS complex [20] or sodium boranate in combination with a transition metal [21]. Stoichiometric metals such as zinc [22], aluminum [23], and lithium [24] or electrochemical methods have also been reported [25].

9.2.2
Transition Metal-Free Alternatives for Amine–Halogen Exchange

Of course, arylamines had long before been prepared from aryl halides without the use of a transition metal or other catalyst, with the first examples dating from 1891. Only a set of highly activated examples and harsh conditions have been published, however, only strongly electron-withdrawing substituents furnish "mobile" halogen leaving groups in all these cases, such as in *o*- and *p*-chloronitrobenzene, chlorodinitrobenzene, and chloro-trinitrobenzene [26–29]. Even though this method is too limited in scope for a general approach, numerous applications using similar conditions were published later [30, 31].

9.2.2.1 **Metal-Free Replacement of Halogens with Amines**

One example is Kym's 1895 synthesis of ditolylphenyldiamine, probably by an arine mechanism. Through treatment of dibromobenzene with *p*-toluidine in the presence of sodium lime ($NaOH/Ca(CH)_2$) as a base, ditolylphenyldiamine can be isolated in high yield [32] (Scheme 9.7).

Another example, this time from 1934, is Foohey's synthesis of 3-chloro-6-nitroaniline from 2,4-dichloronitrobenzene [33, 34] (Scheme 9.8).

In some cases shifts of the newly introduced functionalities have been observed, easily explained by the intermediacy of an arine discussed above [35]. A good demonstration of this theory is the following example. In 1945, Gilman showed that when *o*-chloroanisole is treated with sodium amide, *m*-aminoanisole can be isolated in 45% yield (Scheme 9.9). Since the methoxy group of the intermediate methoxyamine induces a meta-directing effect on the addition of the amine.

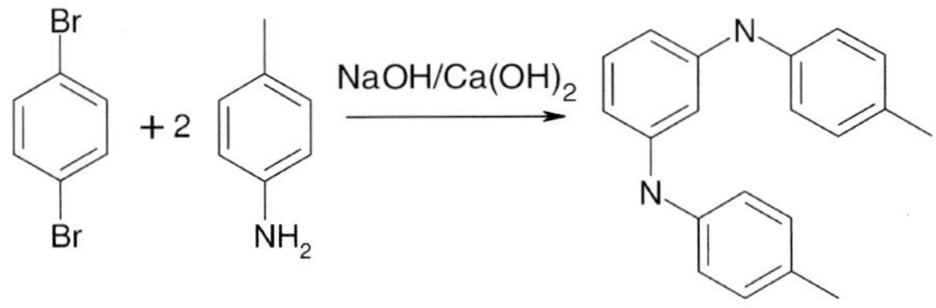

Scheme 9.7 Synthesis of ditolylphenyldiamine.

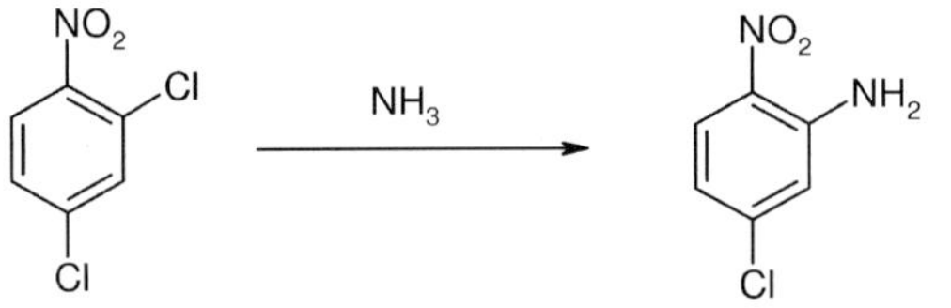

Scheme 9.8 Synthesis of 3-chloro-6-nitroaniline without copper.

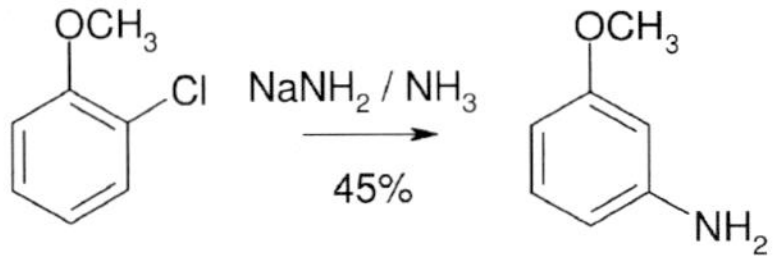

Scheme 9.9 Synthesis of *m*-aminoanisol.

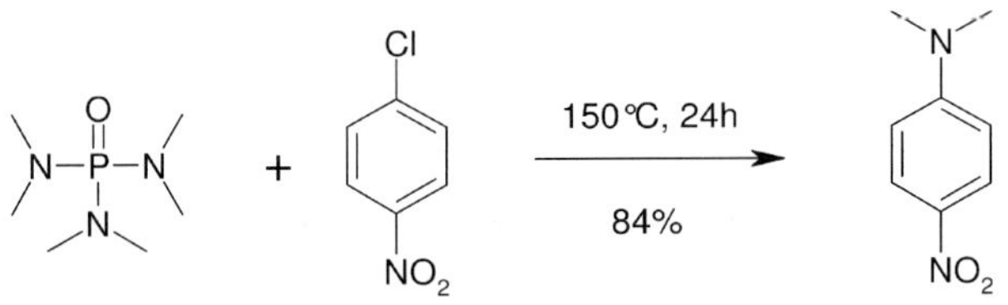

Scheme 9.10 Synthesis of 4-nitro-*N,N*-dimethylaniline.

The amine nucleophile can also be introduced in the form of a phosphoramidite such as HMPT. *N,N*-Dimethyl-4-nitroaniline, for example, can be synthesized very effectively from *p*-chloronitrobenzene and HMPT at 150 °C, as shown by Idoux in 1982 [36] (Scheme 9.10).

9.2.2.2 **The Chichibabin Reaction**

An important alternative arrived in 1914 with Chichibabin's attempts to introduce amines into aromatic substrates directly by the use of strong bases [37, 38], although his methodology was again, usually limited to activated aromatic carbon hydrogen bonds. These are to be found in heterocyclic compounds [39], such as pyridine, so the textbook example of the Chichibabin reaction is the synthesis of 2-aminopyridine (Scheme 9.11).

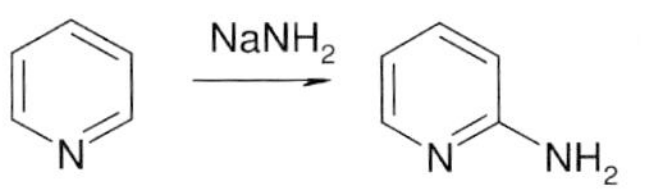

Scheme 9.11 Synthesis of aminopyridine.

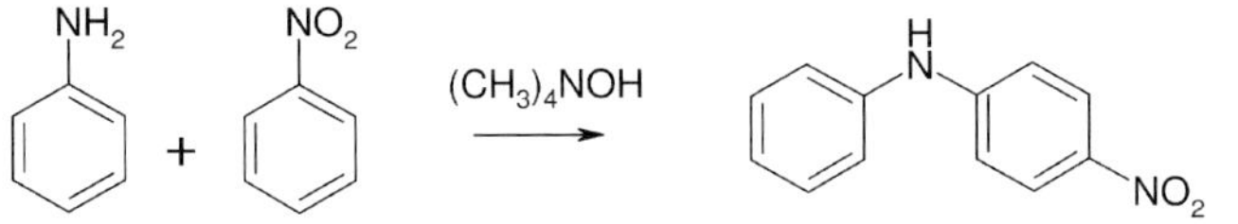

Scheme 9.12 Commercial route to the synthesis of 4-nitrodiphenylamine.

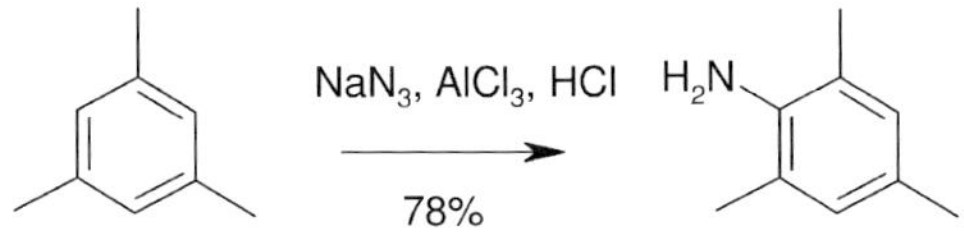

Scheme 9.13 Synthesis of 2,4,6-trimethylaniline.

The Chichibabin reaction offers advantages in cases in which it is difficult to generate the corresponding aryl nitro compound and to reduce the nitro group afterwards, since these are methodologies subject to complementary directing effects. However in comparison with a halide–amine exchange reaction on aromatic compounds, both are nucleophilic processes, and therefore follow similar rules, so the obvious limitations of the Chichibabin approach often outweigh the advantages and the reaction is rarely used these days.

9.2.2.3 **The Nucleophilic Aromatic Substitution of Hydrogen (NASH Reaction)**

Even with these obvious disadvantages, however, direct amination can in some cases be the method of choice. In 1942 Bergstrom published the synthesis of a nitrotriarylamine in low yield on treatment of nitrobenzene with the sodium salt of diphenylamine [40]. While this is not a very attractive process in itself, a similar method, using a strongly basic phase-transfer catalyst instead, is today used on a multi thousand ton scale for the synthesis of rubber antioxidants (Scheme 9.12) [41–43]. This reaction is sometimes referred to as the NASH reaction.

9.2.2.4 **Aromatic Amination by Use of Azides**

Also an alternative, again very limited in scope and selectivity, is direct amination with use of azides and aluminum chloride. These have proven their value when selectivity is not important, such as in the direct amination of mesitylene [44], in which, according to the literature, the aminating intermediate is $NH_2^+AlCl_4^-$, which initially forms from sodium azide and aluminium chloride [45] (Scheme 9.13).

Last but not least, direct amination can also be achieved by use of azides, trifluoroacetic acid, and irradiation with light. In this way, 2,4,6,4′-tetramethyldiphenylamine can be synthesized from *p*-tolylazide and mesitylene in an 84% yield, as shown by Sundberg in 1973 [46]. A similar reaction in more modern guise was presented by Knochel in 2006 in the form of the amination of functionalized arylazo-tosylates using dialkylzinc reagents and subsequent N–N bond cleavage with Raney nickel [47].

9.2.2.5 The Minisci Reaction

Very similar from a mechanistic point of view is direct amination of arenes with chloroamine derivatives. Several variations are known, in terms both of the amine part and of the conditions under which the reactive species is generated. Therefore, trichloroamine or dichloroamine in combination with aluminum chloride [48–51] or with dialkylchloroamines in combination with iron(II), chromium(II), copper(I), or titanium(III) salts in sulfuric acid are useful conditions for aryl amination. The reaction is often referred to as the Minisci reaction [52, 53]. In some cases activation by irradiation with light has been described. While not suitable as a general approach, this methodology has to be recognized for some starting materials as a very straightforward and flexible approach, as demonstrated by the synthesis of *p*-(piperidino)acetanilide [54, 55] (Scheme 9.14).

9.2.2.6 The Bucherer Reaction

The famous transformation of β-naphthols into β-aminonaphthalenes was first described by Bucherer in 1904 [56] and further investigated afterwards [57]. The popularity of this reaction is easily understandable, since various naphthalene amine derivatives have gained tremendous industrial importance as bulk intermediates (Scheme 9.15).

The hydroxy group is replaced via an addition product of sulfurous acid. While the Bucherer reaction allows the easy introduction of amino groups in the naphthalene β position, which is often difficult to achieve otherwise, it also limits this reaction type to very few aromatic systems, mostly naphthalenes. Alternative reagents, such as ammonium sulfite solutions (i.e., SO_2 in aqueous ammonia), have also been investigated. In this way the Bucherer reaction has also been used to introduce groups other than NH_2, such as by the use of a methylammonium sulfite solution. The reaction is still sometimes of practical importance, when

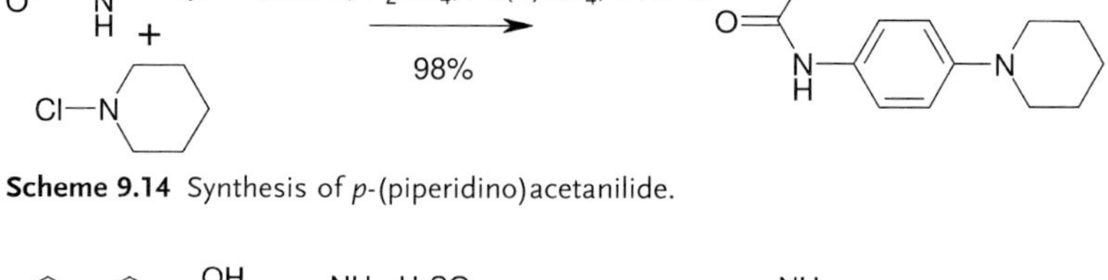

Scheme 9.14 Synthesis of *p*-(piperidino)acetanilide.

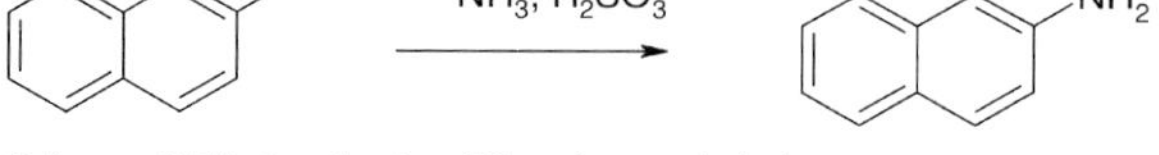

Scheme 9.15 Synthesis of β-aminonaphthalene.

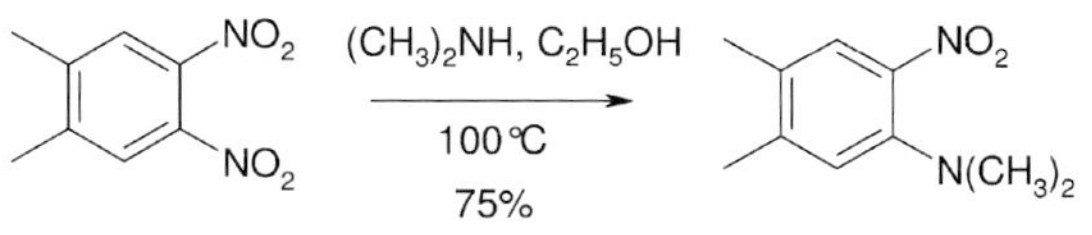

Scheme 9.16 Synthesis of 4,5-dimethyl-2-nitro-*N,N*-dimethylaniline.

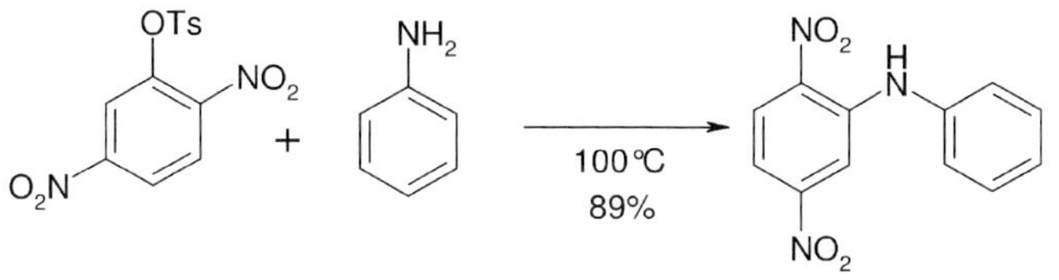

Scheme 9.17 Ullmann's synthesis of 2,4-dinitrodiphenylamine.

helped, for example, with modern techniques such as the use of microwaves as heat source [58].

9.2.2.7 **Metal-Free Replacement of Nitro Groups by Amines**

The replacement of nitro groups by amines was first described by Laubenheimer in 1876 [59, 60]. It is often applicable when dinitroarenes are used as starting materials. The reaction is especially charming, since the nitration of aromatics is one of the best understood organic transformations, so there is a vast abundance of starting materials available. The reaction has a certain popularity in dye chemistry, most probably for the same reasons. An example is Rudy's 1939 synthesis of 4,5-dimethyl-2-nitro-*N,N*-dimethylaniline [61] (Scheme 9.16).

9.2.2.8 **Metal-Free Replacement of Sulfonic Acid Esters by Amines**

As described below, highly activated sulfonic acid esters can serve well in both copper- and palladium-catalyzed amine arylation. It is interesting to note that the metal-free version of this coupling was described by Ullmann as early as 1908 [62] (Scheme 9.17). Of course, this replacement works only with very electron-poor arylating agents, such as toluene-4-sulfonic acid 2,5-dinitrophenyl ester, and therefore lacks a broader scope.

9.3 The Quest for Industrial Applications of Transition Metal-Catalyzed Arylamination

For most new chemical technologies their first industrial applications constitute a breakthrough in their importance. In very few cases, however, is this fact acknowledged by the chemical community. One of the exceptions is the Haber–Bosch process, at first glance an obvious chemical reaction for the generation of industrially important ammonia from the abundant nitrogen in the Earth's atmosphere. On closer inspection, however, this process – a cascade of difficult catalytic transformations, in combination with ingenious engineering skills – has to considered one of the major milestones in modern chemical technology [63, 64].

9.3.1
Industrial-Scale Halogen–Amine Exchanges

Encouraged by the success story of this process, chemists and engineers also tried to apply the principles of Ullmann's metal-catalyzed aryl amine synthesis in the form of an industrial process [65]. The rather labile chlorine of *p*-chloronitrobenzene had been observed to be replaceable with ammonia without the necessity to use a catalyst, as mentioned above. A temperature of 175 °C, high ammonia pressure, and a reaction time of 16–19 h directly furnish *p*-nitroaniline [66]. To improve the economics of the process, most of the excess ammonia is recycled by distillation after venting, the reaction mixture is then filtered, and the highly pure product separates from the aqueous phase.

An obvious goal, of course, is to extend the ammonolysis of highly activated halogenated arenes to industrially easily accessible compounds such as chlorobenzene, which would represent an alternative to the large-scale reduction of nitrobenzene. Although described as early as 1907 in the case of chlorobenzene, this reaction intrigued industrial chemists in the first third of the 20th century and engendered a plethora of patents [67–80], mostly concerned with the catalytic activity of different copper catalysts and the difficult handling of such complicated intermediates on industrial scales. A simple mechanism for the reaction was proposed in 1934 by Woroshzow (Scheme 9.18) [81].

Attempts to introduce organic additives such as glucose as cocatalysts were investigated first, but several authors quickly found that only a temperature of >200 °C allowed for acceptable reaction times. At this temperature, however, the competing formation of phenols and diphenyl ethers become a serious problem. The formation of these side products can be partly suppressed by increasing the excess of ammonia to 6 : 1 and by increasing the pressure [81]. At the same time, use of a lower ammonia excess can result in increased formation of ammonium chloride, a salt that is known to deactivate the copper catalyst. Since there is strong economic competition between aniline synthesis by ammonolysis and by reduction of nitrobenzene, as small an excess as possible is favorable for ammonolysis. The economic production of aniline by ammonolysis also needs a complicated general infrastructure with regard to more than just the transformation alone. For example, the coupling of aniline synthesis by ammonolysis with bulk chlorine

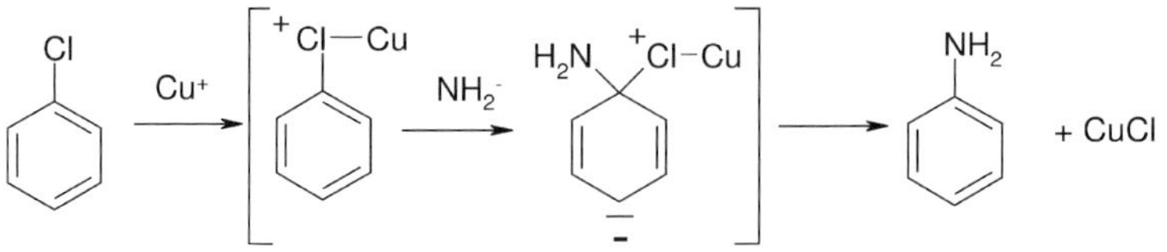

Scheme 9.18 Proposed mechanism of chloro-amine exchange.

production, including the manufacture of other chlorinated aromatics, is advantageous to make the process more competitive as a whole for a bulk chemical product. Again, the recycling of the majority of the ammonia upon venting is necessary to reduce costs once more. The crude product is then treated with alkali to free excess ammonia and to precipitate the copper catalyst and the alkali salt of the phenol byproducts. Steam distillation is then used to separate the volatile compounds – ammonia, chlorobenzene, aniline, and diphenylamine – from one another. The amount of heat necessary for the steam distillation can be reduced by running the reaction in an alternative mode, in which about 90% of the aniline separates from the aqueous phase upon venting. This aniline-rich phase is separated, then made basic, and finally distilled in a process that needs far less energy than a steam distillation and is therefore more economical. It is interesting to note that the copper catalyst can be reused after washing without loss of activity. Of course, all side products are isolated by distillation in pure form and can be sold separately.

To improve the economics of the ammonolysis process further, a continuously run process was investigated in 1947 by Dow chemists [82]. One of the major obstacles to overcome was the enormous corrosion from the reaction slurry in the tubular reactor observed at higher flow rates. To limit the corrosion, a setup consisting of a premixer, which preheated the reaction mixture to 180–220 °C, in combination with a plug flow reactor gave best results. In addition to corrosion, the Dow group also observed the precipitation of copper scales. The copper scales could be removed by purging the reaction stream with water and also by re-feeding new copper catalyst into the reactor. Slight adjustments of the catalyst by addition of the oxides of calcium, tin, lead, arsenic, and antimony gave a system that showed neither corrosion, nor copper scales, nor hydrolysis of the product [83].

A second variation of a continuous process to produce aniline from chlorobenzene was demonstrated by Hughes and Veatch in 1949 [84]. The reaction was run in such a way that two liquid phases, one organic and one aqueous, were mixed. The organic phase was partly removed from the system and the aniline from this phase was removed by steam distillation. The organic phase was then re-fed into the reactor. The aqueous stream was also separated, set basic, and the freed ammonia was recycled into the reactor. At the same time the copper catalyst precipitates and can be re-fed into the reactor to keep the catalytic activity constant.

As a second industrial process using a transition metal catalyst, the production of diphenylamine from chlorobenzene and ammonia can be mentioned, though stoichiometric amounts of rather expensive potassium hydroxide are usually required. A mixture of cheaper but otherwise ineffective sodium hydroxide with traces of potassium salts and copper oxide provided the basis for another industrially important copper-catalyzed ammonolysis reaction (Scheme 9.19) [85].

When ammonia is replaced by other N-nucleophiles such as aniline, the copper-catalyzed coupling of chloronitrobenzene with aniline as run by Lanxess serves as a multi thousand ton example [86]. The catalyst consists of copper oxide in combination with potassium carbonate as a stoichiometric base. The catalytic activity

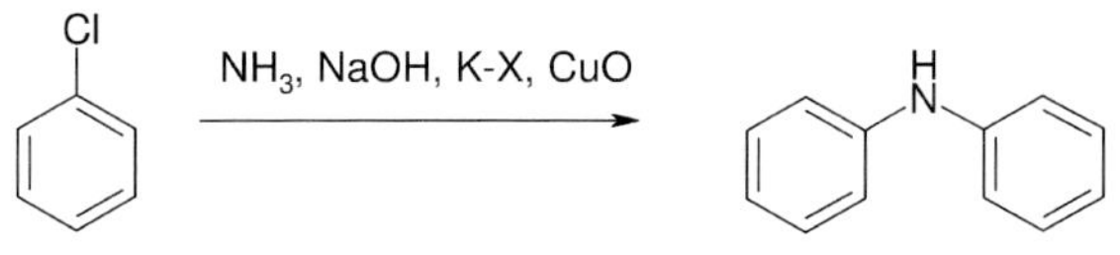

Scheme 9.19 Industrial scale diphenylamine synthesis.

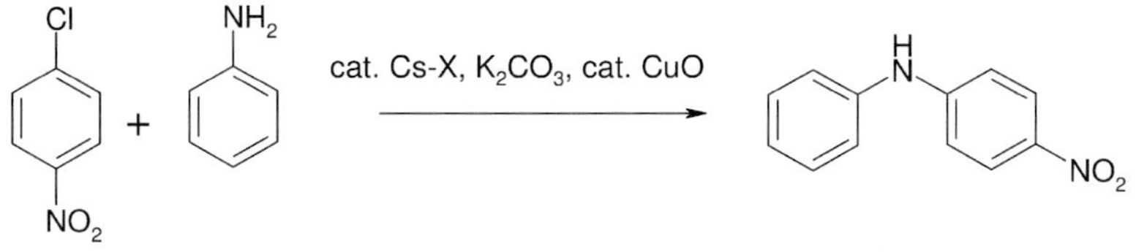

Scheme 9.20 Industrial scale production of 4-nitrodiphenylamine.

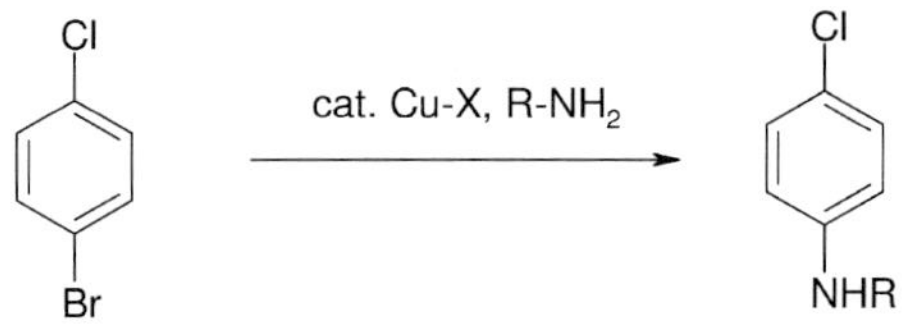

Scheme 9.21 Industrial synthesis of *p*-chloroaniline.

was augmented by the addition of traces of cesium salts, but in this very economic process a relatively large amount of nitrated triarylamine byproduct – the product of a coupling of one aniline molecule with two chloronitrobenzene molecules – is formed (Scheme 9.20).

It is very interesting to note that the same company patented a palladium-catalyzed process to manufacture 4-nitrodiphenylamine in 2001, probably the first large-scale process to produce a bulk intermediate by the Buchwald–Hartwig reaction [87].

In some cases industrially important processes, such as the replacement of bromine with ammonia for the synthesis of aminoanthraquiones, have been optimized to low temperatures rather uncommon for copper-catalyzed industrial processes. In the case mentioned, the reaction runs at 80 °C [88].

The greater ease of replacement of bromine over chlorine can result in industrial processes in which it can even be advantageous to use bromoarenes, often stigmatized as expensive intermediates, as starting materials. For the synthesis of *p*-chloroaniline from bromochlorobenzene, use of a mild 120 °C, ammonia or methylamine, and a copper chloride catalyst yields the product very selectively (Scheme 9.21), thus alleviating the purification in a fashion that for high quality chloroanilines makes this process competitive with the reduction of the corresponding chloro-nitroarenes [89].

9.3.2
Transition Metal-Catalyzed Direct Amination of Aromatic Compounds

Without a doubt, one of the dream reactions for bulk-scale chemistry is the direct amination of aromatic compounds, such as the direct synthesis of aniline from benzene and ammonia, first described by Wibaut in 1917 [90].

The concept evolved over the next decades but remained largely in the realms of heterogeneous catalysis, but no system has ever yet proven promising enough for industrial application [91–98]. The most promising catalysts so far have been nickel-based, and were published for the first time by Dupont in 1975. The catalyst is a metal oxide mixture of Ni, NiO, and ZrO_2 [99–102] (Scheme 9.22).

High ammonia pressures of 300 bar are necessary to yield conversions of almost 14% in a discontinuous setup. The mechanism of the reaction has been subject to a recent investigation by Hoffmann [103]. While with a 14% conversion a continuous process would already be in reach for an industrial scale up, it seems that this reaction is still not economically compatible with the classical synthesis of aniline from nitrobenzene, phenol, or chlorobenzene.

9.3.3
Industrial-Scale Aminolysis of Phenols

Even today, the aminolysis of phenols is still one of the large-scale processes for the synthesis of aniline. In these cases, however, activated aluminium oxide catalysts, high temperatures, pressures, and large excesses of the amine are necessary to yield sometimes only low conversions from phenol to aniline; in Briner's 1924 synthesis of aniline from phenol at normal pressure, for example, only a 12% conversion was observed [104] (Scheme 9.23), though when higher pressures were applied the conversion could be shifted into the >90% region [105]. Modern catalytic systems are slightly different and contain niobium oxides [106] or palladium/tin combinations [107].

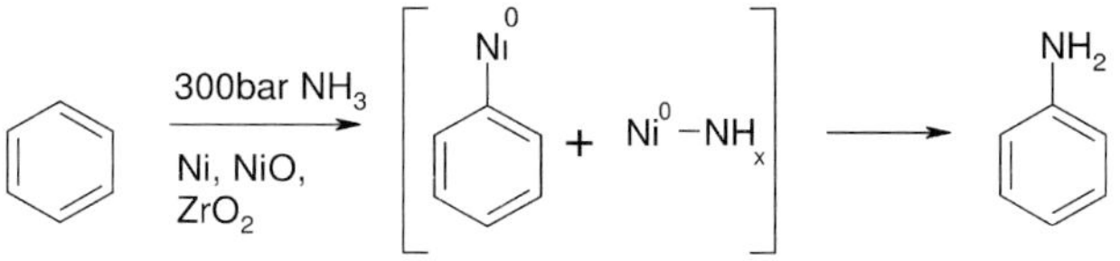

Scheme 9.22 Direct amination of benzene.

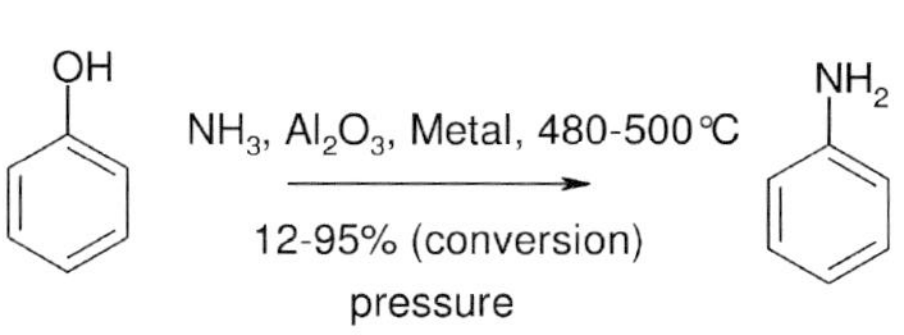

Scheme 9.23 Technical synthesis of aniline from phenol.

The method is also used for the industrial synthesis of other anilines from phenols, such as resorcines, cresols and several others.

9.4
Copper-Catalyzed Processes – More Recent Developments

After these first steps in copper-catalyzed aromatic amination, elaboration of the methods gave rise to various improvements in the conditions, allowing wider and wider scope and, of course, less harsh reaction conditions [108].

9.4.1
Alternative Arylating Agents

Alongside the evolution of the classical Goldberg conditions, under which halogenated aromatics were used as a abundant resource as arylating agents, some groups went different ways. The copper-catalyzed coupling of arylbismuthanes with secondary amines as shown by Dodonov in 1985, for example (Scheme 9.24) [109], offers an unconventional approach to arylamines.

Its large excesses of the amine and extremely long reaction times are disadvantageous to the popularity of the method, but the example shows that room temperature copper-catalyzed amination can, in principle, work.

Interesting selectivities, more appealing conditions, and the use of sterically hindered anilines were reported by, for example, Esteves in 2001 [110] (Scheme 9.25).

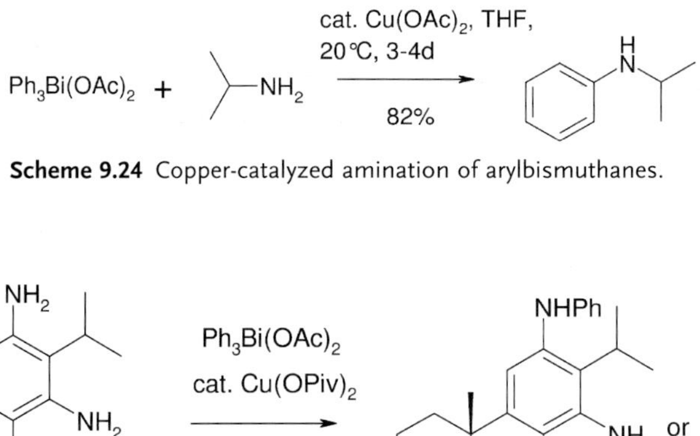

Scheme 9.24 Copper-catalyzed amination of arylbismuthanes.

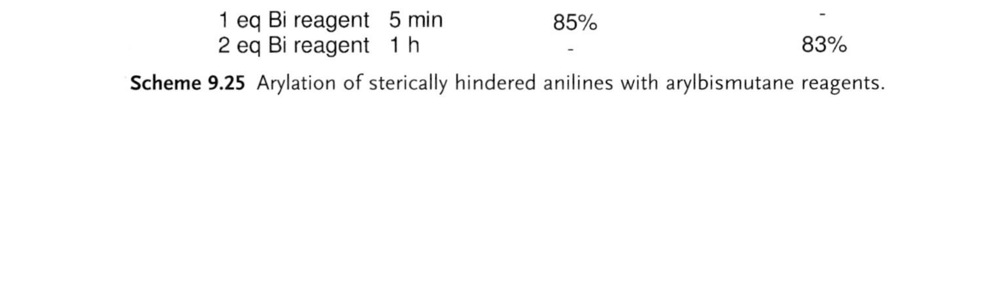

Scheme 9.25 Arylation of sterically hindered anilines with arylbismutane reagents.

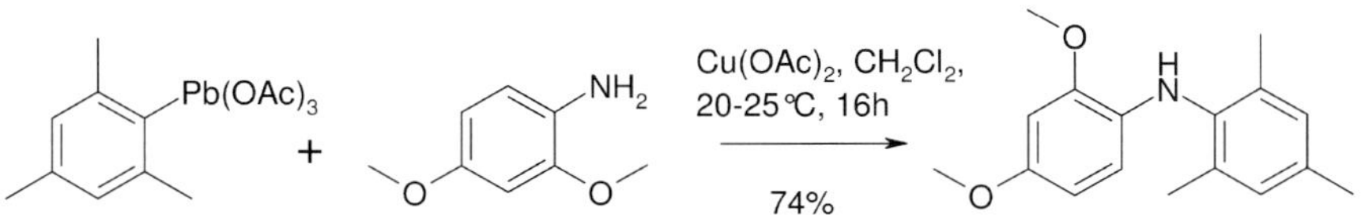

Scheme 9.26 Copper-catalyzed amination of arylplumbanes.

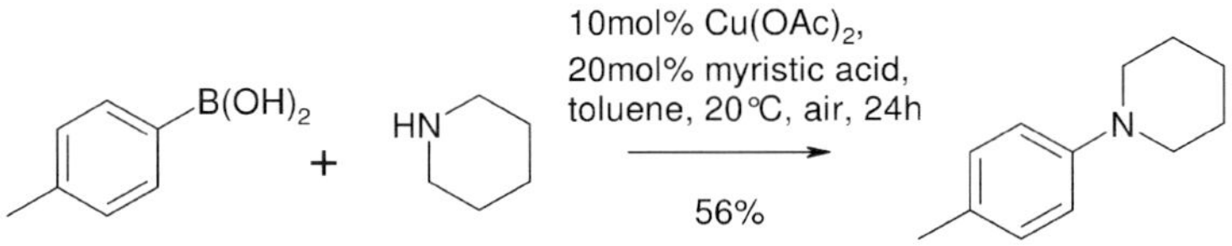

Scheme 9.27 Synthesis of *N*-(*p*-tolyl)-piperidine.

A similar approach was pursued by Barton in 1989, in which the group showed that arylplumbanes could be aminated with use of a copper catalyst in several cases, again at room temperature [111] (Scheme 9.26).

Again, limited scope, unfavorable reaction intermediates, and toxic byproducts are probably responsible for rare mentioning of this method in the literature.

Other arylating agents such as siloxanes [112] or arylstannanes [113] have also been identified, though with use of stoichiometric amounts of copper acetate only.

A breakthrough was certainly achieved in 1998, when Lam and Chan reported a generally applicable method for the synthesis of arylamines from arylboronic acids under mild conditions [114–116]. Thanks to the popularity of the Suzuki methodology for the synthesis of biphenyls, arylboronic acids had become widespread intermediates. Availability of the arylating agent is of course a major prerequisite for a procedure to be applied by chemists all over the world. This method was developed into a room temperature version with wide applicability by Buchwald in 2001 [117], with catalytic amounts of myristic acid and 2,6-lutidine as base being used. Again this procedure works well for the synthesis of diarylamines, and alkylarylamines can also be synthesized in acceptable to good yields with that method (Scheme 9.27).

A review article dealing with this variety of arylating agents, summarizing more examples of alternative arylating agents, was published by Finet in 2002 [118].

9.4.2
Catalyst Tuning

Since the early publications of Ullmann and Goldberg the synthesis of diarylamines from aryl halides and anilines has constantly evolved towards more and more robust, milder, more practical, and “greener” processes. One such example was supplied by Pellon in 1993, demonstrating the use of water as solvent for Goldmann-type couplings [119] (Scheme 9.28).

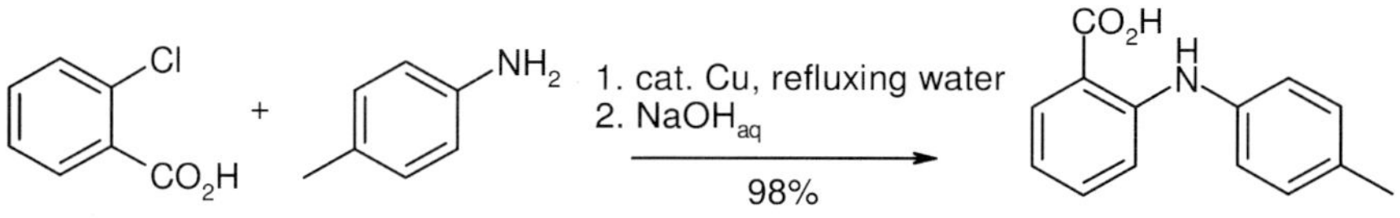

Scheme 9.28 Ullmann coupling in water.

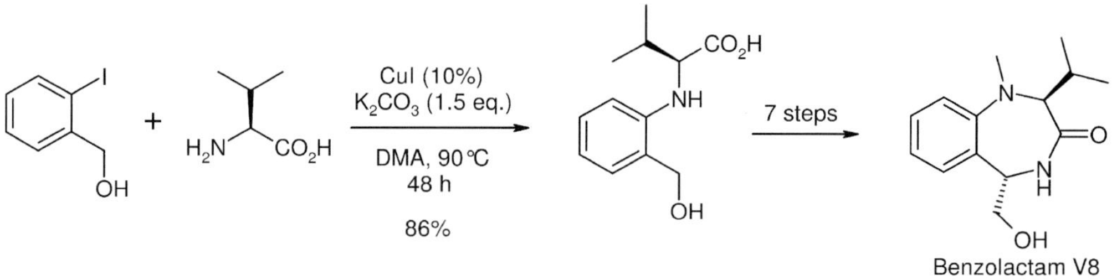

Scheme 9.29 Ma's benzolactam V8 synthesis of 1998.

It is interesting to note that at the end of the 20th century the number of publications focused on the development of more robust and widely applicable copper-catalyzed protocols made a tremendous jump. Optimization cycles similar to those observed just a few years previously in the field of palladium-catalyzed amine arylation were undertaken. The result was the development of catalysts that today allow the coupling of most aryl halides with amines under very mild conditions. A major advantage of the new methods is the fact that the new catalysts also allow the coupling of aliphatic amines to aryl halides, a reaction type that is rather difficult to succeed with under the classical conditions of Ullmann and Goldberg.

As a first step towards modern copper-catalyzed aryl aminations, Ma's 1998 benzolactam V8 synthesis gave a first view of the potential of copper catalysis [120]. This group revisited the old procedures and recognized that α-amino acids can be good substrates for amine arylation, since they exhibit the same chelating nature and act as ligands for copper just as in Goldberg's original work [5] (Scheme 9.29).

Continuing this work on the selective coupling of amino acids, Ma reported the use of β-amino acids as intramolecular coupling partners with aryl iodides in the 2001 total synthesis of Lotrafiban [121] (Scheme 9.30). The synthesis of the same target structure was also demonstrated by Hayes in the same year, though in an intermolecular version from L-aspartic acid – an amino acid containing both an α-amino acid and a β-amino acid substructure – as starting material [122] (Scheme 9.30).

Shortly thereafter, Buchwald presented a procedure for the arylation of β-amino alcohols. This reaction serves as a convenient alternative to the opening of epoxides with anilines. By use of different reaction conditions, both N and C selectivities can be achieved. The group had to identify an additive for this reaction, since the amino alcohol alone – valinol alone, say – did not show an accelerating effect in

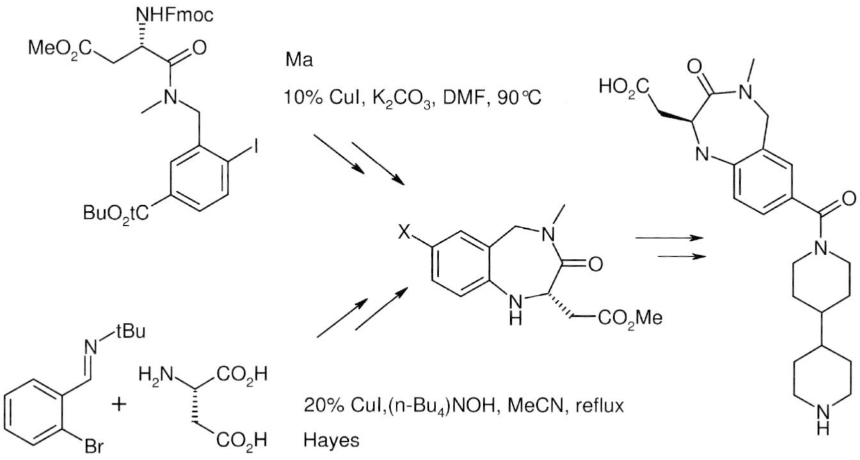

Scheme 9.30 Overview of the two lotrafiban syntheses by Ma and Hayes.

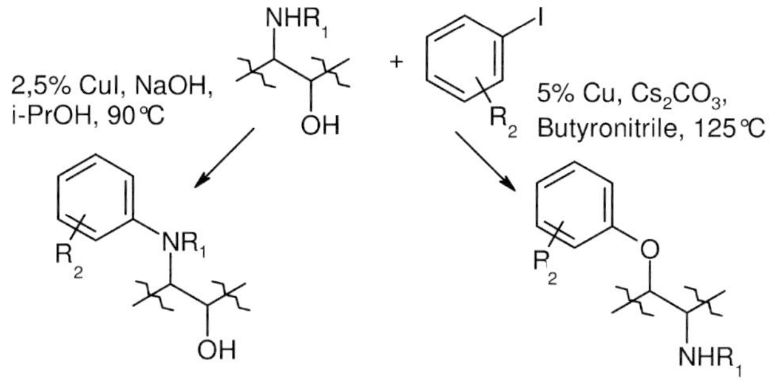

Scheme 9.31 Selective *N*- and *O*-arylation of aminoalcohols.

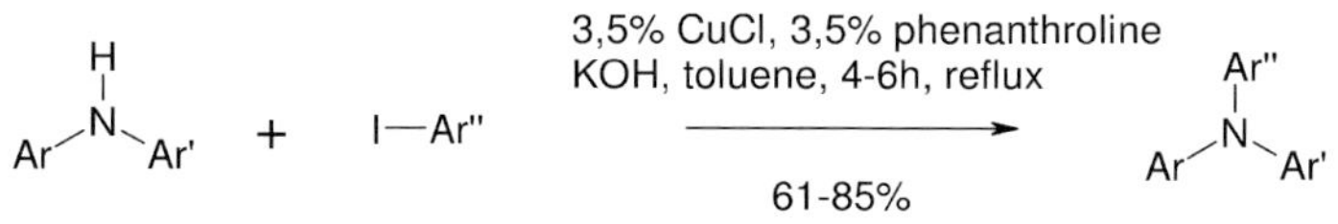

Scheme 9.32 Triarylamine synthesis by Goodbrand.

the reaction. *N*-Arylation proved to be possible through the use of glycol as ligand, while *O*-arylation was successful at higher temperatures with use of other bases [123] (Scheme 9.31).

The idea of using chelating ligands as substitutes for chelating substrates to broaden the scope of copper-catalyzed arylations of amines, amides, and related structures seems to be rather new. Early attempts can be observed in the 1999 work of Goodbrand, in which phenanthroline was extensively used for the synthesis of triarylamine-based photoreceptors (Scheme 9.32) [124].

In the same year Buchwald presented a mild procedure for the arylation of N-heterocycles such as imidazole, by use of stoichiometric phenanthroline and catalytic amounts of dibenzylideneacetone as additives in combination with a copper catalyst and cesium carbonate [125] (Scheme 9.33).

In 2001 Venkataraman extended the applicability of copper-phenanthroline or neocuproine complexes with a procedure that used very mild conditions [126]. These complexes again show superior activity in the synthesis of triarylamines. Especially versatile is the use of these catalysts as isolated complexes, since they are easy to prepare and stable to air. Evindar also showed that, in the intramolecular guanidinylation of aryl bromides, copper phenanthroline catalysts are superior even to palladium catalysts (Scheme 9.34) [127].

The arylation of aliphatic cyclic and acyclic amines with aryl iodides can be achieved very efficiently by a procedure published by Buchwald in 2003. Again, the use of cheap glycol as a ligand in twofold stoichiometric amounts adds to the attractiveness of this method, together with its mild temperatures, broad scope and good to excellent yields (Scheme 9.35) [128].

Also focused on the arylation of aliphatic amines, especially primary amines, with aryl bromides was a paper presented by Buchwald in 2003, in which commercially available diethylsalicylamide is used as a chelating ligand [129]. Together with the mild conditions and a cheap catalytic system, the group also investigated a solvent-free version of the reaction (Scheme 9.36).

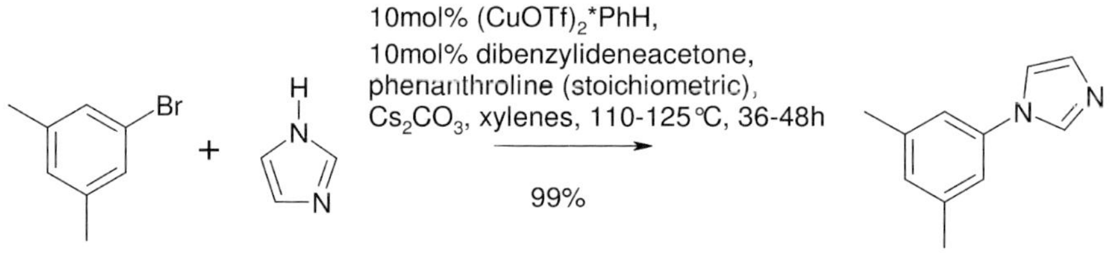

Scheme 9.33 Buchwald's 1999 Ullmann-type imidazole arylation procedure.

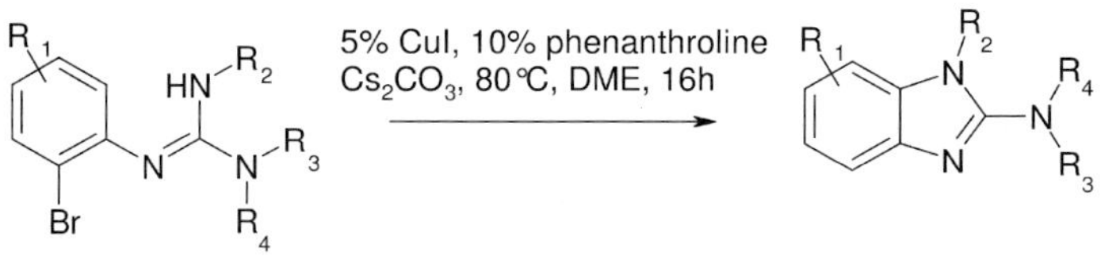

Scheme 9.34 Guanidinylation of aryl bromides by Evindar.

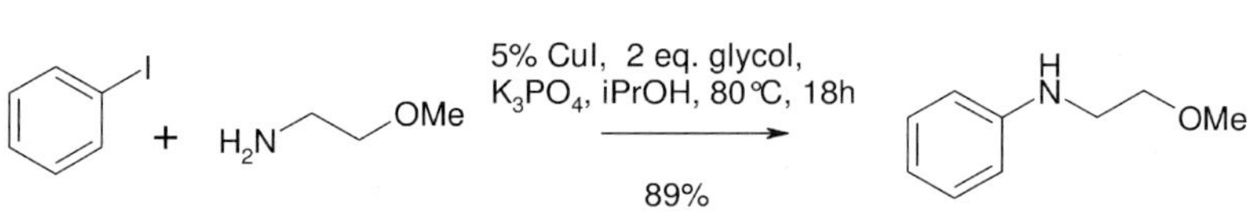

Scheme 9.35 The use of glycol as chelating additive.

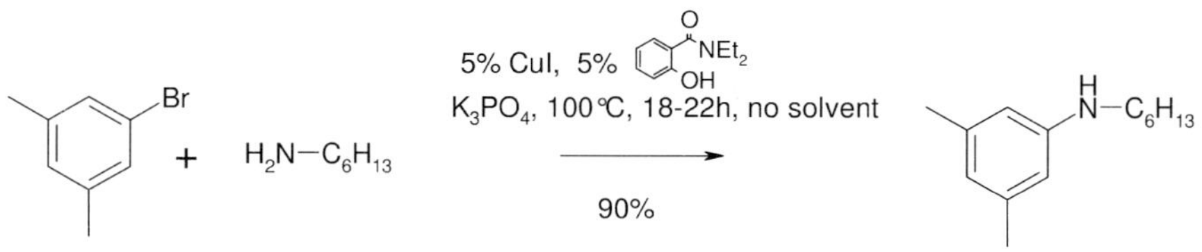

Scheme 9.36 Diethylsalicylicamide as ligand.

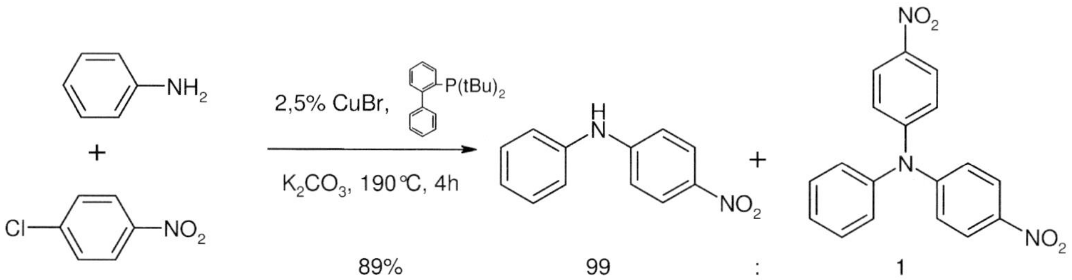

Scheme 9.37 Selective monoarylation of aniline.

One of the first ligand screenings for the copper-catalyzed synthesis of triarylamines by double arylation of anilines was published by Chaudhari in 2002; interestingly to note, no ligand at all showed the best reactivity [130]. Improved conditions, however, also feasible for a broader scope of aryl halides, were demonstrated by the same group two years afterwards [131]. In this case 2,6-diphenylpyridine as ligand in combination with CuI in toluene and KOtBu as base gave optimal yields. A procedure for the opposite selectivity (i.e., the selective monoarylation of anilines) was demonstrated by Scholz in 2004 [132–136] (Scheme 9.37).

The coupling of alternative N-nucleophiles was the subject of a publication by Buchwald in 2001, in which a procedure for effective arylation of N-Boc-hydrazine [137] was presented.

Since coupling approaches generally offer a strategic advantage over alternative methods for the synthesis of arylated N-heterocycles, a series of publications has added tremendously to the scope of copper-catalyzed arylations. They all deal with the *N*-arylation of various N-heterocycles. Buchwald started in 2001 with a very general family of protocols based on copper-*N*,*N*-dimethylated 1,2-diamine catalysts (Scheme 9.38). Even aryl chlorides can be used as arylating agents with this procedure [138]. The group extended the scope of the reaction with a systematic study on the influence of the ligand structure on the catalytic activity [139].

Specifically optimized conditions were then published in 2002 [140, 141]. The system was still optimized to yield a procedure with increased functional group tolerance (aldehydes, for example) and broader applicability (such as the arylation of pyrrazoles, pyrazoles, indazoles, and triazoles) [142] (Scheme 9.39).

Kinetic studies suggested that one of the key roles of the ligand is to prevent multiple ligation of the substrates [143].

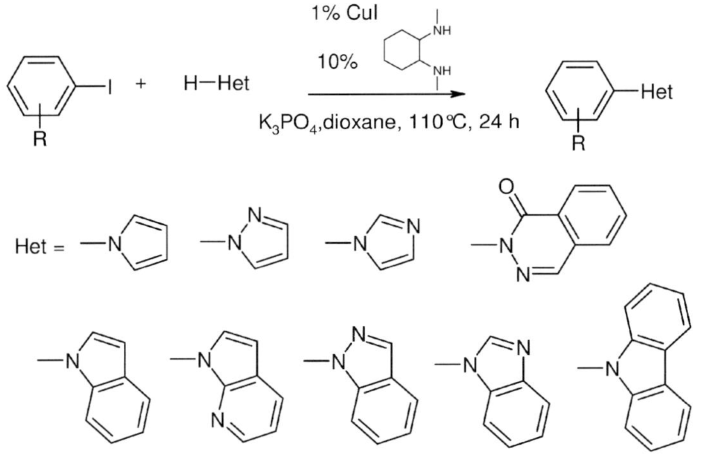

Scheme 9.38 N-Arylation of N-heterocyles by Buchwald.

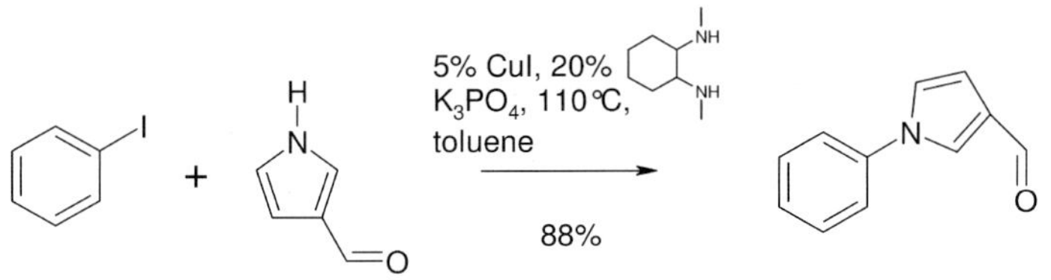

Scheme 9.39 General arylation of N-heterocycles.

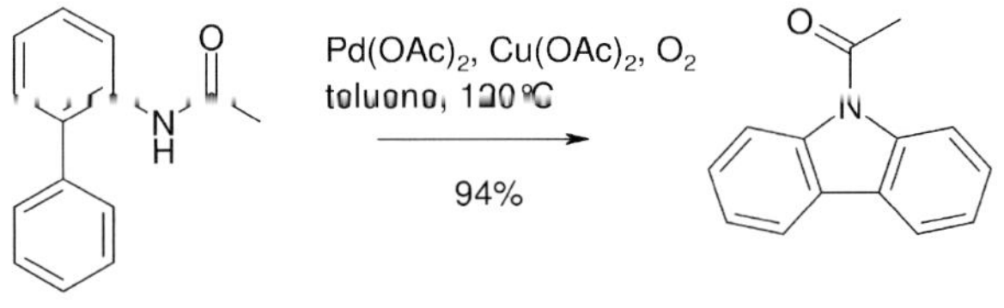

Scheme 9.40 One pot carbazole synthesis.

Other groups also added to the versatility of copper-catalyzed arylation of amides or heterocycles; Kang et al. in 2002, for example, promoted ethylenediamine as a ligand for copper [144]. In the same year, Padwa used a copper *N,N,N,N*-tetramethylethylenediamine (TMEDA) complex for the amidation of furans and thiophenes [145].

A new development in copper-catalyzed arylation reactions is the combination of catalytic steps, such as the initial use of a palladium-catalyzed C–H oxidative activation and then of a copper-catalyzed amidation reaction, as presented by Buchwald in 2005. With this method carbazoles can be synthesized from 2-phenylacetanilides in a one-pot procedure in excellent yields [146] (Scheme 9.40).

9.5 Palladium-Catalyzed Processes

Palladium-catalyzed arylation of amines has been the subject of multiple review articles over the past few years [147–160], so this section focuses mainly on the historical order of the developments and some general statements about the reaction. Different variations have become so popular among chemists that it has been promoted to the rank of a name reaction. It is usually referred to as the Buchwald–Hartwig amination, in acknowledgement of the two authors who independently reported on the reaction in 1995 [161–163]. Since the readers of the mentioned review articles with no doubt get the impression that there is no pair of substrates that cannot be coupled by this methodology, it is more valuable to focus on the concepts of the reaction than on single examples.

9.5.1 Early Developments

Most authors agree that the first milestone in palladium-catalyzed aryl amination was the palladium-catalyzed reaction between tin amides and aryl bromides, sometimes referred to as the Migita reaction (Scheme 9.41). This reaction was described for the first time in 1983 [164].

This transformation elegantly arylates amines with nonactivated aryl halides, with no apparent intermediacy of arynes. However, both the high toxicities of organotin compounds and the relatively high synthetic effort needed to generate and to isolate them have kept this reaction hidden in the realms of laboratories specialized in organometallic transformations, far away from practical organic synthesis.

As an improvement in terms of toxic intermediates, Boger demonstrated the possibility of direct amination of a nonactivated aryl bromide in the presence of stoichiometric amounts of tetrakis-triphenylphosphin-palladium complex in 1984 [165]. While this was the first example of a palladium-mediated direct replacement of bromine by aniline, it only worked as an intramolecular reaction. This reaction, while already elegantly showing the concept (Scheme 9.42), therefore lacks certain aspects of practicability, notwithstanding the fact that expensive transition metal-phosphine complexes are not usually used in stoichiometric amounts.

Another important contribution was the replacement of the toxic tin amides with magnesium amides as reported by Dhzemilev in 1987 (Scheme 9.43). Again,

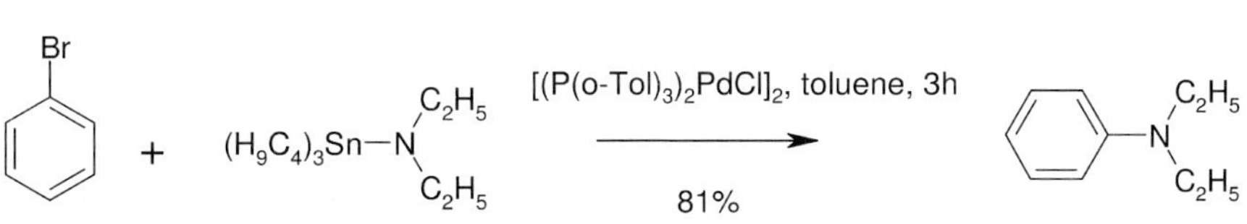

Scheme 9.41 The Migita reaction.

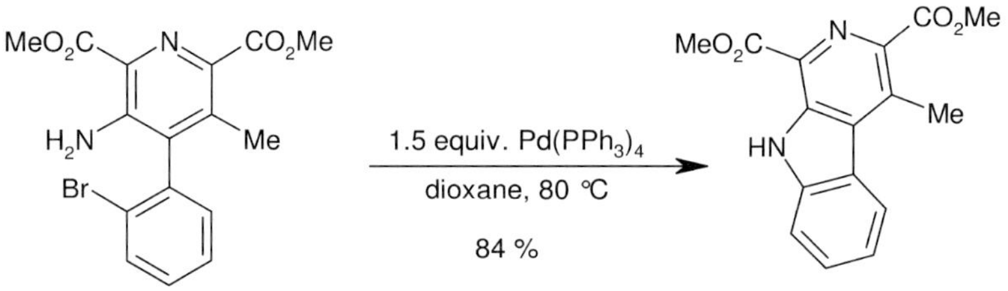

Scheme 9.42 Boger's intramolecular amination of aryl bromides.

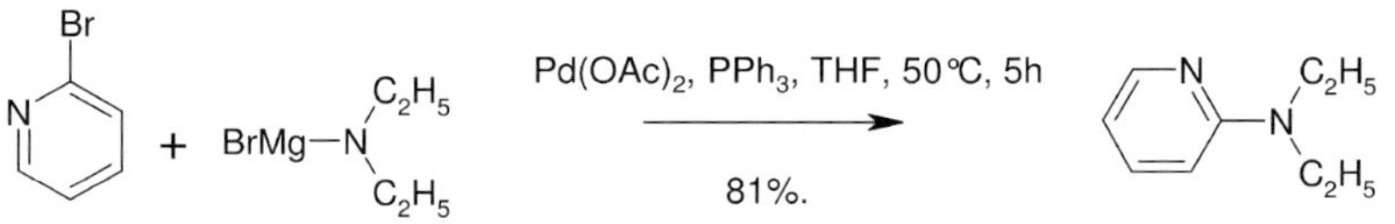

Scheme 9.43 The Dhzemilev reaction of 1987.

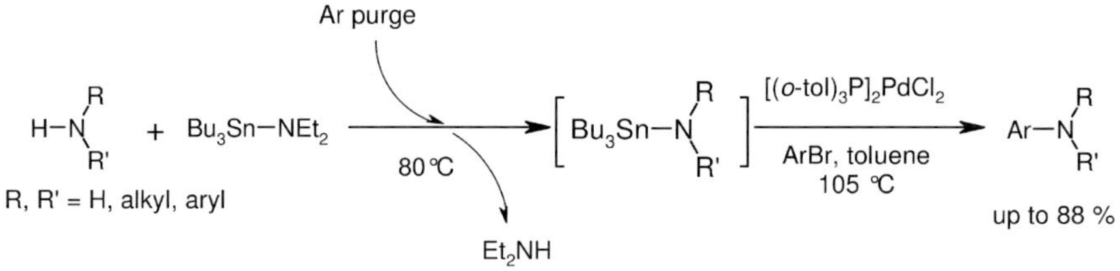

Scheme 9.44 Buchwald's in situ method for amination with tin amides.

it is a big improvement in comparison with the tin amide coupling, but still not as practical as the developments that followed in the 1990s [166].

The two pioneers of palladium-catalyzed arylation of amines with aryl halides, Buchwald and Hartwig, both entered the field with their first publications in 1994. Buchwald demonstrated a more practical approach to the palladium-catalyzed amination with tin amides, which were prepared in situ by metal–hydrogen exchange. The group mixed different high-boiling primary and secondary amines with tributyltin diethylamide at 80 °C, a constant argon purge being necessary to drive the formation of the new tin amide to completion by removal of the volatile diethylamine. This new tin amide could then be used in situ as an aminating agent (Scheme 9.44) [167].

This was another improvement on Migita's original procedure, but the majority of the drawbacks of the old method still prevailed.

While the one group was busy trying to improve Migita's procedure, Hartwig in the same year published an insightful study on the catalytic intermediates of the Migita coupling [168]. His results suggested transmetalation of the amide and the reductive elimination of the arylamine as key steps in the catalytic cycle. This deeper understanding of the mechanism, albeit still with tin amides, surely has to be regarded as the solid knowledge base that allowed for the tremendous improvements in the reaction in the following years.

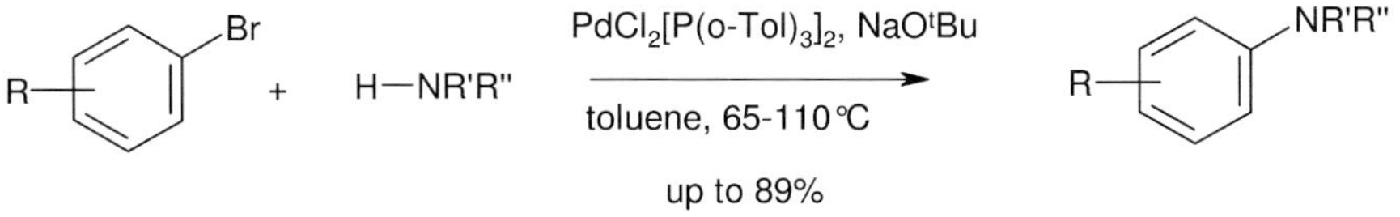

Scheme 9.45 Buchwald's amination of aryl bromides (1995).

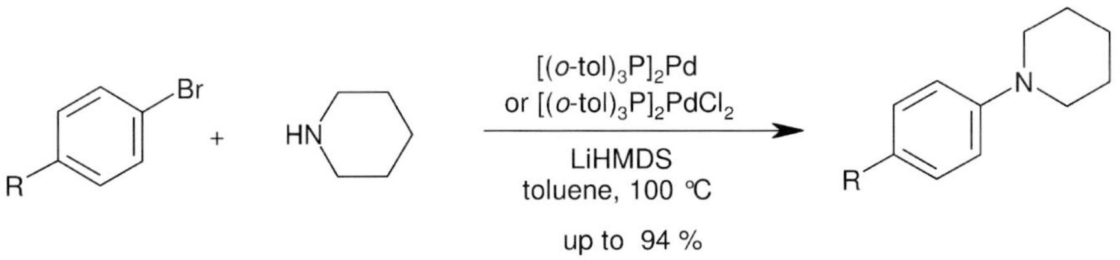

Scheme 9.46 Hartwig's amination of aryl halides (1995).

In 1995 both groups then came up with already very practical procedures for palladium-catalyzed amination of aryl halides. Buchwald demonstrated the amination of aryl bromides with different amines in the presence of a palladacyclic catalyst and sodium *tert*-butoxide as a strong base that allowed for the closing of the catalytic cycle (Scheme 9.45) [161, 162].

In a very similar procedure with a similar catalyst, Hartwig demonstrated the use of LHMDS as base (Scheme 9.46) [163, 169]. Once again, mechanistic considerations are already presented in detail in Hartwig's papers, suggesting a mechanism of the catalytic cycle that is partly still accepted today.

With these contributions published, there followed an explosive development of the technology that allows organic chemists to use this method today in a "shake and bake" fashion. The transformation has to be regarded as nearly failure-proof, robust, usually high-yielding, and in many cases as an economical and convergent method for the preparation of highly functionalized anilines. The development was again driven by Buchwald and Hartwig, but more and more other groups picked up the technology, not only demonstrating new applications in pharmaceutical and material science fields, but also contributing by introducing new catalysts or ligands. The development of the reaction parameters is summarized below.

9.5.2
Ligand Developments

Buchwald and Hartwig quickly came to understand that their original palladium-phosphine complex was far from optimal. Therefore, after 1995, chelating phosphines were often used to improve catalyst activity and scope of the reaction?

The ligand BINAP (Scheme 9.47) is often referred to as a privileged structure, since this chelating phosphine has shown superior activity in countless applications, mostly chiral transformations, of course. For palladium-catalyzed aryl

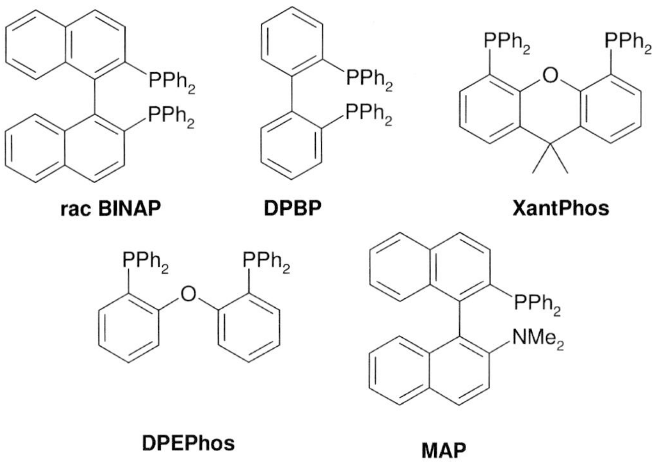

Scheme 9.47 Chelating biarylphosphines.

amination, however, the racemic mixture is preferred, most probably thanks to its better solubility in organic media. BINAP has become very popular, especially in industrial groups, since it is one of the highly active ligand systems that is both available on commercial scale and has slightly more limited patent protection. To name a few examples of its appearance in C–N couplings, BINAP has been used in various applications [170–173], including in combination with NaOMe as base [174], in coupling of primary [175] and acyclic secondary amines [170], in coupling with aryl iodides at room temperature [176], in coupling with chiral amines without racemization [177, 178], in coupling of aryl triflates [179–182], usually in combination with Cs_2CO_3, in selective arylation of primary amines in the presence of secondary amines [183], and in polymerization reactions [184]. Its combination with cheap palladium chloride [185] has also been demonstrated, as well as special applications for the synthesis of benzothiophenes [186]. The palladium-BINAP catalytic system has also been investigated quite extensively from a mechanistic point of view [187–189].

As another chelating ligand, XantPhos (Scheme 9.47) is also popular in areas including special applications in diphenylamine synthesis [185], coupling of alkylarylamines [171], coupling with N-heterocycles [190], and coupling of these with halogenated heterocycles [191].

MAP [192], DPBP [193], and DPEPhos [185, 194] (Scheme 9.47) are also frequently used chelating bisphosphines and have been applied by different groups with good results.

As well as this family of ligands, ferrocene-based phosphines have also shown good activity in palladium-catalyzed C–N coupling (Scheme 9.48).

Especially important to mention is the chelating phosphine DPPF (Scheme 9.48). This ligand arrived for C–N coupling at the same time as the privileged role of BINAP came to be understood, with similar good results [195]. Several applica-

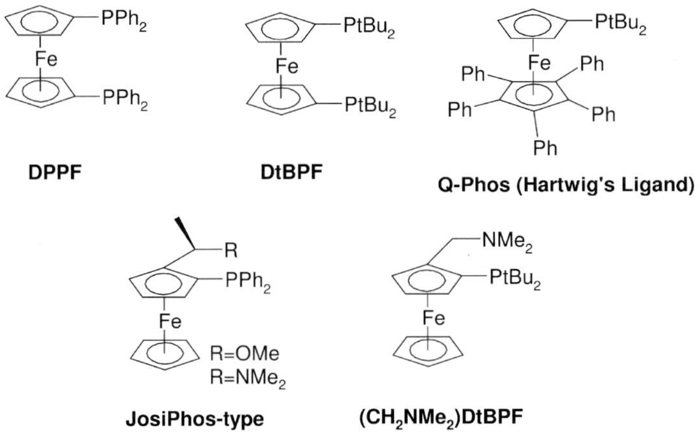

Scheme 9.48 Ferrocene-based phosphines.

tions in triarylamine synthesis have been reported [195–202], together with aryl triflate couplings [203], nonaflate couplings [197], coupling with primary amines [195], and selective monoarylation of primary diamines [204].

JosiPhos-type ligands such as PPF-OMe (R = OMe, Scheme 9.48) work particularly well with acyclic secondary amines [205]; PPF-OMe was the first ligand to be used in combination with the mild base Cs_2CO_3 [206]. With PPF-A (R = NMe_2, Scheme 9.48), certain couplings of dibutylamine run with very little dehalogenation of the aryl halide, and even electron-rich aryl halides can be converted at room temperature [205].

1-(*N*,*N*-Dimethylaminomethyl)-2-(di-*tert*-butylphosphino)ferrocene ((CH_2NMe_2)DtBPF; Scheme 9.48) has been used extensively for the coupling of halogenated indoles with cyclic amines [207, 208].

Hartwig's Q-Phos ligand (Scheme 9.48) has shown superior activity for C–O couplings, but also works quite well for C–N coupling [209]. The DtBPF ligand (Scheme 9.48) also shows good results with cyclic amines and aryl chlorides [210].

$P(tBu)_3$, or Tosoh's ligand (Scheme 9.49), was introduced by that company in 1997. It is one of the first high-activity phosphine ligands that is still intensely used today, and is the template for almost all high-performance ligands. The main disadvantage of this ligand is its sensitivity to air, usually circumvented on industrial scales by use of 10% solutions of the ligand in organic solvents. Another possibility is to follow Fu's idea of using the tetrafluoroborate salt of the ligand and to free the phosphine in situ by addition of an organic base [211]. For an industrial company, Tosoh showed relatively high speed in developing the invention of Buchwald and Hartwig into practice, presenting widely patentable results approximately one year after the first reports [212] and thereby reserving themselves a seat in the front row of contributors to palladium-catalyzed aryl amination [213–217]. Their ligand works well in most applications, such as for the arylation

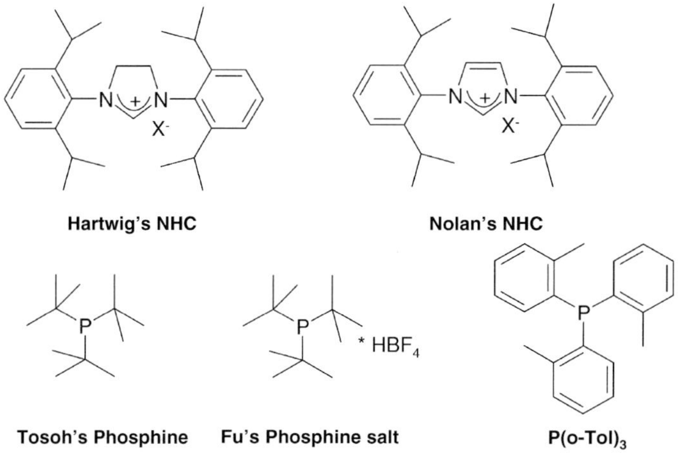

Scheme 9.49 Various C-N ligands.

of piperazines [218], coupling of secondary amines at room temperature [219], triarylamine synthesis [196–198, 219–221], and coupling of secondary amines with aryl chlorides [222]. The great popularity of the ligand can also be explained, since it quickly became the favorite ligand of the Hartwig group, so many very important papers concerning mechanistic studies, scope, and limitations were focused on the use of this ligand.

The ligand used in the first publications by Buchwald and Hartwig, $P(oTol)_3$ (Scheme 9.49) can be used nicely for triarylamine synthesis [195–202] and aryl-iodide couplings [223, 224], but in general other ligands are preferred by most users, including the two authors mentioned above.

Nolan's N-heterocyclic carbene ligand (Scheme 9.49) represents a phosphine-free ligand class and their equivalence to phosphines, with stronger ligation of course, has been shown in various examples by Herrmann [225] and others. Nolan's ligand works for arylation of piperidine [226] and coupling of aryl chlorides with other secondary amines. Hartwig's saturated N-heterocyclic carbene (Scheme 9.49) showed that this is possible even at room temperature [227].

Buchwald's biphenyl-dialkylphosphino class ligand DavePhos (Scheme 9.50) shows very high reactivity in combination with mild bases and secondary amines [228, 229], coupling of aryl chlorides [226] (even at room temperature [228, 229]), or for the preparation of aza crown ethers [230] and can be prepared in a surprisingly straightforward manner [231, 232].

JohnPhos (Scheme 9.50) works well for room temperature reactions of cyclic amines [233], coupling of acyclic secondary amines [228], and for triarylamine synthesis [228, 233].

Cy-JohnPhos (Scheme 9.50) again has shown wide applicability for aryl chloride aminations [228].

X-Phos (Scheme 9.50) is probably the most active and versatile of the Buchwald ligands for C-N coupling [234]. In the author's experience, no ligand at all matches

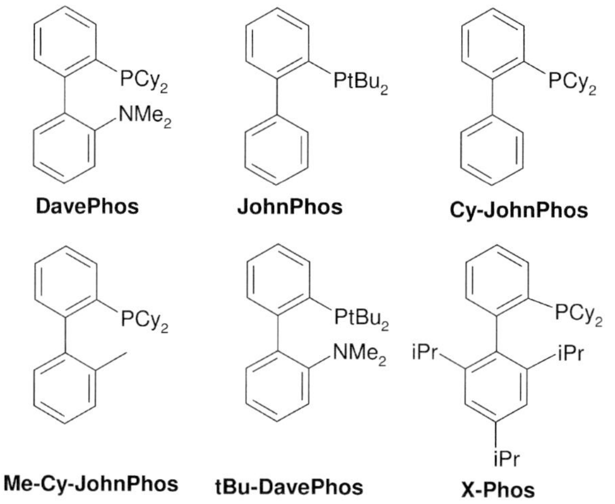

Scheme 9.50 Members of the Buchwald dialkylbiphenylphosphine ligand family.

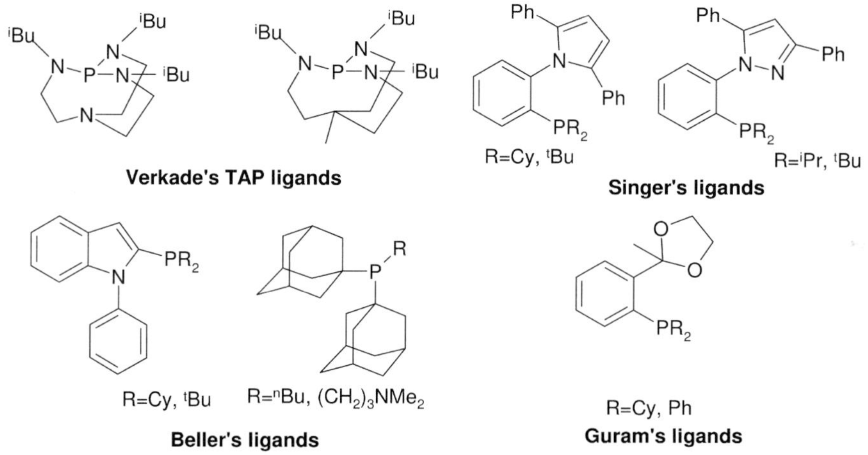

Scheme 9.51 Alternative ligand developments.

the versatility of X-Phos, one of the very few ligands that also work with weakly activated (i.e., non-fluorinated) sulfonic esters [235] such as aryl mesylate esters. These are very attractive starting materials from an economical point of view.

To mention also a few of the ligand developments less frequently used than those discussed above, Scheme 9.51 summarizes different representatives. Guram's P–O ligands (Scheme 9.51) have been used, for example, for the coupling of bromacetophenones with piperidine [236].

Verkade's TAP ligands (bicyclic *tri-amino-phosphines*; Scheme 9.51) [237–240] have shown wide scope in general coupling of halogenated arenes with all types of amines. Their obvious advantage is their simple preparation.

Singer introduced heterocyclic phosphines (Scheme 9.51) as a ligand system that is simple to prepare and shows interesting activity in various applications [241].

Beller's adamantyl and indole-type ligands [242–245] (Scheme 9.51) have shown results comparable to those obtained with the structurally similar Singer ligands and $P(tBu)_3$ and so represent a nice alternative to those systems.

As well as isolated ligands, different groups also liked the idea of providing stable palladium-ligand or precursor ligand complexes. This of course has advantages in everyday applications, especially if the thus formed complex is more stable to air and moisture than the ligand or palladium precursor alone. Various examples are summarized in Scheme 9.52.

The palladacycle catalyst (Scheme 9.52) was the first system to be used as a preformed complex for C–N coupling, working well for the amination of aryl chlorides with cyclic secondary amines [246]. Other examples followed, often stressing the potential for use of these complexes as stable solids, easy to store and usually with no sensibility to air. Buchwald proposed a preformed complex of his JohnPhos ligand with palladium [247], Solvias went one step ahead and used a palladacyclic complex with a similar structure as a precatalyst [248], while Hartwig prepared a dimeric palladium $P(tBu)_3$ complex and showed its superior properties for rapid low temperature C–N couplings [249]. Nolan also presented a preformed version of his catalyst showing high activity for aryl amination [250], and two new catalyst concepts were presented by Li, who used dialkylphosphineoxides as ligands [251], and by Bedford, who used an amine-phosphine-palladacycle complex [252]. These two in particular are attractive from an industrial point of view, since their production on large scales should be much cheaper then that of complicated phosphines.

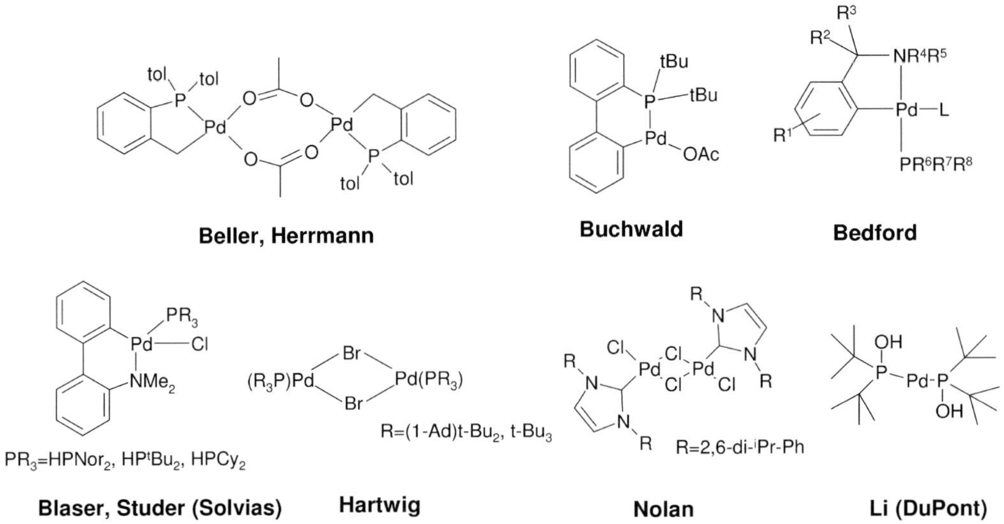

Scheme 9.52 Preformed palladium ligand complexes.

9.5.3
Other Components of the Reaction

The other key components for the successful amination of aryl halides are the stoichiometric base that is used, together with the choice of the right solvent. Originally, the amination procedures worked only with very strong bases such as NaO*t*Bu or LiHMDS, and most side reactions observed can be attributed to these strong bases. So far these bases are still the most versatile ones and published examples with high catalytic turnovers are still using them. Additionally, some of the strong alkoxide bases are soluble in media that are compatible with C–N coupling and can therefore be handled as liquids, a form that is mostly welcome on industrial scales. The stronger alkyl-lithium bases have not so far been utilized in C–N coupling, because of their alkylating and dehydrogenating nature.

Modern synthesis, however, requires the applicability of a method even with highly functionalized, delicate, and late-stage intermediates. To meet these requirements, the introduction of very mild bases, especially in combination with low temperatures, have to be considered an important milestone in palladium-catalyzed aryl amination. The most important developments in the utilization of mild bases are K_3PO_4 and Cs_2CO_3 [253–255], while it is also important to note the development of sodium hydroxide as an extremely cheap base [235, 256]. Several systems have been found also to work at low temperatures, including Hartwig's dimeric Pd-Br-P(tBu$_3$) complex (Scheme 9.52) [249] or catalysts with carbene ligands [226, 227].

9.6
Nickel-Catalyzed Processes

The high activity of nickel towards oxidative addition to aryl chlorides had already been converted into a nickel-catalyzed aryl amination by 1950 by Hughes, $NiCl_2$ being used as an effective catalyst for the amination of chlorobenzene with methyl amine [257]. The general conditions, however, cannot be considered to be very practical. The scope of the reaction was broadened 25 years later, in 1975 [258]. Cramer and Coulson of Du Pont improved understanding of several reaction parameters by means of small-scale screening reactions. They were already aware that cyclic secondary amines converted best to the arylamines and used large excesses of the amine to improve the reaction (Scheme 9.53).

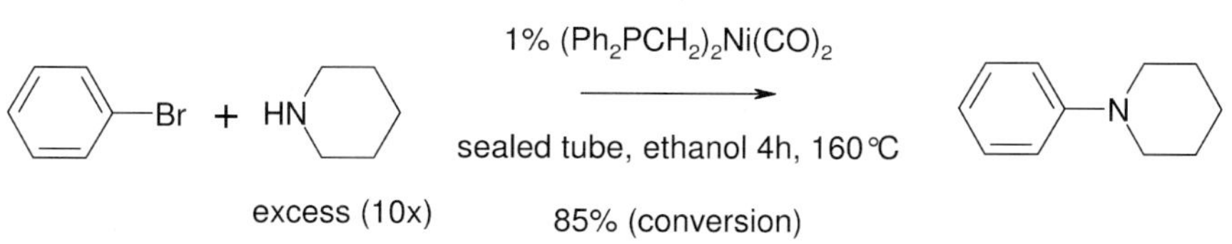

Scheme 9.53 Nickel-catalyzed aryl amination by Cramer (1975).

Again, their work only shows the concept of nickel-catalyzed aryl aminations. Another 22 years had to pass for the first practical approach to the same reaction, when Buchwald demonstrated the first reliable protocol for a nickel-catalyzed aryl amination on preparative scale in 1997 [259]. The major difference from his predecessors is the use of a stoichiometric amount of strong base, no doubt a lesson learned from palladium-catalyzed aryl amination. The reaction conditions are somewhat similar to those of the palladium-catalyzed alternative: temperatures around 100 °C, strong bases, nonpolar solvents, and a nickel complex as catalyst. In 1997 the coupling of aryl chlorides in the palladium-catalyzed version worked only with highly activated aryl chlorides, so the use of a catalyst that actually preferred aryl chlorides over bromides must have seemed a worthwhile endeavor (Scheme 9.54).

The first procedures included coupling of aryl chlorides with anilines, cyclic secondary amines, and primary aliphatic amines (the last of these, though, only in higher yields when electron-poor aryl halides were used as starting materials). NaO*t*Bu as base and 1,10-phenanthroline or dppf as co-ligands together with the common nickel precursor $Ni(COD)_2$ complete the set of reagents necessary to run these reactions, usually in toluene.

With the advent of more active Pd-phosphine catalysts, the obvious disadvantages of nickel (i.e., toxicity) apparently motivated more groups to continue to work on palladium-catalyzed aryl amination, but some groups still kept developing the nickel version.

In 1998, shortly after Buchwald, Fort's group made mechanistic speculations involving the conversion of classical nickel(0) catalysts from C–C coupling to C–N coupling and worked on a nickel-bipyridine catalyst in combination with a mixture of sodium hydride and sodium amylate [260]. They added a compound that is easily hydrogenated, such as styrene, to limit the excessive formation of benzene side products to below 20% (Scheme 9.55).

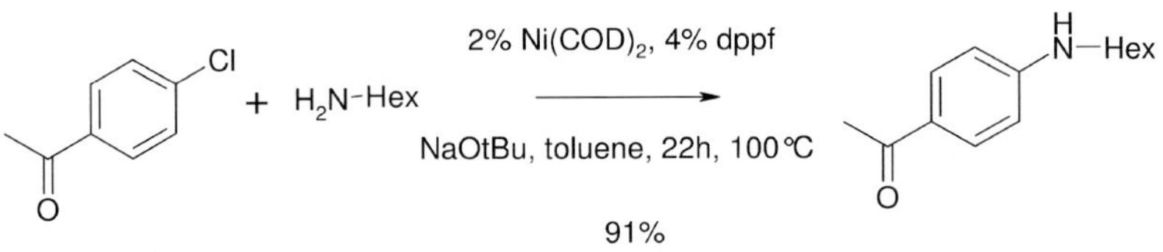

Scheme 9.54 Nickel-catalyzed amination of aryl chlorides by Buchwald (1997).

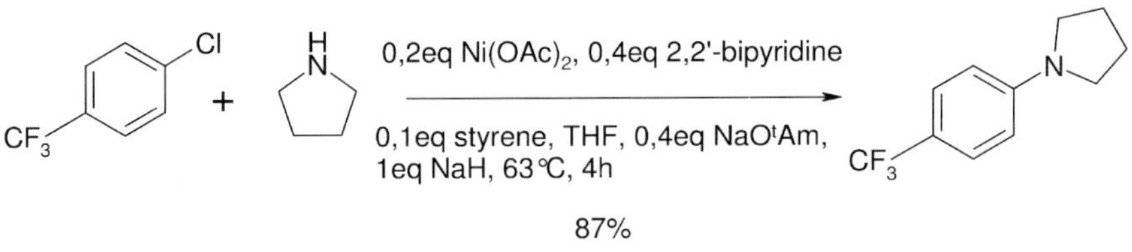

Scheme 9.55 Nickel-catalyzed amination of aryl chlorides with a styrene additive by Fort (1998).

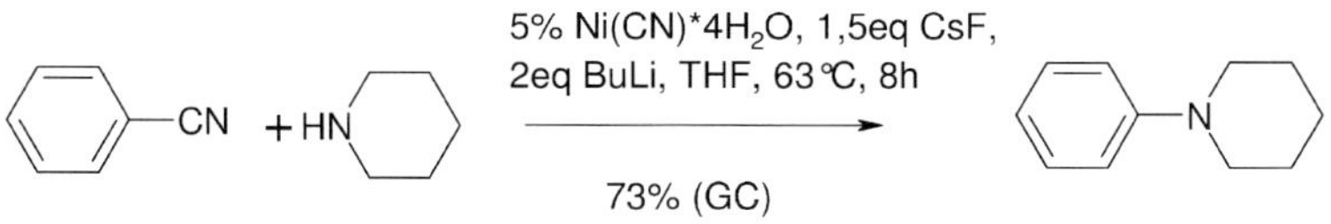

Scheme 9.56 Nickel-catalyzed amine-nitrile exchange by Miller (2003).

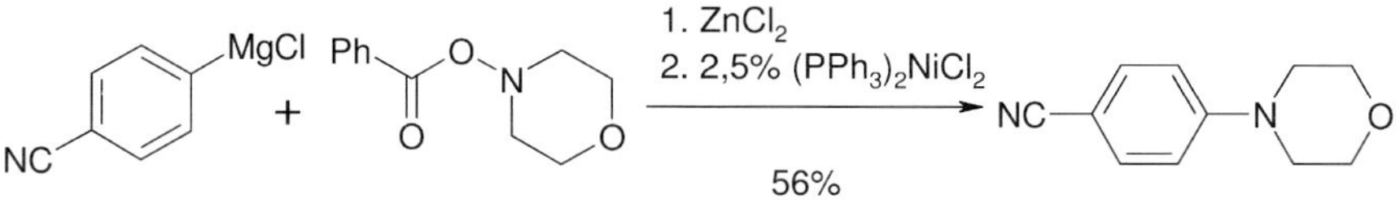

Scheme 9.57 Nickel-catalyzed amination of aryl zinc halides by Berman (2005).

In the subsequent years the group extended their method with selective mono-amination of dichlorobenzenes [261, 262]. In 2001 they also introduced saturated N-heterocyclic carbenes in combination with nickel acetylacetonate and sodium hydride/sodium *tert*-butoxide in THF at 63 °C [263, 264]. This time a low catalytic load (2% of nickel) matched Buchwald's results from 1997. Nolan published an improved procedure for a similar catalyst synthesis in 2005 [265].

In 2003 a procedure for intramolecular amination was presented [266].

Also in 2003, Lipshutz introduced a very practical method using his nickel on charcoal catalyst in combination with phosphine ligands such as dppf or PPh_3 and lithium *tert*-butoxide for effective aryl amination [267]. In 2005, Ackermann reported that his new class of diamino- and dioxophosphine-ligands that work well for palladium-catalyzed amination also work with nickel [268].

Following another approach in 2003, Miller demonstrated successful amine-nitrile exchange catalyzed by nickel [269] (Scheme 9.56), though this method still has room for improvement before it can be called a practical approach.

A last method to mention is the amination of organozinc halides by use of nickel-phosphine catalysts as presented by Berman in 2005 [270]. This procedure, however, is still a complicated organochemical transformation that cannot compete with the established copper- and palladium-catalyzed approaches and only works with benzoyloxy-protected amines. (Scheme 9.57).

9.7 Summary

From the author's point of view, the development of transition metal-catalyzed aryl aminations has furnished two major highlights: Buchwald–Hartwig amination and Ullmann amination. These two methods have reduced a formerly extremely difficult transformation into a trivial task with lots of possibilities for fine tuning for a perfect fit to a given system. Modern synthetic targets can now be produced in an even more highly convergent manner, giving chemists worldwide the opportunity to come up with compounds with new and improved properties. Work on alternatives is still ongoing, however, and so far history has shown that continuous doubt on the completeness of an approach has always been helpful for its perfection, so let's keep our fingers crossed.

References

1 Ullmann, F.; Bielecki, J. Synthesis in the biphenyl series. I. *Berichte der Deutschen Chemischen Gesellschaft 34,* 2174–2185. 1901.

2 Ullmann, F. A new path for preparing diphenylamine derivatives. *Berichte der Deutschen Chemischen Gesellschaft 36,* 2382–2384. 1903.

3 Ullmann, F.; Sponagel, P. Phenylation of phenols. *Berichte der Deutschen Chemischen Gesellschaft 38,* 2211–2212. 1905.

4 Goldberg, I. Phenylation in the presence of copper as catalyst. *Berichte der Deutschen Chemischen Gesellschaft 39,* 1691–1692. 1906.

5 Goldberg, I.; Genf; Ullman, F. Arylanthranilic acids. DE 173523, 1905.

6 Gowda, D. C.; Gowda, S. Formic acid with 10% palladium on carbon: a reagent for selective reduction of aromatic nitro compounds. *Indian Journal of Chemistry, Section B: Organic Chemistry Including Medicinal Chemistry 39B*[9], 709–711. 2000.

7 Gowda, D. C.; Mahesh, B. Catalytic transfer hydrogenation of aromatic nitro compounds by employing ammonium formate and 5% platinum on carbon. *Synthetic Communications 30*[20], 3639–3644. 2000.

8 Haber, F. *Zeitschrift für Elektrochemie und Angewandte Physikalische Chemie 4,* 506. 1897.

9 Haber, F. *Zeitschrift für Elektrochemie und Angewandte Physikalische Chemie 5,* 77. 1898.

10 Haber, F. *Zeitschrift für Physikalische Chemie 32,* 271. 1900.

11 Blout, E. R.; Silverman, D. C. Catalytic reduction of nitrocinnamic acids and esters. *Journal of the American Chemical Society* 66, 1442–1443. 1944.

12 de Traz, C. Two trichloroanisidines. *Helvetica Chimica Acta 30,* 232–236. 1947.

13 Romeo, A. A method of preparation of 4-amino-2-hydroxybenzoic acid. *Ricerca sci. 18,* 1057–1058. 1948.

14 Galatis, L. Derivatives of o-aminophenol. II. *Journal fuer Praktische Chemie (Leipzig) 151,* 331–341. 1938.

15 Kanth, S. R.; Reddy, G. V.; Rao, V. V. N.; Maitraie, D.; Narsaiah, B.; Rao, P. S. A simple and convenient method for the reduction of nitroarenes. *Synthetic Communications 32*[18], 2849–2853. 2002.

16 Hodgson, H. H.; Ward, E. R. Reactions of aromatic nitro compounds with alkaline sulfides. II. The three dinitrobenzenes. *Journal of the Chemical Society,* 1316–1317. 1949.

17 Steck, E. A.; Hallock, L. L.; Holland, A. J. Quinolines. IV. Some bz-iodo-3-methyl-4-(1-methyl-4-diethylaminobutylamino)quinolines. *Journal of the American Chemical Society* 68, 1241–1243. 1946.

18 Schröter, R. Amine durch Reduktion. In *Houben-Weyl – Methoden der organischen Chemie,* IV ed.; Müller, E., Ed.; Georg Thieme Verlag, Stuttgart, 1957; p 455.

19 Gowda, S.; Gowda, B. K. K.; Gowda, D. C. Hydrazinium monoformate: A new hydrogen donor. Selective reduction of nitrocompounds catalyzed by commercial zinc dust. *Synthetic Communications 33*[2], 281–289. 2003.

20 Cha, J. S.; Jeong, M. K.; Kwon, O. O.; Lee, K. D.; Lee, H. S. Reaction of diisobutylaluminum hydride-dimethyl sulfide complex with selected organic compounds containing representative functional groups. Comparison of the reducing characteristics of diisobutylaluminum hydride and its dimethyl sulfide complex. *Bulletin of the*

Korean Chemical Society 15[10], 873–881. 1994.

21 Petrini, M.; Ballini, R.; Rosini, G. Reduction of aliphatic and aromatic nitro compounds with sodium borohydride in tetrahydrofuran using 10% palladium-on-carbon as catalyst. *Synthesis* [8], 713–714. 1987.

22 Weil, H.; Traun, M.; Marcel, S. Reduction of substituted salicylic acids. *Ber. 55B*, 2664–2674. 1922.

23 Morgan, G. T.; Harrison, H. A. Acenaphthene series. V. *Journal of the Society of Chemical Industry*, London *49*, 413–21T. 1930.

24 Benkeser, R. A.; Lambert, R. F.; Ryan, P. W.; Stoffey, D. G. Reduction of organic compounds by lithium in low-molecular weight amines. IV. The effect of nitro and amino groups on the course of the reduction. *Journal of the American Chemical Society 80*, 6573–6577. 1958.

25 Noel, M.; Ravichandran, C.; Anantharaman, P. N. An electrochemical technique for the reduction of aromatic nitro compounds in H_2SO_4 medium on thermally coated Ti/TiO_2 electrodes. *Journal of Applied Electrochemistry 25*[7], 690–698. 1995.

26 Turpin, G. S. *Journal of the Chemical Society 69*, 714. 1891.

27 Turpin, G. S. *Berichte der Deutschen Chemischen Gesellschaft 24*, 949. 1891.

28 Pisani, F. *Comptes Rendus Hebdomadaires des Seances de l'Academie des Sciences 39*, 852. 1854.

29 Romburgh, P. v. *Recueil des Travaux Chimiques des Pays-Bas 2*, 105. 1883.

30 Thiess, K.; Deicke, B. DE 507831, 1928.

31 Hentrich, W.; Stroebel, R.; Tietze, E. DE 541567, 1930.

32 Kym, O. *Journal fuer Praktische Chemie 2* (51), 325. 1895.

33 Foohey, W. L.; Peck, F. W. US 2048790, 1933.

34 Kremer, C. B. *Journal of the American Chemical Society 61*, 1321. 1939.

35 Gilman, H.; Avakian, S. Dibenzofuran. XXIII. Rearrangement of halogen compounds in amination by sodamide. *Journal of the American Chemical Society 67*, 349–351. 1945.

36 Idoux, J. P.; Gupton, J. T.; Colon, C. Aromatic nucleophilic substitution. II. The reaction of chlorobenzonitriles and chloronitrobenzenes with HMPA. *Synthetic Communications 12*[12], 907–914. 1982.

37 Chichibabin, A. E.; Zeide, O. A. New reaction for compounds containing the pyridine nucleus. *Zhurnal Russkago Fiziko-Khimicheskago Obshchestva 46*, 1216–1236. 1914.

38 Chichibabin, A. E. Formation of γ-amino derivatives in the preparation of amino compounds of pyridine. *Zhurnal Russkago Fiziko-Khimicheskago Obshchestva 47*, 835–838. 1915.

39 Leffler, M. T. *Organic Reactions 1*, 91. 1942.

40 Bergstrom, F. W.; Granara, I. M.; Erickson, V. *Journal of Organic Chemistry 7*, 98. 1942.

41 Stern, M. K.; Cheng, B. K. Amination of nitrobenzene via nucleophilic aromatic substitution for hydrogen: direct formation of aromatic amide bonds. *Journal of Organic Chemistry 58*[24], 6883–6888. 1993.

42 Stern, M. K. Method of preparing 4-aminodiphenylamine. US A 5117063, 19920526.

43 De Vera, A. L. Zeolite support loaded with a base for coupling of aniline and nitrobenzene. WO A2 20fd01098252, 20011227.

44 Mertens, A.; Lammertsma, K.; Arvanaghi, M.; Olah, G. A. *Journal of the American Chemical Society 105*, 5657. 1983.

45 Borodkin, G. I.; Shubin, V. G. Nitrenium ions and problem of direct electrophilic amination of aromatic compounds. *Russian Journal of Organic Chemistry 41*[4], 473–504. 2005.

46 Sundberg, R. J.; Sloan, K. B. Acid-promoted aromatic substitution processes in photochemical and thermal decompositions of aryl azides. *Journal of Organic Chemistry 38*[11], 2052–2057. 1973.

47 Sinha, P.; Kofink, C. C.; Knochel, P. Preparation of aryl-alkylamines via electrophilic amination of functionalized arylazo tosylates with alkylzinc reagents. *Organic Letters 8*[17], 3741–3744. 2006.

48 Kovacic, P.; Levisky, J. A.; Goralski, C. T. Amination of alkylbenzenes with trichloramine-aluminum chloride. Synthetic utility and theoretical aspects. *Journal of the American Chemical Society 88*[1], 100–103. 1966.

49 Kovacic, P.; Lange, R. M.; Foote, J. L.; Goralski, C. T.; Hiller, J. J., Jr.; Levisky, J. A. New method of aromatic substitution yielding unusual orientation. Amination with N-haloamines catalyzed by aluminum chloride. *Journal of the American Chemical Society 86*[8], 1650–1651. 1964.

50 Kovacic, P.; Lange, R. M.; Goralski, C. T.; Hiller, J. J., Jr.; Levisky, J. A. *Journal of the American Chemical Society 87*, 1262. 1965.

51 Kovacic, P.; Lowery, M. K.; Field, K. W. *Chemical Reviews* (Washington, DC, United States) 70, 639. 1970.

52 Stella, L.; Raynier, B.; Surzur, J. M. Homolytic synthesis of potential analgesics: 2-substituted 4-arylpiperidines and benzomorphans. *Tetrahedron 37*[16], 2843–2854. 1981.

53 Minisci, F. Recent aspects of homolytic aromatic substitutions. *Topics in Current Chemistry 62*[Synth. Mech. Org. Chem.], 1–48. 1976.

54 Minisci, F.; Galli, R.; Cecere, M.; Quilico, A. Dialkylamination of aromatic amines. FR 93635, 19690425.

55 Minisci, F.; Galli, R.; Cecere, M. Radical amination of active aromatic compounds: acetamides. New process for the synthesis of p-amino-N,N-dialkylaniline. *Chemistry & Industry* (London, United Kingdom) *48*[12], 1324–1326. 1966.

56 Bucherer, H. T. *Journal fuer Praktische Chemie 69*, 49. 1904.

57 Cowdrey, W. A. The mechanism of the Bucherer reaction. IV. The kinetics of the conversion of naphthols to naphthylamines. *Journal of the Chemical Society*, 1046–1050. 1946.

58 Canete, A.; Melendrez, M. X.; Saitz, C.; Zanocco, A. L. Synthesis of aminonaphthalene derivatives using the Bucherer reaction under microwave irradiation. *Synthetic Communications 31*[14], 2143–2148. 2001.

59 Laubenheimer, A. *Berichte der Deutschen Chemischen Gesellschaft 9*, 1826. 1876.

60 Laubenheimer, A. *Berichte der Deutschen Chemischen Gesellschaft 11*, 1155. 1878.

61 Rudy, H.; Cramer, K. E. Homologs of alloxandimethylaminoanil (dimethylaminobarbiturylideneaniline) and (barbiturylideneiminodimethylamino-phenyl)dialuric acids. *Ber. 72B*, 227–248. 1939.

62 Ullmann, F.; Nadai, G. Preparation of o-Nitrated Amines from the Corresponding Phenol Derivatives (I). *Ber. 41*, 1870–1878. 1908.

63 Mittasch, A. *Geschichte der Ammoniaksynthese*, Verlag Chemie, Weinheim, 1951.

64 Schlögl, R. *Katalytische Ammoniaksynthese – eine „unendliche Geschichte. Angewandte Chemie 115*[18], 2050–2055. 2003.

65 In *Encyclopedia of Chemical Technology*, 2 ed., Wiley, New York, 1963, pp 332–373.

66 Groggins, P. H. Amination by Ammonolysis. In *Unit Processes in Organic Synthesis*, 5 ed., McGraw Hill Publishing Co., New York, 1958; p 388.

67 Hale, W. J.; Britton, J. W. US 1607824, 1925.

68 Hale, W. J.; Cheney, G. H. US 1729775, 1926.

69 Hale, W. J. US 1764869, 1926.

70 Hale, W. J. US 1804466, 1926.

71 Hale, W. J. US 1932518, 1928.

72 Britton, J. W.; Williams, W. H. US 1726170, 1926.

73 Britton, J. W.; Williams, W. H. US 1726171, 1927.

74 Britton, J. W.; Williams, W. H. US 1726172, 1927.

75 Britton, J. W.; Williams, W. H. US 1726173, 1927.

76 Britton, J. W. US 1823025, 1929.

77 Britton, J. W.; Williams, W. H.; Putnam, M. E. US 1823026, 1932.

78 DE 204951, 1907.

79 Williams, W. H. US 1775360, 1927.

80 Putnam, M. E. US 1885625, 1928.

81 Woroshzow, N. N.; Kobelew, V. A. *Zurnal Obscej Chimii (Journal of General Chemistry) 3*, 111–114. 1934.

82 Williams, W. H.; Holmes, R. D.; Fruehoff, H. F. US 2432551, 1947.

83 Williams, W. H.; Holmes, R. D.; Widiger, A. H. US 2432552, 1947.

84 Hughes, E. C.; Veatch, F. US 2490813, 1949.

85 Widiger, A. H. US 2476170, 1949.

86 Mueller, W. 4-Nitrodiphenylamines. DE A1 3246151, 19840614.

87 Giera, H.; Lange, W.; Pohl, T.; Sicheneder, A.; Schild, C. Procedure for the production of (N-nitrophenyl)(N-phenyl)amines from nitrohalobenzenes and aniline in the presence of ground bases. DE C1 19942394, 2001.

88 Krock, F. W.; Neett, R. EP 31783, 1982.

89 Mills, L. E. US 1935515, 1933.

90 Wibaut, J. P. *Berichte der Deutschen Chemischen Gesellschaft 50*, 541. 1917.

91 Axon, S. A. WO 9910311, 1999.

92 Hoelderich, W. F.; Becker, J. *Catalysis Letters 54*, 125. 1998.

93 Desrosiers, P.; Guan, S.; Hagemeyer, A.; Lowe, D. M.; Lugmair, C.; Poojary, D. M.; Turner, H.; Weinberg, H.; Zhou, X.; Armbrust, R.; Fengler, G.; Notheis, U. *Catalysis Today 81*, 319. 2003.

94 Desrosiers, P.; Guan, S.; Hagemeyer, A.; Lowe, D. M.; Poojary, D. M.; Turner, H.; Weinberg, H.; Zhou, X.; Armbrust, R.; Fengler, G.; Notheis, U.; Borade, R. *Applied Catalysis 227*, 43. 2002.

95 Schmerling, L. US 2948755, 1960.

96 Stitt, H. E. WO 0009473, 2000.

97 Thomas, C. I. CA 553988, 1958.

98 Yasushi, H. JP 6293715, 1994.

99 DelPesco, T. W. US 4001260, 1977.

100 DelPesco, T. W. US 4031106, 1977.

101 Squire, E. N. US 3929889, 1975.

102 Squire, E. N. US 3919155, 1975.

103 Hoffmann, N.; Muhler, M. *Catal. Lett. 103*, 155. 2005.

104 Briner, E. *Helvetica Chimica Acta 7*, 282. 1924.

105 Fischer, F.; Bahr, T.; Wiedeking, K. *Brennstoffchemie 15*, 101. 1934.

106 Mori, Y.; Noro, H.; Hara, Y.; Washama, T. Preparation of aromatic amines from phenols. JP A2 06184062, 19940705.

107 Liu, Z.; Xie, Z.; Zhang, H.; Chen, Q. Catalyst for preparing aniline and alkylaniline. CN A 1381440, 20021127.

108 Kunz, K.; Scholz, U.; Ganzer, D. Renaissance of Ullmann and Goldberg reactions – progress in copper catalyzed C-N-, C-O- and C-S-coupling. *Synlett* [15], 2428–2439. 2003.

109 Dodonov, V. A.; Gushchin, A. V.; Brilkina, T. G. Some catalytic reactions of triphenylbismuth diacetate in the presence of copper salts. *Zhurnal Obshchei Khimii 55*[11], 2514–2519. 1985.

110 Esteves, M. A.; Narander, M.; Marcelo-Curto, M. J.; Gigante, B. *Journal of Natural Products 64*, 761–766. 2001.

111 Barton, D. H. R.; Donnelly, D. M. X.; Finet, J. P.; Guiry, P. J. Arylation of amines by aryllead triacetates using copper catalysis. *Tetrahedron Letters 30*[11], 1377–1380. 1989.

112 Lam, P. Y. S.; Deudon, S.; Averill, K. S.; Li, R.; He, M. Y.; DeShong, P.; Clark, C. G. *Journal of the American Chemical Society 122*, 7600–7601. 2000.

113 Lam, P. Y. S.; Averill, K. S.; Subern, S.; Clark, C. G.; Adams, J.; Chan, D. M. T.; Combs, A. *Synlett 5*, 674–676. 2000.

114 Chan, D. M.; Monaco, K. L.; Wang, R. P.; Winters, M. P. *Tetrahedron Letters 39*, 2933–2936. 1998.

115 Evans, D. A.; Katz, J. L.; West, T. R. *Tetrahedron Letters 39*, 2937–2940. 1998.

116 Lam, P. Y. S.; Clark, C. G.; Saubern, S.; Adams, J.; Winters, M. P. *Tetrahedron Letters 39*, 2941–2944. 1998.

117 Antilla, J. C.; Buchwald, S. L. Copper-Catalyzed Coupling of Arylboronic Acids and Amines. *Organic Letters 3*[13], 2077–2079. 2001.

118 Finet, J. P.; Fedorov, A. Y.; Boyer, G. *Current Organic Chemistry 6*, 597–626. 2002.

119 Pellon, R. F.; Carrasco, R.; Rodes, L. Synthesis of N-phenylanthranilic acid using water as solvent. *Synthetic Communications 23*[10], 1447–1453. 1993.

120 Ma, D.; Zhang, Y.; Yao, J.; Wu, S.; Tao, F. Accelerating effect induced by the structure of α-amino acid in the copper-catalyzed coupling reaction of aryl halides with α-amino acids. Synthesis of benzolactam-V8. *Journal of the American Chemical Society 120*[48], 12459–12467. 1998.

121 Ma, D.; Xia, C. CuI-catalyzed coupling reaction of β-amino acids or esters with aryl halides at temperature lower than that employed in the normal Ullmann reaction. Facile synthesis of SB-214857. *Organic Letters 3*[16], 2583–2586. 2001.

122 Clement, J. B.; Hayes, J. F.; Sheldrake, H. M.; Sheldrake, P. W.; Wells, A. S. Synthesis of SB-214857 using copper catalyzed amination of aryl bromides with L-aspartic acid. *Synlett* [9], 1423–1427. 2001.

123 Job, G. E.; Buchwald, S. L. Copper-Catalyzed Arylation of β-Amino Alcohols. *Organic Letters 4*[21], 3703–3706. 2002.

124 Goodbrand, H. B.; Hu, N. X. Ligand-Accelerated Catalysis of the Ullmann Condensation: Application to Hole Conducting Triarylamines. *Journal of Organic Chemistry 64*[2], 670–674. 1999.

125 Kiyomori, A.; Marcoux, J. F.; Buchwald, S. L. An efficient copper-catalyzed coupling of aryl halides with imidazoles. *Tetrahedron Letters 40*[14], 2657–2660. 1999.

126 Gujadhur, R. K.; Bates, C. G.; Venkataraman, D. Formation of Aryl-Nitrogen, Aryl-Oxygen, and Aryl-Carbon Bonds Using Well-Defined Copper(I)-Based Catalysts. *Organic Letters 3*[26], 4315–4317. 2001.

127 Evindar, G.; Batey, R. A. Copper- and Palladium-Catalyzed Intramolecular Aryl Guanidinylation: An Efficient Method for the Synthesis of 2-Aminobenzimidazoles. *Organic Letters 5*[2], 133–136. 2003.

128 Kwong, F. Y.; Klapars, A.; Buchwald, S. L. Copper-Catalyzed Coupling of Alkylamines and Aryl Iodides: An Efficient System Even in an Air Atmosphere. *Organic Letters 4*[4], 581–584. 2002.

129 Kwong, F. Y.; Buchwald, S. L. Mild and Efficient Copper-Catalyzed Amination of Aryl Bromides with Primary Alkylamines. *Organic Letters 5*[6], 793–796. 2003.

130 Kelkar, A. A.; Patil, N. M.; Chaudhari, R. V. Copper-catalyzed amination of aryl halides: single-step synthesis of triarylamines. *Tetrahedron Letters 43*[40], 7143–7146. 2002.

131 Patil, N. M.; Kelkar, A. A.; Chaudhari, R. V. Synthesis of triarylamines by copper-catalyzed amination of aryl halides. *Journal of Molecular Catalysis A: Chemical 223*[1–2], 45–50. 2004.

132 Scholz, U.; Kunz, K.; Gaertzen, O.; Benet-Buchholz, J.; Wesener, J. Copper complexes of phosphorus-containing ligands and their use as coupling catalysts. EP A1 1437356, 20040714.

133 Kunz, K.; Scholz, U.; Gaertzen, O.; Ganzer, D.; Wesener, J. Preparation of copper carbene complexes and their use as coupling reaction catalysts. EP A1 1437355, 20040714.

134 Kunz, K.; Haider, J.; Ganzer, D.; Scholz, U.; Sicheneder, A. Process for the preparation of nitrodiphenylamines. EP A1 1437341, 20040714.

135 Scholz, U.; Kunz, K.; Gaertzen, O.; Benet-Buchholz, J.; Wesener, J. Copper complexes of phosphorus-containing ligands and their use as coupling catalysts. EP A1 1437356, 20040714.

136 Haider, J.; Kunz, K.; Scholz, U. Highly selective copper-catalyzed monoarylation of aniline. *Advanced Synthesis & Catalysis 346*[7], 717–722. 2004.

137 Wolter, M.; Klapars, A.; Buchwald, S. L. Synthesis of N-Aryl Hydrazides by Copper-Catalyzed Coupling of Hydrazides with Aryl Iodides. *Organic Letters 3*[23], 3803–3805. 2001.

138 Klapars, A.; Antilla, J. C.; Huang, X.; Buchwald, S. L. A general and efficient copper catalyst for the amidation of aryl halides and the N-arylation of nitrogen heterocycles. *Journal of the American Chemical Society 123*[31], 7727–7729. 2001.

139 Klapars, A.; Huang, X.; Buchwald, S. L. A General and Efficient Copper Catalyst for the Amidation of Aryl Halides. *Journal of the American Chemical Society 124*[25], 7421–7428. 2002.

140 Antilla, J. C.; Klapars, A.; Buchwald, S. L. The Copper-Catalyzed N-Arylation of Indoles. *Journal of the American Chemical Society 124*[39], 11684–11688. 2002.

141 Buchwald, S. L. Copper-catalyzed formation of carbon-heteroatom and carbon-carbon bonds by arylation and vinylation of amines, amides, hydrazides, heterocycles, alcohols, enolates, and malonates, using aryl, heteroaryl, and vinyl halides and analogs. WO A1 2002085838, 20021031.

142 Antilla, J. C.; Baskin, J. M.; Barder, T. E.; Buchwald, S. L. Copper-diamine-catalyzed N-arylation of pyrroles, pyrazoles, indazoles, imidazoles, and triazoles. *Journal of Organic Chemistry* *69*[17], 5578–5587. 2004.

143 Strieter, E. R.; Blackmond, D. G.; Buchwald, S. L. The Role of Chelating Diamine Ligands in the Goldberg Reaction: A Kinetic Study on the Copper-Catalyzed Amidation of Aryl Iodides. *Journal of the American Chemical Society 127*[12], 4120–4121. 2005.

144 Kang, S. K.; Kim, D. H.; Park, J. N. Copper-catalyzed N-arylation of aryl iodides with benzamides or nitrogen heterocycles in the presence of ethylenediamine. *Synlett* [3], 427–430. 2002.

145 Crawford, K. R.; Padwa, A. Copper-catalyzed amidations of bromo substituted furans and thiophenes. *Tetrahedron Letters 43*[41], 7365–7368. 2002.

146 Tsang, W. C. P.; Zheng, N.; Buchwald, S. L. Combined C-H functionalization/C-N bond formation route to carbazoles. *Journal of the American Chemical Society 127*[42], 14560–14561. 2005.

147 Guram, A. S.; Rennels, R. A.; Buchwald, S. L.; Barta, N. S.; Pearson, W. H. Palladium-catalyzed amination of aryl halides with amines. *Chemtracts: Inorganic Chemistry* 8[1], 1–5. 1996.

148 Hartwig, J. F. Palladium-catalyzed amination of aryl halides. Mechanism and rational catalyst design. *Synlett* [4], 329–340. 1997.

149 Wolfe, J. P.; Wagaw, S.; Marcoux, J. F.; Buchwald, S. L. Rational Development of Practical Catalysts for Aromatic Carbon-Nitrogen Bond Formation. *Accounts of Chemical Research 31*[12], 805–818. 1998.

150 Hartwig, J. F. Carbon-Heteroatom Bond-Forming Reductive Eliminations of Amines, Ethers, and Sulfides. *Accounts of Chemical Research 31*[12], 852–860. 1998.

151 Hartwig, J. F. Transition metal catalyzed synthesis of arylamines and aryl ethers from aryl halides and triflates: scope and mechanism. *Angewandte Chemie, International Edition 37*[15], 2046–2067. 1998.

152 Yang, B. H.; Buchwald, S. L. Palladium-catalyzed amination of aryl halides and sulfonates. *Journal of Organometallic Chemistry 576*[1–2], 125–146. 1999.

153 Muci, A. R.; Buchwald, S. L. Practical palladium catalysts for C-N and C-O bond formation. *Topics in Current Chemistry 219*[Cross-Coupling Reactions], 131–209. 2002.

154 Hartwig, J. F. Palladium-catalyzed amination of aryl halides and sulfonates. *Modern Arene Chemistry*, 107–168. 2002.

155 Hartwig, J. F. Palladium-catalyzed amination of aryl halides and related reactions. *Handbook of Organopalladium Chemistry for Organic Synthesis 1*, 1051–1096. 2002.

156 Prim, D.; Campagne, J. M.; Joseph, D.; Andrioletti, B. Palladium-catalyzed reactions of aryl halides with soft, non-organometallic nucleophiles. *Tetrahedron 58*[11], 2041–2075. 2002.

157 Littke, A. F.; Fu, G. C. Palladium-catalyzed coupling reactions of aryl chlorides. *Angewandte Chemie, International Edition 41*[22], 4176–4211. 2002.

158 Hartwig, J. F. Metal complexes as catalysts for carbon-heteroatom cross-coupling reactions. *Comprehensive Coordination Chemistry II 9*, 369–398. 2004.

159 Schlummer, B.; Scholz, U. Palladium-catalyzed C-N and C-O coupling-A practical guide from an industrial vantage point. *Advanced Synthesis & Catalysis 346*[13–15], 1599–1626. 2004.

160 Hartwig, J. F. Discovery and understanding of transition-metal-catalyzed aromatic substitution reactions. *Synlett* [9], 1283–1294. 2006.

161 Buchwald, S. L. Process and catalysts for the preparation of arylamines. US A 5576460, 19961119.

162 Guram, A. S.; Rennels, R. A.; Buchwald, S. L. A simple catalytic method for the conversion of aryl bromides to arylamines. *Angewandte Chemie, International Edition in English* *34*[12], 1348–1350. 1995.
163 Louie, J.; Hartwig, J. F. Palladium-catalyzed synthesis of arylamines from aryl halides. Mechanistic studies lead to coupling in the absence of tin reagents. *Tetrahedron Letters* *36*[21], 3609–3612. 1995.
164 Kosugi, M.; Kameyama, M.; Migita, T. Palladium-catalyzed aromatic amination of aryl bromides with N,N-diethylaminotributyltin. *Chemistry Letters* [6], 927–928. 1983.
165 Boger, D. L.; Duff, S. R.; Panek, J. S.; Yasuda, M. *Journal of Organic Chemistry* *50*, 5782. 1985.
166 Dzhemilev, U. M.; Ibragimov, A. G.; Minsker, D. L.; Muslukhov, R. R. Novel reaction of magnesium amines with allylic electrophiles catalyzed by palladium complexes. *Izvestiya Akademii Nauk SSSR, Seriya Khimicheskaya* [2], 406–409. 1987.
167 Guram, A. S.; Buchwald, S. L. Palladium-Catalyzed Aromatic Aminations with in situ Generated Aminostannanes. *Journal of the American Chemical Society* *116*[17], 7901–7902. 1994.
168 Paul, F.; Patt, J.; Hartwig, J. F. Palladium-catalyzed formation of carbon-nitrogen bonds. Reaction intermediates and catalyst improvements in the hetero cross-coupling of aryl halides and tin amides. *Journal of the American Chemical Society* *116*[13], 5969–5970. 1994.
169 Paul, F.; Patt, J.; Hartwig, J. F. Structural Characterization and Simple Synthesis of {Pd[P(o-Tol)3]2}. Spectroscopic Study and Structural Characterization of the Dimeric Palladium(II) Complexes Obtained by Oxidative Addition of Aryl Bromides and Their Reactivity with Amines. *Organometallics* *14*[6], 3030–3039. 1995.
170 Wolfe, J. P.; Buchwald, S. L. Scope and limitations of the Pd/BINAP-catalyzed amination of aryl bromides. *Journal of Organic Chemistry* *65*[4], 1144–1157. 2000.
171 Harris, M. C.; Geis, O.; Buchwald, S. L. Sequential N-arylation of primary amines as a route to alkyldiarylamines. *Journal of Organic Chemistry* *64*[16], 6019–6022. 1999.
172 Laufer, R. S.; Dmitrienko, G. I. *Journal of the American Chemical Society* *124*[9], 1854–1855. 2002.
173 Wolfe, J. P.; Wagaw, S.; Buchwald, S. L. An Improved Catalyst System for Aromatic Carbon-Nitrogen Bond Formation: The Possible Involvement of Bis(Phosphine) Palladium Complexes as Key Intermediates. *Journal of the American Chemical Society* *118*[30], 7215–7216. 1996.
174 Prashad, M.; Hu, B.; Lu, Y. M.; Draper, R.; Har, D.; Repic, O.; Blacklock, T. J. *Journal of Organic Chemistry* *68*, 2612–2614. 2000.
175 Prashad, M.; Hu, B.; Lu, Y. S.; Draper, R.; Har, D.; Repic, O.; Blacklock, T. *Journal of Organic Chemistry* *65*, 2612–2614. 2000.
176 Wolfe, J. P.; Buchwald, S. L. Room temperature catalytic amination of aryl iodides. *Journal of Organic Chemistry* *62*[17], 6066–6068. 1997.
177 Marinetti, A.; Hubert, P.; Genêt, J. P. *European Journal of Organic Chemistry*, 1815. 2000.
178 Wagaw, S.; Rennels, R. A.; Buchwald, S. L. Palladium-Catalyzed Coupling of Optically Active Amines with Aryl Bromides. *Journal of the American Chemical Society* *119*[36], 8451–8458. 1997.
179 Wolfe, J. P.; Buchwald, S. L. Palladium-Catalyzed Amination of Aryl Triflates. *Journal of Organic Chemistry* *62*[5], 1264–1267. 1997.
180 Ahman, J.; Buchwald, S. L. An improved method for the palladium-catalyzed amination of aryl triflates. *Tetrahedron Letters* *38*[36], 6363–6366. 1997.
181 Demadrille, R.; Moustrou, C.; Samat, A.; Guglielmetti, R. *Heterocyclic Communications* *5*, 123. 1999.
182 Wentland, M. P.; Xu, G.; Cioffi, C. L.; Ye, Y.; Duan, W.; Cohen, D. J.; Colasurdo,

A. M.; Bidlack, J. M. *Bioorganic & Medicinal Chemistry Letters 10*, 183. 2000.

183 Hong, Y.; Senanayake, C. H.; Xiang, T.; Vandenbossche, C. P.; Tanoury, G. J.; Bakale, R. P.; Wald, S. A. *Tetrahedron Letters 39*, 3121. 1998.

184 Kanbara, T.; Izumi, K.; Nakadani, Y.; Narise, T.; Hasegawa, K. *Chemistry Letters*, 1185. 1997.

185 Zhang, X. X.; Harris, M. C.; Sadighi, J. P.; Buchwald, S. L. The use of palladium chloride as a precatalyst for the amination of aryl bromides. *Canadian Journal of Chemistry 79*[11], 1799–1805. 2001.

186 Ferreira, I. C. F. R.; Queiroz, M. J. R. P.; Kirsch, G. *Tetrahedron 59*, 975–981. 2003.

187 Alcazar-Roman, L. M.; Hartwig, J. F.; Rheingold, A. L.; Liable-Sands, L. M.; Guzei, I. A. Mechanistic Studies of the Palladium-Catalyzed Amination of Aryl Halides and the Oxidative Addition of Aryl Bromides to $Pd(BINAP)_2$ and $Pd(DPPF)_2$: An Unusual Case of Zero-Order Kinetic Behavior and Product Inhibition. *Journal of the American Chemical Society 122*[19], 4618–4630. 2000.

188 Singh, U. K.; Strieter, E. R.; Blackmond, D. G.; Buchwald, S. L. Mechanistic Insights into the Pd(BINAP)-Catalyzed Amination of Aryl Bromides: Kinetic Studies under Synthetically Relevant Conditions. *Journal of the American Chemical Society 124*[47], 14104–14114. 2002.

189 Shekhar, S.; Ryberg, P.; Hartwig, J. F.; Mathew, J. S.; Blackmond, D. G.; Strieter, E. R.; Buchwald, S. L. Reevaluation of the Mechanism of the Amination of Aryl Halides Catalyzed by BINAP-Ligated Palladium Complexes. *Journal of the American Chemical Society 128*[11], 3584–3591. 2006.

190 Yin, J.; Zhao, M. M.; Huffman, M. A.; McNamara, J. M. *Organic Letters 4*[20], 3481–3484. 2002.

191 Urgaonkar, S.; Xu, J. H.; Verkade, J. G. *Journal of Organic Chemistry 68*, 8416–8423. 2003.

192 Vyskocil, S.; Smrcina, M.; Kocovsky, P. Synthesis of 2-amino-2′-diphenylphosphino-1,1′-binaphthyl (MAP) and its accelerating effect on the Pd(0)-catalyzed N-arylation. *Tetrahedron Letters 39*[50], 9289–9292. 1998.

193 Ogasawara, M.; Yoshehida, K.; Hayashi, T. *Organometallics 19*, 1567. 2000.

194 Sadighi, J. P.; Harris, M. C.; Buchwald, S. L. A highly active palladium catalyst system for the arylation of anilines. *Tetrahedron Letters 39*[30], 5327–5330. 1998.

195 Driver, M. S.; Hartwig, J. F. A Second-Generation Catalyst for Aryl Halide Amination: Mixed Secondary Amines from Aryl Halides and Primary Amines Catalyzed by $(DPPF)PdCl_2$. *Journal of the American Chemical Society 118*[30], 7217–7218. 1996.

196 Louie, J.; Hartwig, J. F.; Fry, A. J. Discrete High Molecular Weight Triarylamine Dendrimers Prepared by Palladium-Catalyzed Amination. *Journal of the American Chemical Society 119*[48], 11695–11696. 1997.

197 Louie, J.; Hartwig, J. F. *Macromolecules 31*, 6737. 1998.

198 Goodson, F. E.; Hartwig, J. F. Regiodefined Poly(N-arylaniline)s and Donor-Acceptor Copolymers via Palladium-Mediated Amination Chemistry. *Macromolecules 31*[5], 1700–1703. 1998.

199 Hauck, S. I.; Lakshmi, K. V.; Hartwig, J. F. *Organic Letters 1*, 2057. 1999.

200 Tew, G. N.; Pralle, M. U.; Stupp, S. I. *Angewandte Chemie, International Edition 39*, 517. 2000.

201 Braig, T.; Muller, D. C.; Gross, M.; Meerholz, K.; Nuyken, O. *Macromolecular Rapid Communications 21*, 583. 2000.

202 Thelakkhat, M.; Hagen, J.; Haarer, D.; Schmidt, H. W. *Synthetic Methods 102*, 1125. 1999.

203 Louie, J.; Driver, M. S.; Hamann, B. C.; Hartwig, J. F. Palladium-Catalyzed Amination of Aryl Triflates and Importance of Triflate Addition Rate. *Journal of Organic Chemistry 62*[5], 1268–1273. 1997.

204 Beletskaya, I. P.; Bessmertnykh, A. G.; Guilard, R. *Synlett*, 1459. 1999.

205 Buchwald, S. L.; Marcoux, J. F.; Wagaw, S. *Journal of Organic Chemistry 62*, 1568. 1997.

206 Buchwald, S. L.; Wolfe, J. P. *Tetrahedron Letters 38*, 6359. 1997.

207 Watanabe, M.; Yamamoto, T.; Nishiyama, M. Preparation of 1-aminoindoles and 1-aminoquinolines. EP A2 1035114, 20000913.

208 Watanabe, M.; Yamamoto, T.; Nishiyama, M. A new palladium-catalyzed intramolecular cyclization: synthesis of 1-aminoindole derivatives and functionalization of their carbocyclic rings. *Angewandte Chemie, International Edition 39*[14], 2501–2504. 2000.

209 Kataoka, N.; Shelby, Q.; Stambuli, J. P.; Hartwig, J. F. Air stable, sterically hindered ferrocenyl dialkylphosphines for palladium-catalyzed C-C, C-N, and C-O bond-forming cross-couplings. *Journal of Organic Chemistry 67*[16], 5553–5566. 2002.

210 Hamann, B. C.; Hartwig, J. F. Sterically Hindered Chelating Alkyl Phosphines Provide Large Rate Accelerations in Palladium-Catalyzed Amination of Aryl Iodides, Bromides, and Chlorides, and the First Amination of Aryl Tosylates. *Journal of the American Chemical Society 120*[29], 7369–7370. 1998.

211 Netherton, M. R.; Fu, G. C. Air-Stable Trialkylphosphonium Salts: Simple, Practical, and Versatile Replacements for Air-Sensitive Trialkylphosphines. Applications in Stoichiometric and Catalytic Processes. *Organic Letters 3*[26], 4295–4298. 2001.

212 Nishiyama, M.; Koie, Y. Process for producing heterocyclic aromatic amine or arylamine. EP A1 802173, 1997.

213 Yamamoto, T.; Nishiyama, S.; Koie, Y. One-pot preparation of tertiary arylamines from primary amines. JP A2 10310561, 19981124.

214 Yamamoto, T.; Nishiyama, S.; Koie, Y. Manufacture of polyarylamines. JP A2 11080346, 19990326.

215 Watanabe, M.; Nishiyama, M.; Yamamoto, T.; Koie, Y. Palladium/P(t-Bu)3 – a superior amination catalyst. *Speciality Chemicals 18*[10], 445–446, 448, 450. 1998.

216 Yamamoto, T.; Nishiyama, M.; Koie, Y. Palladium-catalyzed synthesis of triarylamines from aryl halides and diarylamines. *Tetrahedron Letters 39*[16], 2367–2370. 1998.

217 Nishiyama, M.; Yamamoto, T.; Koie, Y. Synthesis of N-arylpiperazines from aryl halides and piperazine under a palladium tri-tert-butylphosphine catalyst. *Tetrahedron Letters 39*[7], 617–620. 1998.

218 Nishiyama, M.; Yamamoto, T.; Koie, Y. *Tetrahedron Letters 39*, 617. 1998.

219 Hartwig, J. F.; Kawatsura, M.; Hauck, S. I.; Shaughnessy, K. H.; Alccazar-Roman, L. M. Room-Temperature Palladium-Catalyzed Amination of Aryl Bromides and Chlorides and Extended Scope of Aromatic C-N Bond Formation with a Commercial Ligand. *Journal of Organic Chemistry 64*[15], 5575–5580. 1999.

220 Yamamoto, T.; Nishiyama, M.; Koie, Y. *Tetrahedron Letters 39*, 2367. 1998.

221 Goodson, F. E.; Hauck, S. I.; Hartwig, J. F. Palladium-Catalyzed Synthesis of Pure, Regiodefined Polymeric Triarylamines. *Journal of the American Chemical Society 121*[33], 7527–7539. 1999.

222 Reddy, N. P.; Tanaka, M. *Tetrahedron Letters 38*, 4807. 1997.

223 Wolfe, J. P.; Buchwald, S. L. Palladium-Catalyzed Amination of Aryl Iodides. *Journal of Organic Chemistry 61*[3], 1133–1135. 1996.

224 Zhao, S. H.; Miller, A. K.; Berger, J.; Flippin, L. A. *Tetrahedron Letters 37*, 4463. 1996.

225 Herrmann, W. A.; Kocher, C. Essays on organometallic chemistry. 9. N-Heterocyclic carbenes. *Angewandte Chemie, International Edition in English 36*[20], 2162–2187. 1997.

226 Nolan, S. P.; Grasa, G.; Huang, J. *Organic Letters 1*, 1307. 1999.

227 Stauffer, S. R.; Lee, S.; Stambuli, J. P.; Hauck, S. I.; Hartwig, J. F. High Turnover Number and Rapid, Room-Temperature Amination of Chloroarenes Using Saturated Carbene Ligands. *Organic Letters 2*[10], 1423–1426. 2000.

228 Wolfe, J. P.; Tomori, H.; Sadighi, J. P.; Yin, J.; Buchwald, S. L. Simple, efficient

catalyst system for the palladium-catalyzed amination of aryl chlorides, bromides, and triflates. *Journal of Organic Chemistry* 65[4], 1158–1174. 2000.

229 Buchwald, S. L.; Wolfe, J. P.; Old, D. W. *Journal of the American Chemical Society 120*, 9722. 1998.

230 Zhang, X. X.; Buchwald, S. L. Efficient Synthesis of N-Aryl-Aza-Crown Ethers via Palladium-Catalyzed Amination. *Journal of Organic Chemistry* 65[23], 8027–8031. 2000.

231 Tomori, H.; Fox, J. M.; Buchwald, S. L. *Journal of Organic Chemistry 65*, 5334. 2000.

232 Buchwald, S. L.; Mauger, C.; Mignani, G.; Scholz, U. Industrial-scale palladium-catalyzed coupling of aryl halides and amines – a personal account. *Advanced Synthesis & Catalysis 348*[1 + 2], 23–39. 2006.

233 Wolfe, J. P.; Buchwald, S. L. *Angewandte Chemie, International Edition 38*, 2413. 1999.

234 Strieter, E. R.; Blackmond, D. G.; Buchwald, S. L. Insights into the Origin of High Activity and Stability of Catalysts Derived from Bulky, Electron-Rich Monophosphinobiaryl Ligands in the Pd-Catalyzed C-N Bond Formation. *Journal of the American Chemical Society 125*[46], 13978–13980. 2003.

235 Huang, X.; Anderson, K. W.; Zim, D.; Jiang, L.; Klapars, A.; Buchwald, S. L. Expanding Pd-Catalyzed C-N Bond-Forming Processes: The First Amidation of Aryl Sulfonates, Aqueous Amination, and Complementarity with Cu-Catalyzed Reactions. *Journal of the American Chemical Society 125*[22], 6653–6655. 2003.

236 Guram, A. S.; Bei, X. H.; Uno, T.; Norris, J.; Turner, H. W.; Weinberg, W. H.; Peterson, J. L. *Organometallics 18*, 1840. 1999.

237 Urgaonkar, S.; Xu, J. H.; Verkade, J. G. Application of a new bicyclic triaminophosphine ligand in Pd-catalyzed Buchwald-Hartwig amination reactions of aryl chlorides, bromides, and iodides. *Journal of Organic Chemistry* 68[22], 8416–8423. 2003.

238 Urgaonkar, S.; Nagarajan, M.; Verkade, J. G. P[N(i-Bu)CH2CH2]3N: A Versatile Ligand for the Pd-Catalyzed Amination of Aryl Chlorides. *Organic Letters* 5[6], 815–818. 2003.

239 Urgaonkar, S.; Verkade, J. G. Scope and limitations of Pd2(dba)3/P(i-BuNCH2CH2)3N-catalyzed Buchwald–Hartwig amination reactions of aryl chlorides. *Journal of Organic Chemistry* 69[26], 9135–9142. 2004.

240 Urgaonkar, S.; Nagarajan, M.; Verkade, J. G. P(i-BuNCH$_2$CH$_{2)3}$N: An Effective Ligand in the Palladium-Catalyzed Amination of Aryl Bromides and Iodides. *Journal of Organic Chemistry* 69[26], 9323. 2004.

241 Singer, R. A.; Caron, S.; McDermott, R. E.; Arpin, P.; Do, N. M. Alternative biarylphosphines for use in the palladium-catalyzed amination of aryl halides. *Synthesis* [11], 1727–1731. 2003.

242 Beller, M.; Ehrentraut, A.; Fuhrmann, C.; Zapf, A. Preparation of novel phosphine ligands and use in catalytic reactions. WO A1 2002010178, 20020207.

243 Ehrentraut, A.; Zapf, A.; Beller, M. A new improved catalyst for the palladium-catalyzed amination of aryl chlorides. *Journal of Molecular Catalysis A: Chemical 182–183*, 515–523. 2002.

244 Harkal, S.; Rataboul, F.; Zapf, A.; Fuhrmann, C.; Riermeier, T.; Monsees, A.; Beller, M. Dialkylphosphinoimidazoles as new ligands for palladium-catalyzed coupling reactions of aryl chlorides. *Advanced Synthesis & Catalysis 346*[13–15], 1742–1748. 2004.

245 Rataboul, F.; Zapf, A.; Jackstell, R.; Harkal, S.; Riermeier, T.; Monsees, A.; Dingerdissen, U.; Beller, M. New ligands for a general palladium-catalyzed amination of aryl and heteroaryl chlorides. *Chemistry – A European Journal 10*[12], 2983–2990. 2004.

246 Beller, M.; Riermeier, T. H.; Reisinger, C. P.; Herrmann, W. A. *Tetrahedron Letters 38*, 2073. 1997.

247 Zim, D.; Buchwald, S. L. An Air and Thermally Stable One-Component Catalyst for the Amination of Aryl

Chlorides. *Organic Letters* 5[14], 2413–2415. 2003.

248 Schnyder, A.; Indolese, A. F.; Studer, M.; Blaser, H. U. A new generation of air stable, highly active Pd complexes for C-C and C-N coupling reactions with aryl chlorides. *Angewandte Chemie, International Edition* 41[19], 3668–3671. 2002.

249 Stambuli, J. P.; Kuwano, R.; Hartwig, J. F. Unparalleled rates for the activation of aryl chlorides and bromides: coupling with amines and boronic acids in minutes at room temperature. *Angewandte Chemie, International Edition* 41[24], 4746–4748. 2002.

250 Viciu, M. S.; Kissling, R. M.; Stevens, E. D.; Nolan, S. P. An Air-Stable Palladium/N-Heterocyclic Carbene Complex and Its Reactivity in Aryl Amination. *Organic Letters* 4[13], 2229–2231. 2002.

251 Li, G. Y.; Zheng, G.; Noonan, A. F. Highly Active, Air-Stable Versatile Palladium Catalysts for the C-C, C-N, and C-S Bond Formations via Cross-Coupling Reactions of Aryl Chlorides. *Journal of Organic Chemistry* 66[25], 8677–8681. 2001.

252 Bedford, R. B.; Cazin, C. S. J. Preparation of ortho-cyclometalated palladium aryl complexes for use as catalysts in bond-forming reactions. GB A 2376946, 20021231.

253 Jiang, L.; Buchwald, S. L. Palladium-catalyzed aromatic carbon-nitrogen bond formation. *Metal-Catalyzed Cross-Coupling Reactions* (2nd Edition) 2, 699–760. 2004.

254 Yang, B. H.; Buchwald, S. L. Development of Efficient Protocols for the Palladium-Catalyzed Cyclization Reactions of Secondary Amides and Carbamates. *Organic Letters* 1[1], 35–37. 1999.

255 Stauffer, S. R.; Hartwig, J. F. Fluorescence Resonance Energy Transfer (FRET) as a High-Throughput Assay for Coupling Reactions. Arylation of Amines as a Case Study. *Journal of the American Chemical Society* 125[23], 6977–6985. 2003.

256 Kuwano, R.; Utsunomiya, M.; Hartwig, J. F. Aqueous Hydroxide as a Base for Palladium-Catalyzed Amination of Aryl Chlorides and Bromides. *Journal of Organic Chemistry* 67[18], 6479–6486. 2002.

257 Hughes, E. C.; Veatch, F.; Elersich, V. N-Methylaniline from chlorobenzene and methylamine. *Journal of Industrial and Engineering Chemistry* (Washington, D.C.) 42, 787–790. 1950.

258 Cramer, R.; Coulson, D. R. Nickel-catalyzed displacement reactions of aryl halides. *Journal of Organic Chemistry* 40[16], 2267–2273. 1975.

259 Wolfe, J. P.; Buchwald, S. L. Nickel-Catalyzed Amination of Aryl Chlorides. *Journal of the American Chemical Society* 119[26], 6054–6058. 1997.

260 Brenner, E.; Fort, Y. New efficient nickel(0) catalyzed amination of aryl chlorides. *Tetrahedron Letters* 39[30], 5359–5362. 1998.

261 Desmarets, C.; Schneider, R.; Fort, Y. Nickel-catalysed synthesis of 3-chloroanilines and chloro aminopyridines via cross-coupling reactions of aryl and heteroaryl dichlorides with amines. *Tetrahedron Letters* 42[2], 247–250. 2001.

262 Desmarets, C.; Schneider, R.; Fort, Y. Nickel-catalyzed sequential amination of aryl- and heteroaryl di- and trichlorides. *Tetrahedron* 57[36], 7657–7664. 2001.

263 Gradel, B.; Brenner, E.; Schneider, R.; Fort, Y. Nickel-catalyzed amination of aryl chlorides using a dihydroimidazoline carbene ligand. *Tetrahedron Letters* 42[33], 5689–5692. 2001.

264 Desmarets, C.; Schneider, R.; Fort, Y. Nickel of (0)/Dihydroimidazol-2-ylidene Complex Catalyzed Coupling of Aryl Chlorides and Amines. *Journal of Organic Chemistry* 67[9], 3029–3036. 2002.

265 Kelly, R. A., III; Scott, N. M.; Diez-Gonzalez, S.; Stevens, E. D.; Nolan, S. P. Simple Synthesis of CpNi(NHC)Cl Complexes (Cp = Cyclopentadienyl; NHC = N-Heterocyclic Carbene). *Organometallics* 24[14], 3442–3447. 2005.

266 Omar-Amrani, R.; Thomas, A.; Brenner, E.; Schneider, R.; Fort, Y. Efficient Nickel-Mediated Intramolecular

Amination of Aryl Chlorides. *Organic Letters* 5[13], 2311–2314. 2003.

267 Tasler, S.; Lipshutz, B. H. Nickel-on-Charcoal-Catalyzed Aromatic Aminations and Kumada Couplings: Mechanistic and Synthetic Aspects. *Journal of Organic Chemistry* 68[4], 1190–1199. 2003.

268 Ackermann, L.; Born, R. Modular diamino- and dioxophosphine oxides and chlorides as ligands for transition-metal-catalyzed C-C and C-N couplings with aryl chlorides. *Angewandte Chemie, International Edition* 44[16], 2444–2447. 2005.

269 Miller, J. A.; Dankwardt, J. W.; Penney, J. M. Nickel-catalyzed cross-coupling and amination reactions of aryl nitriles. *Synthesis* [11], 1643–1648. 2003.

270 Berman, A. M.; Johnson, J. S. Nickel-catalyzed electrophilic amination of organozinc halides. *Synlett* [11], 1799–1801. 2005.

Index

Amino Group Chemistry. From Synthesis to the Life Sciences. Edited by Alfredo Ricci

ISBN: 978-3-527-31741-7

i

k

l